Progress in Molecular and Subcellular Biology

Series Editors Werner E.G. Müller
Philippe Jeanteur, Robert E. Rhoads, Đurđica Ugarković,
Márcio Reis Custódio

52

Volumes Published in the Series

Progress in Molecular and Subcellular Biology

Subseries: Marine Molecular Biotechnology

Volume 35
RNA Trafficking and Nuclear Structure Dynamics
Ph. Jeanteur (Ed.)

Volume 36
Viruses and Apoptosis
C. Alonso (Ed.)

Volume 38
Epigenetics and Chromatin
Ph. Jeanteur (Ed.)

Volume 40
Developmental Biology of Neoplastic Growth
A. Macieira-Coelho (Ed.)

Volume 41
Molecular Basis of Symbiosis
J. Overmann (Ed.)

Volume 44
Alternative Splicing and Disease
Ph. Jeanteur (Ed.)

Volume 45
Asymmetric Cell Division
A. Macieira Coelho (Ed.)

Volume 48
Centromere
Đurđica Ugarković (Ed.)

Volume 49
Aestivation
C.A. Navas and J.E. Carvalho (Eds.)

Volume 50
miRNA Regulation of the Translational Machinery
R.E. Rhoads (Ed.)

Volume 51
Long Non-Coding RNAs
Đurđica Ugarkovic (Ed.)

Volume 52
Molecular Biomineralization
W.E.G. Müller (Ed.)

Volume 37
Sponges (Porifera)
W.E.G. Müller (Ed.)

Volume 39
Echinodermata
V. Matranga (Ed.)

Volume 42
Antifouling Compounds
N. Fusetani and A.S. Clare (Eds.)

Volume 43
Molluscs
G. Cimino and M. Gavagnin (Eds.)

Volume 46
Marine Toxins as Research Tools
N. Fusetani and W. Kem (Eds.)

Volume 47
Biosilica in Evolution, Morphogenesis, and Nanobiotechnology
W.E.G. Müller and M.A. Grachev (Eds.)

Werner E.G. Müller
Editor

Molecular Biomineralization

Aquatic Organisms Forming Extraordinary Materials

Springer

Editor
Prof.Dr. Werner E.G. Müller
Universität Mainz
Institute for Physiological Chemistry
University Medical Center of the
Johannes Gutenberg University Mainz
Duesbergweg 6
55128 Mainz
Germany
wmueller@uni-mainz.de

ISSN 0079-6484
ISBN 978-3-642-21229-1 e-ISBN 978-3-642-21230-7
DOI 10.1007/978-3-642-21230-7
Springer Heidelberg Dordrecht London New York

Library of Congress Control Number: 2011936517

Printed on acid-free paper

Springer is part of Springer Science+Business Media (www.springer.com)

Preface

Living beings, in particular aquatic organisms are capable of synthesizing a high diversity of biominerals, ranging from silica, calcium carbonate, calcium phosphate to metallic, e.g. iron oxide, biominerals. Some of these biominerals, e.g. calcium carbonate, can be present in various phases, regulated by certain organic macromolecules, and they are found both in prokaryotic and eukaryotic organisms. This book of the series Progress in Molecular and Subcellular Biology gives a survey on the most recent developments in the field of Molecular Biomineralization highlighting the importance and the mechanisms of this process occurring at the interface between the inorganic and the organic world.

Part I on Metallic Biominerals describes the surprising ability of certain bacteria (magnetotactic bacteria) to biomineralize magnetic crystals in their "magnetosomes", the synthesis of ferric oxide biominerals in protein (ferritin) nanocages, the oxidation of manganese by bacteria, as well as the contribution of microorganisms to the biogenic formation of mineral deposits in manganese nodules and seamount crusts. Part II on Biocalcium illustrates the molecular mechanisms of formation of calcium-based biominerals, including the calcium carbonate precipitation by bacteria and the formation of calcium carbonate and calcium phosphate biominerals in a variety of aquatic (invertebrate and vertebrate) organisms. Special emphasis is on the role of organic matrix proteins in the biomineralization of the Echinoderm calcite endoskeleton and the role of skeletogenic genes in the regulation of biocalcification in sea urchin. The main focus of Part III on Biosilica is on the unique enzyme, silicatein, which forms the biosilica skeleton of the siliceous sponges (demosponges and hexactinellids). The extraordinary properties of this biomaterial, an inorganic-organic nanocomposite with the capability of "bio-sintering", but also its bioactivity, in particular its ability to stimulate bone hydroxyapatite formation and to modulate the expression of certain cytokines involved in pathogenesis of osteoporosis have attracted increasing interest in its possible application in nanotechnology and nanobiomedicine. Part IV on Nacre will attract the attention of the reader on the intriguing function of matrix proteins in the calcification and decalcification of the hard cuticle in Crustaceans. The most recent research advances in the formation of molluscan shell nacreous layers, the control of the nucleation and growth of aragonitic crystals as well as the function of extracellular matrix macromolecules in these biomineralization processes will be delineated.

This book hopes to contribute to our present understanding of the role of organic proteins and matrices in skeletal formation, one fundamental process of life, and biogenic mineral deposition in aqueous environments as a base for the biomimetic design of novel functional materials for future biotechnological and biomedical applications.

Werner E.G. Müller
Heinz C. Schröder
Institute for Physiological Chemistry
University Medical Center of the Johannes Gutenberg
University Mainz

Contents

Contributors

P.A. Loka Bharathi National Institute of Oceanography (Council of Scientific and Industrial Research), Dona Paula, Goa, India

Jens Baumgartner Department of Biomaterials, Max Planck Institute of Colloids and Interfaces, Potsdam, Germany

Loes E. Bevers Council for BioIron, CHORI (Children's Hospital Oakland Research Institute), Oakland, CA, USA

Rosa Bonaventura Consiglio Nazionale delle Ricerche, Istituto di Biomedicina e Immunologia Molecolare "Alberto Monroy", Palermo, Italy

Ailin Chen Institute for Physiological Chemistry, University Medical Center of the Johannes Gutenberg University Mainz, Mainz, Germany; Yunnan Key Laboratory for Palaeobiology, Yunnan University, Kunming, China

Caterina Costa Consiglio Nazionale delle Ricerche, Istituto di Biomedicina e Immunologia Molecolare "Alberto Monroy", Palermo, Italy

Damien Faivre Department of Biomaterials, Max Planck Institute of Colloids and Interfaces, Potsdam, Germany

Qingling Feng Department of Materials Science and Engineering, Tsinghua University, Beijing, China

Lu Gan Yunnan Institute of Geological Sciences, Kunming, China

P.U.P.A. Gilbert Department of Physics, University of Wisconsin-Madison, Madison, WI, USA

Shixue Hu Yunnan Key Laboratory for Palaeobiology, Yunnan University, Kunming, China

Konstantinos Karakostis Consiglio Nazionale delle Ricerche, Istituto di Biomedicina e Immunologia Molecolare "Alberto Monroy", Palermo, Italy

Frédéric Marin UMR CNRS 5561 "Biogéosciences", Université de Bourgogne, Dijon, France

Giorgio Mastromei Department of Evolutionary Biology "Leo Pardi", University of Florence, Firenze, Italy

Valeria Matranga Consiglio Nazionale delle Ricerche, Istituto di Biomedicina e Immunologia Molecolare "Alberto Monroy", Palermo, Italy

Sébastien Motreuil UMR CNRS 5561 "Biogéosciences", Université de Bourgogne, Dijon, France

Werner E.G. Müller Institute for Physiological Chemistry, University Medical Center of the Johannes Gutenberg University Mainz, Mainz, Germany; Nanotec-MARIN GmbH, Mainz, Germany

Hiromichi Nagasawa Department of Applied Biological Chemistry, Graduate School of Agricultural and Life Sciences, The University of Tokyo, Bunkyo, Tokyo, Japan

Prabakaran Narayanappa UMR CNRS 5561 "Biogéosciences", Université de Bourgogne, Dijon, France

Brunella Perito Department of Evolutionary Biology "Leo Pardi", University of Florence, Firenze, Italy

Annalisa Pinsino Consiglio Nazionale delle Ricerche, Istituto di Biomedicina e Immunologia Molecolare "Alberto Monroy", Palermo, Italy

Roberta Russo Consiglio Nazionale delle Ricerche, Istituto di Biomedicina e Immunologia Molecolare "Alberto Monroy", Palermo, Italy

Ute Schloßmacher Institute for Physiological Chemistry, University Medical Center of the Johannes Gutenberg University Mainz, Mainz, Germany

Heinz C. Schröder Institute for Physiological Chemistry, University Medical Center of the Johannes Gutenberg University Mainz, Mainz, Germany; Nanotec-MARIN GmbH, Mainz, Germany

P.P. Sujith National Institute of Oceanography (Council of Scientific and Industrial Research), Dona Paula, Goa, India

Elizabeth C. Theil Council for BioIron, CHORI (Children's Hospital Oakland Research Institute), Oakland, CA, USA; Department of Nutritional Sciences and Molecular Toxicology, University of California-Berkeley, Berkeley, CA, USA

Xiaohong Wang Institute for Physiological Chemistry, University Medical Center of the Johannes Gutenberg University Mainz, Mainz, Germany; National Research Center for Geoanalysis, Beijing, China

Matthias Wiens Institute for Physiological Chemistry, University Medical Center of the Johannes Gutenberg University Mainz, Mainz, Germany; NanotecMARIN GmbH, Mainz, Germany

Fred H. Wilt Molecular and Cell Biology Department, University of California, Berkeley, CA, USA

Li-ping Xie Protein Science Laboratory of the Ministry of Education, Institute of Marine Biotechnology, School of Life Sciences, Tsinghua University, Beijing, P. R. China

Chao Yang Protein Science Laboratory of the Ministry of Education, Institute of Marine Biotechnology, School of Life Sciences, Tsinghua University, Beijing, P. R. China

Rong-qing Zhang Protein Science Laboratory of the Ministry of Education, Institute of Marine Biotechnology, School of Life Sciences, Tsinghua University, Beijing, P. R. China

Yu-juan Zhou Protein Science Laboratory of the Ministry of Education, Institute of Marine Biotechnology, School of Life Sciences, Tsinghua University, Beijing, P. R. China

Fang-jie Zhu Protein Science Laboratory of the Ministry of Education, Institute of Marine Biotechnology, School of Life Sciences, Tsinghua University, Beijing, P. R. China

Francesca Zito Consiglio Nazionale delle Ricerche, Istituto di Biomedicina e Immunologia Molecolare "Alberto Monroy", Palermo, Italy

Part I
Metallic Biominerals

Chapter 1
Magnetite Biomineralization in Bacteria

Jens Baumgartner and Damien Faivre

Contents

Abstract Magnetotactic bacteria are able to biomineralize magnetic crystals in intracellular organelles, so-called "magnetosomes." These particles exhibit species- and strain-specific size and morphology. They are of great interest for biomimetic nanotechnological and biotechnological research due to their fine-tuned magnetic properties and because they challenge our understanding of the classical principles of crystallization. Magnetotactic bacteria use these highly optimized particles, which form chains within the bacterial cells, as a magnetic field actuator, enabling them to navigate. In this chapter, we discuss the current biological and chemical knowledge of magnetite biomineralization in these bacteria. We highlight the extraordinary properties of magnetosomes and some resulting potential applications.

D. Faivre (✉)
Department of Biomaterials, Max Planck Institute of Colloids and Interfaces, Wissenschaftspark Golm, Potsdam 14424, Germany
e-mail: damien.faivre@mpikg.mpg.de

W.E.G. Müller (ed.), *Molecular Biomineralization*, Progress in Molecular and Subcellular Biology 52, DOI 10.1007/978-3-642-21230-7_1,

1.1 Introduction

Iron biominerals are formed by a broad range of terrestrial and aquatic organisms, in which they serve various functions. The best known function is magnetoreception, i.e., the ability to detect a magnetic field (Johnsen and Lohmann 2005). Magnetite-based magnetoreception has been recognized in birds and fish, a typical example being the homing pigeon (Winklhofer et al. 2001). Further functions typically include the strengthening of tissues (Frankel and Blakemore 1991) and hardening of teeth (Lowenstam 1967).

One of the most intriguing examples for the aquatic biomineralization of iron oxides and biomineral formation, in general, is represented by the synthesis of magnetic minerals in prokaryotes. Biomineralization has been divided into two distinct fields: Extracellularly bio-induced formation (Frankel and Bazylinski 2003) or intracellularly bio-controlled formation, as in the bacterial magnetosomes (Faivre and Schüler 2008).

It is remarkable that such simple organisms can induce the extracellular formation of minerals, but that this can also be performed under highly controlled conditions with high reproducibility and perfection is even more impressive. The formation of magnetosomes is a fascinating example of how supposedly primitive organisms can translate genetic blueprint information into complex inorganic and cellular structures. Because many of the fundamental mechanisms of biomineralization are found in bacterial magnetosome formation, magnetotactic bacteria (hereafter referred to as MTB) may serve as an accessible and relatively simple model for studying and understanding biomineralization processes in general.

Magnetite crystals from MTB are formed in the so-called magnetosomes (Gorby et al. 1988), which are specialized organelles synthesized by the cells for the purpose of geomagnetic navigation in their aquatic habitats (Bazylinski and Frankel 2004). The magnetosomes comprise membrane-enveloped, nano-sized crystals of either the magnetic iron oxide magnetite, Fe_3O_4 (Frankel et al. 1979) or the magnetic iron sulfide greigite, Fe_3S_4 (Farina et al. 1990; Mann et al. 1990). The magnetosomes are arranged in one or more intracellular chains by a recently discovered molecular mechanism (Komeili 2007), which enables the cell to align and swim along external magnetic fields, a behavior known as "magnetotaxis" (Blakemore 1975).

Magnetotaxis facilitates the search for growth-favoring microoxic zones within chemically stratified natural waters (Frankel et al. 2007). The synthesis of bacterial magnetosomes involves iron uptake from an environmental source, iron transport to the deposition site via possible storage and precursor compounds, and finally mineral formation in the dedicated magnetosome organelle, which is achieved by a high degree of control over the biomineralization of perfectly shaped and sized magnetic crystals. Moreover, in MTB this process also includes the assembly into hierarchically structured chains to serve most efficiently as a magnetic field actuator. The unique characteristics of magnetosome biomineralization have attracted multidisciplinary interest and might be exploited for a variety of applications, specifically in bio- and nanotechnologies (Lang et al. 2007; Matsunaga and Arakaki 2007).

Although several early reports by Salvatore Bellini described the observation of bacteria, in which the swimming direction was apparently affected by magnetic fields, this research was only translated and published very recently (Bellini 2009a, b). Therefore, it was Richard Blakemore's report of MTB that initiated the research activities during the last few decades (Blakemore 1975). Since then, the subject of magnetosome biomineralization has evolved into an interdisciplinary and unique field of research. The aim of this review is to give a broad overview of the current state of knowledge on the bacteria with an emphasis on the materials that are formed. Thus, we will point out the given properties that make this material so special, and explain the molecular processes that enable such a property. Finally, we will discuss what we believe to be possible new perspectives and directions for future studies.

1.2 Magnetotactic Bacteria

All MTB are motile aquatic prokaryotes that intracellularly biomineralize magnetite or greigite in specific organelles termed "magnetosomes." The magnetosomes are confined by a protein-containing lipid bilayer membrane (Balkwill et al. 1980; Gorby et al. 1988) and are typically aligned along the longitudinal axis of the cell, although sometimes clusters of magnetic particles have also been observed, as shown in Fig. 1.1 (Sparks et al. 1986). The alignment of magnetic crystals provides bacteria with the ability to orient along magnetic field lines and thus allows for a defined axial motion in the water column by utilizing the Earth's magnetic field. This phenomenon called "magnetotaxis" has been observed in bacteria with different cell morphologies and from different phyla (Bazylinski and Frankel 2004).

In the following sections, we will shortly discuss their habitats, magnetotactic bacterial diversity, and the basics of the currently known genetics and cell biology responsible for biomineralization and magnetotaxis.

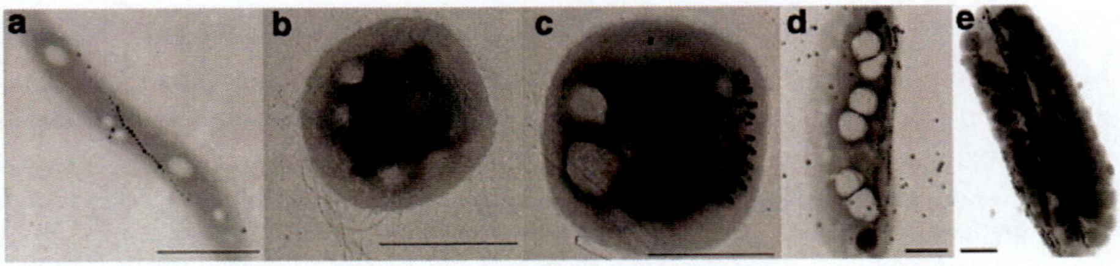

Fig. 1.1 Transmission electron micrographs of various magnetotactic bacterial species: (**a**) a spirillum with single chain of cuboctahedral magnetosomes, (**b**) coccus with double chains of elongated magnetosomes, (**c**) coccus with clustered elongated magnetosomes, (**d**) vibrio with two chains, and (**e**) a rod-shaped bacterium with multiple chains of bullet-shaped magnetosomes. Scale bars represent 1 μm

1.2.1 Ecology

MTB have been ubiquitously found in aquatic environments such as fresh- and marine water columns or the upper layers of the respective sediments. Their habitats are usually chemically stratified with respect to oxygen, which diffuses from the water–air surface downward, and sulfide generated by sulfate-reducing bacteria, which diffuses upward from the anaerobic zone. This stratification establishes a respective double gradient within the water column. MTB preferably move to and inhabit the so-called oxic–anoxic transition zone (OATZ), where they find optimal conditions and can reach cell densities of 10^5–10^6 per mL (Blakemore 1982). It is thought that magnetotaxis aids them in finding the OATZ by simplifying their search problem to one dimension, as the magnetic field lines of Earth are inclined in the northern and southern hemispheres and align bacteria mostly vertically in the water column.

All MTB discovered thus far are microaerophiles and/or anaerobes, and the biomineralization of intracellular magnetite crystals is favored under the low oxygen concentrations found in their optimal habitats (Heyen and Schüler 2003; Flies et al. 2005). Typical iron concentrations in such environments are around 0.01–1 mg/L, i.e., 0.2–20 μM, and higher abundance in lab experiments does not lead to increasing numbers of bacteria but rather has been shown to be toxic (Schüler and Baeuerlein 1996). MTB have also been detected in highly alkaline and saline environments where available iron is even scarcer due to its low solubility at high pH (Nash 2004). Thus, they must have developed means to accumulate iron against a large concentration gradient. Studied strains have been isolated by taking sediment samples from aquatic habitats, storage in microcosms, and subsequent cell harvest with a magnet (Schleifer et al. 1991). Only a few strains have been obtained in pure culture so far, which is probably due to difficulties in providing similarly complex environmental conditions in the lab as MTB inhabit in nature.

1.2.2 Diversity

MTB are polyphyletic, meaning they represent a heterogeneous group of species as determined by 16 S rRNA analysis. Known strains have very different morphologies including rods, vibrios, spirilla, cocci and more complex forms such as giant, barbell-shaped, and multicellular bacteria (see Fig. 1.1). They have been attributed to the Gram-negative *Alpha-*, *Gamma-*, *Deltaproteobacteria*, and the *Nitrospira* phylum (Amann et al. 2007). Best studied is the *Magnetospirillum* genus as several species can be cultured in the laboratory, namely, *M. magnetotacticum* (Blakemore et al. 1979), *M. magneticum* (Matsunaga et al. 1991), and *M. gryphiswaldense* (Schleifer et al. 1991). The vibrio strain MV-1 from the *Alphaproteobacteria*, *Desulfovibrio magneticus* (Sakaguchi et al. 2002) from the

Table 1.1 Cultured and some unusual examples of magnetotactic bacteria, their respective cell morphologies, and magnetosome properties

Species	Cell morphology	Magnetosome size	Magnetosome morphology	Magnetosome number
Magnetospirillum magneticum	helicoidal	~50 nm	cuboctahedral	>15
Magnetospirillum magnetotacticum	helicoidal	~40 nm	cuboctahedral	Up to ~40
Magnetospirillum gryphiswaldense	helicoidal	~40 nm	cuboctahedral	Up to ~60
Magnetococcus (marinus) MC-1	coccus	80–120 nm	elongated pseudo-hexagonal prismatic	<15
Magnetovibrio MV-1	vibrio	40–60 nm	elongated pseudo-hexagonal prismatic	~10
Desulfovibrio magneticus	vibrio	~40 nm	bullet-shaped	~10–15
Magnetobacterium bavaricum	rod	110–150 nm	hook-shaped	~1,000
Coccoid MTB	coccus	~250 nm	elongated	<10
Multicellular magnetotactic prokaryote (MMP)	ovoid	~90 nm	irregular	Up to ~65 per cell

Deltaproteobacteria, and the coccus MC-1 (Frankel et al. 1997) are also available in pure culture and have been studied in more detail. All greigite-forming MTB or other unusual types such as the multicellular magnetotactic prokaryote (MMP), which consists of an assembly of multiple flagellated bacterial cells containing magnetosome chains or the very large *Magnetobacterium bavaricum* have not yet been obtained in axenic culture and are thus less well investigated (Vali et al. 1987; Farina et al. 1990; Rodgers et al. 1990) (Table 1.1).

1.2.3 Genetics

To understand what distinguishes magnetotactic bacteria from non-magnetotactic species, their molecular biology and genetics have been studied in recent years, providing new insights into biomineralization. Genetic information has been obtained from the freshwater species *M. magneticum* (Matsunaga et al. 2005), *M. gryphiswaldense* (Ullrich et al. 2005), *M. magnetotacticum* (http://genome.jgi-psf.org/magma/magma.home.html), the marine species *Magnetococcus* MC-1 (Schubbe et al. 2009), the vibrio strain MV-1, and the *Deltaproteobacterium Desulfovibrio magneticus* (Nakazawa et al. 2009). Comparative genome analyses have identified a common MTB-specific subset of genes, which was termed "magnetosome island" (MAI), and which presumably provides the magnetite-biomineralizing capability and potentially also magnetotactic functionalities

(Ullrich et al. 2005; Richter et al. 2007). The most recently published genome analysis of *Desulfovibrio magneticus* RS-1 contains so far the lowest number of MAI genes common to all sequenced magnetotactic bacterial genomes and thus gives insights into which genes might be essential for intracellular magnetite formation. This reduced subset of genes consists of *mamA*, *mamB*, *mamE*, *mamK*, *mamM*, *mamO*, *mamP*, *mamQ*, and *mamT*. *mamA* has so far been described in *M. magneticum* as required for functional magnetosome vesicle formation (Komeili et al. 2004), and *mamB* and *mamM* are putative iron transporter genes, which will be discussed later. *mamK* is involved in magnetosome chain formation as a filamentous actin-like protein (Komeili et al. 2006). The roles of the other MAI genes remain unclear. *Magnetospirillum* spp. genomes comprise several further so-called *mam* or *mms* genes (referring to "*ma*gnetosome *m*embrane" and "*m*agnetosome *m*embrane *s*pecific"), which are unique to MTB and have been shown to be somehow involved in magnetite biomineralization, as will be discussed later. Because not all magnetotactic species are closely related and various transposases were detected along with MAI genes, it has been proposed that the respective genes have spread by horizontal gene transfer (Jogler et al. 2009). A further distinct feature of magnetotactic bacterial genomes is the presence of numerous regulatory and signaling genes similar to chemotaxis genes that thus might be involved in magnetotaxis (Jogler et al. 2009; Nakazawa et al. 2009).

1.2.4 Cell Biology

Structurally, MTB possess two cell membranes, the outer and inner membrane separating periplasm from cytoplasm, as in all bacteria. Magnetosomes form as invaginations of the inner cell membrane and possibly completely detach from it to form isolated organelles within the cell (Komeili et al. 2006; Faivre et al. 2007). Biochemical analysis of the lipid bilayer membrane (Gorby et al. 1988; Grünberg et al. 2004), electron microscopic (Balkwill et al. 1980; Komeili et al. 2004), tomographic (Komeili et al. 2006; Scheffel et al. 2006), and Mössbauer spectroscopy (Faivre et al. 2007) studies have led to this current model. The lipid composition of both magnetosome and cytoplasmic membrane are virtually identical (Gorby et al. 1988; Grünberg et al. 2004). The protein content of each, however, differs dramatically; proteomic analyses showed that the proteins mainly encoded in the MAI are exclusively found in or attached to the magnetosome membranes of *M. magneticum* and *M. gryphiswaldense* (Grünberg et al. 2004; Matsunaga et al. 2005). The proteins have been partly attributed to known families according to sequence homologies. Among them are proteins involved in transport such as the generic transporters and cation diffusion facilitators (CDF) MamB and MamM, proteases (HtrA-like serine proteases) MamE and MamO, the actin-like filamentous structural protein MamK, proteins putatively involved in protein–protein interactions, and complex scaffolding containing PDZ and TPR (tetracopeptide repeat) domains, and a number of proteins without similarities to any other proteins with known function (Richter et al. 2007). Generic transporters and cation diffusion facilitators control influx and efflux of diverse solutes necessary for metabolism and in this case, might possibly transport iron for the biomineralization of magnetite (Schüler 2008; Jogler and Schüler 2009). HtrA-like proteases are normally found in the periplasm where they degrade misfolded proteins (Pallen and Wren 1997). Magnetosome-related proteins can be proteolytically cleaved and could be potential targets of these respective proteases (Arakaki et al. 2003). The development of genetic systems for some MTB provided the base to selectively determine in vivo the function of involved proteins (Matsunaga et al. 1992; Schultheiss and Schüler 2003). MamG, F, D, and C have been shown to have an influence on crystal size and possibly morphology (Scheffel et al. 2008). MamK and MamJ have been made responsible for chain alignment of magnetosomes along the cell axis, as will be explained in Sect. 3.2.4 (Komeili et al. 2006; Pradel et al. 2006; Scheffel et al. 2006). Fusions with fluorescent reporters helped determine the localization of some proteins, including MamA, MamC, MamJ, and MamK (Komeili et al. 2004; Komeili et al. 2006; Scheffel et al. 2006; Lang and Schüler 2008). Additionally, recombinant expression of magnetosome-related proteins has shed some light on their possible functions (Arakaki et al. 2003; Pradel et al. 2006; Scheffel and Schüler 2007; Taoka et al. 2007). For example, Mms6 has been proposed as a potential magnetite crystal nucleator or growth modulator by face-selective adhesion (Arakaki et al. 2003). All these approaches give us a first glimpse in comprehending the molecular basis of magnetite biomineralization, but many details remain poorly understood.

1.3 Magnetosomes

As only magnetite-forming bacteria have been obtained in axenic culture so far, our knowledge about the chemistry, molecular biology, and genetics has mainly been obtained from these strains. Future studies on greigite formers will hopefully provide insights on their respective differences. In the following sections, we will present the current knowledge on how magnetite is biomineralized in MTB and which biological and chemical factors influence the specific properties of the magnetosome particles.

1.3.1 Magnetite Biomineralization Pathway

The intracellular formation of magnetite nanocrystals requires the accumulation of substantial amounts of iron from the surrounding environment and a precisely coordinated biological machinery to transport and deposit it under the right conditions for the formation of the magnetite phase. This chemistry must be highly controlled by MTB as other potentially formed iron oxide phases do not provide the magnetic properties required for magnetotaxis.

1.3.1.1 Iron Source

Ferrous and ferric iron oxides are sparingly soluble in water around neutral pH and thus the total soluble iron without any complexing or reducing agents at pH 4–10 lies below micromolar concentrations (Cornell and Schwertmann 2003). Most bacteria, also non-magnetotactic organisms, require micromolar levels for growth, yet many aquatic environments provide much less. Thus, organisms have developed means to accumulate the necessary amounts of iron from their respective environments (Sandy and Butler 2009). Natural MTB habitats are freshwater or marine sediments with typically micromolar concentrations of soluble iron (Flies et al. 2005). Due to the biomineralization of magnetite, MTB acquire several orders of magnitude more iron than non-magnetotactic organisms, reaching up to 4% of their cell dry weight (Schüler and Baeuerlein 1996). Cultivated MTB strains are grown in medium supplied with iron (typically as citrate or quinate) in similar amounts as within their natural habitats. It has been demonstrated that they can only tolerate concentrations up to the millimolar range. Higher quantities were shown to be toxic rather than beneficial for growth (Nakamura et al. 1993; Schüler and Baeuerlein 1996; Schüler and Baeuerlein 1998; Faivre et al. 2007).

1.3.1.2 Iron Uptake and Transport

MTB have been proclaimed to possess specific iron uptake systems coupled to magnetosome synthesis in order to obtain the necessary quantities for biomineralization. Iron must be taken up from the surrounding environment and transported across the outer membrane into the periplasm. Two alternative routes of further processing are possible as it has not been fully resolved yet whether magnetosomes remain interconnected with the periplasm or are completely detached (Komeili et al. 2006; Faivre et al. 2007). Iron must be transported either directly from periplasm into the magnetosome vesicle or across the cytoplasmic membrane and subsequently the magnetosome membrane in order to be concentrated and form magnetite (Fig. 1.2). Both proposed pathways would involve different transporters, proteins that provide energy, and regulators that control uptake and transport.

It was shown that MTB are able to take up iron in both its divalent and trivalent (ferrous and ferric) forms (Schüler and Baeuerlein 1996; Faivre et al. 2007), and this uptake involves siderophores in some cases, similarly to non-magnetotactic organisms. Siderophores are low-molecular-weight iron chelators that function to enable uptake of ferric iron (Sandy and Butler 2009). They are produced and released by bacterial cells into the surrounding medium to complex available ferric iron. Iron–siderophore complexes are then internalized by specific transporters such as TonB-dependent receptors. Intracellular iron release is enabled by reduction to

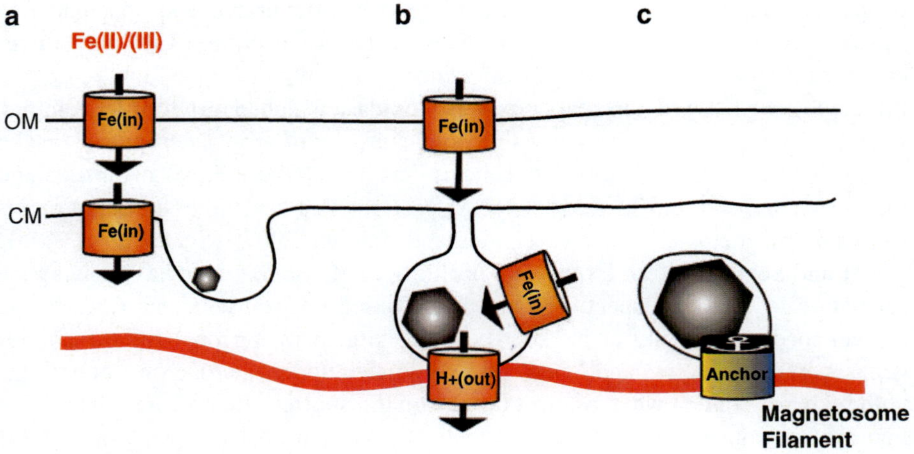

Fig. 1.2 Iron uptake and intravesicular magnetite formation; (**a**) Fe(II) and Fe(III) are taken up by MTB through respective transporter proteins; (**b**) it is yet unclear whether iron is transported into magnetosome vesicles from the periplasm and/or from cytoplasm; (**c**) mature magnetosomes might be detached from the cytoplasmic membrane, alignment is provided by anchoring to the magnetosome filament; OM: outer membrane, CM: cytoplasmic membrane

ferrous iron and thus a change in the complex-binding constant or decomposition of the siderophore. *M. magnetotacticum* (Paoletti and Blakemore 1986), MV-1 (Dubbels et al. 2004) and *M. magneticum* have been reported to secrete siderophores in amounts dependent on the medium iron concentrations and cellular demand. The latter species was shown to secrete hydroxamates and the catechol siderophore 3,4-dihydroxybenzoic acid for Fe^{III} complexation (Calugay et al. 2003; Calugay et al. 2006). *M. magnetotacticum* has been reported to show unusual siderophore production behavior, as production increased with higher ferric iron concentrations (Paoletti and Blakemore 1986) instead of downregulation when supply is high. It has been speculated, however, that very rapid assimilation of dissolved iron and the subsequent low concentration levels triggering siderophore production might account for this effect (Calugay et al. 2003; Calugay et al. 2004). Surprisingly, no siderophore production has been detected in cultures of *M. gryphiswaldense* although spent media fluid increased ferric iron uptake which was also shown to be energy coupled in this organism and might indicate the existence of other potential chelators involved in ferric iron uptake. Ferrous iron acquisition seemed to be diffusion based and not coupled to energetically driven uptake (Schüler and Baeuerlein 1996; Schüler and Baeuerlein 1998). Thus, a general role of siderophores related to iron uptake necessary for magnetosome formation is still debated and might be species dependent.

No mechanism common to all MTB has been revealed yet, but a number of proteins involved in general iron uptake have been identified. It is not clear whether these proteins are part of biochemical pathways involved in biomineralization or in simply providing essential iron for various metabolic processes. In MV-1, a major copper-containing periplasmic protein (ChpA) was identified and speculated to function as an iron-uptake regulator analogously to the protein Ctr1 in a three-component iron-uptake system in *Saccharomyces cerevisiae*. In this yeast, the protein is known to provide copper to an Fe^{II} oxidase which in turn forms a complex with an iron permease for activation and subsequent iron–siderophore uptake (Dubbels et al. 2004). Genome sequencing has revealed a number of ferrous and ferric iron transporters in magnetotactic bacteria that resemble similar proteins from non-magnetotactic species (Jogler and Schüler 2007; Richter et al. 2007; Jogler and Schüler 2009). Expression profiling in *M. magneticum* has shown gene regulation of several respective transporters based on iron concentrations in the growth medium (Suzuki et al. 2006). In this organism, ferrous iron transporter genes, namely, *ftr1*, *tpd*, and *feoAB* are upregulated under microaerobic conditions with high Fe^{II} content whereas most ferric iron transporters are downregulated. The Feo system consists of three components: FeoA, a probably cytoplasmic SH-3 domain protein; FeoB, the inner membrane Fe^{II} permease possessing an N-terminal G-protein domain; and only found in *Deltaproteobacteria*, FeoC, an [Fe-S]-dependent transcriptional repressor (Cartron et al. 2006). Furthermore, *M. magnetotacticum* grown in ferrous iron-rich medium upregulates expression of FeoB1 and the protein was found to be localized at the cytoplasmic membrane (Taoka et al. 2009). *M. gryphiswaldense* seems to behave differently, as *feoAB1* gene transcription is downregulated with higher ferrous iron concentration (Rong et al. 2008). As uptake in this species apparently does not depend on siderophores, other transporters might be involved in the process, possibly explaining this difference. However, it was experimentally demonstrated in *M. gryphiswaldense* that the Feo system plays at least an accessory role in iron uptake related to magnetosome formation, since a deletion mutant (Δ*feoB1*) forms smaller and fewer magnetite particles (Rong et al. 2008). A cytoplasmic ATPase identified in *M. magneticum* was shown to be involved in ferrous iron uptake, most probably by energizing membrane transporters such as FeoB. Its target transporter, however, is still unknown (Suzuki et al. 2007). Disruption of a *fur*-like gene in *M. gryphiswaldense* resulted in low intracellular iron levels and inhibition of magnetosome formation, thereby indicating its involvement in biomineralization-related iron uptake (Yijun et al. 2007). Fur (ferric uptake regulator) controls iron homeostasis in bacteria and was shown in *Escherichia coli* to act by regulating gene expression according to intracellular iron concentrations (Escolar et al. 1999). It can bind Fe^{II} and the resulting complex acts as a transcriptional repressor for genes related to iron uptake. When intracellular iron concentrations are low, the Fe^{II}–Fur complex dissociates and thereby loses its ability to bind DNA. The respective genes are then transcribed and expressed.

Final iron transport into magnetosomes is poorly understood. In *M. magneticum*, a possible proton/iron antiporter protein, MagA, has been described as being involved in the process. The protein was detected at both cell and magnetosome membrane (Nakamura et al. 1995). However, no experimental evidence from other MTB exists to confirm an essential role of MagA in magnetite biomineralization. Proteomic analysis has also identified two cation diffusion facilitators located exclusively at the magnetosome membrane, namely, MamB and MamM (Grünberg et al. 2004). These proteins have been attributed to the so-called CDF3 subfamily which is supposed to comprise iron transporters (Nies 2003). Deletion mutants of these genes in *M. gryphiswaldense* are nonmagnetic, supporting the hypothesis of their involvement in the magnetite biomineralization process (Schüler, personal communication). Both genes are found in genomes of all other sequenced magnetotactic bacterial species as well, emphasizing their potentially essential role. So far, various proteins have been shown to be involved in iron uptake, but it remains unclear which are related to biomineralization and whether separate and specific uptake mechanisms exist besides the one for metabolic needs. Only the proteins MagA, MamB, and MamM seem to be specific to MTB and thus are candidates for further studies which will hopefully lead to insights into how iron transport is coupled to biogenic magnetite synthesis.

1.3.1.3 Magnetite Formation

Magnetite formation in *Magnetospirillum* spp. requires microaerobic or anaerobic conditions and higher oxygen levels can suppress its biomineralization (Heyen and Schüler 2003). This is consistent with abiotic synthesis conditions, which require low or no oxygen to prevent oxidation of ferrous to ferric iron and thus formation of oxidized iron oxide phases such as maghemite, hematite, or goethite. Generally, magnetite forms under slightly reducing conditions ($E_h \sim -0.2$ to -0.4 V) and alkaline pH (>8) (Winklhofer and Petersen 2007; Faivre and Schüler 2008). Thus, MTB have to form specific intracellular compartments for the synthesis of the magnetite phase as normal physiological conditions do not allow for the formation of this iron oxide phase. The pH level present within magnetosome vesicles is still unclear; however, foraminifera, marine calcite-forming unicellular organisms have been shown to be able to regulate pH above 9 within intracellular vesicles (de Nooijer et al. 2009). This finding suggests that other single-cell aquatic organisms, and therefore MTB, may be able to regulate pH within this range.

Two alternative magnetosome formation pathways have been proposed since the discovery of bacterial magnetite biomineralization. One suggestion involves invagination and detachment of magnetosome vesicles from the cytoplasmic membrane before nucleation and growth of crystals. Another proposed mechanism involves the formation of small crystal nuclei at the cytoplasmic membrane and simultaneous or consecutive vesicle invagination (Komeili 2007). Various magnetite mineral precipitation mechanisms have also been proposed. Initial assumptions after the discovery of magnetite biomineralization involved an amorphous or low crystallinity mineral precursor such as an amorphous ferrous oxyhydroxide or ferrihydrite, which is subsequently transformed into magnetite by oxidation or reduction, respectively (Frankel et al. 1983). Recent findings, however, seem to refute the existence of such precursors as they have not been detectable in cultivable MTB. A possible mechanism without mineral precursors is the coprecipitation of Fe^{II} and Fe^{III} ions under alkaline conditions within the magnetosome vesicle

(Faivre et al. 2007). Both iron species would have to be supplied to the process in the correct stoichiometry, i.e., in the ratio $Fe^{III}/Fe^{II} = 2$. This could be obtained through precise tuning of transport rates of the respective species into the magnetosome vesicle or controlled intravesicular redox reactions. In any scenario, iron has to be concentrated to supersaturating conditions in order to form nuclei that can then grow into mature crystals. For a coprecipitation-like process, results obtained from abiotic chemical synthesis indicate that a lower concentration limit exists in the range of 30 mM iron to form the magnetite phase. Lower concentrations favored formation of poorly crystalline iron oxides, hydroxides, and goethite (Faivre et al. 2004). Supersaturating conditions could be provided by two different means, regulated iron transport into the magnetosome vesicle by proteins such as MamB and MamM, and/or localized supersaturation by iron-binding entities such as the lipid membrane or attached proteins.

Growth kinetics has been studied to some extent by induction experiments. Some MTB can be grown in culture with very low iron supply and/or high oxygen concentrations, suppressing magnetite biomineralization. Addition of iron under microaerobic conditions then leads to induced crystal formation and gives insights into magnetosome growth behavior (Komeili et al. 2004; Faivre et al. 2007). *M. magneticum* has been shown to form arranged magnetite crystals within 2 h after iron addition with increasing size and number over 21 h. Magnetite crystallized simultaneously within several vesicles of the same chain (Komeili et al. 2004). In induction experiments with iron-starved *M. gryphiswaldense,* transmission electron microscopy showed magnetite crystals approximately an hour after iron addition. Particles reached full size and number over a period of 6 h. Mössbauer spectroscopy detected the first magnetite-related signals only 20 min after induction, indicating an even faster nucleation. Analysis of cell fractions suggests that very small (<5 nm) magnetite particles are formed at the cytoplasmic membrane, in accordance with the model that proposes nucleation in undetached magnetosomes and subsequent growth in detached vesicles (Faivre et al. 2007; Faivre et al. 2008).

In a recent paper by Staniland et al., the authors proposed rapid formation of full chains with mature crystals within 15 min after iron induction, challenging the slower growth observed in *M. magneticum* and *M. gryphiswaldense* before (Staniland et al. 2007). X-ray magnetic circular dichroism results suggested the existence of an α-Fe_2O_3 precursor phase during the first 30 min after induction. It is still unclear which experimental details led to these substantial reported differences. However, experiments were conducted with different iron sources in the medium, potentially influencing uptake rates. Furthermore, reported C_{mag} values, a measure of cell magnetization based on a light scattering assay are significantly different. C_{mag} is zero for nonmagnetic cells and increases with magnetization. Thus, in induction experiments, C_{mag} should begin at zero and increase over time with progressing magnetosome formation. Reported C_{mag} values in the latter experiment were marginally above zero at $t = 0$, indicating slight magnetization of the cells already before induction and possibly explaining differences in described magnetosome formation rates (Staniland et al. 2007).

Except for twinning in some cases, only single crystals are found within magnetosomes, indicating a single nucleation event (Devouard et al. 1998; Faivre and Schüler 2008). This finding suggests the existence of a specific nucleation site putatively provided by a single protein or complex. Several proteins have been proposed to play roles in the nucleation and later growth process, but in vivo proof remains sparse. The small Mms6 protein was described in *M. magneticum* as tightly bound to the magnetite crystal (Arakaki et al. 2003). The protein is processed proteolytically and bears a leucine-glycine (LG)-rich motif which is also conserved in other magnetosome proteins and might mediate its aggregation. Its C-terminus is acidic and thought to provide its iron-binding functionality as shown in vitro; however, thus far the lack of *mms6* mutants leaves open the question of whether it acts as a nucleator in vivo or performs other necessary functions. The absence of *mms6* in the genome of *D. magneticus* also suggests that this gene might not be essential for magnetite formation in MTB (Nakazawa et al. 2009).

During the formation of magnetite, protons are released and have to be translocated from the magnetosome vesicles to maintain pH within the stable regime for the magnetite phase. It has been speculated that MamN might provide this H^+ efflux function in *Magnetospirillum* because it shows similarities to respective known proteins (Jogler and Schüler 2007). MamT, thought to be a cytochrome *c* heme-binding protein, has been speculated to be involved in possible redox reactions involved in magnetite formation (Jogler and Schüler 2007). However, for both proteins no experimental data is available that proves or disproves these postulated roles.

1.3.2 Why Are Magnetosomes an Extraordinary Material?

MTB have evolved means to optimize the structure of magnetosomes to work as a magnetic field actuator. This optimization can be observed on at least three hierarchical levels from the Ångström to the submicrometer length scale, i.e., from the atomic structure up to filaments within the bacterial cell. In the following sections, we describe these levels of structural optimization with respect to function and how they are achieved.

1.3.2.1 Magnetosome Structure

Magnetite ($Fe^{III}[Fe^{II}Fe^{III}]O_4$) can easily be oxidized to maghemite ($Fe^{III}[Fe_{5/3}{}^{III}[\ ]_{1/3}]O_4$) even at low temperatures, and naturally occurring magnetite is typically partially oxidized in its equilibrium state. This oxidation from stoichiometric magnetite to maghemite leads to a decrease in magnetic saturation moment that is disadvantageous for its magnetic properties required for magnetotaxis. The transformation between both iron oxide phases is indicated by a gradual decrease in its lattice parameter *a* from 8.397 Å to 8.347 Å, respectively. Recent findings in our group obtained by high-resolution synchrotron X-ray diffraction show that both cultivated *M. gryphiswaldense* and *M. magneticum* form stoichiometric magnetite and magnetosome particles seem to be protected by the cells from oxidation (Fischer et al. 2010, submitted manuscript). Most magnetosomes are also free of imperfections except for frequently observed twinning along the [111] direction. However, this kind of twinning does not necessarily affect the magnetization of magnetosome crystals, as it is oriented along the easy axis (<111>) of magnetite (Winklhofer 2007). Also, no dislocation lines which might indicate a screw dislocation mechanism for growth in elongated directions have been observed by transmission electron microscopy (Devouard et al. 1998). Thus, MTB seem to have optimized iron usage and crystal growth with respect to magnetosome formation and their magnetic properties on the atomic structural level. Additionally, magnetosome crystals were initially thought to be chemically pure, i.e., no traces of other elements are normally incorporated. However, recent literature has shown that in certain cases, elements such as manganese (Keim et al. 2009) and cobalt (Staniland et al. 2008) might be incorporated under specific conditions when highly abundant in the growth medium. This has created enormous interest as doping of magnetite can be used to increase the coercivity and hence its magnetic hardness. Co-doped magnetosomes were shown to have an up to 49% higher coercivity than their non-doped counterparts (Staniland et al. 2008). However, it remains to be shown whether cobalt or other metals can be incorporated within the whole magnetosome crystal or if doping is more prominent on its surface. It also remains to be verified whether cobalt uptake is an active process or unspecific diffusion. Thus, future experiments will show whether MTB can be used to form other ferrite phases with interesting properties and how other metals than iron affect the biology of MTB.

1.3.2.2 Magnetosome Dimensions

Mature magnetosome crystals are typically in the single-magnetic-domain (SD) size range of 35–120 nm in diameter and have narrow asymmetric size distributions with sharp cutoffs toward larger sizes (Devouard et al. 1998). Particles of smaller size are superparamagnetic at ambient temperature, meaning that due to thermal fluctuations they have no remanent magnetization and cannot provide cells with a magnetic dipole sufficient for alignment in Earth's magnetic field. Larger particles form multiple domains with separate magnetic moments orienting antiparallel. This results in less magnetic remanence per unit volume and therefore less efficiency with respect to iron usage (Dunlop and Özdemir 1997; Muxworthy and Williams 2006; Winklhofer 2007; Muxworthy and Williams 2009). It is thus thought that MTB have evolved highly species-specific means to genetically control the size of magnetosome crystals to tune their magnetic properties. Typical sizes for *magnetospirilla* are in the 30–50 nm size range (Devouard et al. 1998). *D. magneticus* crystals are about 40 nm long (Pósfai et al. 2006), MV-1 and MC-1 produce elongated particles with sizes of 40–60 nm and 80–120 nm, respectively

(Devouard et al. 1998). Unusually large magnetosomes of 250 nm have been reported for uncultured magnetotactic cocci. These particles are exceptional because they still exhibit SD properties although their size is outside the typical SD size range (Lins et al. 2005). It has been speculated that magnetosome crystal size is restricted by the vesicles themselves, as transmission electron micrographs of thin sections prepared by cryo-ultramicrotomy suggest that preformed vesicles are of a similar size as the mature crystals (Komeili et al. 2004).

Furthermore, as shown in *M. gryphiswaldense*, various membrane proteins seem to be able to influence size and potentially morphology. A deletion mutant lacking the whole *mamGFDC* operon produced magnetite crystals of only 75% the size of wild-type cells (Scheffel et al. 2008). The resulting crystals were also less regular in morphology and chain alignment. Thus, the precise roles of the proteins are unclear, i.e., which effects cause the respective phenotypical outcomes. Complementation experiments revealed that they might share redundant functions, as single gene complements led to recovery of wild-type size. Reestablishing the full operon even caused the particles to exceed their original size. All four proteins account for around 35% of total protein content in the magnetosome membrane fraction of *M. gryphiswaldense* and seem to be located there exclusively. *mamG* has so far only been found in spirilla, but *mamD*, *mamF*, and *manC* are found in all sequenced genomes of magnetotactic *Alphaproteobacteria*, although their organization in operons may differ (Schübbe et al. 2009). The *Deltaproteobacterium D. magneticus* lacks all respective genes (Nakazawa et al. 2009). MamC and MamF are the most abundant proteins, and the latter has been shown to form highly stable oligomers. Mam12, the ortholog of MamC in *M. magnetotacticum*, has also been identified as a magnetosome membrane protein in this organism (Taoka et al. 2006). MamD and MamG share similarities with the aforementioned Mms6 protein, as they also contain leucine–glycine repeats putatively responsible for aggregation of multimeric complexes.

Mms6 of *M. magneticum* was shown to exert effects on crystal size and morphology under certain conditions in in vitro syntheses of magnetite using recombinant protein as an additive. In syntheses by coprecipitation of Fe^{II} and Fe^{III} with sodium hydroxide, larger particles of around 30 nm were claimed to be found as compared to a control without the additive (Arakaki et al. 2003). In another experiment, magnetite nanoparticles were formed through an oxidative route at elevated temperature (90°C) (Amemiya et al. 2007). Under these conditions, Mms6 restricted crystallite size and seemed to favor the formation of cuboctahedral rather than octahedral particles, indicating a possible role as crystal face recognizing growth regulator. However, these conditions are not physiological and due to the lack of *mms6* mutants so far, it is difficult to draw reliable conclusions.

Changes in magnetsome size and morphology have also been reported for *M. magnetotacticum* grown in media supplied with high amounts of various trace metals such as zinc and nickel, but these changes have not been characterized in detail (Kundu et al. 2009). It can be speculated that respective metal ions compete with iron uptake and thus limit its supply for magnetosome mineralization; however, this has not yet been investigated systematically.

1.3.2.3 Magnetosome Morphologies

Many different crystal morphologies have been observed in magnetosomes and are typically combined forms with cubic, octahedral, and dodecahedral faces which can be distorted and elongated (Devouard et al. 1998) (Fig. 1.3). Furthermore, non-isometric morphologies such as bullet- or tooth-shaped crystals have been described (Mann et al. 1987a, b; Spring et al. 1993; Taylor and Barry 2004; Isambert et al. 2007). Shape anisotropy in magnetic single-domain particles influences their coercivity with elongations along the easy magnetization axis leading to a decrease in spontaneous reversal of the magnetization (Vereda et al. 2009). Thus, in some cases, MTB have optimized particle morphology with respect to properties which are advantageous for magnetotaxis.

Generally, isometric cuboctahedral structures, as found in magnetospirilla, are also obtained by various abiotic syntheses, suggesting that in these organisms no or little genetically encoded morphological control is necessary. In contrast, symmetry-breaking anisotropic habits, such as elongated forms along one of the equivalent [111] faces and bullet- and tooth-shaped crystals, are not in agreement with the symmetry of the crystal structure and their formation is not understood. Iron transporters could possibly be localized or active at specific sites in the magnetosome membrane and could favor crystal growth and elongation by a directed iron flow and thus localized saturation. Another possibility is the existence of face-selective adhesion by certain biomolecules on the particles, preventing growth at blocked crystal sites. For neither hypothesis has experimental evidence been obtained, and this might prove difficult due to resolution problems with potentially useful microscopy techniques. Indirect evidence could be obtained

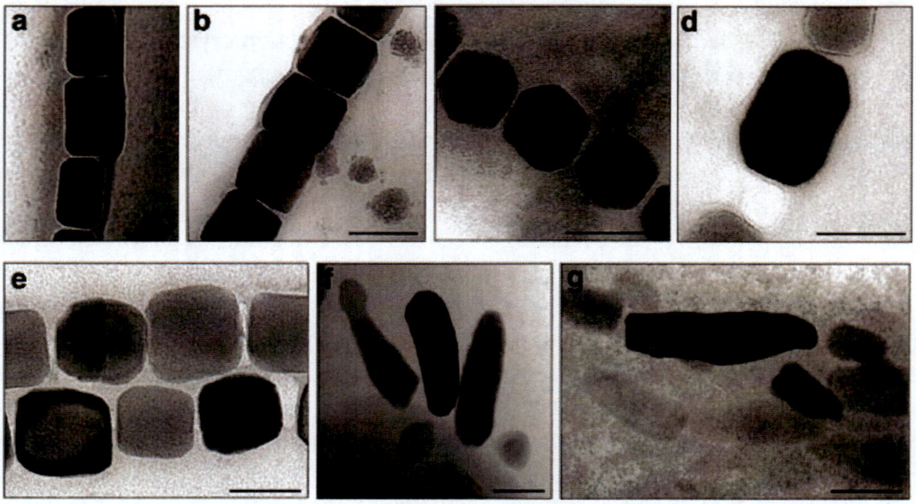

Fig. 1.3 Examples of different magnetosome morphology: (**a**), (**b**), (**c**) elongated, (**d**) cubic, (**e**) octahedral, and (**f**) and (**g**) irregularly shaped habits. Scale bar represents 50 nm

in vitro with magnetotactic bacterial proteins such as Mms6 or peptides which selectively recognize crystal faces and thus influence crystal morphology during growth. An amorphous precursor which is later transformed into magnetite would more probably allow shaping of more extraordinary crystals forms such as bullet or tooth shapes, a phenomenon known from other biomineralizing systems such as calcium carbonate in various higher organisms (Politi et al. 2008). So far, explanations for non-isometric crystal forms in magnetosomes have remained purely speculative and will need verification by experimental evidence in vivo and vitro.

1.3.2.4 Magnetosome Chain Assembly and Chain Function

Magnetosome crystals are arranged in chain structures within the bacterial cell to increase the magnetic dipole for enhanced functionality (Dunin-Borkowski et al. 1998). By assembly, the individual magnetic dipoles of magnetosome particles are summed up, providing cells with the ability to orient in a magnetic field. *Magnetospirilla* contain a single chain that is oriented along the longitudinal cell axis. Some uncultured magnetotactic bacteria possess multiple chains with sometimes more complex assemblies (Schüler 2008). Colloidal magnetic particles tend to aggregate or form so-called flux closure rings, a feature which can also be observed with isolated magnetosomes (Philipse and Maas 2002; Xiong et al. 2007). Induction experiments in *M. gryphiswaldense* have shown that magnetite particles form at various sites and start to align during further growth at midcell (Faivre et al. 2007). A filamentous structure that would prevent collapse of magnetosome chains into clusters or rings by connecting it to cellular structures had been proposed and was later confirmed by cryo-electron tomography in *M. gryphiswaldense* and *M. magneticum* (Komeili et al. 2006; Scheffel et al. 2006). The two proteins MamK and MamJ have been shown to be involved in this chain assembly. MamK is a bacterial actin-like protein assembling into a filamentous structure along the magnetosome chain as visualized by fluorescent microscopy with a GFP-fusion protein. The phenotype of a *M. magneticum* MamK deletion mutant shows magnetosomes dispersed within the bacterial cell (Komeili et al. 2006). Recombinant expression of MamK in *E. coli* led to the formation of detectable filaments within this host and the isolated protein has also been polymerized in vitro into filamentous bundles of 100 μm length and around 100 nm width, with single filaments of around 6 nm in width (Pradel et al. 2006; Taoka et al. 2007). The polymerization has been shown to be ATP dependent and could possibly drive magnetosome arrangement by a treadmilling-like mechanism known from actin and other filamentous protein polymers. Furthermore, the acidic protein MamJ, which is only found in *magnetospirilla*, is involved in magnetosome chain formation in these species. A *M. gryphiswaldense* deletion mutant (Δ*mamJ*) no longer produced chains of magnetosomes but clustered aggregates. Complementation of the *mamJ* gene in the deletion mutant could restore chain formation. Two-hybrid experiments in *E. coli* suggest that both proteins interact and that MamJ might serve as an anchor for magnetosome vesicles to the MamK filament (Scheffel et al. 2006; Scheffel and Schüler 2007). However, this proposed model is partly contradicted by the different phenotypes of Δ*mamK* and Δ*mamJ* mutants, as one would expect similarities between both, assuming an anchoring function on a filamentous structure. It is unclear whether discrepancies are due to experimental differences, species-dependent differences, or because the proposed model is oversimplified (Schüler 2008). Furthermore, MamJ has so far only been identified in *magnetospirilla*, leaving open the question of how other magnetotactic organisms achieve chain formation.

1.4 Applications of Magnetosomes

Synthetic magnetic iron oxide nanoparticles are used or in development for a variety of technical applications, particularly in the biotechnological and medical field. Examples are the magnetic separation of various biomolecules, magnetic resonance imaging (MRI), hyperthermia treatment of cancer cells and other possible applications (Laurent et al. 2008). Due to their monodisperse size in the magnetic single domain range, magnetosomes are particularly interesting with respect to these potential applications.

Synthetic particles can easily be produced in larger quantities, but for a wide array of applications, a coating of the particle surface is necessary to obtain stable colloidal solutions in water and to provide an anchor to add further functionality. Several synthetic routes have been developed to address these issues, e.g., coating with charged, aliphatic molecules or SiO_2.

Magnetosomes are themselves provided with colloidal stability in aqueous solution as they possess a lipid membrane preventing crystallite aggregation and transmembrane proteins which can be used as anchors for diverse modifications, introduced either chemically or genetically (Lang et al. 2007). Functionalizations have been chemically introduced using standard methods for covalent modifications of proteins by, e.g., glutaraldehyde cross-linking or NHS-esters (Matsunaga and Kamiya 1987). Secondary modifications have been produced using biotin–streptavidin links (Amemiya et al. 2005; Ceyhan et al. 2006). Genetically introduced modifications have been obtained by fusing proteins of interest to magnetosome membrane proteins.

First attempts were carried out with the MagA protein and the myristoyl-anchored Mms16 in *M. magneticum* (Nakamura et al. 1995; Matsunaga et al. 2000; Yoshino et al. 2004). A luciferase-based assay later identified Mms13 as better anchor in magnetosomes of this organism, putatively by a crystal-binding mechanism (Yoshino and Matsunaga 2006). More recently, to detect the best potential membrane proteins as anchors in *M. gryphiswaldense*, several candidates were fused to the enhanced green fluorescence protein (EGFP) and analyzed according to resulting magnetosome particle fluorescence by flow cytometry and fluorescence microscopy. As biomineralization of magnetite is favored under anaerobic and microaerobic conditions, whereas the final maturation of GFP requires oxygen, growth parameters had to be optimized to perform both tasks. It was found that the most abundant magnetosome membrane protein, MamC, which is the ortholog to Mms13, serves best as anchor protein for possible fusions (Lang and Schüler 2008).

As proofs-of-principle, several biotechnologically relevant assays based on magnetosomes as magnetic carrier colloids have been developed, such as immunoassays for pollutants, hormones and toxic detergents, ligand-receptor binding assays, and target cell separation methods (Tanaka and Matsunaga 2000; Matsunaga et al. 2003; Kuhara et al. 2004; Yoshino et al. 2004). Furthermore, modified magnetosomes have been used to extract DNA and to discriminate single-nucleotide polymorphisms by automated systems (Tanaka et al. 2003; Yoza et al. 2003).

More recently, magnetosomes have been tested as potential drug carriers for antitumor treatments (Sun et al. 2007; Sun et al. 2008) and as a contrast agents for MRI (Lisy et al. 2007). Particularly for the latter application, magnetosomes are becoming interesting, because magnetic single-domain particles with anisotropical shapes elongated along an easy magnetization axis show longer relaxation times which results in higher imaging contrast (Vereda et al. 2009).

1.5 Conclusion and Outlook

In this chapter, we described magnetotactic bacteria and their remarkable capability to biomineralize magnetite nanoparticles, so-called magnetosomes. This material is extraordinary because it has been optimized with respect to its function as a magnetic field actuator. The crystallites are made of stoichiometric magnetite, which is the most efficient iron oxide phase with respect to magnetization and iron usage. Particles are in the single-magnetic-domain size range and this feature is optimal for their function. They are sometimes elongated along the easy magnetization axis, improving the stability of the magnetic dipole against thermal fluctuations. Finally, chain formation of magnetosomes adds magnetic dipoles of single particles and enables the bacteria to align along the field lines of Earth's magnetic field.

Many aspects of this biomineralization process remain unclear. Iron uptake and particularly transport into the intracellular vesicles has still to be resolved. Furthermore, it is not understood how MTB are able to form the appropriate iron phase under the very soft physiological chemical conditions. Compartmentalization might allow them to control local pH and redox potential in order to precipitate the magnetite phase. However, it has not been shown which conditions exist within the magnetosome organelles. Proteins that are involved in all the stages of magnetite formation and particle chain alignment have been identified to some extent, but their specific roles remain widely unclear. Concerning the chemistry, one particularly interesting fact is that MTB are able to produce particle morphologies that break the crystal symmetry of magnetite. This has thus far not been achieved synthetically under soft chemical conditions. Learning from MTB, chemists and nanotechnologists might be able to produce similar crystals of any shape and thus with interesting magnetic properties using environmentally friendly methods.

Acknowledgments Prof. Müller is acknowledged for inviting us to contribute to this book. We thank Nicolas Menguy for the TEM images of magnetosomes in Fig. 1.3. The authors want to thank Prof. Fratzl for offering them the opportunity to join his department. Discussions with current and older group members were appreciated. Corrections and suggestions on the chapter by Kevin Eckes and Matthew Harrington are acknowledged. Research in the laboratory is supported by the Deutsche Forschungsgemeinschaft (DFG), the European Union, and the Max Planck Society.

References

Amann R, Peplies J, Schüler D (2007) Diversity and taxonomy of magnetotactic bacteria. In: Schüler D (ed) Magnetoreception and magnetosomes in bacteria. Springer, Heidelberg

Amemiya Y, Tanaka T, Yoza B, Matsunaga T (2005) Novel detection system for biomolecules using nano-sized bacterial magnetic particles and magnetic force microscopy. J Biotechnol 120:308–314

Amemiya Y, Arakaki A, Staniland SS, Tanaka T, Matsunaga T (2007) Controlled formation of magnetite crystal by partial oxidation of ferrous hydroxide in the presence of recombinant magnetotactic bacterial protein Mms6. Biomater 28:5381–5389

Arakaki A, Webbs J, Matsunaga T (2003) A novel protein tightly bound to bacterial magnetite particles in *Magnetospirillum magnetotacticum* strain AMB-1. J Biol Chem 278:8745–8750

Balkwill D, Maratea D, Blakemore RP (1980) Ultrastructure of a magnetotactic spirillum. J Bacteriol 141:1399–1408

Bazylinski DA, Frankel RB (2004) Magnetosome formation in prokaryotes. Nat Rev Microbiol 2:217–230

Bellini S (2009a) Further studies on "magnetosensitive bacteria". Chi J Oceanogr Limnol 27:6–12

Bellini S (2009b) On a unique behavior of freshwater bacteria. Chi J Oceanogr Limnol 27:3–5

Blakemore RP (1975) Magnetotactic bacteria. Science 190:377–379

Blakemore RP (1982) Magnetotactic bacteria. Ann Rev Microbiol 36:217–238

Blakemore RP, Maratea D, Wolfe RS (1979) Isolation and pure culture of freshwater magnetic spirillum in chemically defined medium. J Bacteriol 140:720–729

Calugay RJ, Miyashita H, Okamura Y, Matsunaga T (2003) Siderophore production by the magnetic bacterium Magnetospirillum magneticum AMB-1. FEMS Microbiol Let 218: 371–375

Calugay RJ, Okamura Y, Wahyudi AT, Takeyama H, Matsunaga T (2004) Siderophore production of a periplasmic transport binding protein kinase gene defective mutant of Magnetospirillum magneticum AMB-1. Biochem Biophys Res Comm 323:852–857

Calugay RJ, Takeyama H, Mukoyama D, Fukuda Y, Suzuki T, Kanoh K, Matsunaga T (2006) Catechol siderophore excretion by magnetotactic bacterium *Magnetospirillum magneticum* AMB-1. J Biosci Bioeng 101:445–447

Cartron ML, Maddocks S, Gillingham P, Craven CJ, Andrews SC (2006) Feo – transport of ferrous iron into bacteria. Biometals 19:143–157

Ceyhan B, Alhorn P, Lang C, Schüler D, Niemeyer CM (2006) Semisynthetic biogenic magnetosome nanoparticles for the detection of proteins and nucleic acids. Small 2:1251–1255

Cornell RM, Schwertmann U (2003) The Iron Oxides. Wiley-VCH Verlag GmBH & Co. KGaA, Weinheim

de Nooijer LJ, Toyofuku T, Kitazato H (2009) Foraminifera promote calcification by elevating their intracellular pH. Proc Natl Acad Sci U S A 106:15374–15378

Devouard B, Pósfai M, Hua X, Bazylinski DA, Frankel RB, Buseck PR (1998) Magnetite from magnetotactic bacteria: Size distributions and twinning. Am Miner 83:1387–1398

Dubbels BL, DiSpirito AA, Morton JD, Semrau JD, Neto JNE, Bazylinski DA (2004) Evidence for a copper-dependent iron transport system in the marine, magnetotactic bacterium strain MV-1. Microbiology 150:2931–2945

Dunin-Borkowski RE, McCartney MR, Frankel RB, Bazylinski DA, Pósfai M, Buseck PR (1998) Magnetic microstructure of magnetotactic bacteria by electron holography. Science 282:1868–1870

Dunlop DJ, Özdemir O (1997) Rock magnetism: fundamentals and frontiers. Cambridge University Press, Cambridge

Escolar L, Perez-Martin J, De Lorenzo V (1999) Opening the iron box: transcriptional metalloregulation by the fur protein. J Bacteriol 181:6223–6229

Faivre D, Schüler D (2008) Magnetotactic bacteria and magnetosomes. Chem Rev 108:4875–4898

Faivre D, Agrinier P, Menguy N, Zuddas P, Pachana K, Gloter A, Laval J-Y, Guyot F (2004) Mineralogical and isotopic properties of inorganic nanocrystalline magnetites. Geochim Cosmochim Acta 68:4395–4403

Faivre D, Böttger LH, Matzanke BF, Schüler D (2007) Intracellular magnetite biomineralization in bacteria proceeds by a distinct pathway involving membrane-bound ferritin and an iron(II) species. Angew Chem Int Ed 46:8495–8499

Faivre D, Menguy N, Pósfai M, Schüler D (2008) Effects of environmental parameters on the physical properties of fast-growing magnetosomes. Am Mineral 93:463–469

Farina M, Esquivel DMS, Lins de Barros H (1990) Magnetic iron-sulphur crystals from a magnetotactic microorganism. Nature 343:256–258

Fischer A, Schmitz M, Aichmayer B, Fratzl P, Faivre D (2011) Structural purity of magnetite nanoparticles in magnetotactic bacteria. J R Soc Interface 8:1011–1018

Flies CB, Jonkers HM, de Beer D, Bosselmann K, Böttcher ME, Schüler D (2005) Diversity and vertical distribution of magnetotactic bacteria along chemical gradients in freshwater microcosms. FEMS Microbiol Ecol 52:185–195

Frankel RB, Bazylinski DA (2003) Biologically induced mineralization by bacteria. Rev Mineral Geochem 54:95–114

Frankel RB, Blakemore RP (1991) Iron Biominerals. Plenum Press, New York and London

Frankel RB, Blakemore R, Wolfe RS (1979) Magnetite in freshwater magnetotactic bacteria. Science 203:1355–1356

Frankel RB, Papaefthymiou GC, Blakemore RP, O'Brien W (1983) Fe_3O_4 precipitation in magnetotactic bacteria. Biochim Biophys Acta 763:147–159

Frankel RB, Bazylinski DA, Johnson MS, Taylor BL (1997) Magneto-aerotaxis in marine coccoid bacteria. Biophy J 73:994–1000

Frankel RB, Williams TJ, Bazylinski DA (2007) Magneto-Aerotaxis. In: Schüler D (ed) Magnetoreception and magnetosomes in bacteria. Springer, Heidelberg

Gorby YA, Beveridge TJ, Blakemore R (1988) Characterization of the bacterial magnetosome membrane. J Bacteriol 170:834–841

Grünberg K, Müller EC, Otto A, Reszka R, Linder D, Kube M, Reinhardt R, Schüler D (2004) Biochemical and proteomic analysis of the magnetosome membrane in *Magnetospirillum gryphiswaldense*. Appl Eviron Microbiol 70:1040–1050

Heyen U, Schüler D (2003) Growth and magnetosome formation by microaerophilic Magnetospirillum strains in an oxygen-controlled fermentor. Appl Microbiol Biotechnol 61:536–544

Isambert A, Menguy N, Larquet E, Guyot F, Valet J-P (2007) Transmission electron microscopy study of magnetites in a freshwater population of magnetotactic bacteria. Am Mineral 92:621–630

Jogler C, Schüler D (2007) Genetic analysis of magnetosome biomineralization. In: Schüler D (ed) Magnetoreception and magnetosomes in bacteria. Springer, Heidelberg
Jogler C, Schüler D (2009) Genomics, genetics, and cell biology of magnetosome formation. Annu Rev Microbiol 63:501–521
Jogler C, Kube M, Schübbe S, Ullrich S, Teeling H, Bazylinski DA, Reinhardt R, Schüler D (2009) Comparative analysis of magnetosome gene clusters in magnetotactic bacteria provides further evidence for horizontal gene transfer. Environ Microbiol 11:1267–1277
Johnsen S, Lohmann KJ (2005) The physics and neurobiology of magnetoreception. Nat Rev Neurosci 6:703–712
Keim CN, Lins U, Farina M (2009) Manganese in biogenic magnetite crystals from magnetotactic bacteria. FEMS Microbiol Lett 292:250–253
Komeili A (2007) Molecular mechanisms of magnetosome formation. Ann Rev Biochem 76:351–366
Komeili A, Vali H, Beveridge TJ, Newman D (2004) Magnetosome vesicles are present prior to magnetite formation and MamA is required for their activation. Proc Natl Acad Sci USA 101:3839–3844
Komeili A, Li Z, Newman DK, Jensen GJ (2006) Magnetosomes are cell membrane invaginations organized by the actin-like protein mamK. Science 311:242–245
Kuhara M, Takeyama H, Tanaka T, Matsunaga T (2004) Magnetic cell separation using antibody binding with protein a expressed on bacterial magnetic particles. Anal Chem 76:6207–6213
Kundu S, Kale AA, Banpurkar AG, Kulkarni GR, Ogale SB (2009) On the change in bacterial size and magnetosome features for Magnetospirillum magnetotacticum (MS-1) under high concentrations of zinc and nickel. Biomater 30:4211–4218
Lang C, Schüler D (2008) Expression of green fluorescent protein fused to magnetosome proteins in microaerophilic magnetotactic bacteria. Appl Environ Microbiol 74:4944–4953
Lang C, Schüler D, Faivre D (2007) Synthesis of magnetite nanoparticles for bio- and nanotechnology: genetic engineering and biomimetics of bacterial magnetosomes. Macromol Biosci 7:144–151
Laurent S, Forge D, Port M, Roch A, Robic C, Vander Elst L, Muller RN (2008) Magnetic iron oxide nanoparticles: synthesis, stabilization, vectorization, physicochemical characterizations, and biological applications. Chem Rev 108:2064–2110
Lins U, McCartney MR, Farina M, Frankel RB, Buseck PR (2005) Habits of magnetosome crystals in coccoid magnetotactic bacteria. Appl Environ Microbiol 71:4902–4905
Lisy MR, Hartung A, Lang C, Schüler D, Richter W, Reichenbach JR, Kaiser WA, Hilger I (2007) Fluorescent bacterial magnetic nanoparticles as bimodal contrast agents. Invest Radiol 42:235–241
Lowenstam HA (1967) Lepidocrocite an apatite mineral and magnetite in teeth of chitons (Polyplacophora). Science 156:1373–1375
Mann S, Sparks N, Blakemore R (1987a) Structure, morphology and crystal growth of anisotropic magnetite crystals in magnetotactic bacteria. Proc R Soc Lond B 231:477–487
Mann S, Sparks N, Blakemore R (1987b) Ultrastructure and characterization of anisotropic magnetic inclusions in magnetotactic bacteria. Proc R Soc Lond B 231:469–476
Mann S, Sparks NHC, Frankel RB, Bazlinski DA, Jannasch HW (1990) Biomineralization of ferrimagnetic greigite (Fe_3S_4) and iron pyrite (FeS_2) in a magnetotactic bacterium. Nature 343:258–261
Matsunaga T, Arakaki A (2007) Molecular bioengineering of bacterial magnetic particles for biotechnological applications. In: Schüler D (ed) Magnetoreception and magnetosomes in bacteria. Springer, Heidelberg
Matsunaga T, Kamiya S (1987) Use of magnetic particles isolated from magnetotactic bacteria for enzyme immobilization. Appl Microbiol Biotechnol 26:328–332
Matsunaga T, Sakaguchi T, Tadokoro F (1991) Magnetite formation by a magnetic bacterium capable of growing aerobically. Appl Microbiol Biotechnol 35:651–655

Matsunaga T, Nakamura C, Burgess JG, Sode K (1992) Gene transfer in magnetic bacteria: transposon mutagenesis and cloning of genomic DNA fragments required for magnetosome synthesis. J Bacteriol 174:2748–2753

Matsunaga T, Togo H, Kikuchi T, Tanaka T (2000) Production of luciferase-magnetic particle complex by recombinant Magnetospirillum sp AMB-1. Biotechnol Bioeng 70:704–709

Matsunaga T, Ueki F, Obata K, Tajima H, Tanaka T, Takeyama H, Goda Y, Fujimoto S (2003) Fully automated immunoassay system of endocrine disrupting chemicals using monoclonal antibodies chemically conjugated to bacterial magnetic particles. Anal Chim Acta 475:75–83

Matsunaga T, Okamura Y, Fukuda Y, Wahyudi AT, Murase Y, Takeyama H (2005) Complete genome sequence of the facultative anaerobic magnetotactic bacterium Magnetospirillum sp strain AMB-1. DNA Res 12:157–166

Muxworthy AR, Williams W (2006) Critical single-domain/multidomain grain sizes in noninteracting and interacting elongated magnetite particles: Implications for magnetosomes. J Geophys Res 111:B12S12

Muxworthy AR, Williams W (2009) Critical superparamagnetic/single-domain grain sizes in interacting magnetite particles: implications for magnetosome crystals. J R Soc Interf 6:1207–1212

Nakamura C, Sakaguchi T, Kudo S, Burgess JG, Sode K, Matsunaga T (1993) Characterization of Iron Uptake in the Mangetic Bacterium *Aquaspirillum* sp. AMB-1. Appl Biochem Biotechnol 39(40):169–176

Nakamura C, Kikuchi T, Burgess JG, Matsunaga T (1995) Iron-regulated expression and membrane localization of the maga protein in magnetospirillum sp strain AMB-1. J Biochem 118:23–27

Nakazawa H, Arakaki A, Narita-Yamada S, Yashiro I, Jinno K, Aoki N, Tsuruyama A, Okamura Y, Tanikawa S, Fujita N, Takeyama H, Matsunaga T (2009) Whole genome sequence of Desulfovibrio magneticus strain RS-1 revealed common gene clusters in magnetotactic bacteria. Genome Res 19:1801–1808

Nash CZ (2004) Magnetic microbes in Mono Lake. Mono Lake Newsletter Fall 2004:14

Nies DH (2003) Efflux-mediated heavy metal resistance in prokaryotes. Fems Microbiol Rev 27:313–339

Pallen MJ, Wren BW (1997) The HtrA family of serine proteases. Mol Microbiol 26:209–221

Paoletti LC, Blakemore RP (1986) Hydroxamate production by Aquaspirillum magnetotacticum. J Bacteriol 167:73–76

Philipse AP, Maas D (2002) Magnetic Colloids from Magnetotactic Bacteria: Chain Formation and Colloidal Stability. Langmuir 18:9977–9984

Politi Y, Metzler RA, Abrecht M, Gilbert B, Wilt FH, Sagi I, Addadi L, Weiner S, Gilbert P (2008) Transformation mechanism of amorphous calcium carbonate into calcite in the sea urchin larval spicule. Proc Natl Acad Sci U S A 105:17362–17366

Pósfai M, Moskowitz BM, Arato B, Schüler D, Flies C, Bazylinski DA, Frankel RB (2006) Properties of intracellular magnetite crystals produced by Desulfovibrio magneticus strain RS-1. Earth Planet Sci Lett 249:444–455

Pradel N, Santini C-L, Bernadac A, Fukumori Y, Wu L-F (2006) Biogenesis of actin-like bacterial cytoskeletal filaments destined for positioning prokaryotic magnetic organelles. Proc Natl Acad Sci USA 103:17485–17489

Richter M, Kube M, Bazylinski DA, Lombardot T, Glockner FO, Reinhardt R, Schüler D (2007) Comparative genome analysis of four magnetotactic bacteria reveals a complex set of group-specific genes implicated in magnetosome biomineralization and function. J Bacteriol 189:4899–4910

Rodgers FG, Blakemore RP, Blakemore NA, Frankel RB, Bazylinski DA, Maratea D, Rodgers C (1990) Intercellular structure in a many-celled magnetotactic prokaryote. Arch Microbiol 154:18–22

Rong CB, Huang YJ, Zhang WJ, Jiang W, Li Y, Li JL (2008) Ferrous iron transport protein B gene (feoB1) plays an accessory role in magnetosome formation in Magnetospirillum gryphiswaldense strain MSR-1. Res Microbiol 159:530–536

Sakaguchi T, Arakaki A, Matsunaga T (2002) Desulfovibrio magneticus sp nov., a novel sulfate-reducing bacterium that produces intracellular single-domain-sized magnetite particles. IntJ Syst Evol Microbiol 52:215–221

Sandy M, Butler A (2009) Microbial iron acquisition: marine and terrestrial siderophores. Chem Rev 109:4580–4595

Scheffel A, Schüler D (2007) The acidic repetitive domain of the magnetospirillum gryphiswaldense mamj protein displays hypervariability but is not required for magnetosome chain assembly. J Bacteriol 189:6437–6446

Scheffel A, Gruska M, Faivre D, Linaroudis A, Plitzko JM, Schüler D (2006) An acidic protein aligns magnetosomes along a filamentous structure in magnetotactic bacteria. Nature 440:110–115

Scheffel A, Gärdes A, Grünberg K, Wanner G, Schüler D (2008) The major magnetosome proteins MamGFDC are not essential for magnetite biomineralization in *Magnetospirillum gryphiswaldense*, but regulate the size of magnetosome crystals. J Bacteriol 190:377–386

Schleifer K-H, Schüler D, Spring S, Weizenegger M, Amann R, Ludwig W, Köhler M (1991) The genus *Magnetospirillum* gen. nov., description of *Magnetospirillum gryphiswaldense* sp. nov. and transfer of *Aquaspirillum magnetotacticum* to *Magnetospirillum magnetotacticum* comb. nov. Syst Appl Microbiol 14:379–385

Schubbe S, Williams TJ, Xie G, Kiss HE, Brettin TS, Martinez D, Ross CA, Schuler D, Cox BL, Nealson KH, Bazylinski DA (2009) Complete genome sequence of the chemolithoautotrophic marine magnetotactic coccus strain MC-1. Appl Environ Microbiol 75:4835–4852

Schübbe S, Williams TJ, Xie G, Kiss HE, Brettin TS, Martinez D, Ross CA, Schüler D, Cox BL, Nealson KH, Bazylinski DA (2009) Complete genome sequence of the chemolithoautotrophic marine magnetotactic coccus strain MC-1. Appl Environ Microbiol 75:4835–4852

Schüler D (2008) Genetics and cell biology of magnetosome formation in magnetotactic bacteria. FEMS Microbiol Rev 32:654–672

Schüler D, Baeuerlein E (1996) Iron-limited growth and kinetics of iron uptake in Magnetospirilum gryphiswaldense. Arch Microbiol 166:301–307

Schüler D, Baeuerlein E (1998) Dynamics of iron uptake and Fe_3O_4 biomineralization during aerobic and microaerobic growth of *Magnetospirillum gryphiswaldense*. J Bacteriol 180:159–162

Schultheiss D, Schüler D (2003) Development of a genetic system for Magnetospirillum gryphiswaldense. Archives Microbiol 179:89–94

Sparks NHC, Courtaux L, Mann S, Board RG (1986) Magnetotactic bacteria are widely distributed in sediments in the U.K. FEMS Microbiol Let 37:305–308

Spring S, Amann R, Ludwig W, Schleifer K-H, van Gemerden H, Petersen N (1993) Dominating role of an unusual magnetotactic bacterium in the microaerobic zone of a freshwater sediment. Appl Environ Microbiol 50:2397–2403

Staniland S, Ward B, Harrison A, van der Laan G, Telling N (2007) Rapid magnetosome formation shown by real-time x-ray magnetic circular dichroism. Proc Natl Acad Sci USA 104:19524–19528

Staniland S, Williams W, Telling N, Van Der Laan G, Harrison A, Ward B (2008) Controlled cobalt doping of magnetosomes in vivo. Nat Nano 3:158–162

Sun J-B, Duan J-H, Dai S-L, Ren J, Zhang Y-D, Tian J-S, Li Y (2007) *In vitro* and in vivo antitumor effects of doxorubicin loaded with bacterial magnetosomes (DBMs) on H22 cells: The magnetic bio-nanoparticles as drug carriers. Cancer Lett 258:109–117

Sun JB, Duan JH, Dai SL, Ren J, Guo L, Jiang W, Li Y (2008) Preparation and anti-tumor efficiency evaluation of doxorubicin-loaded bacterial magnetosomes: magnetic nanoparticles as drug carriers isolated from magnetospirillum gryphiswaldense. Biotechnol Bioeng 101:1313–1320

Suzuki T, Okamura Y, Calugay RJ, Takeyama H, Matsunaga T (2006) Global gene expression analysis of iron-inducible genes in *Magnetospirillum magneticum* AMB-1. J Bacteriol 188:2275–2279

Suzuki T, Okamura Y, Arakaki A, Takeyama H, Matsunaga T (2007) Cytoplasmic ATPase involved in ferrous ion uptake from magnetotactic bacterium Magnetospirillum magneticum AMB-1. FEBS Let 581:3443–3448

Tanaka T, Matsunaga T (2000) Fully automated chemiluminescence immunoassay of insulin using antibody-protein A-bacterial magnetic particle complexes. Anal Chem 72:3518–3522

Tanaka T, Maruyama K, Yoda K, Nemoto E, Udagawa Y, Nakayama H, Takeyama H, Matsunaga T (2003) Development and evaluation of an automated workstation for single nucleotide polymorphism discrimination using bacterial magnetic particles. Biosens Bioelectron 19:325–330

Taoka A, Asada R, Sasaki H, Anzawa K, Wu L-F, Fukumori Y (2006) Spatial localizations of Mam22 and Mam12 in the magnetosomes of magnetospirillum magnetotacticum. J Bacteriol 188:3805–3812

Taoka A, Asada R, Wu LF, Fukumori Y (2007) Polymerization of the actin-like protein MamK, which is associated with magnetosomes. J Bacteriol 189:8737–8740

Taoka A, Umeyama C, Fukumori Y (2009) Identification of iron transporters expressed in the magnetotactic bacterium *Magnetospirillum magnetotacticum*. Curr Microbiol 58:177–181

Taylor AP, Barry JC (2004) Magnetosomal matrix: ultrafine structure may template biomineralization of magnetosomes. J Microsc 213:180–197

Ullrich S, Kube M, Schübbe S, Reinhardt R, Schüler D (2005) A hypervariable 130-kilobase genomic region of Magnetospirillum gryphiswaldense comprises a magnetosome island which undergoes frequent rearrangements during stationary growth. J Bacteriol 187:7176–7184

Vali H, Forster O, Amarantidid G, Petersen H (1987) Magnetotactic bacteria and their magnetofossils in sediments. Earth Planet Sci Lett 86:389–400

Vereda F, de Vicente J, Hidalgo-Alvarez R (2009) Physical properties of elongated magnetic particles: magnetization and friction coefficient anisotropies. ChemPhysChem 10:1165–1179

Winklhofer M (2007) Magnetite-based magnetoreception in higher organisms. In: Schüler D (ed) Magnetoreception and magnetosomes in bacteria. Springer, Heidelberg

Winklhofer M, Petersen N (2007) Paleomagnetism and magnetic bacteria. In: Schüler D (ed) Magnetoreception and magnetosomes in bacteria. Springer, Heidelberg

Winklhofer M, Holtkamp-Rötzler E, Hanzlik M, Fleissner G, Petersen N (2001) Clusters of superparamagnetic magnetite particles in the upper-beak skin of homing pigeons: evidence of a magnetoreceptor? Eur J Mineral 13:659–669

Xiong Y, Ye J, Gu XY, Chen QW (2007) Synthesis and assembly of magnetite nanocubes into flux-closure rings. J Phys Chem C 111:6998–7003

Yijun H, Weijia Z, Wei J, Chengbo R, Ying L (2007) Disruption of a fur-like gene inhibits magnetosome formation in *magnetospirillum gryphiswaldense* MSR-1. Biochem-Moscow 72:1247–1253

Yoshino T, Matsunaga T (2006) Efficient and Stable Display of Functional Proteins on Bacterial Magnetic Particles Using Mms13 as a Novel Anchor Molecule. Appl Environ Microbiol 72:465–471

Yoshino T, Takahashi M, Takeyama H, Okamura Y, Kato F, Matsunaga T (2004) Assembly of G protein-coupled receptors onto nanosized bacterial magnetic particles using Mms16 as an anchor molecule. Appl Environ Microbiol 70:2880–2885

Yoza B, Arakaki A, Matsunaga T (2003) DNA extraction using bacterial magnetic particles modified with hyperbranched polyamidoamine dendrimer. J Biotechnol 101:219–228

Chapter 2
Maxi- and Mini-Ferritins: Minerals and Protein Nanocages

Loes E. Bevers and Elizabeth C. Theil

Contents

Abstract Ferritins synthesize ferric oxide biominerals and are central to all life for concentrating iron and protection against oxidative stress from the ferrous and oxidant chemistry. The ferritin protein nanocages and biomineral synthesis are discussed in terms of wide biological distribution of the maxi-ferritins (24 subunit ± heme) and mini-ferritins (Dps) (12 subunit), conservations of the iron/oxygen catalytic sites in the protein cages, mineral formation (step i. Fe(II) entry and binding, step ii. O_2 or H_2O_2 binding and formation of transition intermediates, step iii. release of differric oxo mineral precursors from active sites, step iv.

E.C. Theil (✉)
Council for BioIron, CHORI (Children's Hospital Oakland Research Institute), 5700 Martin Luther King, Jr. Way, Oakland, CA 94609, USA

Department of Nutritional Sciences and Molecular Toxicology, University of California-Berkeley, Berkeley, CA 94720, USA
e-mail: etheil@chori.org

W.E.G. Müller (ed.), *Molecular Biomineralization*, Progress in Molecular and Subcellular Biology 52, DOI 10.1007/978-3-642-21230-7_2,

nucleation and mineralization) properties of the minerals, and protein control of mineral dissolution and release of Fe(II). Pores in ferritin protein cages control iron entry for mineralization and iron exit after mineral dissolution. The relationship between phosphate or the presence of catalytically inactive subunits (animal L subunits) and ferritin iron mineral disorder is developed based on new information about contributions of ferritin protein cage structure to nucleation in protein cage subunit channels that exit close enough to those of other subunits and exiting mineral nuclei to facilitate bulk mineral formation. How and where protons move in and out of the protein during mineral synthesis and dissolution, how ferritin cage assembly with 12 or 24 subunits is encoded in the widely divergent ferritin amino acid sequences, and what is the role of the protein in synthesis of the bulk mineral are all described as problems requiring new approaches in future investigations of ferritin biominerals.

2.1 Introduction

Ferritins are protein cages that synthesize ferric oxides (hydrated minerals), which are located in a central cavity of the protein, 5–8 nm in diameter (e.g., [Lewin et al. 2005; Liu and Theil 2005]) (Fig. 2.1). Ferrous ions and either dioxygen or hydrogen peroxide are the substrates for protein subunit- based catalytic sites, in the cage, that initiate biomineral synthesis. Iron minerals in ferritin have two major functions. First, ferritin minerals are nutritive iron concentrates that slowly release the iron for use in the synthesis of new iron centers for protein catalysts as in heme, iron–sulfur clusters, or iron bound entirely by protein in amino acid side chains ("non-heme iron"). Iron-containing proteins are key to electron transfer chains in respiration

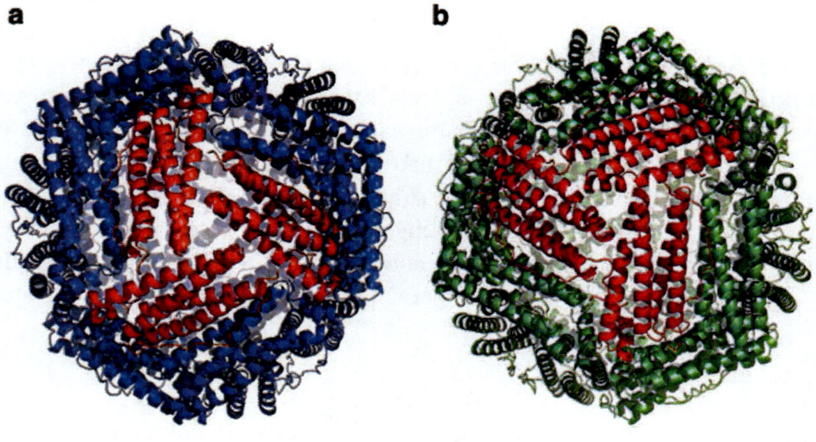

Fig. 2.1 Maxi-ferritin protein cages, which synthesize iron oxide biominerals. View toward the threefold axis: (**a**) *Escherichia coli* bacterioferritin (pdb file: 1BFR, an example of a low phosphate ferritin biomineral (**b**) *Rana catesbeiana* (bullfrog) M-ferritin (pdb file: 1MFR, an example of a low phosphate biomineral. Red-a set of three subunits around one of eight Fe(II) entry/exit pores

(heme proteins) and photosynthesis (iron–sulfur proteins) and play important roles in hydroxylation reactions (e.g., Stiles et al. 2009; Cojocaru et al. 2007) oxygen sensing (Semenza 2009), and reduction, exemplified by the synthesis of deoxyribose nucleotides from ribonucleotides, central to DNA synthesis and replication. Second, ferritins are antioxidants that consume iron and oxygen during mineralization. The two metabolic roles of ferritins are emphasized by genetic regulation with oxidants and iron (Pham et al. 2004; Hintze and Theil 2005; Hintze et al. 2007). In animals, both DNA and mRNA are regulated, and are selectively targeted by oxygen or iron (Theil and Goss 2009). Iron and oxygen consumption to make ferritin minerals is part of a feedback loop where excess iron and oxygen activate ferritin DNA and mRNA to increase ferritin synthesis. Subsequent protein accumulation, which consumes iron and oxygen, decreases the signals and shuts down ferritin synthesis (Theil and Goss 2009).

The protein nanocages of the ferritin superfamily are currently known to occur in two sizes. First, 24 subunit ferritins were observed, in the middle of the twentieth century, in many plant and animals tissues by using electron microscopy with the density of the minerals as a reporter; later, other biophysical and molecular biological methods were used to identify ferritin in bacteria (reviewed in [Theil 1987]). Ferritins occur in all tissues with a wide range of concentrations that reflect environmental and developmental signals. In plants, ferritin is targeted exclusively to cell organelles, whereas in animals, ferritins are in the cytoplasm, and in the lysosomal and mitochondrial organelles. In bacteria, ferritin accumulations are particularly sensitive to the culture phase, environmental iron, and oxidant concentrations; in logarithmic cultures of *Escherichia coli*, e.g., ferritin concentrations are low unless extracellular stressors are added (Nandal et al. 2009). Smaller ferritins or mini-ferritins, discovered as large proteins induced by stress, protect bacterial DNA from damage by consuming ferrous ions and hydrogen peroxide; accordingly they were first named *D*NA *P*rotection *D*uring *S*tress (DPS) proteins (Chiancone 2010). Only later, when protein crystal structures were obtained, was the ferritin family protein cage structure discovered (Grant et al. 1998). However, since the cages are built from only 12, 4 α-bundle subunits, with a smaller cavity that accommodates a smaller mineral (<500 iron atoms), they were called mini-ferritins to distinguish them from the larger, 24 subunit maxi-ferritins. Mini-ferritins, to date, are restricted to bacteria and archaea, contrasting with maxi-ferritins found in bacteria, archaea, and eukaryotes.

2.2 Ferritin Distribution in Organisms of Land and Sea

Ferritins can be identified based on primary sequence in the genomes of archaea, bacteria, and eukaryotes (Fig. 2.2a). Ferritins from organisms of each of the kingdoms have been purified or expressed, heterologously, with variable levels of characterization. Kinetic models and molecular pathways for the uptake of iron, the oxidation from ferrous to ferric, and subsequent reduction and release have been

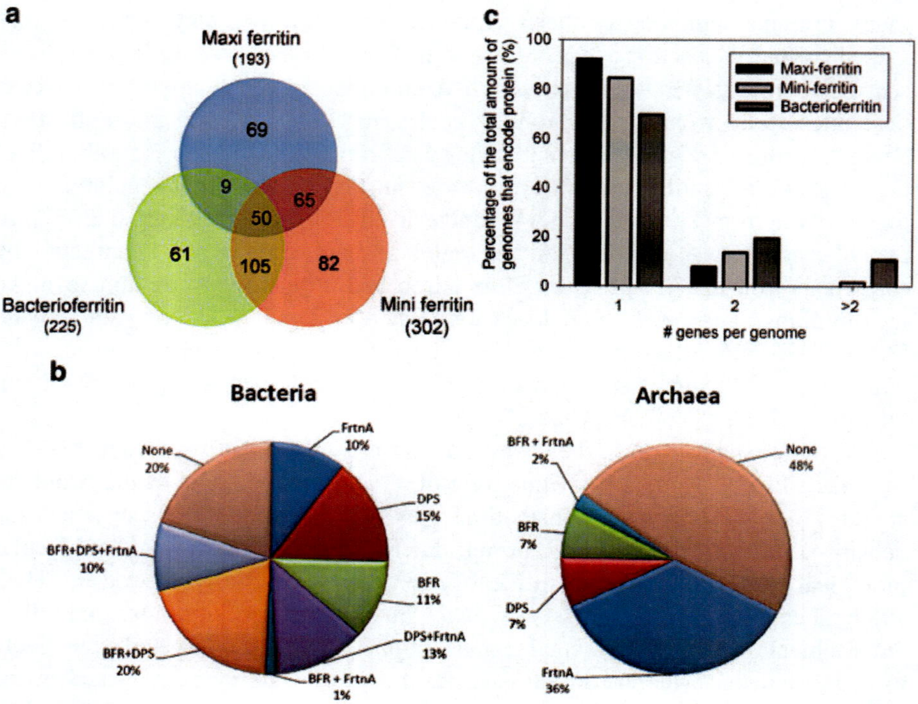

Fig. 2.2 Distribution of the maxi-ferritin (FTNA and BFR) and mini-ferritin (Dps) genes in prokayotes. (**a**). Venn diagram showing the distribution of the three different ferritin genes in 566 sequenced bacterial and archaeal genomes; (**b**). circular diagram showing the distribution for the bacterial (524) and archaeal (Dobson 2001) branches separately. (**c**) The number of genes encoding maxi-ferritins (24 subunits-210), mini-ferritins (Dps)(12 subunits-355), and bacterioferritins (24 subunits; 12 hemes-321) in 441 genomes

Table 2.1 Ferritins with similar protein nanocage structure exhibit a wide range of primary structure (sequence identity/sequence similarity)

	E. coli FTNA	*E. coli* DPS	*E. coli* BFR	Human H	Human L
E. coli FTNA	–				
E. coli DPS	8%/34%	–			
E. coli BFR	12%/41%	9%/37%	–		
Human H	21%/47%	14%/39%	20%/38%	–	
Human L	20%/44%	12%/38%	17%/37%	53%/75%	–

developed over time, especially for animal ferritins and bacterial ferritins (Theil et al. 2008; Chiancone and Ceci 2010; Le Brun et al. 2010). The multiple steps in ferritin biomineralization will be discussed in Sect. 3. In this section, the focus is on the distribution of the different types of ferritin and the conservation of the catalytic sites throughout the tree of life. The conservation of the secondary and quaternary structures, and the primary sequence of the catalytic sites among different ferritins is remarkably constant (Table 2.1), given the wide range of sequence differences.

2.2.1 Ferritin in Prokaryotes

Primary sequences for both maxi- and mini-ferritins are widespread in the genomes of prokaryotes, and even within the genome of one single organism. *E. coli* is an example of an organism that has gene copies for multiple ferritins. *E. coli* mini-ferritin (Dps), maxi-ferritin (FTNA), and bacterioferritin (BFR) were used as queries in BLAST searches in order to study the distribution of recognizable ferritins among prokaryotes (P-value $<5 \times 10^{-5}$), and to generate alignments that provide insights into the conservation of active site residues. The term "recognizable" stresses the limitation of the approach to find homologous proteins based on sequence homology on its own. In the case of ferritin in particular, the three-dimensional protein structure is highly conserved and essential for its function. However, there are no means yet to screen multiple genomes for the presence of specific, quarternary, protein structures. The results of the sequence homology studies cannot be the "final picture" of the ferritin family, because of sequencing gaps. For example, sequenced archaeal genomes, to date, are 12-fold fewer than sequence bacterial genomes (42 vs. 524). We only present a snapshot to illustrate the broad distribution patterns of mini-ferritins (Dps), maxi-ferritins, and bacterioferritins (BFRs), and the conservation of the catalytic sites.

The sequence similarity among microbial maxi-ferritins ranges from 24.2% to 100%. In a total of 566 genomes (524 bacteria and 42 archaea), 321 BFR homologues (sequence identity ranging from 24.2% to 100% and similarity from 42.6% to 100%, P-value $< 3.8 \times 10^{-5}$), 355 mini-ferritin (DPS) homologues (20.9–100%, 41.3–100%, P-value $< 3.4 \times 10^{-5}$), and 210 maxi-ferritin homologues (20.8–100%, 40.9–100%, P-value $< 4.3 \times 10^{-6}$) could be identified.

A group of 27 maxi-ferritin homologues (including *E. coli* FTNB) were discarded in the group when none of the active site residues aligned with the remaining 210 homologues. The sequence conservation in ferritin among all three groups, maxi-ferritins in bacteria + archaea, mini-ferritins in bacteria + archaea, and maxi-ferritins in eukaryotes is enormous and emphasizes that the requirements for forming the assembly of 4 α-helix into a hollow protein nanocage is coded in secondary/tertiary codes still to be defined.

Table 2.2 The distribution of ferritins among bacteria and archaea. Absolute numbers and percentages of the occurrence of recognizable maxi-ferritins, min-ferritins and bacterioferritins are presented from sequences of 524 bacterial genomes and 42 archaeal genomes obtained from BLAST searches against *E. coli* FTNA (ftnA/b-1905), DPS (dps/b-0812), and BFR (bfr/b-3336)

	No. homologous genes	No. genomes	No. bacterial genomes (% of total)	No. archaeal genomes (% of total)
Maxi-ferritin	321	225	221 (42)	4 (10)
Mini-ferritin	355	302	299 (57)	3 (7)
Bacterioferritin	210	193	178 (34)	15 (36)

Using the CMR database (J. Craig Venter Institute)

Over 43% (230 out of 524) of the currently available bacterial genomes have more than one type of recognizable ferritin encoded in the DNA, indicating the importance of these proteins. This percentage is lower in archaea; only 2% of the sequenced genomes encode more than one type of ferritin, which was exclusively a combination of bacterioferritin and maxi-ferritin (Fig. 2.2b). While multiple gene copies of the same type of ferritin occur rarely in archaea (Table 2.2 and Fig. 2.2), many bacteria have multiple genes encoding functional ferritin proteins.

2.2.2 *Ferritin in Eukaryotes*

The distribution of ferritin genes in eukaryotes is fairly different; mini-ferritins and heme-ferritins (BFR) have not been found to date. Studying the distribution of ferritin genes in eukaryotic genomes using BLAST alignments is less straightforward than in prokaryotic genomes for a number of reasons: (1) the presence of introns; (2) large numbers of truncated or pseudogenes; (3) multiple gene annotations. Eukaryotic maxi-ferritins are mixtures of subunits encoded in different genes. In plants, all the genes encode H-type subunits that are catalytically active. The multiple names for the catalytic sites include oxidoreductase, ferroxidase, F_{ox}, FC.

Ferritin cages in animals are co-assembled from active H-type subunits or inactive subunits called L-subunits. Historically, H and L stands for heart, heavy, or higher; L stands for liver, light or lower, referring to the organ where the subunit is dominant or the mass, or migration in an SDS-PAGE gel. However, the only consistent parameter is catalytic activity since H and L combination are found in all tissues, and SDS gel mobility does not always relate to mass. Multiple genes encoding H subunits occur in fish, frogs, mice, humans, maize, soy, etc. where the H subunits are called H and M (frog, fish) (Dickey et al. 1987; Yamashita et al. 1996) or H and mitochondrial (human, mouse) (Corsi et al. 2002; Arosio et al. 2009) or H-1-4/ATF 1–4 (plant ferritins) (Dong et al. 2008; Briat et al. 2009). Some ferritins have been identified from intronless genes that possibly result from the high stability of the animal ferritin mRNA enabling copying by viral reverse transcriptases during infection. In prokaryotes, multiple ferritin genes are expressed at different times in the culture cycle or with different stimuli so that the bacterial ferritin nanocages assemble usually only one type of subunit.

To determine the distribution of eukaryotic ferritin sequences in Nature, human H-ferritin was used in a BLAST search against all eukaryotic curated protein sequences (the UniProtKB/Swiss-Prot database, Expasy) in order to generate alignments of animal and plant ferritins to get insight into the conservation of active site residues among the A and the B site. A total of 55 homologues sequences were identified (sequence identity ranging from 22% to 99.45% and similarity from 40% to 100%, P-value $< 67 \times 10^{-11}$) and the conserved active sites and variants are shown in Table 2.3. Inherent in the analysis is the constraint imposed by the absence of structural information, which will exclude therefore many ferritins. Such limits of the search tools can explain the apparent absence of ferritin in organisms

Table 2.3 Sequence patterns of putative, catalytically relevant Fe^{2+} ligands in mini- and maxi-ferritins. Ferritin homologues were identified by the alignment of prokaryotic (J. Craig Venter Institute) and eukaryotic genomes (Uniprot knowledgebase, Swiss-Prot). Templates: *E. coli* proteins FTNA, DPS, and BFR – prokaryotes and human H-FTNA – eukaryotes

Distribution	Example	Homologues (total)	Fe 1 (Site A)[a]	Fe 2 (Site B)[a]
		Maxi-ferritins		
Prokaryotic FTNA	*E. coli* FTNA	196 (210[b])	E, ExxH	E, E, QxxE
Animal H-ferritin	*Human* H	33 (37[c])	E, ExxH	E, QxxA/S
Plant H-ferritin	*Soybean* H	17 (18[d])	E, ExxH	E, QxxA/S
Prokaryotic BFR	*E. coli* BFR	239 (321[e])	E,ExxH	E, ExxH
		Mini-ferritins		
Prokaryotic DPS	*E. coli* DPS	164 (355[f])	H, DxxxE	HxxxD

[a]Diiron sites in oxygenases and reductases are Fe 1 and Fe 2 (34); ferritin diiron sites were Fe A and Fe B before the similarity to the oxygenases was recognized.

[b]*Prokaryotic FTNA active site variants*: E, (E/K)xxH (Fe 1), E, E, Qxx(Q/K/D/A/A) (Fe 2) in 14/210 homologues.

[c]*Animal FTNA active site variants*: E, (E/G/S)xx(H/RD) (Fe 1), (E/V/K), (Q/V)xx(A/S/D/G/W) (Fe 2) in 4/37 homologues.

[d]*Plant FTN active site variants*: E, HxxH (site A), E, QxxA (Fe 2) in 1/18 homologues.

[e]*Prokaryotic BFR active site variants*: (E/L/H/Q/S/Y/W),(D/E/A/V/K/Q/D) xx (H/A/Y/Q/W/N/E/H/T) (Fe 1), (E/V/M/N/V/Q/G), (T/E/N/K/A/C/S/Q/V/)xx(H/A/Q/Y/I) (Fe2) in 82/ 321 homologues.

[f]*Prokaryotic Dps active site variants*: Fe1 100% conserved, Hxxx(E/M/Q/G) (Fe 2) in 191/ 355 homologues.

such as *Saccharomyces cerevisiae*. The current list of ferritin homologues, nevertheless, demonstrates primary sequence features that are highly conserved, as first observed in the first comparisons of ferritins from bacteria and animals (Grossman et al. 1992).

2.2.3 Conservation of Active Sites in Ferritins of Prokaryotes and Eukaryotes

Catalytic events that initiate biomineral synthesis in ferritins occur in each ferritin subunit. Thus, ferritins have multiple catalytic sites, 24 or 12 in bacteria and archaea, 24 in plants, and up to 24 in animals, depending on the fraction of L subunits; the distribution of H and L subunits in a ferritin nanocage changes the degree of order/crystallinity of the ferritin mineral (St Pierre et al. 1991). At the oxidoreductase sites, ferrous ions transfer electrons to dioxygen or hydrogen peroxide. The presence of ferritins in contemporary organisms that are aerobes, facultative aerobes and anaerobes, as well as the ability of ferritins to use dioxygen and ferrous, or hydrogen peroxide and ferrous, suggest that the antioxidant properties of ferritin mineral synthesis may have contributed to the transition from anaerobic life as earth's atmosphere evolved.

A third metal site near the diiron catalytic centers has been observed in bacterial maxi-ferritins co-crystallized with various metal ions, and name site C or Fe 3 (Stillman et al. 2003; Crow et al. 2009); soaking in iron of co-crystallization with ferrous anaerobically indicates Fe 1, Fe 2, and Fe 3 (Crow et al. 2009). To date, no specific kinetic intermediates have been characterized in prokaryotic maxi-ferritins, contrasting with the well-described differic peroxo intermediate in animal ferritins (Pereira et al. 1998; Moënne-Loccoz et al. 1999; Bou-Abdallah et al. 2002); although changes in heme absorption have been observed during ferrous oxidation of heme-containing BFR (Le Brun et al. 1993). Most likely, Fe 3 represents an alternate iron-binding site related to ferrous ions traveling through the protein cage to the catalytic sites, especially since mutation of the ligands slows down oxidation (Treffry et al. 1998). Alternated conformations of ferritin protein cage side chains have been observed in high-resolution structures of ferritin, dependent on metals present during crystallization (e.g., Trikha et al. 1995; Toussaint et al. 2007) indicate the flexibility of the ligands at and near the active sites needed to move iron through the cage.

The ferritin catalytic sites are diiron sites related to the diiron cofactors sites in dioxygenases by simple, DNA codon differences in two amino acids of the Fe 2 site (Liu and Theil 2005). However, in eukaryotic ferritins, ferrous iron is substrates rather than cofactors, as in the oxygenases. Iron leaves the active sites, after catalytic coupling to dioxygen, as differic oxo products that are mineral precursors (Liu and Theil 2005). In prokaryotes, where primary structure varies as much as 80%, the ferritin protein cages are studded with catalytic sequences within each subunit or at the cavity surface between subunits that bind two iron atoms as cofactors, or are mono or diiron substrates sites (Chiancone and Ceci 2010; Le Brun et al. 2010). Since many of the amino acids at the ferritin catalytic sites are highly conserved, in both eukaryotes and prokaryotes (Table 2.3), the variable amino acids at and around the active sites provide significant tissue and species selectivity (e.g., (Tosha et al. 2008).

At the ferritin catalytic sites, the ligating residues for Fe 1(A) are highly conserved in prokaryotes and eukaryotes (91.3–100%). By contrast, Fe 2 (B) ligands vary considerably (46.2–93.3%) (Table 2.4). In general, iron site 2 has two, nonvariant residues and several variable residues that have specific kinetic effects and are genetically regulated so the abundance differ (Tosha et al. 2008). There appears to be some type of cooperativity among the catalytic sites of multiple ferritin subunits. Since each ferrous ion binds independently at the diiron sites in each ferritin subunit, and the other substrate, dioxygen, is absent, the coooperativity is likely protein: protein (Schwartz et al. 2008).

The diiron sites in maxi-ferritins of both prokaryotes and eukaryotes, Fe 1(A) and Fe 2 (B), are highly conserved (all above 90.9%). Conservation of site Fe 2 of mini-ferritins (46.2%) is low, unless the second most abundant site Fe2 motif is included a consensus sequence; thus mutation of aspartate to the glutamate, in the *E. coli* template, increases the conservation to 92.7%. A similar approach for bacterioferritin increases for Fe 2 ligands from 76.3% to 88.5%. However, in eukaryotic ferritins of organisms with multiple H subunit (catalytically active) genes, Fe 2 site differences as small as changing alanine to serine occur, and in

Table 2.4 Conservation of iron-binding amino acids at the catalytic centers of ferritins. Percentages of conservation of putative Fe^{2+} ligands in site A and B in mini-, maxi-, and bacterioferritin homologues based on the alignment of prokaryotic (J. Craig Venter Institute) and eukaryotic genomes (Uniprot knowledgebase, Swiss-Prot) as presented in Table 2.3

Protein	Fe 1 (Site A)[a]	Fe2 (Site B)[a]
Maxi-ferritin (pro)	99.5	93.3
Maxi-ferritin (euk)	92.7	90.9
Mini-ferritin (Dps)	100.0	46.2
Bacterioferritin	91.3	76.3

[a]Diiron sites at the catalytic centers of oxygenases and reductases, Fe 1 and Fe 2, are homologous to ferritin A and B, which were named before the similarity to the oxygenases and FTNA became clear

recombinant ferritins identical except for the alanine/serine shift, have significant effects on K_{cat} (Tosha et al. 2008). Thus, while the sequence motifs for Fe 2 in both mini-ferritin and bacterioferritin consist of a few common motifs found in the majority of the ferritins, the variations could have functional significance.

2.3 Ferritin Iron Biomineral Formation

Iron biomineralization in ferritins is a multistage process with discretely defined steps such as ferrous substrate binding, oxidation, product (mineral precursor) release and, when the active sites are in the middle of the protein cage, mineral nucleation, as recently observed by 13C–13C and magnetic susceptibility NMR (Turano et al. 2010). The multiphase process is illustrated in progress curves where data are collected at a general wavelength for Fe(III)O species (310–420 nm) as in Fig. 2.3. By contrast, when data are collected at the λmax of the differic peroxo complex (650 nm), a single phase of the reaction is observed (Fig. 2.3).

A goal for the future is deconvolution, both spectroscopic and kinetic, of the multiple intermediates in ferritin iron biomineralization. The different steps in ferritin biomineral synthesis are:

1. Fe (II) enters the protein cage and binds at the active sites
2. Oxygen or hydrogen peroxide binds with oxidoreduction and formation of transition intermediates
3. Release of ferric oxo mineral precursor
4. Nucleation and mineralization

2.3.1 Step i. Fe(II) Entry and Binding

Binding of Fe(II) to the active site has been studied directly by MCD/CD in a eukaryotic maxi-ferritin (Schwartz et al. 2008), and indirectly by soaking Fe(II)

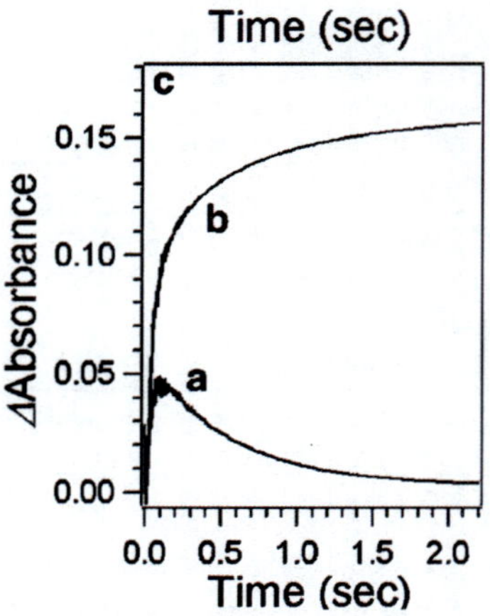

Fig. 2.3 Overall ferritin oxidoreductase activity (Maxi-ferritin – 24 sites). (**a**) Absorbance at 650 nm of the differic peroxo (DFP) catalytic intermediate. (**b**) Absorbance at 350 nm of all Fe(III) species: DFP, differic oxo mineral precursors, other intermediates and mineral itself. The figure is modified from reference 24

into an iron -free protein crystal (Crow et al. 2009), or by tryptophan quenching (Lawson et al. 2009; Bellapadrona et al. 2009). In maxi-ferritin, the two Fe(II) atoms appear to bind independently to Fe sites 1 and 2, with five ligands that include water and a "space" for dioxygen Fe binding. The Fe(II) binding sites in eukaryotic and prokaryotic ferritins are weaker than in BFR (Table 2.3), supporting the cofactor behavior of iron at BFR active sites (Le Brun et al. 2010). In mini-ferritins, the 12 active sites are saturated with 12 Fe(II) atoms (Su et al. 2005) except in proteins that can use dioxygen as the substrate, where 24 Fe(II) bind/cage (Liu et al. 2006). The differential affinities of Fe 1 and Fe 2 sites in mini-ferritins are illustrated by protein crystal structures where the site occupancy of metals varies between one and two depending on the protein (Chiancone and Ceci 2010). In maxi-ferritins, the 24 active sites are saturated with 48 Fe (II), when the formation of the differic peroxo complex is the reporter (reviewed Liu 2005).

2.3.2 Step ii. O_2 or H_2O_2 Binding and Formation of Transition Intermediates

Ferritin catalysis is rapid (msec) and requires stopped flow measurements to monitor the early stages. For maxi-ferritins, formation of the differic peroxo intermediate, well characterized by UV–vis, Mössbauer, resonance Raman, and EXAFS spectroscopies (Pereira et al. 1998; Moënne-Loccoz et al. 1999; Bou-Abdallah et al. 2002), is a convenient spectroscopic probe. However, except for the heme changes during oxidation in BFR, oxidation is measured as Fe(III)O which does not allow separation of catalysis from subsequent mineralization steps, which hampers kinetic analyses. For mini-ferritins that preferentially use hydrogen peroxide as the oxidant (Chiancone and Ceci 2010; Su et al. 2005), the reactions are anaerobic and can involve uncharacterized intermediates with UV absorption properties similar to differic peroxo of maxi-ferritins (Liu et al. 2006).

2.3.3 Step iii. Release of Differric Oxo Mineral Precursors from Active Sites

Release of the Fe(III)O product of ferritin catalysis varies among ferritins. In BFR and most Dps, the ferric oxo product is thought to be released directly into the protein cavity, because of the proximity of the active sites to the cavity surface. In BFR, and possibly *Pyrococcus furiosus* ferritins, where two iron atoms are cofactors, the active sites are proposed to dominate dioxygen reduction while ferrous oxidation occurs in the cavity on protein or mineral surfaces with the released electron used to reduce the cofactor ferric iron and recycle the catalytic center (Le Brun et al. 2010). However, the properties of various ferric species remain uncharacterized. In eukaryotic ferritins, where the differic peroxo intermediate results from dioxygen and ferrous, the reactions produced is a differic oxo complex and multimeric ferric oxo complexes (Jameson et al. 2002). Even though the active sites in eukaryotic ferritin protein cages, like those in prokaryotic ferritins, have direct access to the internal cavity (Tosha et al. 2008), recent NMR evidence reveals that the differic oxo products of ferritin protein catalysts remain inside the protein cage and move along an interior trajectory of 20 Å through the protein cage to the cavity surface. During passage through the protein, reactions occur among the ferric oxo products of multiple catalytic cycles (Turano et al. 2010) (Fig. 2.4), leading to the formation of multinuclear ferric oxide complexes inside the protein cage itself. Whether this feature is an evolutionary advance in eukaryotic ferritins or a property yet to be detected in prokaryotic and archaeal ferritins, remains unknown.

2.3.4 Step iv. Nucleation and Mineralization

Nucleation and mineralization in prokaryotic ferritins, BFR, and mini-ferritins is considered to be essentially inorganic hydrolytic chemistry of ferric oxo complexes where the effective concentration of iron is increased by the volume of the cavity. Ferrous ions are propelled into the cavity from the exterior of the protein cages by high-affinity binding sites (active sites) that harvest ferrous ions moving into the cage through pores lined with negatively charged amino acids (Theil et al. 2008; Bellapadrona et al. 2009); in recombinant ferritins with only animal L subunits and no catalytic sites, oxidation and mineralization is ~1,000 times slower. Pores with negative charges are formed around the threefold axes in both mini-ferritins and maxi-ferritins; highly conserved residues control pore folding/unfolding which influences iron mineral dissolution more than mineral synthesis.

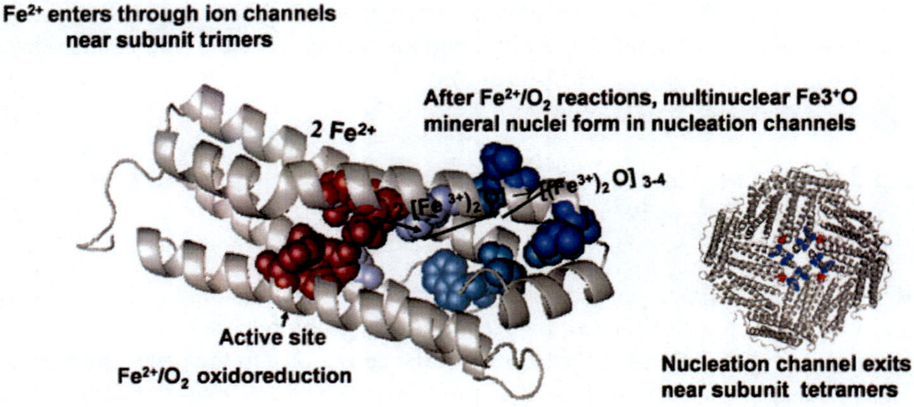

Fig. 2.4 Movement iron premineral through a eukaryotic ferritin protein cage. The data are from 13C–13C solution NOESY NMR obtained after the addition of 2 Fe (II)/F_{ox} site, for each of four catalytic turnovers. Modified from reference 35

Eukaryotic ferritin protein cages appear to be much more active in mineralization than those in prokaryotes, which could mean either a different evolutionary path to achieve the quaternary protein cage structure, or a lag in functional studies for prokaryotes. (See also comments in ref. Le Brun et al. 2010). In a recent surface, where the position of iron oxo mineral precursors was monitored as the ferric-induced broadening of resonance in 13C–13C solution NMR spectra, the first two ferrous ions oxidized (48 Fe/cage) appeared as a dimer (magnetic susceptibility) distal to the active sites and away from the cavity surfaces (Turano et al. 2010). The addition of saturating Fe(II) for second, third, and fourth oxidoreductase cycles revealed the ferric oxo species moving through about 20 Å channels along the long axes of each four-helix bundle subunit to the inner edge of the protein cages. The ferric oxo mineral precursor that is produced by the multiple catalytic sites emerged into the cavity as multimers (mineral nuclei) of 4–8 Fe atoms. Since each subunit channel exit is near that of three others (near the cage fourfold axes), the ferritin protein cage of eukaryotes facilitates biomineral growth.

2.4 Ferritin Iron Biominerals

Iron biominerals in the center of ferritin protein cages of animals, plants, bacteria, and archaea are hydrated ferric oxides. The microcrystalline materials have been observed by electron microscopy, X-ray powder diffraction, Mössbauer spectroscopy, and more recently by scanning electron microscopy (SEM) (TEM) (Rodriguez et al. 2005; Dobson 2001). Mineralized iron in ferritins varies in the maximum number of mineralized iron atoms/cage varying from 500 in mini-ferritins to >4,000 in maxi-ferritins. However, in natural tissues, the average number of mineralized iron atoms in maxi-ferritins is much less, from 800 to

1,500, with a distribution from 1,000 to 3 or 4,000 Fe atoms/cage; ferritins reconstituted in the laboratory have much more homogeneous mineral sizes see review [Theil 1987]. When the mineral content of a ferritin protein cage is very high, damage occurs to the cage, likely from side reactions of repeated oxidoreductase catalysis, and an insoluble mass of damaged protein forms, called hemosiderin. In living systems, multiple mechanisms regulate both the cellular influx of iron and the synthesis of ferritin protein cages, so that hemosiderin only forms with abnormal regulation (disease). In the laboratory, the iron content of ferritin protein cages is controlled by the amount of ferrous iron added, which is usually up 240 for mini-ferritins and up to 480–1,000 for maxi-ferritins (Theil 1987; Arosio et al. 2009).

All ferritin minerals contain some phosphate, but the amount varies. For example, in animals the phosphorous content is about 12% of the iron, whereas in plants and bacteria the phosphorous content approached 100% that of the iron (Lewin et al. 2005). The higher phosphate content in ferritin minerals minimizes order and leads to a more amorphous ferritin mineral. When iron minerals were reconstituted in empty ferritin protein cages with high phosphate (P: Fe = 4:1) and low phosphate (P: Fe = 0:1) in the solution, the phosphate-containing minerals were more disordered when analyzed by EXAFS, TEM, electron diffraction, and Mössbauer spectroscopy, and were independent of the proteins cages, which were bacterial (*Azotobacter vinelandii* or *Pseudomonas aeruginosa*) or animal (horse spleen) in origin (Mann et al. 1987; Rohrer et al. 1990).

The phosphate content of prokaryote cytoplasm is higher and more variable than the cytoplasm of eukaryotic cells, which explains the high phosphate content and greater disorder of ferritin biominerals in bacteria (Mann et al. 1987; Rohrer et al. 1990; Wade et al. 1993; Waldo et al. 1995). In plants, the plastid is evolutionarily related to prokaryotes and thus, even though ferritin protein is encoded in a eukaryotic nuclear gene, it is targeted to the plastid. The high phosphate content of plant ferritin minerals, then, is explained if the plant ferritin mineral is synthesized by the nanocage inside the plastid (Waldo et al. 1995), rather than in the cytoplasm where the nanocage is synthesized before transport to plastids. Whether the phosphate content of ferritin minerals is simply a reflection of the ambient concentration where the mineral is formed or includes phosphate storage function, is unknown.

Order in ferritin minerals varies from microcrystalline to amorphous. The most dramatic effect is the effect of phosphate that distinguishes animal ferritin minerals from those of bacteria and plants, where the ferritin minerals appear to form in the prokaryotic-derived plastids. The properties of the mineral in mitochondrial ferritins of animal, only recently discovered and studied (Corsi et al. 2002) only at the level of the protein, is unknown. In animal tissues, the crystallinity of the ferritin mineral varies in different tissues and disease states (St Pierre et al. 1991). One contributing factor is the presence of catalytically inactive subunits, the animal-specific L subunit. Ferritins with a high L subunit content have more disordered minerals than those with few L subunits (St Pierre et al. 1991), which may relate to the recently detected role of the catalytically active subunits in directing nucleation and mineralization of ferritin minerals (Turano et al. 2010) (Fig. 2.4). The underlying mechanism of ferritin biomineral synthesis is ferric hydrolysis. Apparently, ferritin protein cages contribute to a much larger fraction of the mineralization process than previously thought, but the fraction of ferritin biominerals laid down without passage nucleation in the protein cage remains unknown.

2.5 Ferritin Iron Biomineral Dissolution

Recovering iron from the ferritin mineral core is crucial for the biosynthesis of heme and iron-sulfur cluster-containing proteins that are important for all forms of life. However, the exit of iron from the ferritin cage, which requires the delivery of electrons and protons to dissolve the mineral, is much less studied than entry/oxidation/mineralization. The reduction and subsequent release of iron from the ferritin cavity can be measured in an assay as first described by Jones et al. (1978), and has been used to study maxi-ferritins and mini-ferritins from plants, animals, bacteria, and cyanobacteria (Bellapadrona et al. 2009; Takagi et al. 1998; Richards et al. 1996; Castruita et al. 2007). Usually, NADH/FMN, or sodium dithionite is the source of electrons, added to mineralized ferritin protein in neutral buffer (Bellapadrona et al. 2009), with a chelator such as bipyridyl, as the reporter for iron exit the pink Fe(II)-chelate (A 522 nm); other chelators such as desferrioxamine B (DFO) are also used. Curves for the formation of the Fe–chelator complexes are multiphasic and complex due to multiple sequential and simultaneous reactions; iron reduction, hydration of ferric oxo bridges in the mineral, transport of iron and reductants through the protein cage in the case of Fe(III) chelators, the competition between binding to the chelator or to the catalytic sites of the protein (Liu et al. 2007).

In both mini- and maxi-ferritins, iron leaves the mineral cage via pore structures in the protein that connect the inner cavity with the outside of the protein (Liu and Theil 2005; Bellapadrona et al. 2009). These funnel-shaped, hydrophilic pores are localized at the threefold symmetry axes by junctions of three subunits (Takagi et al. 1998). There are eight of these hydrophilic pores in maxi-ferritin and four in mini-ferritin. Experiments with the H-frog maxi-ferritin, using mutagenesis, protein crystallography, helix content (CD spectroscopy) have shown that the helix content around ferritin pores is sensitive to slight changes in temperature, ionic strength, or by the presence of low concentrations (1–10 mM) of chaotropes like urea or guanidine (Liu et al. 2003). Partial unfolding of these pore structures was observed simultaneously with an increase in rate of iron reduction and iron release from the proteins inner core illustrating the regulatory role of ferritin cage pores in controlling mineral/reductant access. Similar effects of threefold pore mutations were observed *Listeria innocua* mini-ferritin where replacement of conserved, negatively charged residues with neutral amino acids changed the initial rate of iron release from the mineral core. Even though the amino acid sequences of mini-ferritin and maxi-ferritin subunits vary considerably (up to 80%), quaternary structures have pores, at the junctions of matched helix turns in three subunits, which control reductant/chelator access/Fe (II) exit during demineralization of both maxi-ferritins and mini-ferritins. The turnover rate of ferritin minerals in vivo is not known but studies with natural maxi-ferritin minerals suggest it could be significant (Castruita et al. 2007).

Dissolution of ferritin minerals has a fast, pH-dependent initial phase and a slower, pH-independent phase in ~100–200 irons per cage are removed. These iron atoms may be anchored in the protein and represent iron still in transit through the proteins (Turano et al. 2010). The phosphate content of ferritin minerals also influences dissolution of mineral dissolution rates. Rates in the early phase of mineralization decreases when phosphate is high, whereas the later, slow phase is independent of phosphate (Richards et al. 1996). Such observations support the notion that phosphate replaces coordinated water and thereby blocks hydrolysis and formation ferric oxo bridges, which would have a larger effect on mineral nuclei than on bulk mineral.

What keeps ferritin minerals dissolving at a rapid rate? The cytoplasm of most cells is thought to be "reducing," i.e., full of reductants. Increased rates of iron mineral dissolution occurred when ferritin threefold pores were unfolded, as observed in a crystal structure of a ferritin folded normally (Takagi et al. 1998). Thus, in normal ferritin cages, the folded pores keep reductant away from the ferric mineral. A search for other amino acid residues that stabilize pore folding in ferritin protein nanocages was performed using frog H maxi-ferritin, the protein in which pore unfolding was observed first (Fig. 2.5). (Takagi et al. 1998; Jin et al. 2001). Three highly conserved residues near the pores in three dimensions, but with no assigned function, were substituted with a variety of amino acids, each of which increased rates of mineral reduction and release of iron compared to the wild-type ferritin, identifying a set of amino acids that control pore opening. Even in wild-type protein, pore unfolding, measured as increased rates of reducing/dissolving ferritin minerals, is increased by low heat and low concentrations of chaotropes (both in the range of normal physiological conditions). Protein pore function was

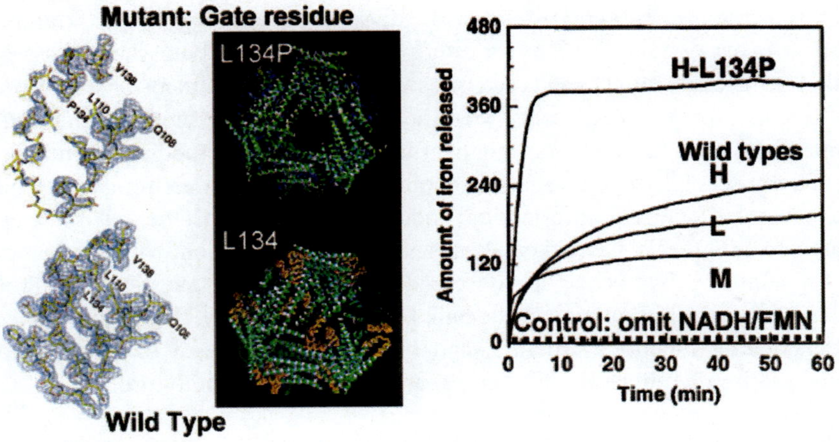

Fig. 2.5 Ferritin threefold pore influences ferritin demineralization rates. The eukaryotic model shows the mid-point constriction in the pore. There are eight ferritin pores in maxi-ferritins (Fig. 2.1) and four functionally analogous pores in mini-ferritins (Tosha et al. 2008)

further probed with five peptides identified from a combinatorial phage display library of 10^9 peptides that bound to wild-type ferritin (Liu et al. 2007). One of the five binding peptides increased iron release as if it were unfolding the pores. Such results lead to the hypothesis that cellular proteins control folding/unfolding of ferritin pores and, thereby, control ferritin mineral/reductant/chelator interactions in vivo (Liu et al. 2003).

2.5.1 *Perspectives*

Ferritin protein nanocages, which synthesize hydrated iron oxide biominerals and protect them from uncontrolled cell reductants, are ubiquitous. Recognizable sequences are found in archaea, bacteria, plants, and animals, including humans. Ferritins concentrate iron in biominerals containing thousands of iron and oxygen atoms, to match cell need and to act as antioxidants that prevent corrosive ROS (reactive oxygen species) release from iron and dioxygen or hydrogen peroxide chemistry. Ferritin plays a role in normal development and in disease, as pathogens and host battle with iron and inflammatory oxidants. Hallmarks of ferritin family members are a quaternary protein cage structure self-assembled from polypeptide subunits and studded with: (1) catalytic sites that initiate mineralization, (2) pores that control iron entry and mineral reduction/dissolution, and (3) internal channels that facilitate during mineral nucleation. Ferritin subunits fold into 4 α-helix bundles; 12 subunit structures are mini-ferritins (Dps proteins), and 24 subunit structures are maxi-ferritins. The variable phosphate content of ferritin biominerals appears to relate to differences in environmental phosphate during biomineral synthesis. High-phosphate-induced minerals have biological relevance in plants and bacteria while biomineral disorder in low phosphate ion biominerals of animals depends on the relative abundance in the protein cage of subunits that are catalytically inactive (L subunits). The wide divergence among ferritin amino acid sequences has two implications for future research: First, solving the self-assembly code for the cage, pores, and active sites will require new dimensions in informatics. Second, ferritins that escape identification by scanning methods emphasizing primary structure will be trapped by future scanning methods that incorporate features of secondary and tertiary structures to reveal more ferritin genes. Finally, the paucity of current information about the full role of the ferritin cage in biomineral synthesis, beyond catalysis and mineral nucleation, which is hampered by the sequence divergence, has obscured the answer to the question of the role of the protein cage in formation of the bulk biomineral. Clearly, the rapid advances in understanding function, structure, and biological importance of ferritin protein cages and iron biominerals still leave many questions for the future.

Acknowledgment The authors are grateful for the contributions of all the members of the Theil Group. The writing of the manuscript and work described herein were supported by the CHORI Foundation (ECT), the NIH (DK20251) (LB and ECT), and a Rubicon Fellowship from the Netherlands Organization for Scientific Research (NWO), (LB).

References

Arosio P, Ingrassia R, Cavadini P (2009) Ferritins: a family of molecules for iron storage, antioxidation and more. Biochim Biophys Acta 1790:589–599

Bellapadrona G, Stefanini S, Zamparelli C, Theil EC, Chiancone E (2009) Iron translocation into and out of *Listeria innocua* Dps and size distribution of the protein-enclosed nanomineral are modulated by the electrostatic gradient at the 3-fold "ferritin-like" pores. J Biol Chem 284:19101–19109

Bou-Abdallah F, Papaefthymiou GC, Scheswohl DM, Stanga SD, Arosio P, Chasteen ND (2002) μ-1,2-Peroxobridged di-iron(III) dimer formation in human H-chain ferritin. Biochem J 364:57–63

Briat, JF, Duc, C, Ravet, K, and Gaymard, F (2009) Ferritins and iron storage in plants. *Biochim Biophys Acta* 1800:8i06–814

Castruita M, Elmegreen LA, Shaked Y, Stiefel EI, Morel FM (2007) Comparison of the kinetics of iron release from a marine (*Trichodesmium erythraeum*) Dps protein and mammalian ferritin in the presence and absence of ligands. J Inorg Biochem 101:1686–1691

Chiancone E, Ceci P (2010) The multifaceted capacity of Dps proteins to combat bacterial stress conditions: Detoxification of iron and hydrogen peroxide and DNA binding. Biochim Biophys Acta 1800(8):798–805

Cojocaru V, Winn PJ, Wade RC (2007) The ins and outs of cytochrome P450s. Biochim Biophys Acta 1770:390–401

Corsi B, Cozzi A, Arosio P, Drysdale J, Santambrogio P, Campanella A, Biasiotto G, Albertini A, Levi S (2002) Human mitochondrial ferritin expressed in HeLa cells incorporates iron and affects cellular iron metabolism. J Biol Chem 277:22430–22437

Crow A, Lawson TL, Lewin A, Moore GR, Le Brun NE (2009) Structural basis for iron mineralization by bacterioferritin. J Am Chem Soc 131:6808–6813

Dickey LF, Sreedharan S, Theil EC, Didsbury JR, Wang Y-H, Kaufman RE (1987) Differences in the regulation of messenger RNA for housekeeping and specialized-cell ferritin. A comparison of three distinct ferritin complementary DNAs, the corresponding subunits, and identification of the first processed in amphibia. J Biol Chem 262:7901–7907

Dobson, J (2001) On the structural form of iron in ferritin cores associated with progressive supranuclear palsy and Alzheimer's disease. *Cell Mol Biol (Noisy-le-grand)* 47 Online Pub OL49–50

Dong X, Tang B, Li J, Xu Q, Fang S, Hua Z (2008) Expression and purification of intact and functional soybean (Glycine max) seed ferritin complex in *Escherichia coli*. J Microbiol Biotechnol 18:299–307

Grant RA, Filman DJ, Finkel SE, Kolter R, Hogle JM (1998) The crystal structure of Dps, a ferritin homolog that binds and protects DNA. Nat Struct Biol 5:294–303

Grossman MJ, Hinton SM, Minak-Bernero V, Slaughter C, Stiefel EI (1992) Unification of the ferritin family of proteins. Proc Natl Acad Sci USA 89:2419–2423

Hintze KJ, Theil EC (2005) DNA and mRNA elements with complementary responses to hemin, antioxidant inducers, and iron control ferritin-L expression. Proc Natl Acad Sci USA 102: 15048–15052

Hintze KJ, Katoh Y, Igarashi K, Theil EC (2007) Bach1 repression of ferritin and thioredoxin reductase1 is heme-sensitive in cells and in vitro and coordinates expression with heme oxygenase1, beta-globin, and NADP(H) quinone (oxido) reductase1. J Biol Chem 282: 34365–34371

Jameson GNL, Jin W, Krebs C, Perreira AS, Tavares P, Liu X, Theil EC, Huynh BH (2002) Stoichiometric production of hydrogen peroxide and parallel formation of ferric multimers through decay of the diferric-peroxo complex, the first detectable intermediate in ferritin mineralization. Biochemistry 41:13435–13443

Jin W, Takagi H, Pancorbo B, Theil EC (2001) Opening the Ferritin Pore for Iron Release by Mutation of Conserved Amino Acids at Interhelix and Loop Sites. Biochem Biochem 40:7525–7532

Jones T, Spencer R, Walsh C (1978) Mechanism and kinetics of iron release from ferritin by dihydroflavins and dihydroflavin analogues. Biochemistry 17:4011–4017

Lawson TL, Crow A, Lewin A, Yasmin S, Moore GR, Le Brun NE (2009) Monitoring the iron status of the ferroxidase center of *Escherichia coli* bacterioferritin using fluorescence spectroscopy. Biochemistry 48:9031–9039

Le Brun NE, Wilson MT, Andrews SC, Guest JR, Harrison PM, Thomson AJ, Moore GR (1993) Kinetic and structural characterization of an intermediate in the biomineralization of bacterioferritin. FEBS Lett 333:197–202

Le Brun NE, Crow A, Murphy ME, Mauk AG, Moore GR (2010) Iron core mineralisation in prokaryotic ferritins. Biochim Biophys Acta 1800(8):732–744

Lewin, A, Moore, GR, and Le Brun, NE (2005) Formation of protein-coated iron minerals. *Dalton Trans* 3597–3610

Liu X, Theil EC (2005) Ferritin: dynamic management of biological iron and oxygen chemistry. Acc Chem Res 38:167–175

Liu X, Jin W, Theil EC (2003) Opening protein pores with chaotropes enhances Fe reduction and chelation of Fe from the ferritin biomineral. *Proc.* Natl. Acad. Sci. U. S. A. Proc Natl Acad Sci 100:3653–3658

Liu X, Kim K, Leighton T, Theil EC (2006) Paired *Bacillus anthracis* Dps (mini-ferritin) have different reactivities with peroxide. J Biol Chem 28:27827–27835

Liu XS, Patterson LD, Miller MJ, Theil EC (2007) Peptides Selected for the Protein Nanocage Pores Change the Rate of Iron Recovery from the Ferritin Mineral. J Biol Chem J Biol Chem 282:31821–31825

Mann S, Williams JM, Treffry A, Harrison PM (1987) Reconstituted and native iron-cores of bacterioferritin and ferritin. J Mol Biol 198:405–416

Moënne-Loccoz P, Krebs C, Herlihy K, Edmondson DE, Theil EC, Huynh BH, Loehr TM (1999) The ferroxidase reaction of ferritin reveals a diferric m-1,2 bridging peroxide intermediate in common with other O_2-activating non-heme diiron proteins. Biochemistry 38:5290–5295

Nandal A, Huggins CC, Woodhall MR, McHugh J, Rodriguez-Quinones F, Quail MA, Guest JR, Andrews SC (2009) Induction of the ferritin gene (ftnA) of *Escherichia coli* by Fe(2+)-Fur is mediated by reversal of H-NS silencing and is RyhB independent. Mol Microbiol 75 (3):637–657

Pereira A, Small GS, Krebs C, Tavares P, Edmondson DE, Theil EC, Huynh BH (1998) Direct spectroscopic and kinetic evidence for the involvement of a peroxodiferric intermediate during the ferroxidase reaction in fast ferritin mineralization. Biochemistry 37:9871–9876

Pham CG, Bubici C, Zazzeroni F, Papa S, Jones J, Alvarez K, Jayawardena S, De Smaele E, Cong R, Beaumont C, Torti FM, Torti SV, Franzoso G (2004) Ferritin heavy chain upregulation by NF-κB inhibits TNFα-induced apoptosis by suppressing reactive oxygen species. Cell 119:529–542

Richards TD, Pitts KR, Watt GD (1996) A kinetic study of iron release from *Azotobacter vinelandii* bacterial ferritin, J Inorg Biochem 61:1–13

Rodriguez N, Menendez N, Tornero J, Amils R, de la Fuente V (2005) Internal iron biomineralization in *Imperata cylindrica*, a perennial grass: chemical composition, speciation and plant localization. New Phytol 165:781–789

Rohrer JS, Islam QT, Watt GD, Sayers DE, Theil EC (1990) Iron environment in ferritin with large amounts of phosphate, from *Azotobacter vinelandii* and horse spleen, analyzed using extended x-ray absorption fine structure (EXAFS). Biochemistry 29:259–264

Schwartz JK, Liu XS, Tosha T, Theil EC, Solomon EI (2008) Spectroscopic definition of the ferroxidase site in M ferritin: comparison of binuclear substrate vs cofactor active sites. J Am Chem Soc 130:9441–9450

Semenza GL (2009) Involvement of oxygen-sensing pathways in physiologic and pathologic erythropoiesis. Blood 114:2015–2019

St Pierre T, Tran KC, Webb J, Macey DJ, Heywood BR, Sparks NH, Wade VJ, Mann S, Pootrakul P (1991) Organ-specific crystalline strcutures of ferritin cores in beta-thalassemia/hemoglobin E. Biol Met 4:162–165

Stiles AR, McDonald JG, Bauman DR, Russell DW (2009) CYP7B1: one cytochrome P450, two human genetic diseases, and multiple physiological functions. J Biol Chem 284:28485–28489

Stillman TJ, Connolly PP, Latimer CL, Morland AF, Quail MA, Andrews SC, Treffry A, Guest JR, Artymiuk PJ, Harrison PM (2003) Insights into the effects on metal binding of the systematic substitution of five key glutamate ligands in the ferritin of *Escherichia coli*. J Biol Chem 278:26275–26286

Su M, Cavallo S, Stefanini S, Chiancone E, Chasteen ND (2005) The so-called *Listeria innocua* ferritin is a Dps protein. Iron incorporation, detoxification, and DNA protection properties, Biochemistry 44:5572–5578

Takagi H, Shi D, Ha Y, Allewell NM, Theil EC (1998) Localized unfolding at the junction of three ferritin subunits; A mechanism for iron release? J Biol Chem J Biol Chem 273:18685–18688

Theil EC (1987) Ferritin: structure, gene regulation, and cellular function in animals, plants, and microorganisms. Annu Rev Biochem 56:289–315

Theil EC, Goss DJ (2009) Living with iron (and oxygen): questions and answers about iron homeostasis. Chem Rev 109:4568–4579

Theil EC, Liu XS, Tosha T (2008) Gated pores in the ferritin protein nanocage, Inorg Chim Acta 361:868–874

Tosha T, Hasan MR, Theil EC (2008) The ferritin Fe2 site at the diiron catalytic center controls the reaction with O_2 in the rapid mineralization pathway. Proc Natl Acad Sci USA 105:18182–18187

Toussaint L, Bertrand L, Hue L, Crichton RR, Declercq JP (2007) High-resolution X-ray structures of human apoferritin H-chain mutants correlated with their activity and metal-binding sites. J Mol Biol 365:440–452

Treffry A, Zhao Z, Quail MA, Guest JR, Harrison PM (1998) How the presence of three iron binding sites affects the iron storage function of the ferritin (EcFtnA) of *Escherichia coli*. FEBS Lett 432:213–218

Trikha J, Theil EC, Allewell NM (1995) High resolution crystal structures of amphibian red-cell L ferritin: potential roles for structural plasticity and solvation in function. J Mol Biol 248:949–967

Turano P, Lalli D, Felli I, Theil E, Bertini I (2010) NMR reveals pathway for ferric mineral precursors to the central cavity of ferritin. Proc Natl Acad Sci USA 107:545–550

Wade VJ, Treffry A, Laulhere J-P, Bauminger ER, Cleton MI, Mann S, Briat J-F, Harrison PM (1993) Structure and composition of ferritin cores from pea seed (*Pisum sativum*). Biochim Biophys Acta 1161:91–96

Waldo GS, Wright E, Whang ZH, Briat JF, Theil EC, Sayers DE (1995) Formation of the ferritin iron mineral occurs in plastids. Plant Physiol 109:797–802

Yamashita M, Ojima N, Sakamoto T (1996) Molecular cloning and cold-inducible gene expression of ferritin H subunit isoforms in rainbow trout cells. J Biol Chem 271:26908–26913

Chapter 3
Manganese Oxidation by Bacteria: Biogeochemical Aspects

P.P. Sujith and P.A. Loka Bharathi

Contents

Abstract Manganese is an essential trace metal that is not as readily oxidizable like iron. Several bacterial groups posses the ability to oxidize Mn effectively competing with chemical oxidation. The oxides of Mn are the strongest of the oxidants, next to oxygen in the aquatic environment and therefore control the fate of several elements. Mn oxidizing bacteria have a suit of enzymes that not only help to scavenge Mn but also other associated elements, thus playing a crucial role in biogeochemical cycles. This article reviews the importance of manganese and its interaction with microorganisms in the oxidative Mn cycle in aquatic realms.

P.A. Loka Bharathi (✉)
National Institute of Oceanography (Council of Scientific and Industrial Research), Dona Paula, Goa 403 004, India
e-mail: loka@nio.org

W.E.G. Müller (ed.), *Molecular Biomineralization*, Progress in Molecular and Subcellular Biology 52, DOI 10.1007/978-3-642-21230-7_3,

3.1 Introduction

Manganese comprises about 0.1% of the total mass of the Earth (Nealson 1983) and occurs as $MnAl_2O_4$ (Zajic 1969). It is the fifth most abundant transition metal in the Earth's crust (Tebo et al. 2007) and is the second most common trace metal after iron (Tebo et al. 1997). The name manganese is derived from the Greek word mangania, meaning magic (Horsburgh et al. 2002). It occupies the 25th position in the periodic table and belongs to group VII transition elements (Cellier 2002). Mn exist in seven different oxidation states ranging from 0 to +7 and in nature it occurs in +II, +III, and +IV oxidation states (Tebo et al. 1997, 2004). Mn^{2+} has an ionic radius of 0.80 A° and has Gibbs standard energy of $-54.5\ \Delta G°$ in aqueous solutions (Hem 1978). It occurs at a concentration of 100–1000 ppm in river, 1–10 ppm in ground water (Nealson 1983), and averages 8 μg kg^{-1} in freshwater and 0.2 μg kg^{-1} in seawater (Bowen 1979 and Ehrlich 2002a). The concentration of dissolved Mn (Mn^{2+}) in the open ocean ranges from 0.2–3 nmol kg^{-1} of seawater (Glasby 2006). Further details about its distribution and abundance are shown in Fig. 3.1. As Mn exists at a higher redox potential than iron, following comparisons can be made on Mn–Fe relationships. (1) Mn reduces more easily than iron, (2) Mn is harder to oxidize than iron, and (3) Soluble Mn (Mn^{2+}) occurs at a somewhat higher level in the oxygen gradient than iron (Kirchner and Grabowski 1972). Mn enrichment occurs as a result of both artificial and natural processes. The sources of Mn in the ocean are atmospheric input, intense scavenging at mid-depth, and fluxes from reducing shelf and slope sediments and emanations from submarine hydrothermal vents (Saager et al. 1989). This study focuses on the bacterial groups that participate in Mn oxidation and the recent advancements made in the field of metal–microbe interaction. It also delves into genomic and proteomic aspects covering both freshwater and marine systems. Lastly, the review addresses the bacterial contribution to mineral formation and their potential use in biotechnological applications.

Fig. 3.1 Concentrations of manganese in different environments in ppm (*modified from Nealson* 1983)

3.2 Importance of Manganese

Manganese is a critical trace nutrient required for the growth and survival of many living organisms. It is essential for oxygenic photosynthesis in cyanobacteria (Yocum and Pecoraro 1999; Keren et al. 2002; Ogawa et al. 2002), redox reactions, protection from toxic metals, UV light, predation or, viruses, scavenging of micro-nutrient trace metals, breakdown of natural organic matter into metabolizable substrates, maintenance of an electron-acceptor reservoir for use in anaerobic respiration, oxygen production, and protection against oxidative stress in bacteria (Christianson 1997; Spiro et al. 2010). It is important for general metabolism, carbohydrate metabolism, and for both anabolic and catabolic functions in anaerobiosis and aerobiosis (Crowley et al. 2000). It is a part of four metalloenzymes manganese superoxide dismutase (MnSOD), mangani-catalase, arginase, and O-phosphatases (Christianson 1997; Shi 2004). Mn^{2+} containing O-phosphatases are involved in controlling spore formation, stress-response, cell density during stationary phase, carbon and nitrogen assimilation, vegetative growth, development of fruiting bodies, and cell segregation (Shi 2004). Additionally, nonenzymatic Mn^{2+} is crucial for the proper functioning of a variety of bacterial products, including secreted antibiotics (Archibald 1986). It also contributes to the stabilization of bacterial cell walls (Doyle 1989) and plays an important role in bacterial signal transduction (Jakubovics and Jenkinson 2001). Mn^{2+} is required for the stimulation of poly-β-hydroxybutyrate oxidation in *Sphaerotilus discophorous* (Stokes and Powers 1967) and exopolysaccharides (EPS) production in *Rhizobium meliloti* JJ-1 (Appanna 1988). Being a part of an enzyme in glycolysis, Mn^{2+} is required for the activity of 3-phosphoglycerate mutase in several endospore-forming gram-positive bacteria (Chander et al. 1998). Indirectly, Mn functions in controlling nutrient availability in freshwater, most significantly by complexing with iron (Kirchner and Grabowski 1972).

3.3 Biogeochemistry of Manganese

The geochemistry of Mn is a complex pattern of mutually exclusive chemical reactions of oxidation and reduction (Kirchner and Grabowski 1972). The geochemical behavior of Mn differs in environment which shows gradation in oxygen profile (Roitz et al. 2002). Mn occurs as highly soluble Mn^{2+} in oxygen-deficient settings and as insoluble oxyhydroxides under well-oxygenated conditions (Calvert and Pedersen 1996). The concentration of soluble Mn in the environments varies with change in redox condition and the group of microorganisms present. The oxidation of Mn by microorganisms results in decrease in the dissolved Mn^{2+} concentration of metal and increase in the particulate/higher oxidation states (Mn^{3+} and Mn^{4+}) of Mn (Ehrlich 1976, 1978). Redox transitions between soluble Mn^{2+} ions and insoluble Mn^{3+} and Mn^{4+} oxides form the backbone of aquatic biogeochemistry of Mn (Sunda and Huntsman 1990). Mn^{3+} being a strong oxidant and a reductant, it has been largely ignored due to its property to disproportionate to Mn^{2+} and MnO_2 (Johnson 2006). However, recent improvements in the understanding of Mn chemistry indicate that dissolved Mn exist mostly as Mn^{3+} in sub-oxic regions (Trouwborst et al. 2006). Mn being one of the strongest oxidant in the natural environment, eventually participates in redox reactions and due to its sorptive characteristics controls the distributions and bioavailability of several toxic and essential trace elements (Tebo et al. 2004). Microorganisms like bacteria and fungi are known oxidizers of Mn^{2+} and reducers of Mn oxide-containing minerals. They carry out oxidation/reduction of Mn as a way to conserve energy for growth or oxidation of carbon (Nealson and Myers 1992; Tebo et al. 2005). The oxidation of Mn^{2+} under natural conditions is catalyzed only by microbes under pH range of 5.5–8.0, Eh value above +200 mV and oxygen concentration of 3–5 mg L^{-1} (Schweisfurth et al. 1978).

The three representative stoichiometric relationships that typically follow the oxidation of Mn^{2+} depending on the oxide product formed according to Nealson et al. (1988) are

$$Mn^{2+} + \frac{1}{2}O_2 + H_2O \rightarrow MnO_2 + 2H^+$$
$$Mn^{2+} + \frac{1}{4}O_2 + \frac{3}{2}H_2O \rightarrow MnOOH + 2H^+$$
$$3Mn^{2+} + \frac{1}{2}O_2 + 3H_2O \rightarrow Mn_3O_4 + 6H^+$$

A number of bacteria capable of mixotrophic and autotrophic growth could derive useful energy from Mn^{2+} oxidation. The oxidation of Mn^{2+} by bacteria could yield ΔF_r of +2.79 and $\Delta F'_r$ of −16.31 kcal (Ehrlich 1976, 1978) in the reaction

$$Mn^{2+} + \frac{1}{2}O_2 + H_2O \rightarrow MnO_2 + 2H^+$$

and yield ΔG −18.5 kcal in the reverse reaction

$$MnO_2 + 4H^+ + 2e \rightarrow Mn^{2+} + 2H_2O$$

when allowance for physiological pH of 7.0 is made (Ehrlich 1987).

The key steps in the reduction of Mn^{4+} to Mn^{2+} involve the reactions of strongly bound Mn^{3+} $\{\{Mn^{3+}\}\}$ and weakly bound Mn^{2+} $\{Mn^{2+}\}$ on the surface of MnO_2 (Ehrlich 2002b). The following steps explain the process

$$\{Mn^{2+}\} + MnO_2 + 2H_2O \rightarrow 2\{\{Mn^{3+}\}\} + 4OH^-$$
$$2\{\{Mn^{3+}\}\} + 2e^- \rightarrow \{Mn^{2+}\} + Mn^{2+}$$

The reduction of Mn oxide is predicted based on standard redox potential of +1.29 mV (Rusin and Ehrlich 1995) based on the equation

$$\frac{1}{2}MnO_2 + 2H^+ + e^- \rightarrow \frac{1}{2}Mn^{2+} + H_2O$$

In environment where Mn and iron oxides coexist such as in ferromanganese nodules, bacteria preferentially attack manganese oxides. The possible explanation suggested for the preference of Mn over Fe by bacteria was the lower midpoint potential for the Fe(III)/Fe(II) couple relative to the Mn(IV)/Mn(II) couple (Ehrlich 1987). However, the exact reason is yet to be determined.

3.4 Effect of Salinity on Manganese Oxidation

The solubility of Mn is more in freshwater compared to salt water. However, the oxidation rate of Mn^{2+} decreased sixfold in creek water of an English estuary when the salinity of water increased from 1 to 5 (Vojak et al. 1985). Likewise, Spratt et al. (1994b) also demonstrated a fourfold decrease in the rates of manganese oxide production in high marsh sediments than in creek bank sediments. In another study, Spratt et al. (1994a) showed decrease in rate of Mn oxidation in high marsh sediments experimentally exposed to hypersaline (102) conditions. They demonstrated that Mn oxidation in creek bank sediments exhibit much higher rates of oxidation than high marsh sites (2.31 ± 0.28 and 0.45 ± 0.14 nmol mg dwt^{-1} h^{-1}). In mangrove swamp estuary, highest rates of microbial Mn oxide production was encountered in sediments with salinities between 0 and 8 (50–119 pmol mg dwt^{-1} h^{-1}, respectively) compared with sediments from the mouth of the estuary with salinities of 24 and 34 (3–16 pmol mg dwt^{-1} h^{-1}, respectively). Nevertheless, the overall rates of microbial Mn oxide production in salt-marsh sediments were much higher than in mangrove sediments.

3.5 Toxicity of Manganese in the Presence of Other Metals

In an undisturbed marine environment, metal ions are more likely to occur in combinations than in single (Yang and Ehrlich 1976). The microbial response to individual metals may differ from the response to stress from multiple metals. When interactions of one metal have a protective effect on the toxicity of other metal, the resulting effect is referred as antagonistic and the reverse effect where toxicity of one metal is enhanced in the presence of the other metal is synergistic. Third, interactions where the final toxicity is simply a sum of the individual toxicities of the metal ions is called additive (Babich and Stotzky 1983). The effect of Mn, Ni, Cu, and Co in combinations of 10 μg mL^{-1} in seawater enriched with 1% glucose and 0.05%

peptone at 18°C showed different toxic effects (Yang and Ehrlich 1976). Mn and Co when tested alone had no effect on growth but their combination did. The combination of Co and Ni was more toxic than Co alone but less toxic than Ni alone. Cu was comparatively more toxic than other metals. Its toxicity was slightly relieved by Co or by the combination of Co and Mn, or Co, Mn, and Ni, but by no other metal combinations tried. The mixtures of Mn and Ni and of Mn, Co, and Ni were more toxic than their individual components. Likewise, Appanna et al. (1996) demonstrated the influence of Mn, Co, Cs, and Ni on the ability of *Pseudomonas fluorescens* to adapt to and to decontaminate the multiple-metal environment. There the toxicity of metals was comparatively lower in multiple-metal combinations than when individual metals were tested. Mn was the least toxic followed by Cs, Co, and Ni. Cs and Mn did not alter the cellular yield significantly; however, Ni and Co showed marked inhibitory effect on bacterial growth.

A study on redox transformation of Mn in Antarctic lakes (Krishnan et al. 2009) showed maximum stimulation of Mn oxidation (81 ± 57 ppb d^{-1}) in Lister Hooded (LH) strains with Mn–Co combination rather than combinations of Fe (37 ± 16 ppb d^{-1}) and Ni (40 ± 47 ppb d^{-1}), suggesting the role of Co in Mn oxidation. Conversely, maximum stimulation of Mn reduction occurred in combinations of the metals containing Ni. The rates were >50 ppb d^{-1} with strain LH-11 and >125 ppb d^{-1} with strain LH-8 in combinations of Mn–Ni and Mn–Fe–Ni which suggest the critical role of Ni in Mn reduction.

3.6 Manganese Oxidation by Marine Bacteria

The oxidation of Mn^{2+} by marine bacteria is more versatile. Dick et al. (2008a) have shown that *Aurantimonas* sp. Strain SI85-9A1, a marine α-proteobacterium, contain genes for organoheterotrophy, methylotrophy, oxidation of sulfur and carbon monoxide, the ability to grow over a wide range of oxygen concentrations, and the complete Calvin cycle for carbon fixation. In an early study, Ehrlich (1963) observed that bacteria in Mn nodules can enhance the adsorption of Mn^{2+} from seawater in the presence of peptone. He proposed that bacteria play a crucial role in nodule development. Besides, Mn^{2+} oxidation by cell-free extract from a Mn nodule bacterium *Arthrobacter* 37 was found to be mediated by enzyme activity (Ehrlich 1968). The rate of oxidation of free Mn^{2+} by the enzyme depended on the concentration of cell-free extract used. In another study on the effect of temperature and pressure on *Arthrobacter* 37, Ehrlich (1971) demonstrated that at 5°C, temperature optimum for Mn^{2+} oxidation increases with pressure and the effect of pressure on cells can be counteracted by an appropriate increase in temperature. Further, Arcuri and Ehrlich (1979) stated the involvement of cytochrome in Mn^{2+} oxidation by two marine bacteria. With the use of conventional electron transport chain, the bacterium could derive useful energy from Mn^{2+} oxidation by oxidative phosphorylation. By a detailed observation of bacteria that catalyze Mn^{2+} oxidation, Ehrlich (1980) postulated that two groups of Mn^{2+}-oxidizing bacteria exist, one that acts on free Mn (Mn^{2+}) and the other that acts only on Mn^{2+} bound to Mn(IV) oxide and derive energy from the reaction. At the same time, Arcuri and Ehrlich (1980) proposed that cytochromes were involved in Mn^{2+} oxidation by *Oceanospirillum* BIII 45. They could observe that addition of periplasmic/intracellular fraction to membrane fraction was essential for Mn oxidation. Yet in another study on the removal efficiency of Mn^{2+} from seawater by marine sediments and clay minerals with the same organism, Ehrlich (1982) observed that ferric chloride pretreatment of clays is essential when intact cells are used and not when cell-free extracts are used. He remarked that ferric chloride pretreatment was necessary for activating the sediment for bacterial oxidation of sorbed Mn^{2+}.

Another interesting observation on Mn^{2+} oxidation by bacterial isolates from hydrothermal area (Ehrlich 1983) showed that Mn^{2+} oxidation could occur through an inducible enzyme system and an initial fixation of Mn(II) to Mn(IV) oxide was not essential for oxidation to proceed. Our earlier observation on the bacteriology of Fe–Mn nodules from the Indian Ocean region showed that psychrotrophic heterotrophic bacteria were capable of mobilizing and immobilizing Mn. The maximum percentage of Mn oxidizers (32.2%) was restricted to nodule surface and possessed various hydrolytic enzymes (Chandramohan et al. 1987). Later on, Ehrlich and Salerno (1990) demonstrated the coupling of ATP synthesis with that of Mn^{2+} oxidation by a marine bacterial strain SSW22. They proposed chemiosmosis (diffusion of ions across a selectively permeable membrane) as the probable mechanism for energy coupling by intact cells, membrane vesicles, or cell extracts.

In a different study, Rosson and Nealson (1982) observed that live/killed mature spores of *Bacillus* Strain SG-1 could oxidize Mn^{2+} once bound but not when free in solution. They hypothesized that Mn^{2+} may form complex with exosporium or a spore coat protein. In another observation, Kepkay and Nealson (1982) identified that spores rather than vegetative cells are responsible for Mn^{2+} oxidation by SG-1. The adherence of bacterial cells to solid surfaces was found to be essential for proper sporulation and Mn^{2+} oxidation. Using radiotracers, Emerson et al. (1982) explained that Mn^{2+} oxidation in Saanich Inlet is bacterially mediated and the removal of Mn^{2+} is very fast and would occur in a matter of few days on solid surfaces. The study on the role of plasmids in Mn^{2+} oxidation (Lidstrom et al. 1983; Schuett et al. 1986) showed that in marine *Pseudomonads*, increased levels of both binding and oxidation of Mn^{2+} could occur in the presence of plasmids. Using radiotracers, Tebo et al. (1984) provided evidence for Mn^{2+} oxidation with oxygen as the terminal electron acceptor in Saanich Inlet and Framvaren Fjord. The Mn^{2+} oxidation was found to occur faster under air-saturated condition than under conditions of oxygen limitation. Interestingly, Rosson et al. (1984) using poisoned control could differentiate biological from abiotic Mn^{2+} oxidation and demonstrated that bacteria could significantly enhance the rate of Mn^{2+} removal from manganese-rich particulate layer in the water column. Further, by in situ experimental evidence on Mn^{2+} oxidation in Saanich Inlet, Tebo and Emerson (1985) showed that the rate of Mn^{2+} oxidation was limited by both oxygen and the concentration of microbial binding sites in the environment. In a study on Mn^{2+} removal from porewater, Edenborn et al. (1985) observed that Mn^{2+} removal rate is faster in the oxidized surface sediment where the Mn-oxidizing bacteria are abundant.

In an attempt to describe the general mechanism of Mn^{2+} oxidation, Tebo and Emerson (1986) describe a model where they conclude that Mn oxidation rate is not dependent on Mn concentration in the water column but is rather a function of the total number of surface-binding sites available. Similarly, de Vrind et al. (1986b) showed with experimental evidence that mature spores of marine *Bacillus* SG-1 could oxidize Mn^{2+}. On the other hand, de Vrind et al. (1986a) demonstrated that vegetative cells of the same organism could reduce manganese. The reduction of Mn by the vegetative cells was thought to make Mn^{2+} available for sporulation in a manganese-limited environment. Accumulation of MnO_2 by the spore coat prevented further oxidation of bound Mn^{2+} as the active sites get masked when oxygen gets consumed and protons get liberated. In another approach, the marine *Pseudomonas* sp. Strain S-36 when grown in continuous culture was found to obtain energy for CO_2 fixation from Mn^{2+} oxidation (Kepkay and Nealson 1987).

In a different approach, Sunda and Huntsman (1987) used radiotracers to determine the kinetics of particulate Mn formation in seawater. According to them, Mn^{2+} oxidation is microbially catalyzed and the rates depend upon the increase in temperature with respect to the ratio of particulate to dissolved Mn in estuarine water. The indirect process of Mn^{2+} oxidation by *Chlorella* sp at high pH (>9.0) resulting from photosynthesis was reported by Richardson et al. (1988). They demonstrated that growth of photosynthetic organisms as aggregates or as concentrated cell suspension in pelagic waters could generate microenvironments with steep gradients of oxygen and pH conducive for Mn^{2+} oxidation. Radiotracer studies on hydrothermal vent locations showed the scavenging of Mn^{2+} at higher rates under in situ incubations compared to onboard studies (Mandernack and Tebo 1993). Their results could suggest that bacteria not only enhance the scavenging of Mn within vent waters, but also facilitated Mn deposition to the sediments. In another interesting observation by Hansel and Francis (2006), the unrecognized role of *Roseobacter* like planktonic bacteria in Mn^{2+} oxidation and cycling in coastal waters was identified suggesting an alternative means of Mn^{2+} oxidation in the photic zone. The bacterium showed the ability to oxidize Mn^{2+} in the presence of light through photooxidation pathway and by direct enzymatic action in the dark. Based on a kinetic model of the oxidative pathway, Webb et al. (2005) stated that Mn^{3+} is a transient intermediate and the rate-limiting step in the oxidation of Mn^{2+}. They suggested that oxidation of Mn^{2+} could involve a unique multicopper oxidases (MCOs) system capable of two-electron oxidation of its substrate. MCOs are a class of enzymes that have metallocentre assembly containing four Cu atoms (Brouwers et al. 2000b). They couple the four-electron reduction of dioxygen to water with the oxidation of substrate. The well-defined MCOs are laccase, ascorbate oxidase, and ceruloplasmin. The others include phenoxazinone synthase, bilirubin oxidase, dihydrogeodin oxidase, sulochrin oxidase, and FET3 (Solomon et al. 1996).

Our observation on microbially mediated Mn^{2+} oxidation in bacterial isolates belonging to *Halomonas* sp from Carlsberg Ridge (Fernandes et al. 2005) showed that Mn is precipitated extracellularly. Same isolates when grown in the presence of Ni and Co in the absence of Mn^{2+} showed the ability to accumulate these metals both intra- and extracellularly (Sujith et al. 2010; Antony et al. 2010). Further, study from the mangrove sediments Krishnan et al. (2007) offered experimental evidence to demonstrate that both autochthonous autotrophs and heterotrophs work in tandem in reducing Mn^{2+} and other related metal ions in sediments. These processes may indirectly promote more metal oxidation by removing end product inhibition.

3.7 Manganese Oxidation by Freshwater Bacteria

The Mn and Fe oxidizing/depositing bacteria in freshwater habitats belong to *Sphaerotilus*, *Gallionella*, *Leptothrix*, *Pedomicrobium*, *Metallogenium*, *Hyphomicrobium*, *Crenothrix*, *Clonothrix*, and *Cladothrix* groups (Gregory and Staley 1982; Ghiorse 1984). Based on their abundance, Pringsheim (1949) stated that their significance in biochemical processes in rivers must be great but require further investigations to know about their nutritional needs, metabolism, and enzymatic systems. Knowing the importance of Mn^{2+} oxidation by bacteria, Johnson and Stokes (1966) readily stated with experimental evidence that *Sphaerotilus discophorus* belonging to β_1 subdivision of proteobacteria could oxidize Mn^{2+} to dark-brown manganic oxide. They pointed out that cells can lose the Mn^{2+}-oxidizing activity on heating and not poisoned by treatment with $HgCl_2$ suggestive of endogenous Mn^{2+} oxidation catalyzed by an inducible enzyme(s). In continuation of the earlier study with *Sphaerotilus discophorus*, Stokes and Powers (1967) ruled out that endogenous oxidation of Mn^{2+} could be stimulated by the oxidation of poly-β-hydroxybutyrate, a storage product within the cell. Further study by Ali and Stokes (1971) could observe autotrophic growth promotion in *Sphaerotilus discophorus* with Mn^{2+} as the sole source of energy. The results were later on evaluated with evidence that in the late phase of growth, *S. discophorus* do oxidize and accumulate MnO_2 but do not serve as energy source in this organism (Hajj and Makemson 1976). Conversely, Mills and Randles (1979) using electron transport chain inhibitors in their study suggested that Mn^{2+} oxidation in *Sphaerotilus discophorus* could be cytochrome mediated.

In an another study with different Mn-oxidizing filamentous budding bacteria *Pedomicrobium* belonging to α-proteobacteria, Ghiorse and Hirsch (1978) noted that very active Mn-depositing bacterial strains are also very active iron depositors. The presence of budding bacteria in freshwater distribution systems leads to the formation of biofilms heavily encrusted with Mn oxides (Tyler and Marshall 1967; Sly et al. 1988). The depositions of Mn oxides occur in close association with an extracellular matrix of acidic polysaccharides or polymer in these bacterial strains

(Ghiorse and Hirsch 1979; Sly et al. 1990). The mechanism of Mn^{2+} oxidation was found to involve a two-step process composed of rapid binding of Mn^{2+} to EPS followed by the oxidation of Mn^{2+} by an unknown factor (Ghiorse and Hirsch 1979). Incidentally, they could identify the unusual factor, perhaps a protein responsible for Mn^{2+} oxidation associated with the polymer that could not be completely inhibited by glutaraldehyde, $HgCl_2$, or heat. Further findings using inhibitors and cellular fractionation methods (Larsen et al. 1999) showed that heat treatment of cells could enhance Mn^{2+} binding but abolish Mn-oxidizing activity. They could restore the activity of the enzyme upon the addition of Cu in the medium and suggested that Cu-dependent enzyme MCOs catalyze the Mn^{2+} oxidation in *Pedomicrobium* ACM 3067.

Jaquet et al. (1982) indicated that *Metallogenium* plays a key role in Mn cycling in Lake Leman. In contrast, Maki et al. (1987) using ^{54}Mn tracer could not find any significant difference between poisoned and non-oxygen controls in the biological Mn^{2+} oxidation and the number of *Metallogenium* morphotypes in Lake Washington. They suggested that *Metallogenium* plays only a weak role in Mn^{2+} oxidation in Lake Washington. In a discussion on the retention of Mn in the Wahnbach reservoir by bacteria, Herschel and Clasen (1998) state that *Metallogenium personatum* could be a propelling force behind microbially catalyzed transformation of Mn in the reservoir. They explained that increase in dissolved oxygen level in deep water is conducive for Mn^{2+} oxidation. In some earlier observations, Tyler and Marshall (1967) and Tyler (1970) addressed the widespread occurrence of stalked, budding bacteria *Hyphomicrobium* belonging to α-proteobacteria in Mn deposits and suggested that some *Hyphomicrobia* could preferentially oxidize Mn over Fe in hydroelectric pipelines.

In an interesting observation, Uren and Leeper (1978) stated that microbial oxidation of Mn^{2+} in soil could occur at low oxygen pressures provided CO_2 supplied is adequate. With *Arthrobacter* sp from soil, Bromfield and David (1976) had also showed that oxides of Mn could rapidly adsorb manganous ions from aqueous solutions but not in the case of abiotic control. In a kinetic study with cell free extracts of two bacterial isolates *Pseudomonas* III and *Citrobacter freundii* belonging to γ-proteobacteria, Douka (1980) identified that the rate of Mn^{2+} oxidation increased with its concentration, suggesting a strong affinity between the oxidizing system and Mn. Conversely, Chapnick et al. (1982) could show that Mn^{2+} removal from water column and oxidation in Lake Oneida during summer months are mediated by metabolically active Mn-oxidizing bacteria. They also showed that particles in lake water when removed by filtration or killed by ethanol treatment inhibit the activity. Besides, Gregory and Staley (1982) suggested that plasmids may be directly involved in Mn oxidation by providing essential gene products or may act indirectly by altering the microenvironment in a way as to make the chemical oxidation of Mn^{2+} favorable. Using in situ dialysis technique, Kepkay (1985) showed that Mn precipitation in soil could be a microbially mediated process causing a fivefold enhancement of abiotic process such as adsorption. Likewise, Vojak et al. (1985) stated that biological process could be responsible for the change in oxidation state of Mn^{2+} to Mn^{4+}. The rates of Mn^{2+}

oxidation in their study differed with increase in salinity and were depressed in the presence of inhibitors. Similarly, Johnson and Kipphut (1988) showed by in situ incubation technique, the rate of Mn^{2+} oxidation is largely microbially mediated in Toolik lake and is regulated by Mn^{2+} concentration rather than temperature or oxygen concentration. With more detailed experiment on mutants that lack the ability to oxidize Mn^{2+}, Caspi et al. (1998) stated that the Mn-oxidizing ability of *Pseudomonas putida* MnB1 can be recovered by complementation of the mutation in a c-type cytochrome biogenesis-defective mutant. In the kinetic studies on Mn uptake and oxidation by Moy et al. (2003), the uptake of Mn^{2+} by *Rhizobium* sp was greater than the conversion of Mn^{2+} to Mn oxides with significant production of polysaccharides. They suggested that polysaccharides might be involved in the uptake of Mn and in minimizing Mn oxide production. The study on redox transformation of Mn in Antarctic lakes (Krishnan et al. 2009) showed that Co could have a more profound role in Mn^{2+} oxidation and Ni on Mn oxide reduction. Although several studies report the oxidation of Mn^{2+} by bacteria, the identity of Mn-oxidizing bacteria remained undisclosed. Recently, Falamin and Pinevich (2006) determined the phylogenetic position and phenotypic properties of *Pseudomonas siderocapsa* sp.nov. They suggested a mixotrophic mode of nutrition in the strain and deposition of Mn oxides in their capsules rather than in outer membrane as observed in other *Pseudomonas* species.

As an attempt to understand the Mn-oxidizing ability of *Leptothrix discophora* SS-1 Adams and Ghiorse (1986) examined the ultrastructure of the strain by electron microscopy. They could observe extracellular blebs in cells and proposed it as vehicles for Mn-oxidizing protein. Manganese oxidation by sheathless strain of *Leptothrix discophora* SS-1 belonging to β_1 subdivision of proteobacteria (Boogerd and de Vrind 1987) in buffered medium at pH 7.5 showed the release of Mn^{2+}-oxidizing factors in the spent culture medium and was found associated with MnO_2 aggregates. Meanwhile, Adams and Ghiorse (1987) isolated the Mn^{2+}-oxidizing protein from *Leptothrix discophora* SS-1 and characterized the extracellular Mn^{2+}-oxidizing activity. The same authors in 1988 identified the oxidation states of Mn in the Mn oxide produced by *Leptothrix discophora* SS-1. They could identify that the oxidation state of Mn in fresh samples exist as Mn^{3+} and on aging give rise to a mixture of Mn (III,IV) oxides in older samples. Later, study by Corstjens et al. (1997) could identify that gene *mofA* is linked to Mn oxidation. In addition, Nelson et al. (1999) showed that Mn oxides produced by SS-1 can adsorb toxic metal lead. Moreover, Brouwers et al. (2000b) stated that being the core element in putative MCOs, Cu^{2+} could stimulate the oxidation of Mn^{2+}. Investigation on kinetics of Mn^{2+} oxidation by Zhang et al. (2002) explained that at circumneutral pH, at a relatively low numbers of Mn-oxidizing bacteria (*Leptothrix discophora* SS-1), biologically mediated Mn^{2+} oxidation exceeded abiotic oxidation. Interestingly, a recent study by El Gheriany et al. (2009) remarked that Fe is essential for efficient Mn^{2+} oxidation in *Leptothrix discophora* SS-1.

3.8 Manganese Oxidation: A Genomic Perspective

In the recent years, several studies were attempted by researchers to understand the genetics involved in bacterial Mn^{2+} oxidation. Marine bacteria are efficient Mn oxidizers; however, only few studies report the genetic mechanism(s) involved in Mn oxidation. The well-studied Mn-oxidizing bacteria is *Bacillus* sp. Strain SG-1, a marine gram-positive bacterium isolated from shallow marine sediment that produces Mn-oxidizing spores (van Wassbergen et al. 1993, 1996; Francis et al. 2002; Francis and Tebo 2002). This is the only organism for which the direct involvement of MCOs in Mn oxidation is established. They proposed MnxG as one of the first gene products ever shown to be associated with the exosporium possessing oxidase activity. Francis et al. (2002) demonstrated that MnxG is localized to the exosporium of wild-type spores and is absent in the nonoxidizing spores of transposon mutants within the *mnx* gene cluster. Dick et al. (2006) based on phylogenetic analysis of 16S rRNA and *mnxG* genes explained that Mn-oxidizing *Bacillus* sp isolated from Guaymas Basin resembled deep-sea isolates reported earlier from coastal sediments, with few representing novel strains and clusters. Recently, Mayhew et al. (2008) proposed that vertical inheritance and gene loss influenced the distribution of the gene *mnxG* among the *Bacillus* sp.

For the first time, van Waasbergen et al. (1993) identified the genes involved in Mn^{2+} oxidation. They demonstrated that *mnx* region encodes factors that are required for oxidation of Mn^{2+} by SG-1 spores by protoplast transformation and mutagenesis. Later, van Waasbergen et al. (1996) suggested that among the several genes (*mnxA* to *mnxG*) that were earlier proposed to be involved in Mn^{2+} oxidation, the *mnxG* gene product may function like a copper oxidase and would be directly responsible for the oxidation of Mn^{2+} by the bacterial spores. The first direct evidence for the presence of RubisCo genes in a gram-negative Mn-oxidizing bacterium strain S185-9A1 was given by Caspi et al. (1996). The genes were more related to those from non-chlorophyte algal chloroplasts than from bacteria. Dick et al. (2008b) suggested that MnxG catalyzes two sequential one-electron oxidations from Mn^{2+} to Mn^{3+} and from Mn^{3+} to Mn^{4+}, a novel type of reaction for a multicopper oxidase.

Aurantimonas manganoxydans Strain. SI85-9A1 (Dick et al. 2008a, Anderson et al. 2009a, b) and *Erythrobacter* sp Strain. SD-21 (Anderson et al. 2009b) are the two other marine Mn-oxidizing α-proteobacteria known to oxidize Mn^{2+} that have been recently studied in detail. Anderson et al. (2009b) identified five annotated MCOs in the genome sequence of the above strains but none of the MCOs were reported to have any role in Mn^{2+} oxidation. In contrast, they could illustrate the role of heme peroxidase in Mn^{2+} oxidation and tentatively suggested MopA for the putative Ca^{2+} binding heme peroxidase.

Leptothrix discophora SS-1 a freshwater bacterial species that deposits Mn oxides on its extracellular sheath was studied in detail by Corstjens et al. (1997) and Brouwers et al. (2000a) using sophisticated molecular tools. Based on the results, they proposed that MCOs like gene *mofA* (manganese-oxidizing factor) to be involved in Mn^{2+} oxidation and genes *mofB* and *mofC* to be a part of the same *mofA* operon. At the same time, Siering and Ghiorse (1997b) by variable stringency hybridization analysis using digoxigenin-labelled mofA probe of *Leptothrix discophorora* SS-1 showed that Mn-oxidation genes of other *Leptothrix* spp were closely related to one another but were not homologous to the unidentified presumptive Mn oxidation genes from other genera. In the meanwhile, Siering and Ghiorse (1997a) could detect sheathed bacteria (*Leptothrix* spp) in environmental samples based on their designed 16S rRNA-targeted specific probes and proposed its applications in further research.

The other common bacterial genera known for Mn^{2+} oxidation are *Pseudomonas putida* Strains MnB1 and GB-1. There are several reports (Caspi et al. 1996; Caspi et al. 1998; Brouwers et al. 1999; de Vrind et al. 2003) that describe the Mn-oxidizing ability of these strains. But only few (Brouwers et al. 1999, 2000a) attempts have been made to understand the genetic mechanisms involved in Mn^{2+} oxidation. They could identify by molecular analysis that gene *cumA* (copper protein involved in Mn oxidation) participates with MCOs in Mn^{2+} oxidation and *cumB* for optimal growth. Later, Francis and Tebo (2001) surprisingly observed highly conserved *cumA* gene sequences in non-Mn-oxidizing *Pseudomonas* strains. Based on the results, they suggested that *cumA* gene may not be expressed or that it may not be the only gene to confer the ability to oxidize Mn^{2+}. Conversely, they could exert an alternative function in these organisms and the gene could occur in phylogenetically diverse *Pseudomonas* strains.

Pedomicrobium sp. ACM3067 another aquatic bacteria could oxidize Mn^{2+} in close association with an extracellular matrix of acidic polysaccharides or polymer (Ghiorse and Hirsch 1979). Further, understanding about the mechanism involved in Mn^{2+} oxidation (Larsen et al. 1999) showed that Mn^{2+} oxidation is catalyzed by a copper-dependent enzyme in *Pedomicrobium* sp. ACM3067. Recent study by Ridge et al. (2007) provided evidence that *moxA* gene encoding a MCO homolog is essential for both Mn^{2+} oxidation and laccase-like activity in *Pedomicrobium* sp. ACM3067.

Mn-oxidizing genus *Hypomicrobium* is less probed at genetic level. Only one study by Layton et al. (2000) recorded the abundance of *Hypomicrobium* populations in activated sludge based on 16S rRNA analysis. About 5% of 16S rRNA in activated sludge corresponded to *Hypomicrobium* sp. Gregory and Staley (1982) showed experimental evidence that Mn^{2+}-oxidizing ability in the bacterial isolates was lost when maintained in the absence of the metal in the laboratory. They hypothesized that Mn oxidation may be directly related to the presence of plasmids.

3.9 Manganese Oxidation: A Proteomic Perspective

Metal ion efflux systems are central to cellular physiology. The uptake of Mn in bacteria occurs through (1) P-type ATPase (MntA) (Hao et al. 1999), (2) metal binding protein-dependent ABC transport system: Group PsaA, (3) pH-dependent metal ion transporter: MntH Groups A, B, C, and (4) natural resistance-associated macrophage protein (NRAMP) family (Jakubovics and Jenkinson 2001; Cellier 2002; Papp-Wallace and Maguire 2006). Metallochaperones are responsible for the incorporation of correct metal into some of the proteins; however, most metalloproteins acquire their metals directly from cellular pools. In an attempt to understand the cellular mechanisms that govern metal acquisition by most nascent proteins, Tottey et al. (2008) identified the most abundant Cu^{2+}-protein, CucA and the most abundant Mn^{2+}-protein, MncA in the periplasm of cyanobacteria *Synechocystis* PCC 6803. They showed that compartmentalization kept competitive metals out of the wrong nascent proteins during protein folding.

The only Mn-oxidizing proteins identified and characterized so far in bacteria were the MCOs (Brouwers et al. 2000b; Francis and Tebo 2000). Another class of proteins in bacteria rarely known to oxidize Mn is the heme-containing manganese peroxidases (MNPs) (Palma et al. 2000; Anderson et al. 2009b). The known Mn-oxidizing proteins include the MnxG (~138 kDa) of *Bacillus* SG-1 (van Wassbergen et al. 1996; Francis et al. 2002) and MopA of *Aurantimonas manganoxidans* Strain. SI85-9A1 and *Erythrobacter* sp Strain. SD-21[Anderson et al. (2009b)]. In marine α-Proteobacterium SD-21, manganese-oxidizing factors of ≈250 and 150 kDa was observed in the logarithmic phase of growth. However, the expression of Mn(II) oxidase was not completely dependent on Mn^{2+} rather it was required for higher growth yield (Francis et al. 2001). They claimed it as the first group of Mn-containing metalloenzyme in gram-negative marine bacteria. Francis and Tebo (2002) could identify the first active Mn-oxidizing enzymes in spores or gram-positive bacteria. Their study came across proteins of different molecular weights in Mn-oxidizing marine *Bacillus* sp isolated from coastal marine sediment. Based on the inhibition of Mn-oxidizing activity by azide a multicopper oxidase inhibitor suggested that the unidentified proteins belong to the MCO group of enzymes. The role of metalloregulatory protein MntR, a transcriptional regulator of Mn homeostasis, was determined by Lieser et al. (2003). They demonstrated that differences in metal-activated DNA binding could play a role in the mechanism of Mn(II)-selective transcription of factors and the oligomerization of MntR that was metal independent. Further, Huang and Wu (2004) revealed the identity of the genes under control of manganese response regulator ManR in the cyanobacterium, *Anabaena* sp. PCC 7120.

The known Mn-oxidizing proteins include the CumA (50.5 kDa) of *Pseudomonas putida* GB-1 (Brouwers et al. 1999), MofA (~180 kD) of *Leptothrix discophora* SS-1 (Corstjens et al. 1997; Brouwers et al. 2000a), and MoxA (52.47 kDa) of *Pedomicrobium* sp. ACM3067 (Ridge et al. 2007). Several regulatory pathways for Mn in bacteria were investigated by different authors. Que and Helmann (2000), Guedon et al. (2003), and Moore and Helmann (2005) found that MntR, Fur, TnrA, and σ^B regulons regulated Mn uptake in *Bacillus subtilis*. Platero et al. (2004) stated that Fur was involved in manganese-dependent regulation of *mntA*. Patzer and Hantke (2001) and Hohle and O'Brian (2009) provided evidence to show that *mntH* gene encoding NRAMP like Mn^{2+} transporter was repressed by Fur and MntR of the *mntH* gene. Conversely, Kehres et al. (2000) inferred that NRAMP proteins are selective Mn transporters involved in response to reactive oxygen. Diaz-Mireles et al. (2004) affirmed that Fur-like protein Mur (manganese uptake regulator), a Mn^{2+}-responsive transcriptional regulator of *Rhizobium leguminosarum*, differs from Fur that binds Fe^{2+} in γ-proteobacteria and engage in Mn uptake. Groot et al. (2005) identified the expression of three putative Mn transport systems (*mtsCBA*, *mntH1*, *and mntH2*) besides *mntA* in *Lactobacillus plantarum.* They observed the specific derepression or induction of transport systems upon Mn^{2+} limitation, suggesting their role in Mn^{2+} homeostasis. Subsequently, Jakubovics and Valentine (2009) identified the novel Mn^{2+} efflux system MntE in *Streptococcus pneumoniae*. They stated that disruption of the *mntE* gene could lead to widespread transcriptional changes that are distinct from responses to extracellular Mn^{2+}.

The expression of 25 kDa cytoplasmic protein was identified as superoxide dismutase isoenzyme (Mn-SOD) in *Arthrobacter* sp (Ercole et al. 1999). The functioning of the protein under both aerobic and anaerobic conditions in the presence of Mn oxide was found to have additional physiological function. The higher-molecular-weight surface protein (30 kDa) showed no homology with any of the identified proteins and its function is yet to be identified. Jung and Schweisfurth (1979) observed that *Pseudomonas* sp. Strain MnB1 produced a heat labile intracellular Mn-oxidizing protein during stationary phase of growth. Mn-oxidizing protein was not induced by the presence of Mn^{2+}, rather it was particularly dependent on the age of the culture. Likewise, in a comparative study on Mn oxidation using growing and resting cells of a freshwater bacterial isolate strain FMn 1, Zapkin and Ehrlich (1983) observed enzymatic nature of Mn-oxidizing activity in the strain. The activity of the enzyme was inducible. In a review, Shi (2004) stated that protein phosphatases are metalloenzymes with active centers containing two metal ions functioning as cofactors. The Mn-dependent prokaryotic protein O-phosphatases and their function were stated to add new insight into Mn^{2+} homeostasis and protein O-phosphorylation in prokaryotic cells.

3.10 Molecular Biomineralization

Organisms are capable of forming a diverse array of minerals, some of which cannot be formed inorganically in the biosphere. Biogenic minerals may be amorphous, paracrystalline, or crystalline (Lowenstam 1981). The mineralization processes driven by biological activity involving microorganisms constitute biomineralization (Wang and Müller 2009). The microorganisms and their interaction with geologic materials result in geochemical transformations switching between soluble and insoluble phases (White et al. 1997). As a result of close interaction between mineral and bacteria, biomineralization co-occur (Fig. 3.1). It can lead to precipitation of the metal leachate and formation of metal oxide coatings on bacterial wall and other inert surfaces contributing directly as nucleation sites for further mineral formation (Fortin et al. 1995). Microorganisms interact with minerals for creating a more hospitable surrounding by extraction of nutrients and sequestration of toxic substances. Microbes use minerals as sources and sinks of electrons, for coupled oxidation–reduction reactions. Many of these reactions enable the release and capture of energy from unstable or metastable minerals (Shock 2009). Deep-sea minerals in polymetallic nodules, Fe–Mn crusts, and hydrothermal vents are not only formed by abiogenic mineralization but also by free-living and biofilm-forming bacteria which form matrix for Mn deposition. Here the mineralization processes proceed in close association with organic molecules or matrices. It can be an either induced (biological–chemical) or a controlled (enzymatic) process and the details of the processes and the references have been described by Wang and Müller (2009). Besides, in order to understand the biogeochemical phenomena occurring in the Mn-rich marine environments, several microbiological studies have focused on hydrothermal vents and Fe–Mn encrusted seamounts in the recent years (Davis et al. 2009; Emerson 2009; Glazer and Rouxel 2009; Rassa et al. 2009; Sudek et al. 2009). Few extended their research on biologically induced mineralization (Douglas and Beveridge 1998; Wang et al. 2009a, b; Dong 2010). Wang et al. (2009b) examined the biogenic components of the crust and He et al. (2008) examined the microbial community composition of Iron–Manganese nodules using sophisticated analytical tools. They found acidobacteria and proteobacteria in nodules and associated sediments. The firmicutes were restricted to nodules and the soils had more acidobacteria and Verrucomicrobia compared to nodules. Advancement in element-specific mapping of rock surfaces revealed hot spots of Mn accumulation in microbial biofilms (Templeton and Knowles 2009).

The mineral phases produced by bacteria are mostly amorphous and sometimes poorly crystalline. The crystallization process occurs with prolonged incubation time (Tazaki 2005) and the characteristic oxides thus produced are not identical to known synthetic solids possibly, because of solid-phase incorporation of biomolecular constituents (Parikh and Chorover 2005). It is observed that surficial proteins associate with Mn oxidation during the production of a poorly crystalline Mn(IV) phase. The formation of mixed phase minerals like hausmannite (Mn_3O_4), feiknechtite (β-MnOOH), manganite (γ-MnOOH), and Na-buserite following Mn (II) oxidation by *Bacillus* SG-1 was reported by Mann et al. (1988) and Mandernack et al. (1995). Whereas, a todorokite-like mineral was found to be produced by *Leptothrix discophora* Strain SP-6 (Kim et al. 2003) and MnOx produced by *Pseudomonas putida* Strain MnB1 was most similar to "acid" birnessite (Villalobos et al. 2003). Recent understanding about the genes and proteins involved in Mn oxidation help to spread its application in biotechnology. The gene *mnxG* responsible for Mn oxidation in Bacillus SG-1 (van Wassbergen et al. 1996), *cumA* in *Pseudomonas putida* GB-1 (Brouwers et al. 1999), *mofA* in *Leptothrix discophora* SS-1 (Corstjens et al. 1997; Brouwers et al. 2000a), *moxA* in *Pedomicrobium* sp. ACM3067 (Ridge et al. 2007), and *mopA* in *Aurantimonas manganoxidans* Strain. SI85-9A1 and *Erythrobacter* sp Strain. SD-21 (Anderson et al. 2009b) could be cloned and the products expressed under laboratory condition.

3.11 Biotechnological Applications of Manganese Oxidation

Manganese, a comparatively less toxic element, can become toxic to domestic and aquatic lives when its concentration exceeds beyond the EPA permissible levels (0.05 mg L^{-1}). During summer, it is observed that Mn oxides undergo reduction when the oxygen level drops in public and private wells, municipal water supplies, etc. The solubilized Mn is quite stable in the presence of oxygen and therefore can become a health risk for public who consumes the drinking water. The Mn-oxidizing bacterial residents of the Mn^{2+}-rich environments can oxidize Mn^{2+} and reduce its solubility and thereby provide protective mechanism against toxic levels of soluble Mn (Bromfield 1978). An application of Mn oxidizers or their products to such habitats offer chemical/biological solution to the problem on a seasonal/permanent basis (Czekalla et al. 1985). Mn oxides are also excellent electron acceptors for anaerobic respiration (Nealson et al. 1989). The application of Mn-oxidizing bacteria and Mn in sedimentary environments can stimulate respiratory carbon mineralization and could offer a natural system of "Pumping" (via precipitation and sedimentation of Mn oxides) electron-acceptor equivalents into an anaerobic environment (Nealson et al. 1989). It is observed that Mn oxides are potent chelators of several other trace metals, their application has proved to be efficient in the removal of radium from water supplies (Moore and Reid 1973), in retaining heavy metals like Co, Ni, Zn, and others in soil, polymerization of organic compounds, participation in humus formation by oxidation of phenols and quinines (Vodyanitskii 2009).

The removal of Mn^{2+} is conventionally achieved by inorganic oxidation such as chlorination or permanganate oxidation, followed by sand filtration (Miyata et al. 2007). In 1986, Ghiorse proposed the exploitation of Mn-precipitating microorganisms for industrial metal recovery processes. The use of Mn-oxidizing bacteria in treating effluents can minimize the addition of chemical reagents and unwanted by-product formation. The increase in filtration rate and longer runs due to less clogging, savings on wash water, and rapid return to equilibrium following a backwash sequence reduce the operational cost in treatment and maintenance of sludge in biological effluent treatment (Mouchet 1992; Katsoyiannis and Zouboulis 2004; Stembal et al. 2005). The biological processes take advantage of active and passive process in treatment by a variety of mechanisms like adsorption, accumulation, precipitation, and oxidation. The drawback of using bacteria for treating effluent biotechnologically is the slow rate of Mn oxidation. As an early solution to the problem, Stuetz et al. (1996) proposed the usage of combined algal–bacterial Mn oxidation and optimization of bioreactor parameters for treating metal effluents effectively. When using Mn oxide (scavengers of the environment) for treatment of any effluents with unknown composition, precaution needs to be taken as it is known that sometimes interaction of Mn oxide with other elements can result in phase transformations [Se(IV) to Se(VI), Cr(III) to Cr(VI), and As(III) to As(V)], contributing to increase/decrease in metal toxicity (Vodyanitskii 2009; He et al. 2010).

Recent advancement in the understanding of microbial Mn oxidation provides insight into the mechanisms of metal oxidation and the processes involved. This oxidation proceeds at rates up to five orders of magnitude greater than abiotic oxidation (Tebo et al. 1997). The Mn oxides produced by microorganisms are abundant environmental nanoparticles, they have great importance in biotechnology for the removal of heavy metals from aqueous matrices and oxidation of organic micropollutants in wastewater treatment plants (Villalobos et al. 2005b). The higher specific surface area of negatively charged biogenic Mn oxides than synthetic d-MnO_2 and commercially available pyrolusite allow greater sorption of positively charged heavy metals in solution (Hennebel et al. 2009). The bacterial spores from a potent Mn-oxidizing bacteria *Bacillus* sp. SG-1 have found extensive capacity for actively binding and oxidizing Mn and passively binding other metals. Likewise, Mn-oxidizing protein from *Pseudomonas putida* Strains MnB1 and GB-1 as well as sheath of *Leptothrix discophora* is found to have similar function (Francis and Tebo 1999). It was observed by Nelson et al. (2002); Villalobos et al. (2005a) that Mn oxides produced by Leptothrix *discophora* SS-1 and *Pseudomonas putida* MnB1 can adsorb five times more Pb per mole of Mn than abiotic Mn(IV) (hydr) oxide and 500–5,000 times more than pyrolusite oxides, thus stimulating interest in the development of Mn oxides for use in bioremediation. Likewise, Toner et al. (2006) observed a tenfold higher capacity for biogenic Mn oxides in adsorbing Zn than chemically synthesized Mn oxides, and Murray and Tebo (2007) detected seven times higher adsorption of Cr in biogenic Mn oxides produced by *Bacillus* sp. SG-1 than synthetic d-MnO_2. The utilization of these microorganisms in concentrating metal ions from effluents will have intense application in biotechnology for treatment of wastewaters and metal-containing effluents.

It has also been observed recently that biogenic Mn oxides can oxidize 17a-Ethinylestradiol, a potent endocrine-disrupting recalcitrant, and reduce its estrogenic activity to 81.7% (de Rudder et al. 2004). In another recent observation, Forrez et al. (2010) have shown that biogenic Mn oxides can oxidize diclofenac, a nonsteroidal anti-inflammatory drug and can reduce its lethal concentration and toxicity. Similar observations made for triclosan (Zhang and Huang 2003) and ciprofloxacin (Zhang and Huang 2005) with biogenic Mn oxides suggest that biogenic manganese oxide can be a promising polishing technique for sewage treatment plant effluents.

3.12 Conclusion

The current understanding about bacterial Mn oxidation comprises the participation of MCOs, but their direct link to oxidation is emphasized only in *Bacillus* SG-1 and not in other organisms like *Pseudomonas putida* MnB1, GB1, *Leptothrix discophora* SS1, or *Pedomicrobium* sp. ACM3067. The various regulatory mechanisms and transport systems for Mn uptake in bacterial cells are studied but the role of metalloproteins in Mn oxidation or how the proteins select the right metal when competitive metal ions are in excess is still not fully understood. What is known at present is that the compartmentalization of protein during folding regulates the binding of correct metal. Therefore, the production and utilization of biogenic Mn oxide nanoparticle in biotechnology requires further understanding about the molecular mechanism of Mn oxidation.

Acknowledgments We thank the Director, NIO for providing the required facilities to conduct this research work. The work has been carried out under the project "Preliminary exploration of cobalt-rich seamount crusts in the northern Indian Ocean" funded by the Ministry of Earth Sciences (Government of India) lead by Dr. V.K. Banakar. SPP wishes to thank Dr. M.P. Tapaswi for continuous support with required literature for reference and also acknowledges the Council of Scientific and Industrial Research, New Delhi-India, for the award of Senior Research Fellowship. This manuscript has NIO contribution No 5005.

References

Adams LF, Ghiorse WC (1986) Physiology and ultrastructure of *Leptothrix discophora* SS-1. Arch Microbiol 145:126–135

Adams LF, Ghiorse WC (1987) Characterization of an extracellular Mn^{2+}-oxidizing activity and isolation of Mn^{2+}-oxidizing protein from *Leptothrix discophora* SS-1. J Bacteriol 169: 1279–1285

Adams LF, Ghiorse WC (1988) Oxidation state of Mn in the Mn oxide produced by *Leptothrix discophora* SS-1. Geochim Cosmochim Acta 52:2073–2076

Ali SH, Stokes JL (1971) Stimulation of heterotrophic and autotrophic growth of *Sphaerotilus discophorus* by manganous ions. Anton Van Leeuwenhoek 37:519–528

Anderson CR, Dick GJ, Chu ML, Cho JC, Davis RE, Bräuer SL, Tebo BM (2009a) *Aurantimonas manganoxydans*, sp. nov. and *Aurantimonas litoralis*, sp. nov.: Mn(II) oxidizing representatives of a globally distributed clade of alpha-Proteobacteria from the order Rhizobiales. Geomicrobiol J 26:189–198

Anderson CR, Johnson HA, Caputo N, Davis RE, Torpey JW, Tebo BM (2009b) Mn(II) oxidation is catalyzed by heme peroxidases in "*Aurantimonas manganoxydans*" Strain SI85-9A1 and *Erythrobacter* sp. Strain SD-21. Appl Environ Microbiol 75:4130–4138

Antony R, Sujith PP, Fernandes SO, Verma P, Khedekar VD, Loka Bharathi PA (2010) Cobalt immobilization by manganese oxidizing bacteria from Indian Ridge System. Curr Microbiol 62:840–849

Appanna VD (1988) Stimulation of exopolysaccharide production in *Rhizobium meliloti* JJ-1 by manganese. Biotechnol Lett 10:205–206

Appanna VD, Gazso LG, St. Pierre M (1996) Multiple-metal tolerance in *Pseudomonas fluorescens* and its biotechnological significance. J Biotechnol 52:75–80

Archibald F (1986) Manganese: its acquisition by and function in the lactic acid bacteria. Crit Rev Microbiol 13:63–109

Arcuri EJ, Ehrlich HL (1979) Cytochrome involvement in Mn(II) oxidation by two marine bacteria. Appl Environ Microbiol 37:916–923

Arcuri EJ, Ehrlich HL (1980) Electron transfer coupled to Mn(II) oxidation in two deep-sea pacific ocean isolates. In: Trudinger PA, Walter MR, Ralph BJ (eds) Biogeochemistry of ancient and modern environments. Springer, New York, pp 339–344

Babich H, Stotzky G (1983) Influence of chemical speciation on the toxicity of heavy metals to the microbiota. In: Nriagu JO (ed) Aquatic toxicology. Wiley Interscience, New York, pp 1–46

Boogerd FC, de Vrind JPM (1987) Manganese oxidation by *Leptothrix discophora*. J Bacteriol 169:489–494

Bowen HJM (1979) Environmental chemistry of the elements. Academic, London

Bromfield SM (1978) The oxidation of manganous ions under acidic conditions by an acidophilous actinomycete from acid soil. Aust J Soil Res 16:91–100

Bromfield SM, David DJ (1976) Sorption and oxidation of manganous ions and reduction of manganese oxide by cell suspensions of a manganese oxidizing bacterium. Soil Biol Biochem 8:37–43

Brouwers G-J, de Vrind JPM, Corstjens PLAM, Cornelis P, Baysse C, de Vrind-de Jong EW (1999) *CumA*, a gene encoding a multicopper oxidase, is involved in Mn^{2+}-oxidation in *Pseudomonas putida* GB-1. Appl Environ Microbiol 65:1762–1768

Brouwers G-J, Corstjens PLAM, de Vrind JPM, Verkamman A, de Kuyper M, de Vrind-de Jong EW (2000a) Stimulation of Mn^{2+} oxidation in *Leptothrix discophora* SS-1 by Cu^{2+} and sequence analysis of the region flanking the gene encoding putative multicopper oxidase MofA. Geomicrobiol J 17:25–33

Brouwers G-J, Vijgenboom E, Corstjens PLAM, de Vrind JPM, de Vrind-de Jong EW (2000b) Bacterial Mn^{2+} oxidizing systems and multicopper oxidases: An overview of mechanisms and functions. Geomicrobiol J 17:1–24

Calvert SE, Pedersen TF (1996) Sedimentary geochemistry of manganese: implications for the environment of formation of manganiferous black shales. Econ Geol 91:36–47

Caspi R, Haygood MG, Tebo BM (1996) Unusual ribulose-1,5-bisphosphate carboxylase/ oxygenase genes from a marine manganese-oxidizing bacterium. Microbiology 142: 2549–2559

Caspi R, Tebo BM, Haygood MG (1998) c-type cytochromes and manganese oxidation in *Pseudomonas putida* strain MnB1. Appl Environ Microbiol 64:3549–3555

Cellier M (2002) Bacterial genes controlling manganese accumulation. In: Winkelmann G (ed) Microbial Transport Systems. Wiley-VCH Verlag GmbH & Co., KGaA, pp 325–345

Chander M, Setlow B, Setlow P (1998) The enzymatic activity of phosphoglycerate mutase from gram-positive endospore-forming bacteria requires Mn^{2+} and is pH sensitive. Can J Microbiol 44:759–767

Chandramohan D, Loka Bharathi PA, Nair S, Matondkar SGP (1987) Bacteriology of ferromanganese nodules from the Indian Ocean. Geomicrobiol J 5:17–31

Chapnick SD, Mire WS, Nealson KH (1982) Microbially mediated manganese oxidation in a freshwater lake. Limnol Oceanogr 27:1004–11014

Christianson DW (1997) Structural chemistry and biology of manganese metalloenzymes. Prog Biophys Mol Biol 67:217–252

Corstjens PLAM, de Vrind JPM, Goosen T, de Vrind-de Jong EW (1997) Identification and molecular analysis of the *Leptothrix discophora* SS-1 *mofA* gene, a gene putatively encoding a manganese-oxidizing protein with copper domains. Geomicrobiol J 14:91–108

Crowley JD, Traynor DA, Weatherburn DC (2000) Enzymes and proteins containing manganese: an overview. Met Ions Biol Syst 37:209–278

Czekalla C, Mevius W, Hanert H (1985) Quantitative removal of iron and manganese by microorganisms in rapid sand filters. Wat Suppl 3:111–123

Davis RE, Stakes DS, Wheat CG, Moyer CL (2009) Bacterial variability within an iron-silica-manganese-rich hydrothermal mound located off-axis at the cleft segment, Juan de Fuca Ridge. Geomicrobiol J 26:570–580

de Rudder J, Van de Wiele T, Dhooge W, Comhaire F, Verstraete W (2004) Advanced water treatment with manganese oxide for the removal of 17 a-ethynylestradiol (EE2). Water Res 38:184–192

de Vrind JPM, Boogerd FC, de Vrind-de Jong EW (1986a) Manganese reduction by a marine *Bacillus* species. J Bacteriol 167:30–34 de Vrind JPM, de Vrind-de Jong EW, de Voogt J-WH, Westbroek P, Boogerd FC, Rosson RA (1986b) Manganese oxidation by spores and spore coats of a marine *Bacillus* species. Appl Environ Microbiol 52:1096–1100

de Vrind J, de Groot A, Brouwers GJ, Tommassen J, de Vrind-de Jong EW (2003) Identification of a novel Gsp-related pathway required for secretion of the manganese oxidizing factor of *Pseudomonas putida* strain GB-1. Mol Microbiol 47:993–1006

Diaz-Mireles E, Wexler M, Sawers G, Bellini D, Todd JD, Johnston AWB (2004) The Fur-like protein Mur of *Rhizobium leguminosarum* is a Mn^{2+}-responsive transcriptional regulator. Microbiology 150:1447–1456

Dick GJ, Lee YE, Tebo BM (2006) Manganese(II)-oxidizing *bacillus* spores in guaymas basin hydrothermal sediments and plumes. Appl Environ Microbiol 72:3184–3190

Dick GJ, Podell S, Johnson HA, Rivera-Espinoza Y, Bernier-Latmani R, McCarthy JK, Torpey JW, Clement BG, Gaasterland T, Tebo BM (2008a) Genomic insights into Mn(II) oxidation by the marine alphaproteobacterium *Aurantimonas* sp.Strain SI85-9A1. Appl Environ Microbiol 74:2646–2658

Dick GJ, Torpey JW, Beveridge TJ, Tebo BM (2008b) Direct Identification of a bacterial manganese(ii) oxidase, the multicopper oxidase MnxG, from spores of several different marine *Bacillus* species. Appl Environ Microbiol 74:1527–1534

Dong H (2010) Mineral-microbe interactions: a review. Front Earth Sci China 4:127–147

Douglas S, Beveridge TJ (1998) Mineral formation by bacteria in natural microbial communities. FEMS Microbiol Ecol 26:79–88

Douka C (1980) Kinetics of manganese oxidation by cell-free extracts of bacteria isolated from manganese concretions from soil. Appl Environ Microbiol 39:74–80

Doyle RJ (1989) How cell walls of gram-positive bacteria interact with metal ions. In: Beveridge TJ, Doyle RJ (eds) Metal Ions and Bacteria. Wiley, New York, pp 275–293

Edenborn HM, Paquin Y, Chateauneuf G (1985) Bacterial contribution to manganese oxidation in a deep coastal sediment. Estuar Coast Shelf Sci 21:801–815

Ehrlich HL (1963) Bacteriology of manganese nodules. I. Bacterial action on manganese in nodule enrichments. Appl Microbiol 11:15–19

Ehrlich HL (1968) Bacteriology of manganese nodules. II. Manganese oxidation by cell-free extract from a manganese nodule bacterium. Appl Microbiol 16:197–202

Ehrlich HL (1971) Bacteriology of manganese nodules. V. Effect of hydrostatic pressure on bacterial oxidation of Mn(II) and reduction of MnO_2. Appl Microbiol 21:306–310

Ehrlich HL (1976) Manganese as an energy source for bacteria. In: Nriagu JO (ed) Environmental biogeochemistry. Ann Arbor Science, Michigan, pp 633–644

Ehrlich HL (1978) Inorganic energy sources for chemolithotrophic and mixotrophic bacteria. Geomicrobiol J 1:65–83

Ehrlich HL (1980) Different forms of microbial manganese oxidation and reduction and their environmental significance. In: Trudinger PA, Walter MR, Ralph BJ (eds) Biogeochemistry of ancient and modern environments. Springer, New York, pp 327–332

Ehrlich HL (1982) Enhanced removal of Mn^{2+} from seawater by marine sediments and clay minerals in the presence of bacteria. Can J Microbiol 28:1389–1395

Ehrlich HL (1983) Manganese-oxidizing bacteria from a hydrothermally active area on the Galapagos. Rift Ecol Bull 35:357–366

Ehrlich HL (1987) Manganese oxide reduction as a form of anaerobic respiration. Geomicrobiol J 5:423–431

Ehrlich HL (2002a) Geomicrobiology. Marcel Dekker Inc., New York

Ehrlich HL (2002b) How microbes mobilize metals in ores: a review of current understandings and proposals for future research. Miner Metall Proc 19:220–224

Ehrlich HL, Salerno JC (1990) Energy coupling in Mn^{2+} oxidation by a marine bacterium. Arch Microbiol 154:12–17

Emerson D (2009) Potential for iron-reduction and iron-cycling in iron oxyhydroxide-rich microbial mats at Loihi Seamount. Geomicrobiol J 26:639–647

Emerson S, Kalhorn D, Jacobs L, Tebo BM, Nealson KH, Rosson RA (1982) Environmental oxidation rate of manganese (II): Bacterial catalysis. Geochim Cosmochim Acta 46: 1073–1079

Ercole C, Altieri F, Piccone C, Del Gallo M, Lepidi A (1999) Influence of manganese dioxide and manganic ions on the production of two proteins in *Arthrobacter* sp. Geomicrobiol J 16:95–103

Falamin AA, Pinevich AV (2006) Isolation and characterization of a unicellular manganese-oxidizing bacterium from a freshwater lake in Northwestern Russia. Microbiology 75:180–185

Fernandes SO, Krishnan KP, Khedekar VD, Loka Bharathi PA (2005) Manganese oxidation by bacterial isolates from the Indian Ridge System. Biometals 18:483–492

Forrez I, Carballa M, Verbeken K, Vanhaecke L, Schlusener M, Ternes T, Boon N, Verstraete W (2010) Diclofenac oxidation by biogenic manganese oxides. Environ Sci Technol 44:3449–3454

Fortin D, Davis B, Southam G, Beveridge TJ (1995) Biogeochemical phenomena induced by bacteria within sulfidic mine tailings. J Ind Microbiol Biotechnol 14:178–185

Francis CA, Tebo BM (1999) Marine *Bacillus* spores as catalysts for oxidative precipitation and sorption of metals. J Mol Microbiol Biotechnol 1:71–78

Francis CA, Tebo BM (2000) New insights into the diversity of genes and enzymes involved in bacterial Mn(II) oxidation. In: Morgan J (ed) Chemical speciation and reactivity in water chemistry and water technology: a symposium in honor of James. ILSI Press, Washington, DC, pp 488–490

Francis CA, Tebo BM (2001) cumA multicopper oxidase genes from diverse Mn(II)-oxidizing and non-Mn(II)-oxidizing *Pseudomonas* strains. Appl Environ Microbiol 67:4272–4278

Francis CA, Tebo BM (2002) Enzymatic manganese(II) oxidation by metabolically dormant spores of diverse *Bacillus* species. Appl Environ Microbiol 68:874–880

Francis CA, Co E, Tebo BM (2001) Enzymatic manganese(II) oxidation by a marine α-proteobacterium. Appl Environ Microbiol 67:4024–4029

Francis CA, Casciotti KL, Tebo BM (2002) Localization of Mn(II)-oxidizing activity and the putative multicopper oxidase, MnxG, to the exosporium of the marine *Bacillus* sp. strain SG-1. Arch Microbiol 178:450–456

El-Gheriany IA, Bocioaga D, Hay AG, Ghiorse WC, Shuler ML, Lion LW (2009) Iron requirement for Mn(II) oxidation by *Leptothrix discophora* SS-1. Appl Environ Microbiol 75:1229–1235

Ghiorse WC (1984) Biology of iron- and manganese-depositing bacteria. Annu Rev Microbiol 38:515–550

Ghiorse WC, Hirsch P (1978) Iron and manganese deposition by budding bacteria. In: Krumbein WE (ed) Environmental biogeochemistry and geomicrobiology. Ann Arbor Science, Ann Arbor, pp 897–909

Ghiorse WC, Hirsch P (1979) An ultrastructural study of iron and manganese deposition associated with extracellular polymers of *Pedomicrobium*-like budding bacteria. Arch Microbiol 123:213–226

Ghoirse WC (1986) Applications of ferromanganese-depositing microorganisms to industrial metal recovery processes. Biotech Bioeng Symp 16:141–148

Glasby GP (2006) Manganese: predominant role of nodules and crusts. In: Schulz HD, Zabel M (eds) Marine geochemistry. Springer Berlin, Heidelberg, pp 371–427

Glazer BT, Rouxel OJ (2009) Redox speciation and distribution within diverse iron-dominated microbial habitats at Loihi Seamount. Geomicrobiol J 26:606–622

Gregory E, Staley JT (1982) Widespread distribution of ability to oxidize manganese among freshwater bacteria. Appl Environ Microbiol 44:509–511

Groot MNN, Klaassens E, de Vos WM, Delcour J, Hols P, Kleerebezem M (2005) Genome-based in silico detection of putative manganese transport systems in *Lactobacillus plantarum* and their genetic analysis. Microbiology 151:1229–1238

Guedon E, Moore CM, Que Q, Wang T, Ye RW, Helmann JD (2003) The global transcriptional response of Bacillus subtilis to manganese involves the MntR, Fur, TnrA and σ^B regulons. Mol Microbiol 49:1477–1491

Hajj H, Makemson J (1976) Determination of growth of *Sphaerotilus discophorus* in the presence of manganese. Appl Environ Microbiol 32:699–702
Hansel CM, Francis CA (2006) Coupled photochemical and enzymatic Mn(II) oxidation pathways of a planktonic *Roseobacter*-like bacterium. Appl Environ Microbiol 72:3543–3549
Hao Z, Chen S, Wilson DB (1999) Cloning, expression and characterization of cadmium and manganese uptake genes from *Lactobacillus plantarum*. Appl Environ Microbiol 65: 4746–4752
He J, Zhang L, Jin S, Zhu Y, Liu F (2008) Bacterial communities inside and surrounding soil iron–manganese nodules. Geomicrobiol J 25:14–24
He J, Meng Y, Zheng Y, Zhang L (2010) Cr(III) oxidation coupled with Mn(II) bacterial oxidation in the environment. J Soil Sediment 10:767–773
Hem JD (1978) Redox processes at surfaces of manganese oxide and their effects on aqueous metal ions. Chem Geol 21:199–218
Hennebel T, Gusseme BD, Boon N, Verstraete W (2009) Biogenic metals in advanced water treatment. Trends Biotechnol 27:90–98
Herschel A, Clasen J (1998) The importance of the manganese-oxidizing microorganism *Metallogeniurn personaturn* for the retention of manganese in the Wahnbach reservoir. Internat Rev Hydrobiol 83:19–30
Hohle TH, O'Brian MR (2009) The *mntH* gene encodes the major Mn^{2+} transporter in *Bradyrhizobium japonicum* and is regulated by manganese via the Fur protein. Mol Microbiol 72:399–409
Horsburgh MJ, Wharton SJ, Karavolos M, Foster SJ (2002) Manganese: elemental defence for a life with oxygen? Trends Microbiol 10:496–501
Huang W, Wu Q (2004) Identification of genes controlled by the manganese response regulator, ManR, in the cyanobacterium, *Anabaena* sp. PCC 7120. Biotechnol Lett 26:1397–1401
Jakubovics NS, Jenkinson HF (2001) Out of the iron age: new insights into the critical role of manganese homeostasis in bacteria. Microbiology 147:1709–1718
Jakubovics NS, Valentine RA (2009) A new direction for manganese homeostasis in bacteria: identification of a novel efflux system in *Streptococcus pneumonia*. Mol Microbiol 72:1–4
Jaquet JM, Nembrim G, Garcla J, Vernet JP (1982) The manganese cycle in Lac Leman, Switzerland: the role of *Metallogenium*. Hydrobiologia 91:323–340
Johnson KS (2006) Manganese redox chemistry revisited. Science 313:1896–1897
Johnson CG, Kipphut GW (1988) Microbially mediated Mn(II) oxidation in an oligotrophic Arctic lake. Appl Environ Microbiol 54:1440–1445
Johnson AH, Stokes JL (1966) Manganese oxidation by *Sphaerotilus discophorus*. J Bacteriol 91:1543–1547
Jung WK, Schweisfurth R (1979) Manganese oxidation by an intracellular protein of a *Pseudomonas* species. Z Allg Mikrobiol 19:107–115
Katsoyiannis IA, Zouboulis AI (2004) Biological treatment of Mn(II) and Fe(II) containing groundwater: kinetic considerations and product characterization. Water Res 38:1922–1932
Kehres DG, Zaharik ML, Finlay BB, Maguire ME (2000) The NRAMP proteins of *Salmonella typhimurium* and *Escherichia coli* are selective manganese transporters involved in the response to reactive oxygen. Mol Microbiol 36:1085–1100
Kepkay PE (1985) Kinetics of microbial manganese oxidation and trace metal binding in sediments: results from an in situ dialysis technique. Limnol Oceanogr 30:713–726
Kepkay PE, Nealson KH (1982) Surface enhancement of sporulation and manganese oxidation by a marine *Bacillus*. J Bacteriol 151:1022–1026
Kepkay PE, Nealson KH (1987) Growth of a manganese oxidizing *Pseudomonas* sp. in continuous culture. Arch Microbiol 148:63–67
Keren N, Kidd MJ, Penner-Hahn JE, Pakrasi HB (2002) A light-dependent mechanism for massive accumulation of manganese in the photosynthetic bacterium *Synechocystis* sp. PCC 6803. Biochemistry 41:15085–15092
Kim HS, Pasten PA, Gaillard JF, Stair PC (2003) Nanocrystalline todorokite-like manganese oxide produced by bacterial catalysis. J Am Chem Soc 125:14284–14285

Kirchner WB, Grabowski S (1972) Manganese in lacustrine ecosystems: a review. Am Water Resour Assoc 8:1259–1264

Krishnan KP, Fernandes SO, Chandan GS, Loka Bharathi PA (2007) Bacterial contribution to mitigation of iron and manganese in mangrove sediments. Mar Pollut Bull 54:1427–1433

Krishnan KP, Sinha RK, Krishna K, Nair S, Singh SM (2009) Microbially mediated redox transformations of manganese (II) along with some other trace elements: a study from Antarctic lakes. Polar Biol 32:1765–1778

Larsen EI, Sly LI, McEwan AG (1999) Manganese(II) adsorption and oxidation by whole cells and a membrane fraction of *Pedomicrobium* sp. ACM 3067. Arch Microbiol 171:257–264

Layton AC, Karanth PN, Lajoie CA, Meyers AJ, Gregory IR, Stapleton RD, Taylor DE, Sayler GS (2000) Quantification of *Hyphomicrobium* populations in activated sludge from an industrial wastewater treatment system as determined by 16S rRNA analysis. Appl Environ Microbiol 66:1167–1174

Lidstrom ME, Engebrecht J, Nealson KH (1983) Evidence for plasmid-encoded manganese oxidation in a marine pseudomonad. FEMS Microbiol Lett 19:1–6

Lieser SA, Davis TC, Helmann JD, Cohen SM (2003) DNA-binding and oligomerization studies of the manganese(II) metalloregulatory protein MntR from *Bacillus subtilis*. Biochemistry 42:12634–12642

Lowenstam HA (1981) Minerals formed by organisms. Science 211:1126–1131

Maki JS, Tebo BM, Palmer FE, Nealson KH, Staley JT (1987) The abundance and biological activity of manganese-oxidizing bacteria and *Metallogenium-like* morphotypes in Lake Washington, USA. FEMS Microbiol Ecol 45:21–29

Mandernack KW, Tebo BM (1993) Manganese scavenging and oxidation at hydrothermal vents and in vent plumes. Geochim Cosmochim Acta 57:3907–3923

Mandernack KW, Post J, Tebo BM (1995) Manganese mineral formation by bacterial-spores of the marine *Bacillus*, SG-1: evidence for the direct oxidation of Mn(II) to Mn(IV). Geochim Cosmochim Acta 59:4393–4408

Mann S, Sparks NHC, Scott GHE, de Vrind-de Jong EW (1988) Oxidation of manganese and formation of Mn_3O_4 (Hausmannite) by spore coats of a Marine *Bacillus* sp. Appl Environ Microbiol 54:2140–2143

Mayhew LE, Swanner ED, Martin AP, Templeton AS (2008) Phylogenetic relationships and functional genes: distribution of gene (*mnxG*) encoding a putative manganese-oxidizing enzyme in *Bacillus* species. Appl Environ Microbiol 74:7265–7271

Mills VH, Randles CI (1979) Manganese oxidation in *Sphaerotilus discophorus* particles. J Gen Appl Microbiol 25:205–207

Miyata N, Tani Y, Sakata M, Iwahori K (2007) Microbial manganese oxide formation and interaction with toxic metal ions. J Biosci Bioeng 104:1–8

Moore CM, Helmann JD (2005) Metal ion homeostasis in *Bacillus subtilis*. Curr Opin Microbiol 8:188–195

Moore WS, Reid DF (1973) Extraction of radium from natural waters using manganese-impregnated acrylic fibers. J Geophys Res 78:8880–8886

Mouchet P (1992) From conventional to biological removal of iron and manganese in France. J Am Water Works Assoc 84:158–167

Moy YP, Neilan BA, Foster LJR, Madgwick JC, Rogers PL (2003) Screening, identification and kinetic characterization of a bacterium for Mn(II) uptake and oxidation. Biotechnol Lett 25:1407–1413

Murray KJ, Tebo BM (2007) Cr(III) is indirectly oxidized by the Mn(II)-oxidizing bacterium *Bacillus* sp strain SG-1. Environ Sci Technol 41:528–533

Nealson KH (1983) The microbial manganese cycle. In: Krumbein WE (ed) Microbial geochemistry. Blackwell Scientific Publications, Oxford, pp 191–221

Nealson KH, Myers CR (1992) Microbial reduction of manganese and iron: new approaches to carbon cycling. Appl Environ Microbiol 58:439–443

Nealson KH, Tebo BM, Rosson RA (1988) Occurrence and mechanisms of microbial oxidation of manganese. Adv Appl Microbiol 33:279–318

Nealson KH, Rosson RA, Myers CR (1989) Mechanisms of oxidation and reduction of manganese. In: Beveridge T, Doyle R (eds) Metal ions and bacteria. Wiley, New York, pp 383–411

Nelson YM, Lion LW, Ghiorse WC, Shuler ML (1999) Production of biogenic Mn oxides by *Leptothrix discophora* SS-1 in a chemically defined growth medium and evaluation of their Pb adsorption characteristics. Appl Environ Microbiol 65:175–180

Nelson YM, Lion LW, Shuler ML, Ghiorse WC (2002) Effect of oxide formation mechanisms on lead adsorption by biogenic manganese (hydr)oxides, iron (hydr)oxides, and their mixtures. Environ Sci Technol 36:421–425

Ogawa T, Bao DH, Katoh H, Shibata M, Pakrasi HB, Bhattacharyya-Pakrasi M (2002) A two-component signal transduction pathway regulates manganese homeostasis in *Synechocystis* 6803, a photosynthetic organism. J Biol Chem 277:28981–28986

Palma C, Martinez AT, Lema JM, Martinez MJ (2000) Different fungal manganese-oxidizing peroxidases: a comparison between *Bjerkandera* sp. and *Phanerochaete chrysosporium*. J Biotechnol 77:235–245

Papp-Wallace KM, Maguire ME (2006) Manganese transport and the role of manganese in virulence. Annu Rev Microbiol 60:187–209

Parikh SJ, Chorover J (2005) FTIR spectroscopic study of biogenic Mn-oxide formation by *Pseudomonas putida* GB-1. Geomicrobiol J 22:207–218

Patzer SI, Hantke K (2001) Dual repression by Fe^{2+}-Fur and Mn^{2+}-MntR of the *mntH* gene, encoding an NRAMP-like Mn^{2+} transporter in *Escherichia coli*. J Bacteriol 183:4806–4813

Platero R, Peixoto L, O'Brian MR, Fabiano E (2004) Fur is involved in manganese-dependent regulation of mntA (sitA) expression in *Sinorhizobium meliloti*. Appl Environ Microbiol 70:4349–4355

Pringsheim EG (1949) The filamentous bacteria *Sphaerotilus*, *Leptothrix*, *Cladothrix*, and their relation to iron and manganese. Phil Trans R Soc Lond 233:453–482

Que Q, Helmann JD (2000) Manganese homeostasis in *Bacillus subtilis* is regulated by MntR, a bifunctional regulator related to the diphtheria toxin repressor family of proteins. Mol Microbiol 35:1454–1468

Rassa AC, McAllister SM, Safran SA, Moyer CL (2009) Zeta-proteobacteria dominate the colonization and formation of microbial mats in low-temperature hydrothermal vents at Loihi Seamount, Hawaii. Geomicrobiol J 26:623–638

Richardson LL, Aguilar C, Nealson KH (1988) Manganese oxidation in pH and O_2 microenvironments produced by phytoplankton. Limnol Oceanogr 33:352–363

Ridge JP, Lin M, Larsen EI, Fegan M, McEwan AG, Sly LI (2007) A multicopper oxidase is essential for manganese oxidation and laccase-like activity in *Pedomicrobium* sp. ACM 3067. Environ Microbiol 9:944–953

Roitz JS, Flegal AR, Bruland KW (2002) The biogeochemical cycling of manganese in San Francisco Bay: temporal and spatial variations in surface water concentrations. Estuar Coast Shelf Sci 54:227–239

Rosson RA, Nealson KH (1982) Manganese binding and oxidation by spores of a marine *Bacillus*. J Bacteriol 151:1027–1034

Rosson RA, Tebo BM, Nealson KH (1984) The use of poisons in the determination of microbial manganese binding rates in seawater. Appl Environ Microbiol 47:740–745

Rusin P, Ehrlich HL (1995) Developments in microbial leaching-mechanisms of manganese solubilization. In: Fiechter A (ed) Advances in biochemical engineering/biotechnology. Springer-Verlag Berlin, Heidelberg, pp 1–26

Saager PM, De Baar HJW, Burkill PH (1989) Manganese and iron in Indian Ocean waters. Geochim Cosmochim Acta 53:2259–2267

Schuett C, Zelibor JL Jr, Colwell RR (1986) Role of bacterial plasmids in manganese oxidation: evidence for plasmid-encoded heavy metal resistance. Geomicrobiol J 4:389–406

Schweisfurth R, Eleftheriadis D, Gundlach H, Jacobs M, Jung W (1978) Microbiology of the precipitation of manganese. In: Krumbein WE (ed) Environmental biogeochemistry and geomicrobiology. Ann Arbor Science, Ann Arbor, pp 923–928

Shi L (2004) Manganese-dependent protein o-phosphatases in prokaryotes and their biological functions. Front Biosci 9:1382–1397

Shock EL (2009) Minerals as energy source for microorganisms. Econ Geol 104:1235–1248

Siering PL, Ghiorse WC (1997a) Development and application of 16S rRNA-targeted probes for detection of iron- and manganese-oxidizing sheathed bacteria in environmental samples. Appl Environ Microbiol 63:644–651

Siering PL, Ghiorse WC (1997b) PCR detection of a putative manganese oxidation gene (*mofA*) in environmental samples and assessment of *mof*A gene homology among diverse manganese-oxidizing bacteria. Geomicrobiol J 14:109–125

Sly LI, Arunpairojana V, Hodgkinson MC (1988) *Pedomicrobium manganicum* from drinking-water distribution systems with manganese-related "dirty water" problems. Syst Appl Microbiol 11:75–84

Sly LI, Arunpairojana V, Dixon DR (1990) Binding of colloidal MnO_2 by extracellular polysaccharides of *Pedomicrobium manganicum*. Appl Environ Microbiol 56:2791–2794

Solomon EI, Sundaram UM, Machonkin TE (1996) Multicopper oxidases and oxygenases. Chem Rev 96:2563–2605

Spiro TG, Bargar JR, Sposito G, Tebo BM (2010) Bacteriogenic manganese oxides. Acc Chem Res 43:2–9

Spratt HG Jr, Hodson RE (1994) The effect of changing water chemistry on rates of manganese oxidation in surface sediments of a temperate saltmarsh and a tropical mangrove estuary. Estuar Coast Shelf Sci 38:119–135

Spratt HG Jr, Siekmann EC, Hodson RE (1994) Microbial manganese oxidation in saltmarsh surface sediments using leuco crystal violet manganese oxide detection technique. Estuar Coast Shelf Sci 38:91–112

Stembal T, Marinko M, Ribicic N, Briski F, Sipos L (2005) Removal of ammonia, iron and manganese from ground waters of Northern Croatia: pilot plant studies. Process Biochem 40:327–335

Stokes JL, Powers MT (1967) Stimulation of polyhydroxybutyrate oxidation in *Sphaerotilus discophorus* by manganese and magnesium. Arch Microbiol 59:295–301

Stuetz RM, Greene AC, Madgwick JC (1996) The potential use of manganese oxidation in treating metal effluents. Miner Eng 9:1253–1261

Sudek LA, Templeton AS, Tebo BM, Staudigel H (2009) Microbial ecology of Fe (hydr)oxide mats and basaltic rock from Vailulu'u Seamount, American Samoa. Geomicrobiol J 26:581–596

Sujith PP, Khedekar VD, Girish AP, Loka Bharathi PA (2010) Immobilization of nickel by bacterial isolates from the Indian ridge system and the chemical nature of the accumulated metal. Geomicrobiol J 27:424–434

Sunda WG, Huntsman SA (1987) Microbial oxidation of manganese in a North Carolina estuary. Limnol Oceanogr 32:552–564

Sunda WG, Huntsman SA (1990) Diel cycles in microbial manganese oxidation and manganese redox speciation in coastal waters of the Bahama Islands. Limnol Oceanogr 35:325–338

Tazaki K (2005) Microbial formation of a halloysite-like mineral. Clays Clay Miner 53:224–233

Tebo BM, Emerson S (1985) The effect of oxygen tension, Mn(II) concentration and temperature on the microbially catalyzed Mn(I1) oxidation rate in a marine fjord. Appl Environ Microbiol 50:1268–1273

Tebo BM, Emerson S (1986) Microbial manganese(II) oxidation in the marine environment: a quantitative study. Biogeochemistry 2:149–161

Tebo BM, Nealson KH, Emerson S, Jacobs L (1984) Microbial mediation of Mn(II) and Co(II) precipitation at the O_2/H_2S interfaces in two anoxic fjords. Limnol Oceanogr 29:1247–1258

Tebo BM, Ghiorse WC, van Waasbergen LG, Siering PL, Caspi R (1997) Bacterially mediated mineral formation: insights into manganese(II) oxidation from molecular genetic and biochemical studies. In: Banfield JF, Nealson KH (eds) Geomicrobiology: interactions between microbes and minerals. Mineral Soc Am, Washington, DC, pp 225–266

Tebo BM, Bargar JR, Clement BG, Dick GJ, Murray KJ, Parker D, Verity R, Webb SM (2004) Biogenic manganese oxides: properties and mechanisms of formation. Annu Rev Earth Planet Sci 32:287–328

Tebo BM, Johnson HA, McCarthy JK, Templeton AS (2005) Geomicrobiology of manganese(II) oxidation. Trends Microbiol 13:421–428

Tebo BM, Clement BG, Dick GJ (2007) Biotransformations of manganese. In: Hurst CJ, Crawford RL, Garland JL, Lipson DA, Mills AL, Stetzenbach LD (eds) Manual of environmental microbiology. ASM Press, Washington, DC, pp 1223–1238

Templeton A, Knowles E (2009) Microbial transformations of minerals and metals: recent advances in geomicrobiology derived from synchrotron- based X-ray spectroscopy and X-ray microscopy. Annu Rev Earth Planet Sci 37:367–391

Toner B, Manceau A, Webb SM, Sposito G (2006) Zinc sorption to biogenic hexagonal-birnessite particles within a hydrated bacterial biofilm. Geochim Cosmochim Acta 70:27–43

Tottey S, Waldron KJ, Firbank SJ, Reale B, Bessant C, Sato K, Cheek TR, Gray J, Banfield MJ, Dennison C, Robinson NJ (2008) Protein-folding location can regulate manganese binding versus copper- or zinc-binding. Nature 455:1138–1142

Trouwborst RE, Clement BG, Tebo BM, Glazer BT, Luther GW (2006) Soluble Mn(III) in suboxic zones. Science 313:1955–1957

Tyler PA (1970) *Hyphomicrobia* and the oxidation of manganese in aquatic ecosystems. Anton Van Leeuwenhoek 36:567–578

Tyler PA, Marshall KC (1967) Microbial oxidation of manganese in hydro-electric pipelines. Anton Van Leeuwenhoek 33:171–183

Uren NC, Leeper GW (1978) Microbial oxidation of divalent manganese. Soil Biol Biochem 10:85–87

van Waasbergen LG, Hoch JA, Tebo BM (1993) Genetic analysis of the marine manganese oxidizing *Bacillus* sp. strain SG-1: protoplast transformation, Tn917 mutagenesis and identification of chromosomal loci involved in manganese oxidation. J Bacteriol 175:7594–7603

van Waasbergen LG, Hildebrand M, Tebo BM (1996) Identification and characterization of a gene cluster involved in manganese oxidation by spores of the marine *Bacillus* sp. strain SG-1. J Bacteriol 178:3517–3530

Villalobos M, Toner B, Bargar J, Sposito G (2003) Characterization of the manganese oxide produced by *Pseudomonas putida* strain MnB1. Geochim Cosmochim Acta 67:2649–2662

Villalobos M, Bargar J, Sposito G (2005a) Mechanisms of Pb(II) sorption on a biogenic manganese oxide. Environ Sci Technol 39:569–576

Villalobos M, Bargar J, Sposito G (2005b) Trace metal retention on biogenic manganese oxide nanoparticles. Elements 1:223–226

Vodyanitskii YN (2009) Mineralogy and geochemistry of manganese: a review of publications. Eurasian Soil Sci 42:1170–1178

Vojak PWL, Edwards C, Jones MV (1985) Evidence for microbial manganese oxidation in the River Tamar estuary, South West England. Estuar Coast Shelf Sci 20:661–671

Wang X, Müuller WEG (2009) Marine biominerals: perspectives and challenges for polymetallic nodules and crusts. Trends Biotechnol 27:375–383

Wang X, Schloßmacher U, Natalio F, Schröder HC, Wolf SE, Tremel W, Müller WEG (2009a) Evidence for biogenic processes during formation of ferromanganese crusts from the Pacific ocean: implications of biologically induced mineralization. Micron 40:526–535

Wang X, Schröder HC, Wiens M, Schloßmacher U, Müller WEG (2009b) Manganese/ polymetallic nodules: micro-structural characterization of exolithobiontic- and endolithobiontic microbial biofilms by scanning electron microscopy. Micron 40:350–358

Webb SM, Dick GJ, Bargar JR, Tebo BM (2005) Evidence for the presence of Mn(III) intermediates in the bacterial oxidation of Mn(II). Proc Natl Acad Sci USA 102:5558–5563

White C, Sayer JA, Gadd GM (1997) Microbial solubilization and immobilization of toxic metals: key biogeochemical processes for treatment of contamination. FEMS Microbiol Rev 20: 503–516

Yang SH, Ehrlich HL (1976) Effect of four heavy metals (Mn, Ni, Cu and Co) on some bacteria from the deep sea. In: Sharpley JM, Kaplan AM (eds) Proceedings of the third international biodegradation symposium. Applied Science Publishers Ltd, London, pp 867–874
Yocum CF, Pecoraro V (1999) Recent advances in the understanding of the biological chemistry of manganese. Curr Opin Chem Biol 3:182–187
Zajic JE (1969) Microbial biogeochemistry. Academic, New York
Zapkin MA, Ehrlich HL (1983) A comparison of manganese oxidation by growing and resting cells of a freshwater bacterial isolate, strain FMn 1. Z Allg Mikrobiol 23:447–455
Zhang HC, Huang CH (2003) Oxidative transformation of triclosan and chlorophene by manganese oxides. Environ Sci Technol 37:2421–2430
Zhang HC, Huang CH (2005) Oxidative transformation of fluoroquinolone antibacterial agents and structurally related amines by manganese oxide. Environ Sci Technol 39:4474–4483
Zhang J, Lion LW, Nelson YM, Shuler ML, Ghiorse WC (2002) Kinetics of Mn(II) oxidation by *Leptothrix discophora* SS1. Geochim Cosmochim Acta 66:773–781

Chapter 4
Molecular Biomineralization: Toward an Understanding of the Biogenic Origin of Polymetallic Nodules, Seamount Crusts, and Hydrothermal Vents

Xiaohong Wang, Matthias Wiens, Heinz C. Schröder, Ute Schloßmacher, and Werner E.G. Müller

Contents

X. Wang (✉)
National Research Center for Geoanalysis, 26 Baiwanzhuang Dajie, Beijing CHN-100037, China
e-mail: wxh0408@hotmail.com

W.E.G. Müller (✉)
ERC Advanced Grant Research Group at the Institute for Physiological Chemistry, University Medical Center of the Johannes Gutenberg University Mainz, Duesbergweg 6, D-55128 Mainz, Germany
e-mail: wmueller@uni-mainz.de

W.E.G. Müller (ed.), *Molecular Biomineralization*, Progress in Molecular and Subcellular Biology 52, DOI 10.1007/978-3-642-21230-7_4,

Abstract Polymetallic nodules and crusts, hydrothermal vents from the Deep Sea are economically interesting, since they contain alloying components, e.g., manganese or cobalt, that are used in the production of special steels; in addition, they contain rare metals applied for plasma screens, for magnets in hard disks, or in hybrid car motors. While hydrothermal vents can regenerate in weeks, polymetallic nodules and seamount crusts grow slowly. Even though the geochemical basis for the growth of the nodules and crusts has been well studied, the contribution of microorganisms to the formation of these minerals remained obscure. Recent HR-SEM (high-resolution scanning electron microscopy) analyses of nodules and crusts support their biogenic origin. Within the nodules, bacteria with surface S-layers are arranged on biofilm-like structures, around which Mn deposition starts. In crusts, coccoliths represent the dominant biologically formed structures that act as bio-seeds for an initial Mn deposition. In contrast, hydrothermal vents have apparently an abiogenic origin; however, their minerals are biogenically transformed by bacteria. In turn, strategies can now be developed for biotechnological enrichment as well as selective dissolution of metals from such concretions. We are convinced that the recent discoveries will considerably contribute to our understanding of the participation of organic matrices in the enrichment of those metals and will provide the basis for feasibility studies for biotechnological applications.

4.1 Introduction

It is amazing that the composition of elements in the seawater is so different from the (secondary) minerals in the polymetallic nodules, the Co-rich crusts as well as in the hydrothermal vents that accrue in this environment. While in the seawater, the elements Na, Cl, K, and Ca (e.g., in form of ions or salts) are dominant, the marine (secondary) minerals are composed in the first place of Si [silicon] (clay), Fe [iron] (e.g., magnetite/goethide), Mn [manganese] (pyrolusite, braunite), and S (pyrite). Impressive examples are Mn and Fe, that occur in the seawater in only extremely low concentrations (~0.0004 ppm), while they are dominant in the polymetallic nodules or Co-rich crusts (Mero 1962), in which they represent over 30% of the material. Because of the pressing demands for such raw materials, the commercial exploitation of the gigantic occurrences of nodules and crusts on the ocean floors is to be expected (Schrope 2007). Attempts for a sustainable exploitation of Mn and its associated elements/minerals from the marine environment have to rely on the same abiogenic (mineralization) and biogenic (biomineralization) processes/strategies like those implemented in the synthesis of those deposits. Therefore, in this review, the focus is put on the underlying biological/biochemical processes, since the molecules involved are accessible at an "unlimited" scale by molecular biological and cell biological approaches. This concept of molecular biomineralization (nature as a model) will contribute to an understanding of biomineral formation in a causal analytic manner and might allow a sustainable exploitation of those natural resources in an environment-friendly way.

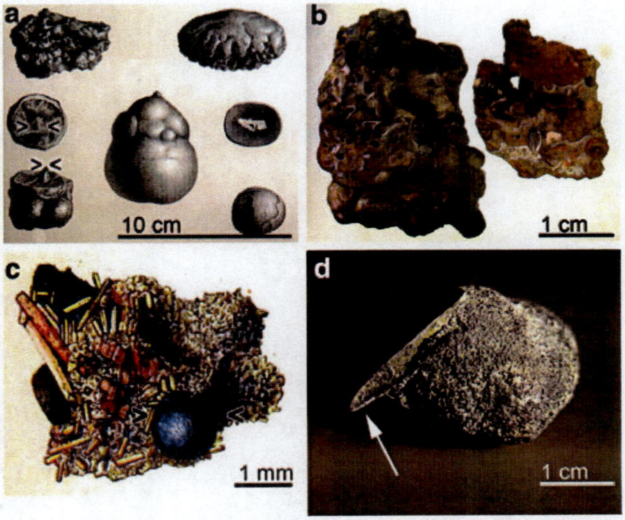

Fig. 4.1 First descriptions of nodules and crusts: (**a**) A close-up of polymetallic nodules, collected in 1891 during the Challenger Expedition (Murray 1891) in the Coral Sea close to New Guinea. The sizes of those nodules varied around 5 cm. Some of the nodules had formed around teeth/bone of sharks (><). (**b**) During the German Deep Sea Expedition "Valdivia" (1898–1899), large deposits (here, 8-cm large samples) of phosphorite crusts have been collected near the Cape of Good Hope (Africa) (Murray and Philippi 1908). (**c**) In those earliest descriptions, it had already been noted that the nodules were composed of smaller entities, metallic corpuscles (><) (Murray 1891). (**d**) A small polymetallic nodule collected recently, formed around a shark tooth (*arrow*)

The discovery of the vast deposits of nodules and crusts goes back to the HMS Challenger Expedition (1872–1876) (Murray 1891) (Fig. 4.1). The hydrothermal vents, with the black smokers as the most prominent examples, have been discovered only more recently in the 1970s (Francheteau et al. 1979). Polymetallic nodules [first described by Murray (1891); Fig. 4.1a] and seamount crusts [Murray and Philippi (1908); Fig. 4.1b] are formed in the interface between the aqueous phase and the seabed in the bathypelagic zone, below 1,000 m. It is remarkable that already in the earliest descriptions of polymetallic nodules and seamount crusts, the underlying principles for a biogenic origin of the minerals have been outlined, e.g., the composition of nodules from smaller entities, called then metallic corpuscles (Murray 1891; Fig. 4.1c). Likewise, the inclusion of teeth from sharks was mentioned at that time (Fig. 4.1a); a recently collected polymetallic nodule from the Clarion/Clipperton zone surrounding a tooth is shown in Fig. 4.1d.

4.2 Discovery

During the famous HMS Challenger Expedition (1872–1876), the first mysteries of the deep-sea deposits were disclosed. The research vessel (length: 61 m; displacement: 2,306 t; Fig. 4.2a) had a crew of 243 seamen, completed with a team of

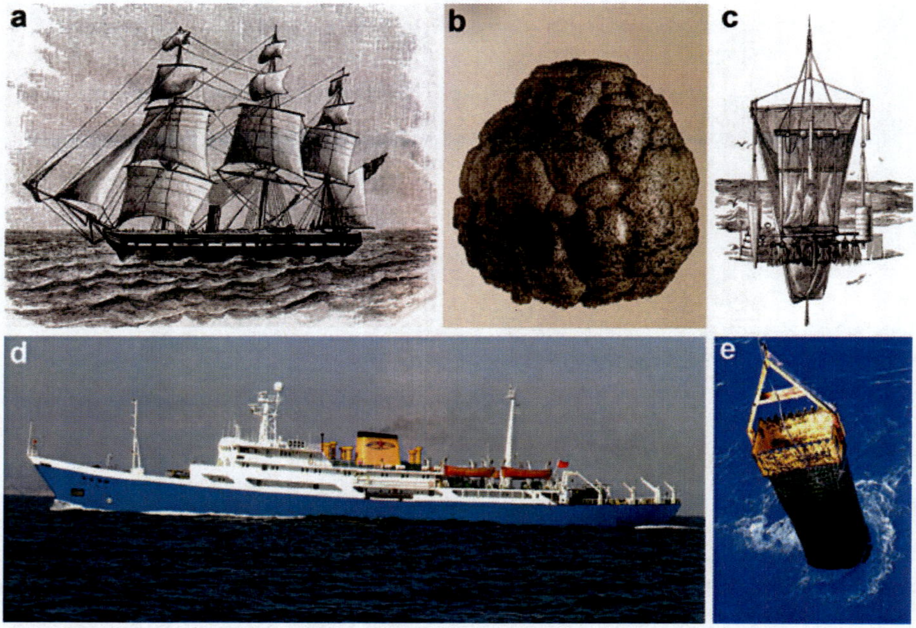

Fig. 4.2 Deep-sea collection of nodules. (**a–c**) Earliest collection during the Challenger Expedition. (**a**) HMS Challenger research vessel (length 61 m). On board, they had a total of 7,315 m hemp lines with a circumference of 5–8 cm to collect the deep-sea deposits, from depths of 8,200 m. (**b**) A large polymetallic nodule (8 cm × 6 cm), collected in the North Pacific Ocean; it contained 28% Mn oxide and 19% Fe oxide. In such nodules, occasionally, organic remains had been identified like algae, diatoms, radiolarians, or fish bones. (**c**) The nodules had been collected with a 62-kg heavy trawl/dredge constructed from iron arms holding a cotton cloth sack. It had been weighed down by five flat-headed swabs. (**d and e**) Geophysical prospecting of nodules with the ship Haiyang 4 in 1991 (length 104 m), using a dredge net as well; (Wang and Müller 2009)

scientists (zoologists, botanists, meteorologists) that were directed by the chief scientist Charles Wyville Thompson. During the 713 days long research journey, 133 bottom dredges were hauled to collect samples of deep-sea animals and plants for analyses of the waters from the surface to the depth of >2,000 m. On board, they had a total of 7,315 m hemp lines with a circumference of 5–8 cm to collect also the deep-sea deposits. A trawl/dredge was 152 cm long, 38 cm wide, and weighed 62 kg (Fig. 4.2c). The polymetallic nodules and the crusts were detected during the same cruises (Murray 1891; Murray and Philippi 1908) but at different depths. While the nodules were found in depths around 3,000–5,000 m, the crusts were collected from depths of only around 500 m. The most prominent samples were the phosphatic concretions detected near the Cape of Good Hope that had sizes of larger than 20 cm with a thickness of more than 10 cm. A characteristic nodule, collected in the North Pacific (between Japan and Hawaii) from a depth of 5,010 m (Fig. 4.1b) contained 28% Mn oxide and 19% Fe oxide. A systematic analysis from 12,000 deposit samples allowed a detailed chemical determination but also an assessment of the organic remains included in them. Murray determined bizarre algae (*Discophaera thomsoni* [Order Coccosphaerales]), diatoms, radiolarians, and a few fish bones. It is remarkable that the author already described that those organic seed structures represent the origin of the deposits. Besides a profound analysis of the nodules, Murray gave also the first descriptions of the phosphatic concretion/crusts, collected during the Challenger Expedition (Murray 1891) and later during the Valdivia Expedition (Murray and Philippi 1908). Already, first reflections on the mode of synthesis of crusts and nodules were outlined including also organic matter and "granular and fragmentary objects" that facilitate the oxidation of Fe(II) to Fe(III) and Mn(II) to Mn(IV), involving also the hydration intermediates. The present-day collection of nodules and crusts for geophysical prospecting, e.g., by the German research vessel "Sonne" or the Chinese "Haiyang 4," follows the same principles and uses almost the same methods (Fig. 4.2d and e).

4.3 Principles of Biomineralization in Marine Ore

Lowenstam and Weiner (Lowenstam 1981; Lowenstam and Weiner 1989; Weiner and Dove 2003) introduced a classification of biomineralization processes (Fig. 4.3). They distinguished two categories of biomineralization; first, the BIOLOGICALLY INDUCED MINERALIZATION (Fig. 4.3d). Biominerals represent genuine composite materials, formed from the inorganic "polymer"/mineral and the organic component (protein, polysaccharide, glycoprotein). The processes of biomineralization occur extracellularly and describe reactions proceeding on the interface between organic membranes and the inorganic environment. Such reactions are frequently observed in the aquatic milieu (within an organism or outside of it) during nucleation of epicellular mineralization processes and are characterized by a certain degree of heterogeneity (Bazylinski and Frankel 2003). The inhomogeneity in the element composition of biominerals is due to the concentrations of the inorganic chemical components in the environment, e.g., in the water, which counteract the selection processes, guided by the organic molecules. Biomineralization processes can be subdivided first into a SEED PHASE, during which the organic matrix functions as a nucleation platform for the deposition of the mineral. In order to distinguish the organic component that initiates a biomineralization process, from an inorganic nucleus that initiates crystallization of a mineral, the term bio-seed is appropriate. While inorganic seed particles allow controlled nucleations and crystal growth in a supersaturated environment of the inorganic precursors, bio-seeds facilitate the initiation of biomineral formation also in non-saturated concentrations of the inorganic precursors. An example is the marine snow, amorphous aggregations of organic particles that display not only adverse effects on the marine biota but have also mineralization activity. Marine snow is a three-dimensional organic/mineral meshwork (Amy et al. 1987; Herndl 1988; Müller et al. 1998; Leppard 1999) that is formed in the photic zone down to 100 m (Cottrell

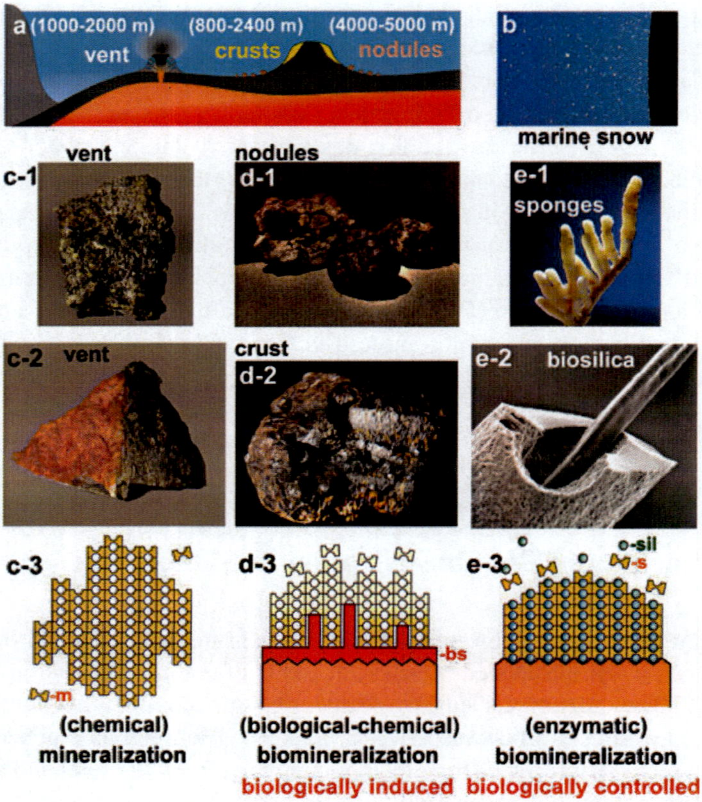

Fig. 4.3 Categories of mineralization–biomineralization. (**a**) Main types of seabed mineral resources and their deposition sites. Biogenic formation of nodules (depths 4,000–5,000 m) and crusts (800–2,400 m) on the seafloor and the seamounts and abiogenic formation of deposits at hydrothermal vents (1,000–2,000 m). (**b**) Organic "marine snow particles" at a depth of 300 m (Zanzibar region), which act as bio-seeds for mineral deposition. (**c**) Formation of vents by chemical mineralization processes driving the deposition of soluble mineral (m), based on chemical and physical reactions only. (**d**) During the biologically induced mineralization (biomineralization), organic molecules act as bio-seeds (bs) to initiate deposition of inorganic minerals. Bacteria act as bio-seeds in nodules and coccoliths as bio-seeds in crusts. (**e**) Biologically controlled mineralization describes the mineralization process which is guided along bio-seeds and organic matrices. A special form is seen in siliceous sponges that synthesize their silica spicules from ortho-silicate (s) by an enzyme (silicatein [sil]) that acts during the seed phase and the subsequent growth phase

et al. 2006), (Fig. 4.3b). It acts as bio-seed and mediates deposition of inorganic materials from an environment that contains the inorganic precursors at non-saturated conditions. The GROWTH PHASE of the biominerals, following the seed phase, is almost exclusively directed by physical or chemical forces. Recent studies characterize polymetallic nodules and crusts as biominerals whose formation is initiated by bio-seeds (Fig. 4.3d-1 and d-2).

The second category of biomineralization, the BIOLOGICALLY CONTROLLED MINERALIZATION, describes the mineralization process which is guided along bio-seeds and organic matrices; those biomolecules control initiation and growth of the biominerals, their morphology, and also the velocity of the mineralization process (Weiner and Dove 2003).

A special form of biologically controlled mineralization, ENZYMATICALLY CONTROLLED MINERALIZATION (Fig. 4.3e), has been described for the biosilicification process in siliceous sponges (see: Müller et al. 2007b; Schröder et al. 2008). In these animals (Demospongiae and Hexactinellida), the enzyme silicatein (Cha et al. 1999; Morse 1999; Krasko et al. 2000; Müller et al. 2008a) is catalytically involved in the formation of bio-silica (Müller et al. 2007a; Wang et al. 2008) and also functions as organic scaffold for the inorganic polysilicate mineral (Müller et al. 2008b; Wang et al. 2008). Hence this enzyme acts as bio-seed and as organic matrix.

In contrast to biomineralization, CHEMICAL MINERALIZATION describes the chemical and physical processes driving accumulation of new inorganic material from solution; Fig. 4.3c. This process is controlled by the initial mineral growth rate, the magnitude of supersaturation of the inorganic precursors as well as the temperature and might be ascribed to a first-order surface reaction kinetics and the respective activation energy of the chemical reaction (Persson et al. 1995). At the present state of knowledge, hydrothermal vents are formed by mineralization processes only (Fig. 4.3c).

4.4 Mineralization/Biomineralization Processes During Formation of Polymetallic Nodules [Mn-Nodules]: Biologically Induced Mineralization

4.4.1 Deposits

The polymetallic nodules are mainly formed in deep water (4,000–5,000 m) within the sediment–water interface. The age of the nodules is about 15 MYR (Somayajulu 2000). In general, they grow extremely slow, one atomic layer per year ([Kerr 1984] ≈ 1 mm Ma^{-1}) and start to form on nuclei/seeds which are not only weathered volcanic rocks or pumice (see: Glasby 2006) but may be also of biogenic origin (bio-seeds) (Wang et al. 2009b). In contrast to the deep-sea polymetallic nodules, the ferromanganese nodules in the shallow marine environments, e.g., in the Baltic Sea (Zhamoida et al. 1996), have a rather high growth rate (8×10^3 mm Ma^{-1} [Anufriev and Boltenkov 2007]). Large deep-sea resources of polymetallic nodules have been localized in the Pacific Ocean (e.g., Clarion/Clipperton zone), the Southern Ocean/Antarctic Convergence, and the Peru Basin (Kawamoto 2008). The marine environment, in which "polymetallic" nodules are formed, contains the constituents of the nodules, the transition metals Mn and Fe as well as trace elements (Halbach et al. 1988; Cronan 2000; Glasby 2006). The nodules, which can reach sizes of 14 cm, often display an alternating growth pattern reflecting a hydrogenous to diagenetic growth (Glasby 2006). The dominant minerals found are todorokite, hydrous MnO_2 (δMnO_2), as well as hydrous iron (Thijssen et al. 1985).

In the deep waters, especially of the Pacific Ocean, the concentration of dissolved oxygen is around 100 μmol kg^{-1} (surface region: > 200 μmol kg^{-1}) and therefore relatively high (Kester 1975). In the depth, the concentration of soluble Mn is 0.1 nmol kg^{-1} [10% of the surface water] and that of Fe is 0.4 nmol kg^{-1} [like on the surface] (Bruland et al. 1994). In the oxygen-rich seafloor layers, Mn and Fe occur in the water mainly as metal oxyhydroxides [MO_xOH_y; as $Mn(IV)O_xOH_y$ or $Fe(III)O_xOH_y$]. The Mn oxyhydroxides [ß-manganite] are stabilized in the seawater by binding to other transition elements and by their fine granulation state (Glasby 1974). Those manganite(III)/manganate(IV) minerals represent the precursors for nodule formation and are formed from Mn(II) via a series of intermediates, partially autocatalytically (Murray and Brewer 1977). The slow oxidation reactions can be accelerated substantially in the presence of inorganic interfaces with Mn(IV)oxide (MnO_2) or Fe(III)hydroxide (FeOOH) (Chukhrov et al. 1976), or on bacterial surfaces (Cowen and Bruland 1985; Hastings and Emerson 1986; Ehrlich 2002). In the nodules, all three major Mn-oxide crystal structures can be identified, ranging from todorokite (10 Å manganite), birnessite (7 Å), and vernadite (δ MnO_2) (Dymond and Eklund 1978; Post 1999; Glasby 2006). During nodule formation, Mn oxides tend to incorporate cations [Ni^{2+}, Cu^{2+}, Zn^{2+}], while the Fe oxyhydroxides scavenge anionic species [P (HPO_4^{2-}), V (HVO_4^{2-}), Mo (MO_4^{2-}), W (WO_4^{2-}), and especially Co (Co_3^{2-}) as well as rare earth elements] (Koschinsky and Halbach 1995).

4.4.2 Growth

The basic reactions responsible for nodule growth are the oxidation of Mn(II) and Fe(II) to the respective oxyhydroxides that are deposited on already existing seed crystals. During this process, other metals are incorporated mainly by scavenging and ionic bond interactions. The Mn and Fe sources are, to a very small extent, weathering products coming from the continental shelf, but mainly originate from pore waters released by diagenetic processes (Bonatti and Nayudu 1965). It is not yet known, if the growth rate on such nodule surfaces that face pore waters is faster than at surfaces that are directed to the open sea (Moore et al. 1981). However, it is established that growth of nodules depends on the concentrations of disequilibria of redox systems of the individual reactants, mainly Mn(II)/Mn(IV) and Fe(II)/Fe(III). The obvious signs of such alterations are ripple marks within the nodules (Halbach et al. 1988). A static view of nodule formation alone does not explain those texture formations and also does not take into consideration the changing chemical reaction conditions on the surfaces of the spheroidal nodules and finally also their mechanical rolling/movement on the seafloor. It is quite conceivable that simple mechanical movement, bioturbation, of the nodules is one determining parameter for the formation of their spheroidal/concretionary shapes (Somayajulu 2000).

Chemical analyses of the nodules did not reveal any marked differences between the center and the surface of the nodules even after their long growth history (Glasby 2006). Likewise, only little information on the dynamics of growth of the spheric nodules (Fig. 4.4a) could be gathered from polished cuts through nodules. They revealed a layered and lamellar structures, but did not allow a resolution of globular units at the μm-scale (Fig. 4.4b–d). Progress came recently from analyses of small, broken, unpolished nodule samples that were studied by high-resolution scanning electron microscopy (HR-SEM) and high-resolution energy dispersive X-ray (HR-EDX) techniques; e.g., samples of polymetallic nodules collected from the Clarion-Clipperton Zone in the Eastern Pacific Ocean basin (see: Wang and Müller 2009). Their fracture planes revealed a distinct sub-composition of the nodules into blackish drops, termed micronodules (Fig. 4.4e–h) that are prominent in the outer lamellae/regions. The diameters of the spheroidal to ellipsoid micronodules vary between 100 and 450 μm (Fig. 4.4d and f). Between the dark-metallic micronodules, an interstitial whitish, non-shiny material is seen which apparently glues the micronodules together (Fig. 4.4e and f). In the nodules studied by us, 2–5 individual lamellae are found, which might reflect the same number of different, consecutively uniform growth periods, each with an own distinct hydrogenetic history (Halbach et al. 1988), as shown in Fig. 4.4b and c. A closer view of the individual lamellae displays their dendritic and ornamental pattern, which is unraveled to single blackish drops, termed micronodules (Fig. 4.4e and f). The micronodules are especially dominant in the surface lamellae. In addition, they are also found, to a lesser degree, in the sublamellar layers (Fig. 4.4g and h).

4.4.3 Microorganisms

HR-SEM analyses were performed to identify bacteria/microorganisms within the Mn nodules. In micronodules, dense accumulations of microorganisms were seen (Fig. 4.5a and d). Only two morphotypes can be distinguished: round-shaped, spherical microorganisms, which were operationally term cocci, and elongated microorganisms, which were termed rods. In most areas, the cocci were predominant with surprisingly uniform diameters of about 2.5–4 μm and a mean of 3.5 μm (Fig. 4.5b). The surface of the cocci was smooth, covered with small-sized platelets, presumably consisting of Mn oxides (Fig. 4.5b). Occasionally, the cocci were arranged in bead-like chains, as in the genus *Streptococcus* (Ryan and Ray 2004); Fig. 4.5c. This chained growth form indicated cellular division along a single axis. The rods on the other side were especially prevalent on the surfaces of the micronodules (Fig. 4.5d). Two growth forms could be distinguished for the rods; (1) either arranged in a palisade-like pattern, in which the rods were attached to

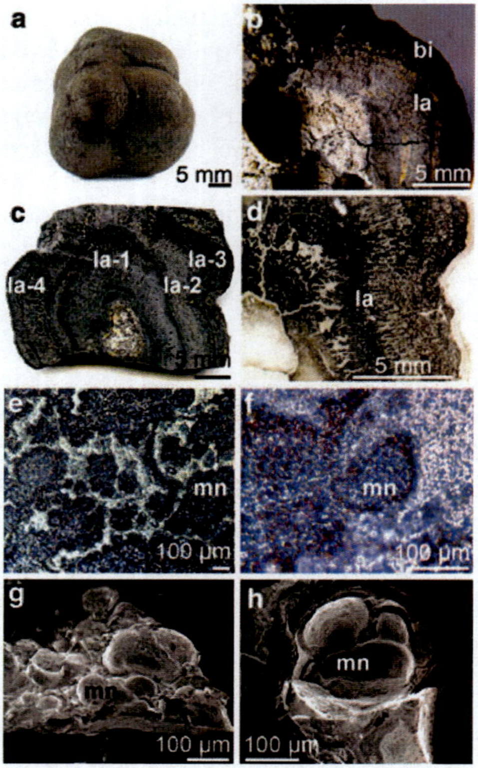

Fig. 4.4 Nodule morphology; light optical analysis (**a–f**) and HR-SEM (**g** and **h**). Nodules were collected from the Clarion-Clipperton Zone in the Pacific Ocean. (**a**) Outer appearance of a Mn nodule, with its smooth surface texture. (**b**) A cross-breakage of a nodule displays the lamellar structure with a pale-gray submetallic luster. The outer lamella (la) is dark brown to black, suggesting a composition of birnessite (bi). (**c** to **f**) Polished sections through nodules. (**c**) This nodule comprises at least four macrolamellae (la-1 to la-4). (**d**) A closer view highlights that the individual lamellae (la) have a slightly different color, indicative for different histories during which the lamellae were formed hydrogenetically. (**e** and **f**) The individual lamellae are separated by a zone that comprises a dendritic, ornamentous pattern. This pattern is assembled and composed of single blackish drops, termed here micronodules (mn). Those micronodules are dominant in the surface lamella. (**g** and **h**) The micronodules (mn) become well distinguishable after breaking the nodules; HR-SEM analyses

each other at their longitudinal surfaces, or (2) linearly arranged as it is often seen in the taxa of Eubacteria (Fig. 4.5e and f). The rods were about 1.5–2.5-µm long and 0.35–0.45-µm thick (Fig. 4.5d and e). From the microscopic images, it appeared that the rods divided medially (Szeto et al. 2001). Such a division pattern is known from bacilli [Streptobacilli (Ryan and Ray 2004)]: The rods appear in chains; a chain with three attached rods is shown in Fig. 4.5e and f, together with the two division septa. It remains unsolved if the rods separated completely after division – which is most likely – or remained fused. The surrounding, adjacent zones of the rods were formed by solid material, which appeared homogenous.

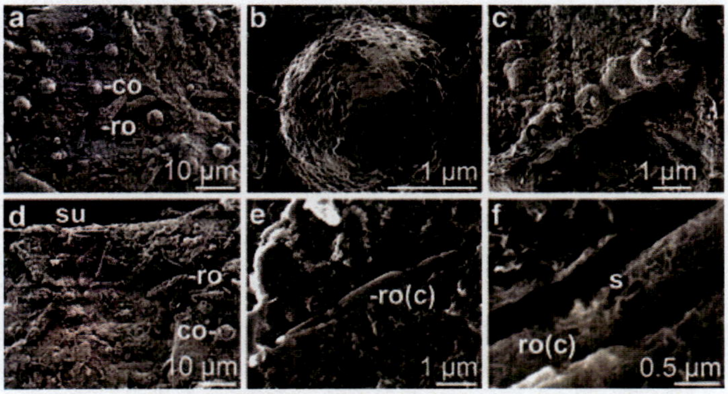

Fig. 4.5 Microorganisms, abundantly present in micronodules; SEM analyses. Within the nodules, both cocci (*co*) and rods (*ro*) are seen. (**a–c**) Cocci. (**b**) An individual coccus (*co*) revealing small-sized platelets on its surface. (**c**) Cocci arranged in bead-like chains, known from the genus Streptococcus. (**d–f**) Rods. (**d**) Rod-like microorganisms, present in micronodules; they are especially abundant close to the surface (*su*) of the nodules. Where present, rods (*ro*) are associated with cocci (*co*) in the micronodules. The rods form either a palisade-like arrangement or are aligned in chains [rod-chain; *ro(c)*] (**e**). (**e** and **f**) The two division septa (*s*) within the linearly oriented rods, rod-chains [*ro(c)*], are enlarged

4.4.4 Bio-seeds

Solid ground for the involvement of bacteria in polymetallic nodule formation had been presented by Ehrlich (see: Ehrlich 2002). However, only recently was it possible to visualize distinct bacteria by in situ HR-SEM analysis within the mineralized deposits of the nodules, and not on their surfaces (Wang and Müller 2009; Wang et al. 2009b, c). Analysis of small, broken splinters revealed two bacterial morphotypes (Wang et al. 2009b); (1) round-shaped spherical microorganisms, operationally term cocci, and (2) rod-shaped bacteria (Fig. 4.5). It is notable that both the cocci and the rods are occasionally arranged in chains, suggesting a distinct pattern of division. This would imply that the bacteria had been buried alive.

Even more surprising are those structures which comprise biofilm-like arrangements of the microorganism (see below). Based on these observations, it became quite certain (Wang et al. 2009b, d) that microorganisms in the micronodules contributed as matrices/seed templates (bio-seeds) to initial mineral deposition, allowing a subsequent concretion of the abiogenic mineral layers. Subsequent EDX analyses disclosed that the bacteria-/microorganism-rich regions of the micronodules contain high levels of Mn and also Fe, while the interstitial zone contains primarily Si and no microorganisms. Hence the following sequence of nodule formation can be formulated (Wang et al. 2009b); formation of micronodules around bio-seeds, the microorganisms, which facilitate deposition of Mn (and Fe). If the micronodules reach sizes of 100–300 µm (Fig. 4.6), they conglomerate together by rotation, as consequences of the water current, or by bioturbation, and form nests (size 3 mm; Fig. 4.6), starting growth of the "composite" nodules.

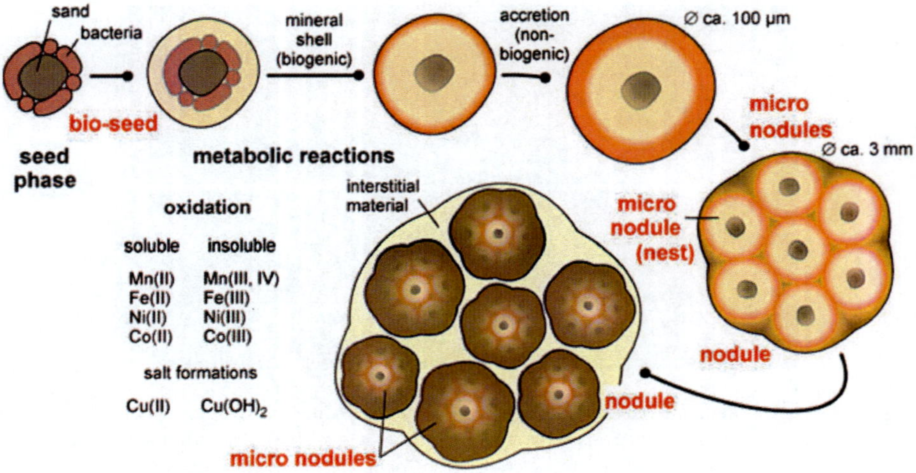

Fig. 4.6 Formation of the polymetallic nodules by aggregating micronodules; scheme. Initially, minute clay/sand aggregates form the substrate for the adhesion of bacteria (bio-seeds). Mediated by microorganisms, soluble elements (Mn, Cu, Fe, and Co) are metabolized by oxidation or salt formation into insoluble minerals. Subsequently, those aggregates function as seed templates and allow a progressing mineralization through abiogenic processes and the formation of micronodules. The micronodules assemble to nests. Finally, nodules are formed under rotating movements on the sea floor, also allowing the inclusion of further inorganic, abiogenic material

4.4.5 Seeds: Bacterial S-Layer

Initially, it had been discussed that the oxidation of Mn is metabolically driven by bacteria (see: Ehrlich 2002), either by direct oxidation of Mn(II) on the surface of the nodules or by an indirect action via a coupled consumption of protons during the ATPase reaction in the respiratory chain. However, with the discovery of imprints of bacteria in the polymetallic nodules that are decorated with S-layer structures (Wang et al. 2009c), a new avenue for the understanding of the biogenic component in nodule formation was opened (Fig. 4.7a–c). S-layers, paracrystalline surface layers that cover both Archea and Bacteria, are 5–25 nm thick and contain ordered repeats of protein(s) or glycoprotein(s) (Sleytr and Messner 1983). The characteristic features of such morphological units of 10–20 nm are; (1) a massive shedding from the bacterial surface (Mengele and Sumper 1992), (2) high capacity of self-assembly to highly organized complex platforms and structures (Sleytr et al. 1999), (3) strong adherence activity (Sleytr and Messner 1983), and (4) surface charge (Schultze-Lam et al. 1993). Due to their regular patterns, with a two-,three-, four-, or sixfold rotational symmetry, the outermost layer of the S-layer is an ideal organic matrix not only to protect the microorganisms against external chemical challenges (Schultze-Lam and Beveridge 1994), but also an anchorage platform for mineral deposits (Fortin et al. 1997). The endolithic microorganisms discovered in the nodules are covered by pillar-shaped protrusions, similar to S-layer units. They measure 75 nm in their longitudinal and 45 nm in

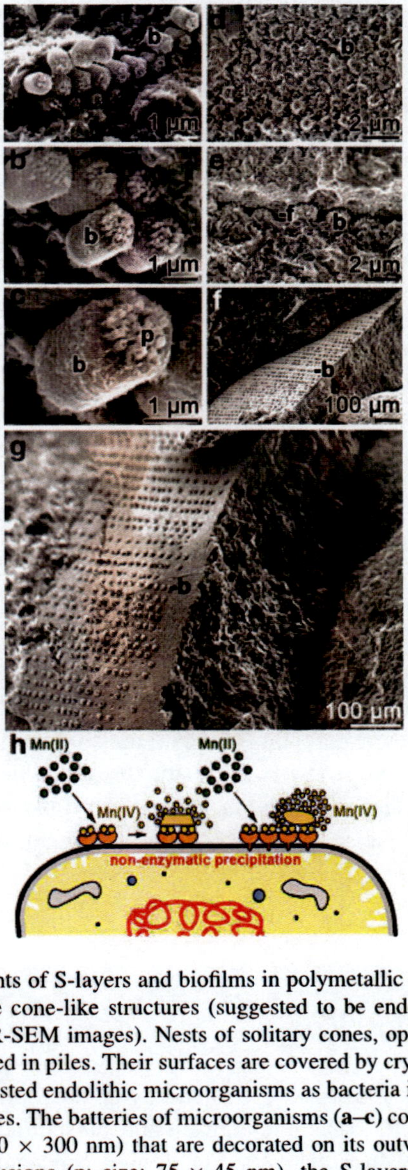

Fig. 4.7 Structural elements of S-layers and biofilms in polymetallic nodules. Bacterial S-layers (**a–c**); identification of the cone-like structures (suggested to be endolithic microorganisms) as bacteria with S-layers (HR-SEM images). Nests of solitary cones, operationally termed bacteria (*b*), exist which are arranged in piles. Their surfaces are covered by crystal-like scale bricks. Final determination of the suggested endolithic microorganisms as bacteria is based on the existence of S-layer crystalline structures. The batteries of microorganisms (**a–c**) consist of individual bacteria-like microfossils (size: 800 × 300 nm) that are decorated on its outward-directed surface, with 20–25 pillar-shaped protrusions (p; size: 75 × 45 nm), the S-layers. The arrangement of the protrusions essentially shows an oblique to square pattern. (**d**) A phalanx of phalanx-like cones, termed bacteria (*b*). Below or above those communities, a fissure (*f*) exists that splits the material along one subsurface structure. One side (*lower*) shows the cones (*b*), whereas the corresponding side comprises the mirror imprints. Biofilm structures (**f** and **g**). Cone-like structures (operationally abbreviated here with (*b*), denoting bacteria, are arranged in a phalanx-like pattern on a surfaces of plane structures within the nodules. The individual cones (*b*), with sizes of 0.8–1.0 μm, are regularly arranged in a pattern with an interspacing of 1.0–1.5 μm. (**h**) Proposed biogenic deposition of minerals onto bacteria, from which the growth of nodules originates (schematic). It is proposed that mineral deposition proceeds nonenzymatically on the enlarged bacterial surface due to the existence of S-layer structures

their transverse dimension (Fig. 4.7d and e). It is assumed that those S-layer structures reduce, due to the enlargement of the bacterial surface and also a particular chemical/physicochemical composition, e.g., on surface wettability, the activation energy for the oxidation processes (Kim et al. 2006); Fig. 4.7h.

4.4.6 *Biofilm Structures in Polymetallic Nodules*

A further input toward an establishment that bio-seeds are causatively involved in nodule formation came from Crerar and Barnes (1974), who proposed that Mn (IV) and Fe(III) deposition also involves autocatalytic processes proceeding on the surfaces of microorganisms which form Mn(IV) oxide from Mn(II). Recently, an in situ detection of distinct bacterial biofilm in nodules has been reported (Wang et al. 2009c); Fig. 4.7f and g. From laboratory studies, it is known that microcolonies in biofilms are surrounded by large amounts of extracellular polymeric substances (Lawrence et al. 1991). Those polymers are composed of exopolysaccharides, carrying functional polyionic groups [anions like carboxylate ($R\text{-}COO^-$), or cations like amino unit ($R\text{-}NH_3^+$)] that allow a spatial organization of the bacteria to cope for optimal supply with nutrients (Wolfaardt et al. 1995); Fig. 4.8a. Metal ions have been found trapped within the biofilm meshwork, and hence are suspected to have been involved in the microbial lithification also during biogeochemical cycles (Dupraz and Visscher 2005). The biofilm structures that were identified in nodules have an intriguingly regular organization. While the rod-like bacteria are arranged in a spatial organization of palisades, the cocci produce a massy extracellular biofilm matrix (Wang et al. 2009c, d); Fig. 4.8d, e.

4.4.7 *Mineral Deposition*

Based on existing data (Zhu et al. 1993) and by application of nuclear microprobe analyses, Marcus et al. (2004) could confirm and substantiate that nodules grow by alternating Fe-rich and Mn-rich layers; Fig. 4.9b–d. Since nodules are formed in the deep-sea within an oxygen-rich bottom zone (Koschinsky and Halbach 1995; Koschinsky et al. 1997), both Mn and Fe occur primarily in their oxidized forms as Mn oxyhydroxides that secondarily associate to colloids (Bau et al. 1996). Those colloidal dispersions carry surface charges, Mn (negative) and Fe (positives), that have the tendency to form first mixed colloids and then coarse agglomerates, a process during which also trace metals are scavenged (Koschinsky and Halbach 1995; Koschinsky and Hein 2003). EDX analysis, coupled with HR-SEM, revealed that the Mn-rich layers in the nodules contain microorganisms, while the Fe-rich layers are almost free of them (Wang et al. 2009a, c).

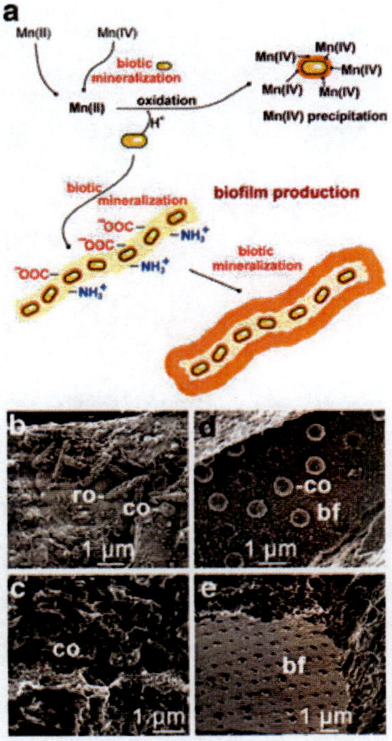

Fig. 4.8 Potential role of bacterial biofilm for biogenic nodule formation. (**a**) Scheme, summarizing the proposed micronodule formation. Initially, bacteria/microorganisms function as bioseeds around which the initial Mn mineral deposition proceeds. Aggregates of bacteria form larger deposits, a process that is augmented by the formation of biofilms. Those Mn deposits grow further through autocatalytic reactions until abiogenically formed Fe oxyhydroxide/colloid is deposited. Finally, the micronodules become surrounded by Si-rich minerals that allow micronodules to form, giving rise to nests and finally to nodules. (**b**) Microorganisms, consisting of rods (*ro*) and cocci (*co*) within the micronodules of a polymetallic nodule. (**c**) Formation of a bacterial phalanx, cocci (*co*), that had been detached for the opposite mineral material by a fission. (**d**) HR-SEM images of a biofilm (*bf*) into which cocci (*co*) are embedded. (**e**) HR-EM image of an empty biofilm (*bf*) from which the bacteria have been detached, leaving behind holes

4.4.8 Approach to Determine Bacteria Species in the Mineralic Material

It is very evident that isolation of DNA from micronodule material and the results obtained after subsequent analyses of sequences from those samples must be taken with greatest caution. For our first approach, pieces from a polymetallic nodule were washed thoroughly with detergent and subsequently ground (Wang et al. 2009d). Then the material was leached with citric acid and the resulting material subjected to DNA isolation followed by PCR analysis. By applying primers specific for 16S ribosomal RNA gene, 28 clones were obtained after PCR reaction. Among those, 19 sequences were identical; they comprised a 1,097-nucleotides-long 16S ribosomal RNA gene sequence (it was termed: AQbac_NOD1_D). This partial

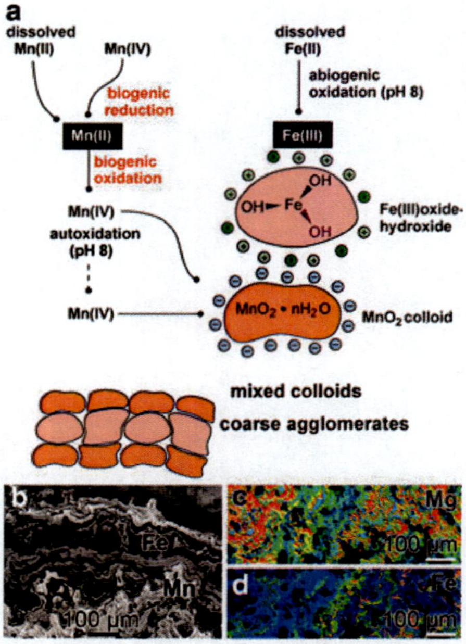

Fig. 4.9 Formation of mixed colloids of Mn- and Fe-oxide-hydroxides through biogenic oxidation of Mn(II) and abiogenic oxidation of Fe(II). (**a**) Schematic illustration of the steps leading to the formation of Mn(II) and Fe(III), to Fe(III)-oxide-hydroxides and to Mn(IV)-oxides and, finally, to their respective colloids. These colloids increase in size until mixed colloids, layers of Fe(III)-colloids (*pink*) and Mn(IV)-colloids (*orange*) are formed. (**b**) A polished cut through a nodule shows the concentric arrangement of Fe-rich and Mn-rich layers. Mn: Mn-rich region; Fe: Fe-rich region. (**c** and **d**) X-ray mapping of a nodule crosscut confirms that separate layers of Mn/(Mn) and Fe exist. *Blue* indicates low levels of Mn or Fe and red indicates high levels of Mn or Fe

16S gene (Fig. 4.10) was found to share high sequence similarity (97%) to the partial 16S ribosomal RNA gene from an uncultured bacterial clone JH-WH45 (AQbac_EF4928; accession number EF492894 [He et al. 2008]) with an E value (maximal score) of 1,991. It is interesting to note that the bacteria that the authors (He et al. 2008) described live in a Mn oxide-rich micro-biotope in soil environment. In the future, more sophisticated and more secure methods with respect to contamination with environmental bacteria will have to be applied in order to draw some conclusions on the possible etiology of modern bacteria with bacteria found as fossils embedded in the minerals of the nodules.

4.4.9 Manganese Depositing Bacteria

As outlined above, the S-layer structures present on the surfaces of many bacteria have been implicated in the nonenzymatic manganese deposition during nodule

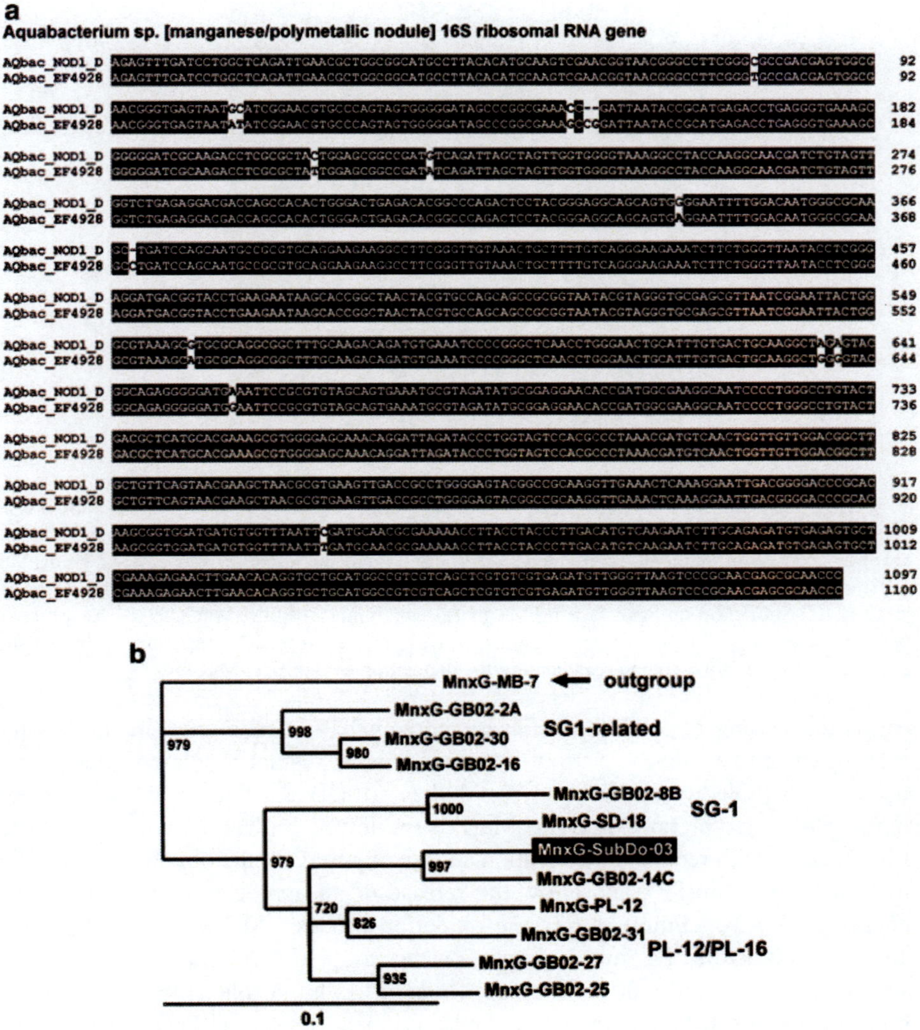

Fig. 4.10 Bacteria, suspected or proven, to be involved in the biomineralization of nodules. **(a)** Uncultured aquabacterium, isolated from the polymetallic nodule. Alignment of the partial 16S ribosomal RNA gene sequence from the Mn-nodule (AQbac_NOD1_D), studied by us, with the corresponding bacterial 16S RNA gene sequence from soil bacterial communities (AQbac_EF4928; accession number EF492894; He et al. 2008). Identical nucleotides are in inverted type and those different between the sequences in black on white. **(b)** Sequence relationship of the *mnxG* gene (termed *mnxG-SubDo-03*) [encoding the bacterial multicopper oxidase-like enzyme], identified in the *Bacillus*-related strain, isolated from the demosponge *S. domuncula*. The deduced polypeptide, MnxG-SubDo-03, was compared with the MnxG proteins from the following *Bacillus* strains: strain PL-12 (MnxG-PL-12; ABP68890.1), strain GB02-31 (MnxG-GB02-31; AAZ31744.1), strain GB02-30 (MnxG-GB02-30; AAZ31743.1), strain GB02-27 (MnxG-GB02-27; AAZ31742.1), strain GB02-25 (MnxG-GB02-25; AAZ31741.1), strain GB02-16 (MnxG-GB02-16; AAZ31739.1), strain GB02-14C (MnxG-GB02-14C; AAZ31738.1), strain GB02-8B (MnxG-GB02-8B; AAZ31736.1), strain SD-18 (MnxG-SD-18; AAL30449.1), strain GB02-8B (MnxG-GB02-8B; AAZ31736), and strain GB02-2A (MnxG-GB02-2A; AAZ31735.1). These

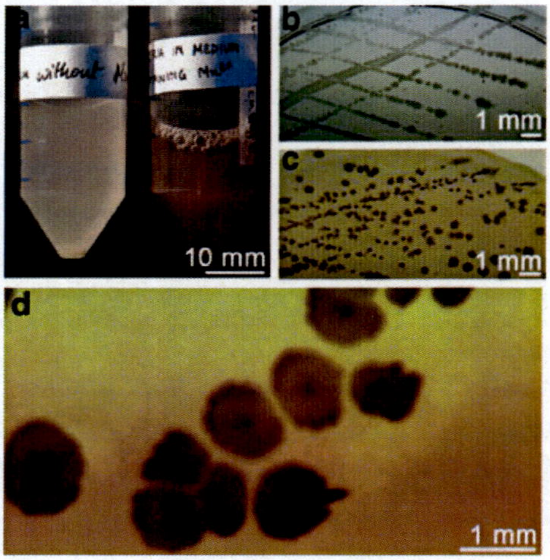

Fig. 4.11 Mn-oxidizing bacterium, isolated from the sponge *S. domuncula* (strain BAC-SubDo-03). Extracts from the sponge were prepared and plated onto Mn-containing agar. Colonies of them were picked and subsequently cultivated (7 d) in liquid; (**a**) this medium was either lacking Mn (*left*) or had been supplemented with Mn [100 μM $MnCl_2$] (*right*). (**b**) Cultivation (for 72 h) of strain BAC-SubDo-03 on agar, lacking Mn or (**c**) containing 100 μM MnCO3. (**d**) At a higher magnification, it is obvious that the cultures growing in the presence of Mn form a brownish-colored area around the colonies, indicating the deposition of oxidized $MnCO_3$

formation (Wang et al. 2009c). Furthermore, it had been proven in thoroughly performed studies that a variety of phylogenetic distantly related, free-living bacteria are able to oxidize and metabolize Mn(II) through their multicopper oxidase-like enzyme(s) (MCO) to Mn(IV) (reviewed in: Tebo et al. 2004). The MCOs are able to reduce the activation energy required for the oxidation of Mn(II) to Mn(III) and Mn(IV), and allow the release of an appreciable amount of free energy (ΔG −50 kJ/mol) during the oxidation process. MCOs use multiple Cu atoms as cofactors that are required for the coupled oxidation of a series of substrates (Tebo et al. 2004). They are thought to play a role in the oxidation of both organic metal-containing compounds and metal ions, e.g., Fe(II) and Mn(II), (Solomon et al. 1996). In extensive studies, the group of Tebo (Dick et al. 2006, 2008a, b) succeeded to provide direct evidence for the role of the MCO during Mn deposition on bacteria, both on enzymic and molecular level.

In a recent study (Wang et al. 2011), we identified for the first time Mn(II)-oxidizing bacteria, a *Bacillus* strain (strain BAC-SubDo-03), in the demosponge *Suberites domuncula* (Figs. 4.10b and 4.11). This sponge species is especially suitable for the identification of potential symbiotic microorganisms, since it can

Fig. 4.10 (continued) sequences had been aligned and the tree was computed and rooted with the sequence from *Bacillus* strain MB-7 (ABP68890). The bootstrap values are based on 1,000 replicates and are indicated at the branch points. The grouping of the sequences has been performed as outlined by Dick et al. (2006)

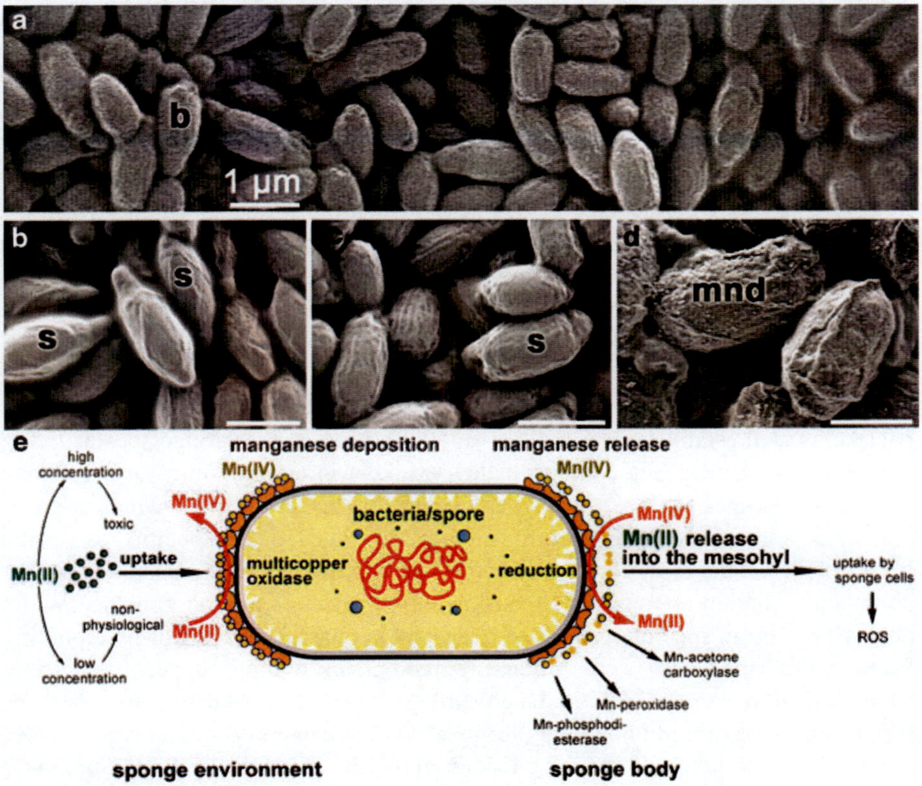

Fig. 4.12 Mn-oxidizing bacteria; SEM images. (**a**) The bacteria (strain BAC-SubDo-03) were cultivated in Mn-containing K-medium. At time 0 h, almost all bacteria (*b*) showed a barrel morphology. However, if they grow longer than 18 h in the Mn-supplemented medium, they change their form and show an elongated spindle-shaped structures (**b** and **c**), which represented endospores (*s*). After a longer incubation period, depositions of Mn onto the outer membranes (*mnd*) of the bacteria can be visualized (**d**), a finding which was supported by EDX analyses as well. All size bars measure 1 µm. (**e**) Schematic representation of the proposed role of the BAC-SubDo-03 *Bacillus* strain, associated with *S. domuncula*, as a Mn store for Mn ions. It is assumed that under high Mn concentrations, the bacteria take up Mn(II) from the environment through the multicopper oxidase (*MCO*) and deposit the ions as insoluble Mn(IV) onto their cell wall. If Mn exists in the environment only at low concentrations, the MCO allows the enrichment of the element to physiological levels. Intracellularly, in the sponge body, Mn is solubilized by reduction from Mn(IV) to Mn(II) released from the cell wall and becomes available as cofactor in a series of essential enzymes, involved in detoxification of reactive oxygen species (*ROS*), or in conversion of acetone to acetoacetate (e.g., Mn–acetone carboxylase) or lipid metabolism (Mn–phosphodiesterase)

be kept under controlled laboratory conditions for over 5 years (LePennec et al. 2003). Our data revealed that those Mn(II)-oxidizing bacteria are highly related to the Mn(II)-oxidizing *Bacillus* strains isolated from Guaymas Basin, a deep-sea hydrothermal vent environment in the Gulf of California (Dick et al. 2006).

On the basis of recently gathered data (Dick et al. 2008a, b), the *mnxG* gene coding for the MCO has been identified by us in the newly discovered *S. domuncula*-associated bacterium (Fig. 4.10b). The sponge-associated bacterium shared the highest sequence similarity only to those bacteria that comprise and express the MCO. The expression of this gene has been found to depend on the presence of Mn in the culture medium. During Mn deposition, the bacteria change their morphology from barrel-like (Fig. 4.12a) to a form which is characterized by distinct elongated projections that we labeled as spores/or spore-like bacteria (Fig. 4.12b–d).

At present, we do not yet know if these Mn-depositing bacteria, associated with *S. domuncula*, have any relationship to those bacteria, which were suspected to be involved in polymetallic nodule formation. Therefore, we attribute those bacteria a crucial role in symbiosis with sponges. The symbiotic coexistence of sponges with bacteria is well established (Althoff et al. 1998). Most of the sponge species living in the Mediterranean Sea contain, if they are kept in an aquarium for more than 2 weeks (LePennec et al. 2003), only a few different bacterial strains which are non-abundant and, with regard to *S. domuncula*, located in specific cells, the bacteriocytes. This is in contrast to a few species, like *Aplysina aerophoba*, that abundantly contain up to 50% of the sponge body mass as microorganisms (Weiss et al. 1996). From specimens that had been kept in quarantine, the Mn-precipitating bacteria described here were isolated. Already, this finding suggests that those bacteria display a crucial role in the physiology/metabolism of the sponge, perhaps supporting our assumption that the Mn-precipitating bacteria act as a reversible Mn store in *S. domuncula*. According to this view, the presence of BAC-SubDo-03 bacteria is required as a protection against higher, toxic concentrations of Mn; after oxidation of Mn(II) to Mn(IV), the ion becomes insoluble. However, the bacteria appear to be also essential for maintaining the physiological concentration in the sponge. Since only minute levels of Mn exist usually in the surrounding seawater, a substantial accumulation of Mn is proposed here to rise onto the bacteria. The release of bacterial-precipitated Mn(IV) is postulated to supply the physiologically needed Mn if needed.

4.5 Mineralization/Biomineralization Processes During Formation of (Co-rich) Polymetallic Crusts

4.5.1 Deposits

The Co-rich crusts, also termed ferromanganese crusts, are formed at depths between 800 and 2,400 m (Bau et al. 1996; Hein et al. 2000). They contain high amounts of Mn oxide (20–30%) and Fe oxide (14–20%) and are rich in Co ($>$1%), Cu, Ni, and Pt (Hein et al. 1997; Mills et al. 2001). Crusts are found mainly in the

Pacific Ocean, as well as in the Indian and Atlantic Oceans (Yihua and Yipu 2002; Glasby 2006). The crusts are slightly older than nodules (can become 20 MYR old). Like the nodules, the Co-rich crusts are formed at low temperatures on the cold deep-sea floor at the interface between the aqueous phase and the seamount slopes mainly basalts being the substrate. The growth rate of the crusts is also slow, 1–2 mm Ma^{-1} (Glasby 2006).

4.5.2 Morphology of the Crusts

Crust samples used in our investigations (Wang et al. 2009a) were collected from the Magellan seamount which is located in the northwest edge of the Mariana basin and represents relatively independent guyots/flat-topped seamounts. The depth of shallowest guyots is about 800–1,500 m and that of the adjacent marine basin is about 5,000–6,500 m. The crust samples have a total thickness of 35 mm with an upper layer of 23 mm, a lower layer of 9 mm, and a separating boundary layer of 3 mm. Since this latter layer has a bright, fatlike appearance, it was termed "speck layer." The overall color of the broken-up crust is dark black (Fig. 4.13a and b).

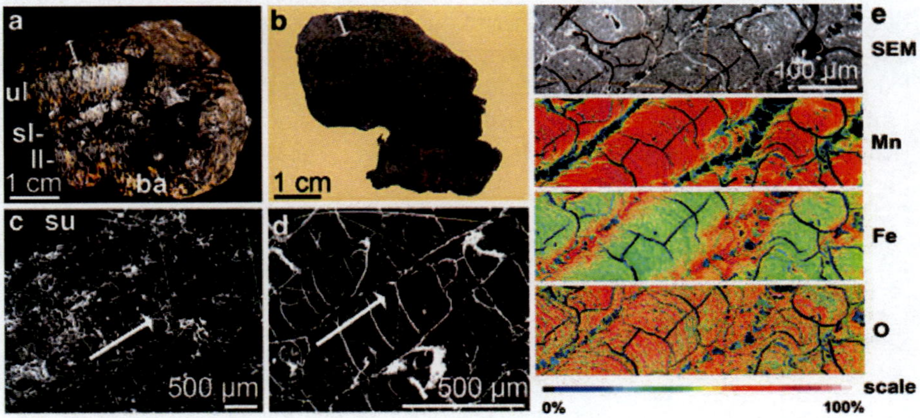

Fig. 4.13 Morphology of the crust sample. (**a**) A vertical fracture site reveals a distinct zonation into *first*, a lower layer (*ll*), *second*, a middle layer ["speck layer" (*sl*)] and *third*, the upper layer (*ul*). The base (*ba*) of the crust, which has contact to the basaltic rock, is characterized by the red encrustations. The *double-headed arrow* marks the band, which has been analyzed. (**b**) Polished vertical slice through the crust. The position/area which has been studied here is marked (*double arrow*). (**c** and **d**) Plane cross section through the upper band/layer of the crust. Individual convex compartments are stacked upon each other in parallel (see direction of *arrow*) to the surface (*su*). (**e**, *right panel*) Element distribution within the upper layer of the crust, as analyzed by EPMA (electron probe microanalyzer). *From top to bottom:* The electronic image shows the characteristic texture, the convex structures displaying wavelike growth lines. The beaded stacks run in parallel to the surface of the crust. Then; elemental mapping through a crust section resolves the concentration changes for the following elements: manganese (Mn), iron (Fe), and oxygen (O). The relative concentrations increase with a change in color from dark blue to yellow to red

The lower part toward the basaltic substrate has a red color, due to the high iron content (Vonderhaar et al. 2000) (Fig. 4.13a). The uppermost 4 mm of the crust has been analyzed in detail. A plane vertical cross section through the crust displays a clear hierarchical structuring (Fig. 4.13c and d). Parallel to the surface of the crust, layers of stacked piles are orderly arranged, which are composed of individual convex units. The width of the stacked piles is approximately 250 μm. HR-SEM displays that these convex structures appear as wavelike growth lines with borders that are approximately 250 μm apart (Wang et al. 2009a). The axis of the convex orientation parallels the surface of the crust (Fig. 4.13c to e).

4.5.3 Growth

Crusts are made from two major building blocks, the negatively charged Mn oxyhydroxides that bind to hydrated cations (e.g., Ca, Ni, Zn, Pb, etc.) and the slightly positively charged Fe hydroxides that complex anions (V, As, P, Zr, etc.) (Halbach et al. 1981; Halbach 1986). Deposition occurs in the mixing zone between the upper oxygen-minimum zone [OMZ] and the lower oxygen-rich bottom zone [ORZ] (Koschinsky and Halbach 1995; Koschinsky et al. 1997). The cold oxygen-rich Antarctic Bottom Water is richer in reduced Mn(II) species than layers under the OMZ and contains the oxidized forms of Mn(IV) (Bruland et al. 1994; Koschinsky and Halbach 1995). In the mixing layer ORZ/OMZ, the crusts are formed from Mn- and Fe oxyhydroxides/hydroxides via a colloidal state (Koschinsky and Halbach 1995; Bau et al. 1996; Koschinsky and Hein 2003); Fig. 4.14.

4.5.4 Coccolithophores

In the photic zone of the oceans, a series calcificating organisms exist, e.g., coccolithophores, foraminifera, pteropods, and calcareous dinophytes; however, most of them dissolve close to their production zone (100 m), while only the coccolithophores reach the deeper zones of the ocean where the crusts are formed (Hay 2004). The presence of coccolithophores in crusts had been described by Cowen et al. (1993). By application of HR-SEM and EDX, those organisms were demonstrated as candidate bio-seed particles in Co-rich crusts from the Magellan seamounts (Wang and Müller 2009). In those deposits, abundantly fossil individual coccolith-plates, shed from single-celled coccolithophores have been detected, that frequently are arranged to coccospheres (Fig. 4.15a–c). It is remarkable

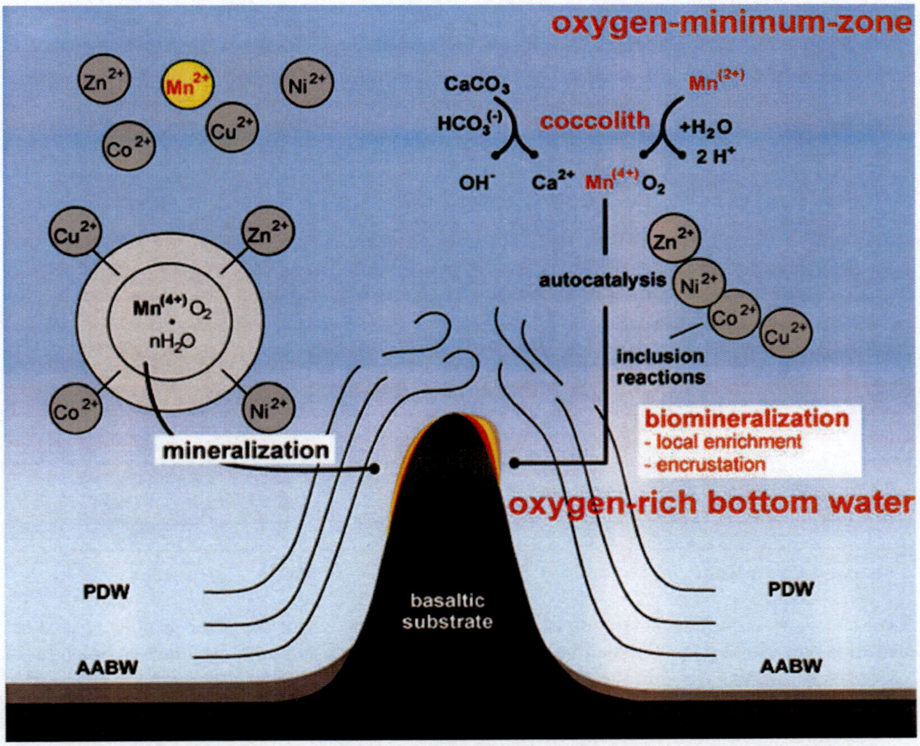

Fig. 4.14 Co-rich crust formation. (**a**) Scheme proposing processes which direct and control deposition of minerals in the hydrogenetic crusts. The crusts are formed on basaltic seamounts in a depth of approximately 1,000–2,000 m. In this depth range, two layers, the upper oxygen-minimum zone and the lower oxygen-rich zone deepwater, mix with each other. The bottom layer originates from two sources, the Pacific Deep Water (*PDW*) and the Antarctic Bottom Water (*AABW*). The formation of the crusts starts at the basaltic seamounts. The left side of the panel sketches the colloid–chemical processes, resulting in the adsorption of heavy metals by the Mn-oxyhydroxide [Mn(IV)] colloids, resulting in the MINERALIZATION at the surface of the seamount (modified after Koschinsky et al. 1997). The *right side* highlights the proposed chemical transformation processes occurring in sinking coccoliths, resulting in a replacement of $CaCO_3$ in their skeleton by Mn oxide [Mn(IV)]. After this step of biologically induced mineralization (BIOMINERALIZATION), the subsequent precipitation of Mn(IV) oxide proceeds "auto" catalytically. As a consequence of the coccolith-dependent biomineralization process, the formed particles attach to the basalt and initiate large-scale encrustation

that within these crust minerals, both the diploid forms of the coccolithophores [heterococcoliths], that propagate asexually by mitotic divisions, and also the haploid forms, the holococcoliths (Fig. 4.15d–f) which likewise reproduce asexually are found (Geisen et al. 2002). Occasionally, meiotic events occur, allowing syngamy.

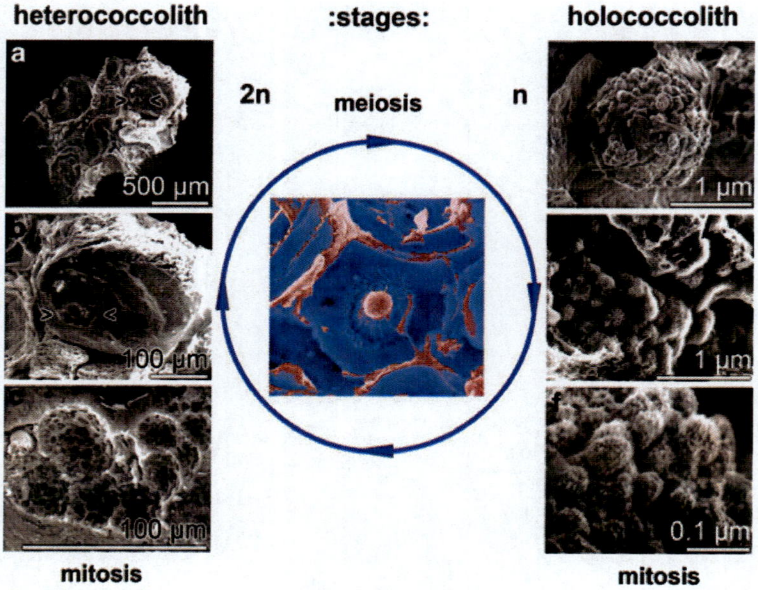

Fig. 4.15 Coccolithophores microfossils in seamount crusts; HR-SEM. In the samples, studied by us (Magellan seamount), both the diploid forms of the coccolithophores [heterococcoliths] **(a–c)**, propagating asexually by mitotic divisions, and also the haploid forms, the holococcoliths **(d–f)** which likewise reproduce asexually, are found frequently. Rarely occurring events link the two forms together and allow syngamy. The *arrows* mark the clusters of coccolithophores

4.5.5 Elemental Mapping

Elemental distribution (mapping) of the crust samples was qualitatively assessed by an energy-dispersive X-ray spectrometer coupled to a focused ion beam (FIB)-assisted SEM (Wang et al. 2009a).

For elemental mapping, we have selected one coccolith and its nearest surrounding to determine the spatial distribution of the elements within a coccolith fragment. For this study, a representative endothecal coccolith, existing within a coccosphere, was selected (Fig. 4.16a); this coccolith was well structured and displayed in the central area the characteristic openings (Fig. 4.16b). Elemental maps of the Ka-lines for Ca (Fig. 165c), Mn (Fig. 4.16d), Si (Fig. 4.16e), O, and Na were recorded using different acquisition channels. The advantage of this technique is the colocalization of the elements with the scanning electronic image. It is evident from the elemental maps that high signals, matching the structure of the coccolith, are seen for Ca and Mn (Fig. 4.16c and d). In contrast, the intensities for Si (Fig. 4.16e), O, and Na were lower at the coccolith structure, compared to the surrounding structures. This series of experiments showed that a high accumulation of both Ca together with Mn exists in the coccoliths, while no Fe can be traced there (to be published). This finding provided the clue to the participation of coccoliths as bio-seeds during Mn precipitation in crusts.

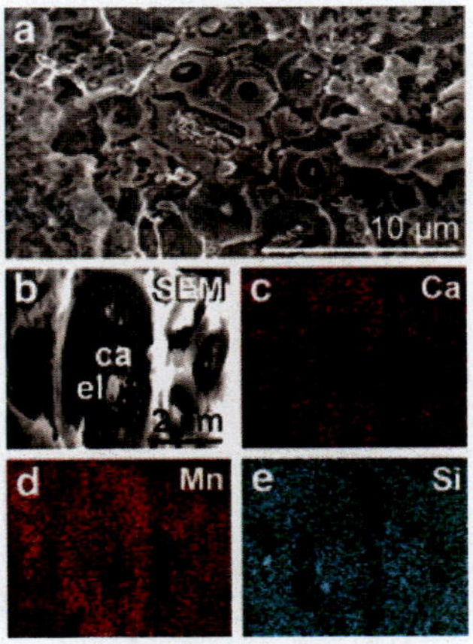

Fig. 4.16 Elemental mapping of an area containing a coccolith existing within a cluster of coccospheres (**a**). (**b**) One representative coccolith displaying its characteristic structures, the central area (*ca*) with the openings, and the element (*el*) is shown. Individual Ka lines of the following elements have been recorded simultaneously using different acquisition channels: calcium (*Ca*) (**c**), manganese (*Mn*) (**d**), and silicon (*Si*) (**e**)

4.5.6 Bio-seeds

In their theoretical considerations, Halbach (1986) and Koschinsky and Halbach (1995) proposed that calcareous particles facilitate the formation of Mn(IV) from Mn(II). They postulated that in the mixing zone [the region in which Mn(II) is oxidized to Mn(III) and Mn(IV)] Mn(II) adsorb physically on $CaCO_3$ surfaces a process that facilitates the oxidation of Mn(II). The two-step process involves first dissolution of calcium carbonate to bicarbonate, a reaction during which free hydroxyl groups are released [$CaCO_3 + H_2O \rightarrow Ca^{2+} + HCO_3^- + OH^-$]. This reaction results in an increase of the pH value during which a microenvironment is created that favors precipitation of MnO_2 (Stumm and Morgan 1996; Glasby and Schulz 1999). Following this coupled chemical reaction, it can be accepted that during $CaCO_3$ dissolution and in the presence of oxygen, Mn(II) is oxidized to Mn(III) and Mn(IV) under simultaneous proton release (Mendhom et al. 2000) [$Mn(II) + H_2O \rightarrow Mn(IV)O_2 + 2H^+$]; Fig. 4.17.

An important step toward an understanding of crust formation, based on principles of inorganic mineralization alone, came from the observation that these deposits occur within the mixing zone between the upper oxygen-minimum zone (OMZ) and the lower ORZ of deep seas (see above). In the mixing zone where the ORZ and OMZ overlap, crusts are formed by precipitation of Mn oxyhydroxides and Fe hydroxides via an intermediate colloidal state. It has been proposed that these colloids adsorb to the hard-rock surfaces, a process that could be promoted, or even catalyzed, by unicellular organisms (Fig. 4.17; left panel). After this initial mineral deposition of the first molecular layers, mineralization proceeds autocatalytically and coincides with co-precipitation of Co (in oxidized

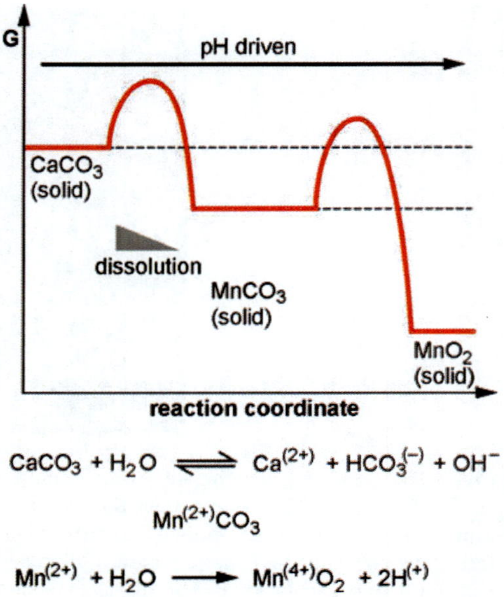

Fig. 4.17 Free energy relationships between $CaCO_3$ dissolution (from coccoliths) via the intermediate $Mn^{(2+)}CO_3$ to $Mn^{(4+)}O_2$ (in crusts). These reactions are driven by pH shifts and concentration differences

form from Co^{2+} to Co^{3+}), Pb, titanium (Ti), thallium (Tl), and cerium (Ce), probably also as a result of oxidation. The concentrations of these minor elements in the minerals reflect their concentrations in the actual environment. Recent results suggested that the growth of crusts might also involve bio-seeds as a biogenic component (Wang et al. 2009a).

4.6 Mineralization/Biomineralization Processes During Formation of Hydrothermal Vents

A new formation of oceanic crust through seafloor spreading, associated with hydrothermal venting and formation of metallic mineral deposits at the seafloor, had been discovered in 1979 with locating the black smokers at the East Pacific Rise (Francheteau et al. 1979; RISE Project 1980). It had been discovered that heat was released by the upwelling of hot magma between the Earth's tectonic plates. The studies revealed that the supply of hydrothermal fluids, released through the basaltic rocks at the ocean floor, originate from the seawater that had been drawn from its vicinity. The water is heated through close contact with crystallizing magma chambers. After enrichment with minerals through interaction with the surrounding rocks, the water is expelled back to the marine environment. Due to intimate fluid-rock interactions during the transport through the crust, the hydrothermal fluids are hot, acidic, highly reduced, and metal-enriched in comparison to ambient seawater (see: Herzig and Hannington 2006). Venting of the hydrothermal fluids at the seafloor takes place at $> 250°C$. During mixing of the high-temperature fluids

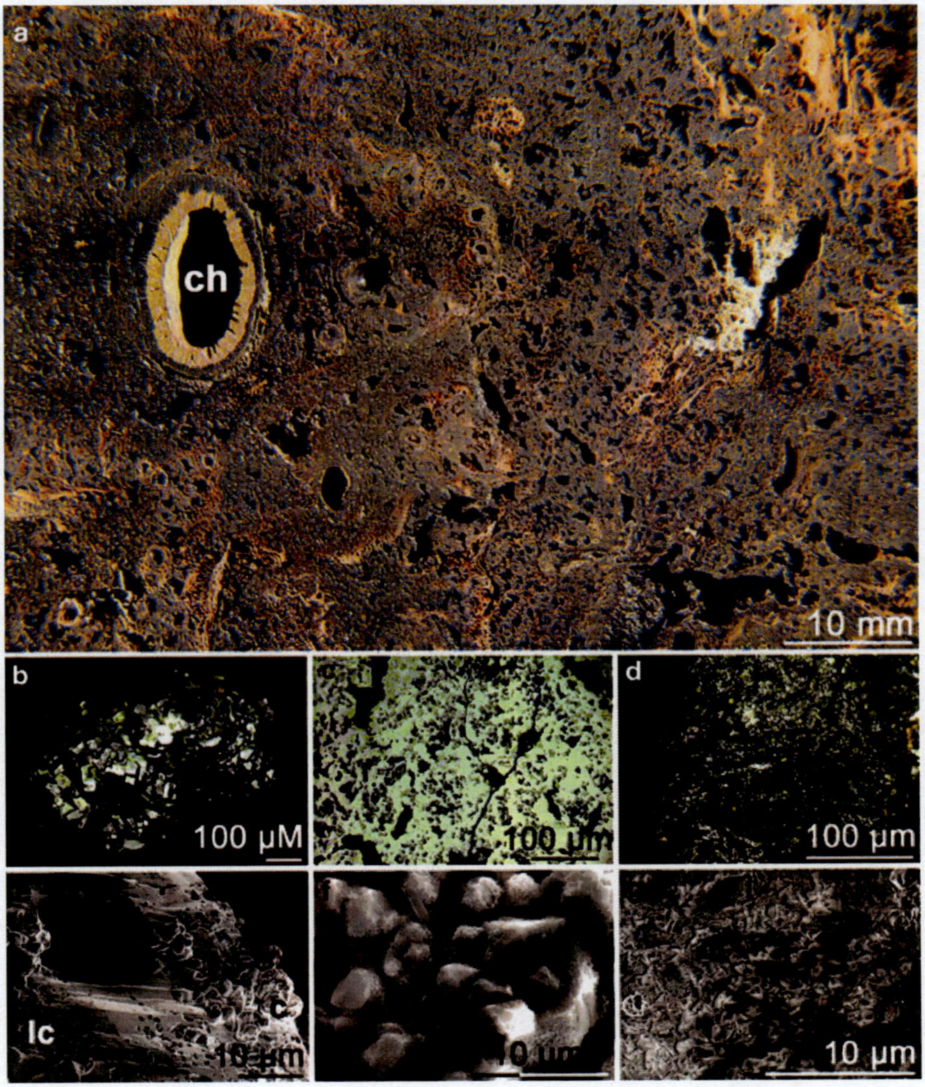

Fig. 4.18 Black smoker. (**a**) Cross section of a black smoker chimney. As hydrothermal fluid vents from the seafloor, different minerals in it precipitate at different temperatures, creating a multilayered black smoker chimney. From the chimney (*ch*), a gradient of chalcopyrite – pyrrhotite:pyrite:sphalerite – silica:barite – anhydrite exists. (**b–g**) A cross fracture through a sample of a black smoker [Indian Ridge collected during the cruise DY105-17 (Lin and Zhang 2006)]; containing high concentrations of sulfur (51%) and iron (45%)], shows (**b**) prominent large pyrite crystals. (**c**) Cross sections show the extensive dimensions of the pyrite (iron sulfide [FeS_2]) and chalcopyrite (copper iron sulfide [$CuFeS_2$]) areas. (**d**) In only μm distances, next to the pyrite crystals, mats of amorphous-looking iron/sulfur-containing areas exist that might be explained as conversion products of bacteria from pyrite. HR-SEM images: (**e**) Transition regions from pyrite crystals (*c*) to less-crystalline (*lc*) Fe-S regions. (**f**) Pyrite crystals; (**g**) less-crystalline Fe-S regions

from the vents with the seawater, a strong precipitation of minerals occurs above the vent; black, cloud-like sulfide minerals are precipitated as microscopic particles in the venting water, giving those vents the name "black smokers." While those black smokers are in direct contact with the magma chambers that are found 1–3 km beneath the seafloor, a second group of hydrothermal vents exists, the Lost City hydrothermal systems (Martin et al. 2008). This circulation system (~200°C) has no intimate connection with the magma chambers, and lacks CO_2 but provides high fluxes of hydrogen and methane at high pH.

The mineralization steps at the black smokers involving chalcopyrite, pyrrhotite, anhydrite, pyrite, and sphalerite are driven by the temperature decrease and can be explained on the basis of chemical reaction processes only, not involving any biogenic component (Halbach et al. 2003; Herzig and Hannington 2006); Fig. 4.18. Likewise, even though vents provide rich substrates for bacterial communities, e.g., chemosynthetic bacteria or a wide array of Metazoa (Suess et al. 1998), no data exist about a potential involvement of microorganisms in vent formation. In contrast, it is more likely that those black smoker-associated microorganisms modify or convert those pyrite derivatives to their oxidation products. One well-studied pyrite-oxidizing bacterium is the gram-negative *Ferrobacillus ferrooxidans*/*Acidithiobacillus ferrooxidans* that obtains energy from the oxidation of ferrous iron or reduced sulfur compounds (Silverman 1967). From this bacterium, it is also known that it produces a biofilm (Mangold et al. 2008). Hence, it can be proposed that *A. ferrooxidans*-related microorganisms (Spiridonova et al. 2006) might exist on the surface of the pyrite crystals and convert them to an accessible area for a wide range of microorganism. First data on such pyrite modifications, from pyrite crystals to amorphous-looking mats composed of Fe and S, had been obtained from black smoker rocks, collected by "Ocean 1" (Lin and Zhang 2006).

4.7 Toward a Molecular Biomineralization

The biomineralization concept provides new avenues for a sustainable exploitation of economically important biominerals, e.g., nodules and vents, since it implements molecular biology as a powerful technique to the understanding of mineralization processes. Application of recent electron microscopic and spectroscopic techniques disclosed the participation of organic molecules in the mineralization processes. Prominent examples are the formation of bio-silica in diatoms or sponges (Müller 2003). In the latter model, even enzymatic activities could be implicated in the mineral deposition. This insight allowed the application of the recombinant technology for the production of inorganic polymers at lower temperature and mild reaction conditions (Müller et al. 2007b; Schröder et al. 2008). Following this strategy, intense efforts yielded the elucidation that defined organisms, bacteria in nodules and coccoliths, in crusts are involved in the deposition of marine minerals, and described in this review. Moreover, exo-lithobiontic microbes have been identified and attributed to the precipitation of minerals (Dupraz and Visscher 2005). Recent studies allowed first identifications of DNA fragments, isolated from within the mineralic part of the polymetallic nodules (Wang et al. 2009d), indicating the existence of endolithobiontic microbes. Focusing on bacteria, strategies had already been developed to utilize microorganisms for biosynthesis and bioleaching of manganese marine deposits (Ehrlich 2001). As a conclusion, it can be expected that in the nearest future, microorganisms will also be isolated that may be used for the accumulation and concentration of trace elements from the marine environment.

Acknowledgment We thank the Research group "Surface Chemistry" of Dr. I. Lieberwirth and Dr M. Kappl (Max Planck Institute for Polymer Research; Mainz) for important contributions in electron microscopic and spectroscopic analyses. This work was supported by grants from the Bundesministerium für Bildung und Forschung Germany [project: Center of Excellence *BIOTEC*marin], and from the International Science & Technology Cooperation Program of China (Grant No. 2008DFA00980).

References

Althoff K, Schütt C, Steffen R, Batel R, Müller WEG (1998) Evidence for a symbiosis between bacteria of the genus *Rhodobacter* and the marine sponge *Halichondria panicea*: harbor also for putatively-toxic bacteria? Mar Biol 130:529–536

Amy PS, Caldwell BA, Soeldner AH, Morita RY, Albright LJ (1987) Microbial activity and ultrastructure of mineral-based marine snow from Howe Sound, British Columbia. Can J Fish Aquat Sci 44:1135–1142

Anufriev G, Boltenkov BS (2007) Ferromanganese nodules of the Baltic Sea: composition, helium isotopes, and growth rate. Lithol Miner Resour 42:240–245

Bau M, Koschinsky A, Dulski P, Hein JR (1996) Comparison of the partitioning behaviours of yttrium, rare earth elements, and titanium between hydrogenetic marine ferromanganese crusts and seawater. Geochim Cosmochim Acta 60:1709–1725

Bazylinski DA, Frankel RB (2003) Biologically controlled mineralization in prokaryotes. Rev Mineral Geochem 54:217–247

Bonatti E, Nayudu YR (1965) The origin of manganese nodules on the ocean floor. Am J Sci 263:17–39

Bruland KW, Orians KJ, Cowen JP (1994) Reactive trace metals in the stratified central North Pacific. Geochim Cosmochim Acta 58:3171–3182

Cha JN, Shimizu K, Zhou Y, Christianssen SC, Chmelka BF, Stucky GD, Morse DE (1999) Silicatein filaments and subunits from a marine sponge direct the polymerization of silica and silicones in vitro. Proc Natl Acad Sci USA 96:361–365

Chukhrov FV, Zvyagin BB, Yermilova LP, Gorshkov AI (1976) Mineralogical criteria in the origin of marine iron-manganese nodules. Miner Deposita (Berl) 11:24–32

Cottrell MT, Mannino A, Kirchman DL (2006) Aerobic anoxygenic phototrophic bacteria in the Mid-Atlantic Bight and the North Pacific Gyre. Appl Environ Microbiol 72:557–564

Cowen JP, Bruland KW (1985) Metal deposits associated with bacteria: implications for Fe and Mn marine geochemistry. Deep-Sea Res 32A:253–272

Cowen JP, DeCarlo EH, McGee DL (1993) Calcareous nannofossil biostratigraphic dating of a ferromanganese crust from Schumann Seamount. Mar Geol 115:289–306

Crerar DA, Barnes HL (1974) Deposition of deep-sea nodules. Geochim Cosmochim Acta 38:279–300

Cronan DS (ed) (2000) Handbook of marine mineral deposits. CRC Press, Boca Raton

Dick GJ, Lee YE, Tebo BM (2006) Manganese(II)-oxidizing *Bacillus* spores in Guaymas Basin hydrothermal sediments and plumes. Appl Environ Microbiol 72:3184–3190

Dick GJ, Podell S, Johnson HA, Rivera-Espinoza Y, Bernier-Latmani R, McCarthy JK, Torpey JW, Clement BG, Gaasterland T, Tebo BM (2008a) Genomic insights into Mn(II) oxidation by the marine alphaproteobacterium *Aurantimonas* sp. strain SI85-9A1. Appl Environ Microbiol 74:2646–2658

Dick GJ, Torpey JW, Beveridge TJ, Tebo BM (2008b) Direct identification of a bacterial manganese(II) oxidase, the multicopper oxidase MnxG, from spores of several different marine bacillus species. Appl Environ Microbiol 74:1527–1534

Dupraz C, Visscher PT (2005) Microbial lithification in marine stromatolites and hypersaline mats. Trends Microbiol 13:429–438

Dymond J, Eklund W (1978) A microprobe study of metalliferous sediment components. Earth Planet Sci Lett 40:243–251

Ehrlich HL (2001) Ocean manganese nodules: biogenesis and bioleaching. In: Kawatra SK, Natarajan KA (eds) Mineral biotechnology: microbial aspects of mineral beneficiation, metal extraction, and environmental control. American Technical Publishers, Littleton, pp 239–252

Ehrlich HL (2002) Geomicrobiology. Marcel Dekker, NY, 768 pp

Fortin D, Ferris FG, Beveridge TJ (1997) Surface-mediated mineral development by bacteria. Rev Mineral 35:161–180

Francheteau J, Needham HD, Choukroune P, Juteau T, Seguret M, Ballard RD, Fox PJ, Normark W, Carranza A, Cordoba D, Guerrero J, Rangin C, Bougault H, Cambon P, Hekinian R (1979) Massive deep-sea sulphide ore deposits discovered on the East Pacific Rise? Nature 277:523–528

Geisen M, Billard C, Broerse ATC, Cros L, Probert I, Young JR (2002) Life-cycle associations involving pairs of holococcolithophorid species: intraspecific variation or cryptic speciation? Eur J Phycol 37:531–550

Glasby GP (1974) Mechanism of incorporation of manganese and associated trace elements in marine manganese nodules. Oceanogr Mar Biol Annu Rev 12:11–40

Glasby GP (2006) Manganese: predominant role of nodules and crusts. In: Schulz HD, Zabel M (eds) Marine geochemistry, 2nd edn. Springer, Berlin

Glasby GP, Schulz HD (1999) EH, pH diagrams for Mn, Fe, Co, Ni, Cu and As under seawater conditions: application of two new types of EH, pH diagrams to the study of specific problems in marine geochemistry. Aquat Geochem 5:227–248

Halbach P (1986) Processes controlling the heavy metal distribution in Pacific ferromanganese nodules and crusts. Geol Rundsch 75:235–247

Halbach P, Scherhag C, Hebisch U, Marchig V (1981) Geochemical and mineralogical control of different genetic types of deep-sea nodules from Pacific Ocean. Miner Deposita 16:59–84

Halbach P, Friedrich G, von Stackelberg U (1988) The manganese nodule belt of the Pacific Ocean. Enke, Stuttgart

Halbach P, Fouquet Y, Herzig P (2003) Mineralization and compositional patterns in deep-sea hydrothermal systems. In: Halbach PE, Tunnicliffe V, Hein JR (eds) Energy and mass transfer in marine hydrothermal systems. Dahlem University Press, Berlin, pp 86–122

Hastings D, Emerson M (1986) Oxidation of manganese by spores of a marine bacillus: kinetics and thermodynamic considerations. Geochim Cosmochim Acta 50:1819–1824

Hay WW (2004) Carbonate fluxes and calcareous nannoplankton. In: Thierstein HR, Young JR (eds) Coccolithophores – from molecular processes to global impact. Springer, Berlin, pp 508–525

He J, Zhang L, Jin S, Zhu Y, Liu F (2008) Bacterial communities inside and surrounding soil iron–manganese nodules. Geomicrobiol J 25:14–24

Hein JR, Koschinsky A, Halbach P, Manheim FT, Bau M, Jung-Keuk K, Lubick N (1997) Iron and manganese oxide mineralization in the Pacific. In: Nicholson K, Hein JR, Bühn B, Dasgupta S (eds) Manganese mineralization: geochemistry and mineralogy of terrestrial and marine deposits. Geological Society Special Publication No 119, pp 123–138

Hein JR, Koschinsky A, Bau M, Manheim FT, Roberts L (2000) Cobalt-rich ferromanganese crusts in the Pacific. In: Cronan DS (ed) Handbook of marine mineral deposits. Boca Raton, FL, USA, pp 239–279

Herndl GJ (1988) Ecology of amorphous aggregations (marine snow) in the Northern Adriatic Sea. 11. Microbial density and activity in marine snow and its implication to overall pelagic processes. Mar Ecol Prog 48:265–275

Herzig PM, Hannington MD (2006) Input from the deep: hot vents and cold seeps. In: Schulz HD, Zabel M (eds) Marine geochemistry. Springer, Berlin, pp 457–479

Kawamoto H (2008) Japan's policies to be adopted on rare metal resources. Quart Rev 27:57–76

Kerr R (1984) Manganese nodules grow by rain from above. Science 223:576–577

Kester DR (1975) Dissolved gases other than CO_2. In: Riley JP, Skirrow G (eds) Chemical oceanography, 2nd edn. Academic, London, pp 498–556

Kim SJ, Bang IC, Buongiorno J, Hu LW (2006) Effects of nanoparticle deposition on surface wettability influencing boiling heat transfer in nanofluids. Appl Phys Lett 89:153–107

Koschinsky A, Halbach P (1995) Sequential leaching of marine ferromanganese precipitates: Genetic implications. Geochim Cosmochim Acta 59:5113–5132

Koschinsky A, Hein JR (2003) Uptake of elements from seawater by ferromanganese crusts: solid-phase associations and seawater speciation. Mar Geol 198:331–351

Koschinsky A, Stascheit A-M, Bau M, Halbach P (1997) Effects of phosphatization on the geochemical and mineralogical composition of marine ferromanganese crusts. Geochim Cosmochim Acta 61:4079–4094

Krasko A, Batel R, Schröder HC, Müller IM, Müller WEG (2000) Expression of silicatein and collagen genes in the marine sponge *Suberites domuncula* is controlled by silicate and myotrophin. Eur J Biochem 267:4878–4887

Lawrence JR, Korber DR, Hoyle BD, Costerton JW, Caldwell DE (1991) Optical sectioning of microbial biofilms. J Bacteriol 173:6558–6567

LePennec G, Perović S, Ammar MSA, Grebenjuk VA, Steffen R, Müller WEG (2003) Cultivation of primmorphs from the marine sponge *Suberites domuncula*: morphogenetic potential of silicon and iron. J Biotechnol 100:93–108

Leppard GG (1999) Structure/function/activity relationships in marine snow. Curr Underst suggested Res thrusts 35:389–395

Lin J, Zhang C (2006) The first collaborative China-international cruises to investigate Mid-Ocean Ridge hydrothermal vents. InterRidge News 15:1–3

Lowenstam HA (1981) Minerals formed by organisms. Science 211:1126–1131

Lowenstam HA, Weiner S (1989) On biomineralization. Oxford University Press, Oxford

Mangold S, Laxander M, Harneit K, Rohwerder T, Claus G, Sand W (2008) Visualization of *Acidithiobacillus ferrooxidans* biofilms on pyrite by atomic force and epifluorescence microscopy under various experimental conditions. Hydrometallurgy 94:127–132

Marcus MA, Manceau A, Kersten M (2004) Mn, Fe, Zn and As speciation in a fast-growing ferromanganese marine nodule. Geochim Cosmochim Acta 68:3125–3136

Martin W, Baross J, Kelley D, Russell MJ (2008) Hydrothermal vents and the origin of life. Nat Rev Microbiol 6:805–814

Mendhom J, Denny RC, Barnes JD, Thomas MJK (2000) Textbook of quantitative chemical analysis. Pearson education, London

Mengele R, Sumper M (1992) Drastic differences in glycosylation of related S-layer glycoproteins from moderate and extreme halophiles. J Biol Chem 267:8182–8185

Mero J (1962) Ocean-floor manganese nodules. Econ Geol 57:747–767

Mills RA, Wells DM, Roberts S (2001) Genesis of ferromanganese crusts from the TAG hydrothermal field. Chem Geol 176:283–293

Moore WS, Ku TL, Macdougall JD, Burns VM, Burns R, Dymond J, Lyle M, Piper DZ (1981) Fluxes of metals to a manganese nodule: radiochemical, chemical, structural, and mineralogical studie. Earth Planet Sci Lett 52:151–171

Morse DE (1999) Silicon biotechnology: harnessing biological silica production to make new materials. Trends Biotechnol 17:230–232

Müller WEG (ed) (2003) Silicon biomineralization: biology, biochemistry, molecular biology, biotechnology. Springer, Berlin

Müller WEG, Steffen R, Kurelec B, Smodlaka N, Puskaric S, Jagic B, Müller-Niklas G, Queric NV (1998) Chemosensitizers of the multixenobiotic resistance in the amorphous aggregates (marine snow): etiology of mass killing on the benthos in the Northern Adriatic? Environ Toxicol Pharmacol 6:229–238

Müller WEG, Eckert C, Kropf K, Wang XH, Schloßmacher U, Seckert C, Wolf SE, Tremel W, Schröder HC (2007a) Formation of the giant spicules of the deep sea hexactinellid *Monorhaphis chuni* (Schulze 1904): electron microscopical and biochemical studies. Cell Tissue Res 329:363–378

Müller WEG, Wang XH, Belikov SI, Tremel W, Schloßmacher U, Natoli A, Brandt D, Boreiko A, Tahir MN, Müller IM, Schröder HC (2007b) Formation of siliceous spicules in demosponges: example *Suberites domuncula*. In: Bäuerlein E (ed) Handbook of biomineralization, vol 1, Biological aspects and structure formation. Wiley, Weinheim, pp 59–82

Müller WEG, Schloßmacher U, Wang XH, Boreiko A, Brandt D, Wolf SE, Tremel W, Schröder HC (2008a) Poly(silicate)-metabolizing silicatein in siliceous spicules and silicasomes of demosponges comprises dual enzymatic activities (silica-polymerase and silica-esterase). FEBS J 275:362–370

Müller WEG, Wang XH, Kropf K, Ushijima H, Geurtsen W, Eckert C, Tahir MN, Tremel W, Boreiko A, Schloßmacher U, Li J, Schröder HC (2008b) Bioorganic/inorganic hybrid composition of sponge spicules: matrix of the giant spicules and of the comitalia of the deep sea hexactinellid *Monorhaphis*. J Struct Biol 161:188–203

Murray J (1891) Report on the scientific results of the voyage of H. M. S. Challenger during the years 1873–76 – deep sea deposits. HMS Stationary Office, London

Murray JW, Brewer PG (1977) Mechanism of removal of manganese, iron and other trace metals from seawater. In: Glasby GP (ed) Marine manganese deposits. Elsevier, Amsterdam, pp 291–325

Murray J, Philippi E (1908) Die Grundproben der Deutschen Tiefsee-Expedition, 1898–99 auf dem Dampfer. In: Valdivia Wiss Ergeb Deutschen Tiefsee–Expedition, vol 10. Gustav Fischer, Jena, pp 77–207

Persson AE, Schoeman BJ, Sterte J, Otterstedt J-E (1995) Synthesis of stable suspensions of discrete colloidal zeolite (Na, TPA)ZSM-5 crystals. Zeolites 15:611–619

Post JE (1999) Manganese oxide minerals: crystal structures and economic and environmental significance. Proc Natl Acad Sci USA 96:3447–3454

RISE Project group: Spiess FN, Macdonald KC, and 20 others (1980) East Pacific Rise: hot springs and geophysical experiment. Science 207:1421–1433

Ryan KJ, Ray CG (eds) (2004) Sherris medical microbiology, 4th edn. McGraw Hill, New York

Schröder HC, Wang XH, Tremel W, Ushijima H, Müller WEG (2008) Biofabrication of biosilica-glass by living organisms. Nat Prod Rep 25:455–474

Schrope M (2007) Digging deep. Nature 447:246–247

Schultze-Lam S, Beveridge TJ (1994) Physicochemical characteristics of the mineral-forming S-layer from the *cyanobacterium Synechococcus* strain GL24. Can J Microbiol 40:216–223

Schultze-Lam S, Thompsom JB, Beveridge TJ (1993) Metal ion immobilization by bacterial surfaces in fresh water environments. Water Pollut Res J Can 28:51–81

Silverman MP (1967) Mechanism of bacterial pyrite oxidation. J Bacteriol 94:1046–1051

Sleytr UB, Messner P (1983) Crystalline surface layers on bacteria. Annu Rev Microbiol 37:311–339

Sleytr UB, Messner P, Pum D, Sara M (1999) Crystalline bacterial cell surface layers (S layers): from supramolecular cell structure to biomimetics and nanotechnology. Angew Chem Int Ed 38:1034–1054

Solomon EI, Sundaram UM, Machonkin TE (1996) Multicopper oxidases and oxygenases. Chem Rev 96:2563–2605

Somayajulu BLK (2000) Growth rates of oceanic manganese nodules: implications to their genesis, palaeo-earth environment and resource potential. Curr Sci 78:300–308

Spiridonova EM, Kuznetsov BB, Pimenov NV, Tourova TP (2006) Phylogenetic characterization of endosymbionts of the hydrothermal vent mussel *Bathymodiolus azoricus* by analysis of the 16S rRNA, *cbb*L, and *pmo*A genes. Microbiology 75:694–701

Stumm W, Morgan JJ (1996) Aquatic chemistry, chemical equilibria and rates in natural waters, 3rd edn. Wiley, New York, p 1022

Suess E, Bohrmann G, von Huene R, Linke P, Wallmann K, Lammers S, Sahling H (1998) Fluid venting in the eastern Aleutian subduction zone. J Geophys Res 103:2597–2614

Szeto J, Ramirez-Arcos S, Raymond C, Hicks LD, Kay CM, Dillon JAR (2001) Gonococcal MinD affects cell division in *Neisseria gonorrhoeae* and *Escherichia coli* and exhibits a novel selfinteraction. J Bacteriol 183:6253–6264

Tebo BM, Bargar JR, Clement BG, Dick GJ, Murray KJ, Parker D, Verity R, Webb SM (2004) Biogenic manganese oxides: properties and mechanisms of formation. Annu Rev Earth Pl Sc 32:287–328

Thijssen T, Glasby GP, Friedrich G, Stoffers P, Sioulas A (1985) Manganese nodules in the Central Peru Basin. Chem Erde 44:1–12

Vonderhaar DL, McMurtry GM, Garbe-Schönberg D, Stüben D, Esser B (2000) Platinum and other related element enrichments in Pacific ferromanganese crust deposits. In: Glenn CR, Prevot-Lucas L, Lucas J (eds) Marine autigenesis: from global to microbial. SEPM Special Publication No 66. Society for Sedimentary Geology, Darlington, pp 287–308

Wang XH, Müller WEG (2009) Contribution of biomineralization during growth of polymetallic nodules and ferromanganese crusts from the Pacific Ocean. Front Mater Sci China FMSC 3:109–123

Wang XH, Boreiko A, Schloßmacher U, Brandt D, Schröder HC, Li J, Kaandorp JA, Götz H, Duschner H, Müller WEG (2008) Axial growth of hexactinellid spicules: formation of cone-like structural units in the giant basal spicules of the hexactinellid *Monorhaphis*. J Struct Biol 164:270–280

Wang XH, Schloßmacher U, Natalio F, Schröder HC, Wolf SE, Tremel W, Müller WEG (2009a) Evidence for biogenic processes during formation of ferromanganese crusts from the Pacific Ocean: implications of biologically induced mineralization. Micron 40:526–535

Wang XH, Schloßmacher U, Wiens M, Schröder HC, Müller WEG (2009b) Biogenic origin of polymetallic nodules from the Clarion-Clipperton zone in the Eastern Pacific Ocean: electron microscopic and EDX evidence. Mar Biotechnol 11:99–108

Wang XH, Schröder HC, Schloßmacher U, Müller WEG (2009c) Organized bacterial assemblies in manganese nodules: evidence for a role of S-layers in metal deposition. Geo-Mar Lett 29:85–91

Wang XH, Schröder HC, Wiens M, Schloßmacher U, Müller WEG (2009d) Manganese/polymetallic nodules: micro-structural characterization of exolithobiontic- and endolithobiontic microbial biofilms by scanning electron microscopy. Micron 40:350–358

Wang XH, Wiens M, Divekar M, Grebenjuk VA, Schröder HC, Batel R, Müller WEG (2011) Isolation and characterization of a Mn(II)-oxidizing bacillus strain from the demosponge *Suberites domuncula*. Marine Drugs 9:1–28

Weiner S, Dove PM (2003) An overview of biomineralization processes and the problem of the vital effect. Rev Mineral Geochem 54:1–29

Weiss B, Ebel R, Elbrächter M, Kirchner M, Proksch P (1996) Defense metabolites from the marine sponge *Verongia aerophoba*. Biochem Syst Ecol 24:1–7

Wolfaardt GM, Lawrence JR, Robarts RD, Caldwell DE (1995) Bioaccumulation of the herbicide diclofop in extracellular polymers and its utilization by biofilm community during starvation. Appl Environ Microbiol 61:152–158
Yihua C, Yipu H (2002) Advances in the study of geochemistry and paleo-oceanography of the Co-rich crust. Mar Sci Bull 4:8–14
Zhamoida VA, Butylin WP, Glasby GP, Popova IA (1996) The nature of ferromanganese concretions from the Eastern Gulf of Finland, Baltic Sea. Mar Georesour Geotec 14:161–176
Zhu J, Wang Y, Legge GJF (1993) Micron scale analysis of deep-sea ferromanganese with nuclear microprobes nodules. Nucl Instrum Methods Phys Res B77:478–483

Part II
Biocalcium

Chapter 5
Molecular Basis of Bacterial Calcium Carbonate Precipitation

Brunella Perito and Giorgio Mastromei

Contents

Abstract Calcium carbonate precipitation is a widespread process, occurring in different bacterial taxonomic groups and in different environments, at a scale ranging from the microscopic one of cells to that of geological formations. It has relevant implications in natural processes and has great potentiality in numerous applications. For these reasons, bacterial precipitation has been investigated extensively both in natural environments and under laboratory conditions. Different mechanisms of bacterial involvement in precipitation have been proposed. There is an agreement that the phenomenon can be influenced by the environmental physicochemical conditions and it is correlated both to the metabolic activity and the cell surface structures of microorganisms. Nevertheless, the role played by bacteria in calcium mineralization remains a matter of debate. This chapter reviews the main mechanisms of the process with particular focus on what is known on

G. Mastromei (✉)
Department of Evolutionary Biology "Leo Pardi", University of Florence, Via Romana 17, 50125 Firenze, Italy
e-mail: giorgio.mastromei@unifi.it

W.E.G. Müller (ed.), *Molecular Biomineralization*, Progress in Molecular and Subcellular Biology 52, DOI 10.1007/978-3-642-21230-7_5,

molecular aspects, and discusses the significance of the precipitation event also from an evolutionary point of view.

Microbiogeochemistry? The proof of the microbiogeochemical process is undeniable, but we have yet to ascertain its magnitude and importance. Is it conceivable that life forms so small could be indispensable for something so enormous? (Beveridge 1989)

5.1 Introduction

Calcium carbonate precipitation (CCP) is a widespread process among bacteria, not restricted to any taxonomic group (Boquet et al. 1973) and common in different environments such as marine waters and sediments, freshwater, and soils (Castanier et al. 1999; Ehrlich 1998). It has drawn much attention in recent decades because of its relevant implications in natural processes and its potentiality in numerous applications. It represents a fundamental part of the calcium biogeochemical cycle as well as the carbon one (Zavarzin 2002), contributing to atmospheric CO_2 fixation and to a vast reservoir of carbon sequestering through formation of calcium carbonate (CC) sediments, deposits, and rocks. Bacterial calcium carbonate precipitation (BCCP) is also involved in the production of pathological concretions such as gallstones and kidney stones in humans and recently the presence of microbial signatures preserved in the rock record has been investigated to assess biogenic origin for speleothems and other carbonates (Barton et al. 2001).

Proposed innovative applications of $CaCO_3$ mineralization by bacteria include biomimetic processes and materials and examples of bioremediation (and stabilization) in several fields ranging from applied environmental microbiology (leaching, solid-phase capture of inorganic contaminants), to civil and environmental engineering (sediment dikes, bioplugging, biogrouting, and self-healing of concrete and limestone structures) and conservation of monumental calcareous stones. For implications and applications of BCCP, see references in Rodriguez-Navarro et al. (2003); Barabesi et al. (2007); De Muynck et al. (2010).

Different mechanisms of bacterial involvement in calcification have been proposed (Ehrlich 1996) and they have been a matter of controversy throughout the last century (von Knorre and Krumbein 2000). The studies made in this field have pointed out the complexity of the phenomenon that can be influenced by the environmental physicochemical conditions and it is correlated both to the metabolic activity and the cell surface structures of microorganisms (Castanier et al. 1999; Beveridge 1989; Fortin et al. 1997).

Although BCCP is a widely occurring natural process and has been investigated extensively both in natural environments and under defined laboratory conditions, the role played by bacteria in calcium mineralization is still debated (von Knorre and Krumbein 2000; Zavarzin 2002). CC deposition by bacteria is a process occurring at a scale ranging from the microscopic one of cells (order of μm) to that of geological formations (order of Km). The mechanisms of precipitation and the function of this process within both the cell physiology and the microbial ecology of the precipitating organism still remains largely unresolved (Yates and Robbins 1999; Rivadeneyra et al. 1994). In particular, very less is known on the molecular basis of the process.

Improving our knowledge on molecular mechanisms of BCCP would be useful for applied purposes.

This chapter starts with a brief overview of the main commonly recognized features of BCCP (Sect. 5.2) followed by the description of interactions between bacterial metabolism and BCCP (Sect. 5.3) and cell surface and BCCP (Sect. 5.4), with particular focus on what is known on molecular aspects.

5.2 General Features of Bacterial Calcium Carbonate Mineralization

5.2.1 Calcium Carbonate Mineralization Terms and Processes

This Section deals with commonly recognized features and mechanisms of bacterial CC mineralization both in natural environments and in the laboratory.

Some introductive clarifications are necessary regarding the terminology. The first one regards the terms precipitation and mineralization. According to Beveridge (1989), there is a subtle but significant difference between precipitation and mineralization. Metal precipitates are hydrous, amorphous aggregates, whereas minerals are much more anhydrous and crystalline. The former are converted to the latter by lithification; water is slowly drawn out by heat and compaction during solidification. In this context, the precipitates found on bacterial surfaces are early stages of mineral development; in a natural setting, bona fide minerals should be long in forming. Nevertheless, in this chapter, the two terms, CC precipitation and CC mineralization, will be considered synonymous.

The second clarification regards the term biomineralization. Published definitions of the term biomineral vary widely (Dupraz et al. 2009) but their use is not banal since their significance is linked to the role of organisms in mineralization. In general terms, biomineralization is the process through which organisms are involved in mineral formation, as a result of cellular activity that fosters the necessary physical and chemical conditions for such a formation and growth to take place (Ben Omar et al. 1997). It refers to a mineral that is produced by living organisms and consists of both mineral and organic components (Weiner and Dove 2003).

From an evolutionary point of view it can be stated, without any doubt, that biomineralization processes originated from bacterial activity (Ben Omar et al. 1997). The majority of the described biominerals contain calcium as their main cation and, in terms of their anions, carbonates are among the most abundant forms. Nevertheless, even if CC mineralization is a widespread process among organisms from bacteria to Chordata, it is generally accepted that the capacities of prokaryotes and eukaryotes in mineralization are different (Zavarzin 2002). Eukaryotes, prevalently tissue-forming multicellular eukaryotes, carry out biologically controlled CC mineralization (Lowenstam and Weiner 1989; Mann 2001). In this case, cellular activity controls the process to a high degree and directs the nucleation, growth, morphology, and final location of the mineral (Decho 2010). The mineral particles formed are synthesized or deposited on or within organic matrices or vesicles in a specific location with regard to the cell and usually intracellularly. CCP is often used for specific purposes and leads to the production of complex and specialized $CaCO_3$ structures (e.g., formation of protozoan and mollusc shells or coral reefs) clearly visible. Every organism synthesizes biogenic minerals in a form that is unique to that species, independently from environmental conditions. Because of these features, both the synthesis and the form of every specific biogenic mineral are thought to be under specific metabolic and genetic control (Bäuerlein 2004). A known example of biologically controlled $CaCO_3$ mineralization by eukaryotic microorganisms is that of the unicellular algae coccolithophores (Bäuerlein 2004).

In contrast, biologically induced mineralization is usually carried out in an open environment and no specialized cell structure or specific molecular mechanism is thought to be involved. Microbially induced mineralization is a specific type of the biologically induced mineralization, referring to precipitation that results from interactions between microbial activities and the environment (Weiner and Dove 2003; Dupraz et al. 2009). With the singular exception of *Achromatium oxaliferum*, CC deposition by bacteria has been generally regarded to be induced and the type of mineral produced largely dependent on environmental conditions (Ben Omar et al. 1997; Brennan et al. 2004; Rivadeneyra et al. 1994). Different mechanisms of bacterial involvement in calcification have been proposed (Ehrlich 1996). However, the role played by bacteria in calcium mineralization is still debated, ranging from passive to active. Some authors emphasize that carbonate precipitation by bacteria is an unwanted by-product of bacterial physiological activities under special environmental conditions, a simple physiological accident. Bacteria would not precipitate carbonate particles by a specific mechanism and the supply of a structure by bacteria would not be necessary (von Knorre and Krumbein 2000). Other authors emphasize that the role of bacteria in CCP can be specific with ecological benefits for the precipitating organisms (McConnaughey and Whelan 1997; Castanier et al. 1999; Barabesi et al. 2007). Castanier et al. (1999) distinguish between passive and active precipitation mechanisms which may occur, often concurrently, in heterotrophic bacteria. Passive precipitation (or passive carbonatogenesis) operates by producing carbonate and bicarbonate ions and inducing chemical modifications in the medium through metabolic pathways (e.g., linked to nitrogen and sulfur cycles, see Sect. 5.3). In active precipitation (or active carbonatogenesis), the carbonate particles would be produced by ionic exchanges through the cell membrane by activation of calcium and/or magnesium ionic pumps or channels, probably coupled with carbonate ion production. Active precipitation would be independent of specific metabolic pathways.

More recently, Dupraz et al. (2009) introduced the term biologically influenced mineralization, distinguished from biologically induced, to refer to passive mineralization of organic matter. While in "biologically induced" precipitation, microbial activities generate biogeochemical conditions that facilitate precipitation, and mineralization results from the interactions between biological activity and the environment, in biologically influenced mineralization, external, environmental parameters, rather than microbial activities are responsible for creating the conditions (e.g., increased alkalinity) for mineral precipitation and the presence of living organisms is not required. An organic matrix (biogenic or abiogenic in origin) is, however, involved in biologically influenced mineralization, influencing the morphology and composition of the crystals through interactions between the mineral and the organic matter (serving as template for precipitation). While both types of precipitation typically occur outside of cells, "biologically influenced" precipitation is a passive process, resulting from interactions between extracellular biopolymers and the geochemical environment.

Moreover, Dupraz et al. (2009) propose the use of the term organomineralization sensu lato (s.l.) as an umbrella term encompassing biologically influenced and biologically induced mineralization. The organomineralization process can be intrinsically (microbial metabolism) or extrinsically driven (e.g., degassing, evaporation).

Microbial precipitation of $CaCO_3$ occurs by organomineralization s.l. through these two major processes: microbiologically induced (active) and microbiologically influenced precipitation (passive). In the geochemical environment near cells, however, both processes occur in spatial and temporal proximity and it is difficult to separate these effects in situ (Weiner and Dove 2003).

Dupraz et al. (2009) proposed that the term biomineral should be used in a restrictive definition for the product of selective uptake of elements, which are incorporated into functional structures under strict biological control. Biominerals are consequently strictly referred as the product of biologically controlled mineralization, which becomes synonymous of biomineralization, while biologically induced minerals are excluded.

For a review on definitions and mechanisms of mineralization, see Dupraz et al. (2009).

The terminology from Dupraz et al. (2009) has the advantage to reconsider in a clear view the debated role of bacteria in $CaCO_3$ mineralization (active vs. passive mineralization requires an active role from bacteria, living and metabolic active cells, independently from the mechanism used to favor precipitation (a specific metabolic pathway, pump, or cell surface properties)). In this sense, the active role of bacteria in CCP is expanded and encompasses any activity fostering CCP by living cells. Urea hydrolysis, sulfate reduction, and denitrification (Sect. 5.3.1) are three examples of metabolic pathways driving this type of precipitation process, while they were considered linked to a passive role by other authors (Castanier et al. 1999). On the contrary, influenced mineralization requires a passive role from bacteria, with cells either living or dead; the organic component required (primarily an esopolymer) may have been produced by living cells and released in the environment or attached to dead or not metabolic active cells.

In this Chapter, we use the classification by Dupraz et al. (2009) and refer to BCCP as a process which bacteria carry out both in microbiologically induced (active) and/or in microbiologically influenced (passive) way.

5.2.2 General Features of Bacterial $CaCO_3$ Mineralization

BCCP was described for the first time by Murray and Irvine (1889–90) who observed the formation of CC crystals after adding seawater to rotting urine. The important geologic role played by bacteria in CCP was suggested since the beginning of the last century and has been suspected for years. In 1903, Nadson showed that calcium carbonate deposits in lake Veisowe in Karkou, Russia, could have originated from bacterial activity. In 1914, Drew called attention to the role of denitrifying bacteria in calcium carbonate deposition in Great Bahama sediments, the most significant calcium carbonate deposits that exist all over the world. Successively, the role of bacteria in $CaCO_3$ deposition was suggested by several authors and it was demonstrated that BCCP was a major biogeochemical process, very diffuse among different taxonomic groups and in different environments such as soils, freshwater, and saline habitats. Boquet et al. (1973) described calcite production by 210 soil bacterial isolates on a solid medium with added calcium and concluded that "under suitable conditions most bacteria form calcite crystals". Different bacterial species are capable of precipitating different amounts, shapes, and types of carbonate crystals from exactly the same synthetic medium, with an apparent occurrence of species- and environment-specific BCCP (Hammes and Verstraete 2002). BCCP has been related to the formation of marine calcareous skeletons, carbonate sediments, and soil carbonate deposits as well as of carbonate rocks, such as travertines and speleothemes (Rivadeneyra et al. 1998; Vasconcelos et al. 1995; Folk 1993).

The phenomenon was investigated both in nature and in the laboratory and an extensive literature is available (see references in Rodriguez-Navarro et al. 2003). Different possible mechanisms of bacterial involvement in calcification were proposed (Ehrlich 1996) and they have been a matter of controversy (von Knorre and Krumbein 2000). They include calcium concentration from the medium by microbial binding, metabolic alteration of the medium that results in changes in bicarbonate concentration and pH, and microbial bodies acting as crystal nucleation sites (Little et al. 1997). From various studies, it emerges that different types of bacteria as well as abiotic factors seem to contribute in a variety of ways to CCP in a wide range of different environments (von Knorre and Krumbein 2000) and that several mechanisms rather than a unique one of CCP could exist in the microbial world.

Bacterial metabolic activities and cell surface structures and their interactions with environmental physicochemical parameters are commonly recognized as the key factors in BCCP (Beveridge 1989; Fortin et al. 1997; Douglas and Beveridge 1998).

Carbonate precipitation is essentially a function of carbonate alkalinity and the availability of free calcium ions (Dupraz et al. 2009; Decho 2010), the two are combined as a saturation index (SI). Concentrations of both free carbonate CO_3^{2-} and Ca^{2+} ions must exceed saturation for precipitation to occur.

According to Hammes and Verstraete (2002), CCP is a rather straightforward chemical process governed by four key factors: (1) the calcium (Ca^{2+}) concentration, (2) the concentration of dissolved inorganic carbon (DIC), (3) the pH, and (4) the availability of nucleation sites. The primary role of bacteria in the precipitation process has been ascribed to their ability to create an alkaline environment by an increase in pH to 8.0 and higher and a DIC increase through various physiological activities (Castanier et al. 1999; Douglas and Beveridge 1998). Bacterial cells are able to induce pH variations in the medium as a consequence of both metabolic processes and ion input, which affect the pH in the surrounding microenvironment (Fortin et al. 1997). Influence of bacterial metabolism on CCP is discussed in Sect. 5.3.1.

Bacterial surfaces also play an important role in calcium precipitation (Fortin et al. 1997). Surface bacterial (macro)molecules can induce CCP, providing a template for carbonate nucleation, both as part of bacterial cell and as cell-free, when released in the environment; in the latter case, primarily as esopolymeric substances. Influence of bacterial surfaces on CCP is discussed in Sect. 5.4.

5.2.3 Bacterial CC Minerals

With the singular exception of *Achromatium oxaliferum*, which precipitates intracellular calcium carbonate crystals (Head et al. 1996), bacteria act upon calcium compounds in the extracellular space and form CC minerals outside their cell. With the exception of several cyanobacteria (Zavarzin 2002), prokaryotes do not build particular inorganic structures of CC, but simple crystals.

Calcite, aragonite, and vaterite are the major crystalline structural polymorphs of $CaCO_3$ in bacterial systems, as well as in all biogenic systems (Ben Omar et al. 1997), with the first two as the most represented isoforms and vaterite as the less stable one. Mineral crystals often contain "substituent elements", i.e., Mg^{2+} ion often substituted for Ca^{2+} in CC precipitates under marine conditions (Decho 2010). Calcium–magnesium carbonates are frequently produced by bacteria, i.e., dolomite (Vasconcelos et al. 1995) or magnesium calcite and high-magnesium calcite (Rivadeneyra et al. 1998; Decho 2010). The isoform produced (vaterite, aragonite, or calcite) depends both on its growing features and the bacterial strain; different bacteria precipitate different types of calcium carbonate, and the most common crystalline forms are either spherical or polyhedral.

The same organism or species can form different kind of minerals in different environmental conditions (Ben Omar et al. 1997; Brennan et al. 2004; Rivadeneyra et al. 1998); e.g., the Mg^{2+}/Ca^{2+} ratio and the Ca^{2+} concentration in the seawater seem to influence biocalcification in cyanobacteria (Arp et al. 2001).

Commonly, carbonate precipitates start to build up and develop on the external surface of bacterial cells by successive stratification, suggesting that the bacterial cell acts as heterogeneous crystallization center, and bacteria can be embedded and calcified in growing carbonate crystals (Castanier et al. 1999; Pentecost and Bauld 1988; Rivadeneyra et al. 1998). Although there is agreement that bacteria become the nucleus of growing crystals (going toward a kind of sacrifice), there is still some uncertainty about bacterial survival inside and some authors suggest that a mineralized coat would protect bacteria from environmental injuries and stress, even if for a short period of time (Zamarreño et al. 2009; Phoenix and Konhauser 2008).

Crystal formation has been studied in natural samples (Castanier et al. 1999) and under laboratory conditions (Rivadeneyra et al. 1998). Castanier et al. (1999) studied the development of mineral products by bacterial carbonatogenesis in eutrophicated karstic originated waters from a natural pool. Solid products were observed at SEM. The first ones were probably amorphous and hydrated at the beginning. They appeared on the surface of the bacterial bodies as patches or stripes that extended and coalesced until forming, in a few steps, a rigid coating as a kind of cocoon (Fig. 5.1). Solid particles formed inside the cellular body and excreted from the cell were also noted (excretates; Fig. 5.1). These tiny particles, including more or less calcified bacterial cells, assembled into biomineral aggregates or buildups which progressively displayed more crystalline structures. Biomineral assemblage appeared to turn into true crystals, either well shaped or poorly organized, which could enclose bacterial cells within the mineral structure (Castanier et al. 1999). The authors propose that different types of crystallogenetic sequences could follow the different metabolic pathways present in the natural sample.

Rivadeneyra et al. (1998) studied the sequence of crystal formation in a liquid culture of the moderately halophilic species *Halomonas eurihalina* under culture conditions in which the mineral phases were aragonite (96%) and magnesium calcite (6%). The sequence of biolith formation basically comprised three phases: (1) calcified filaments or chains, (2) discs, and (3) spherulites (Fig. 5.2).

5.3 Bacterial Metabolism and Precipitation

5.3.1 Role of Bacterial Metabolism on Calcium Carbonate Precipitation

Several authors recognized specific metabolic pathways to be involved in BCCP, with the pH change of the medium as the main mechanism favoring precipitation. In general, metabolic pathways able to shift the environmental pH toward alkalinity

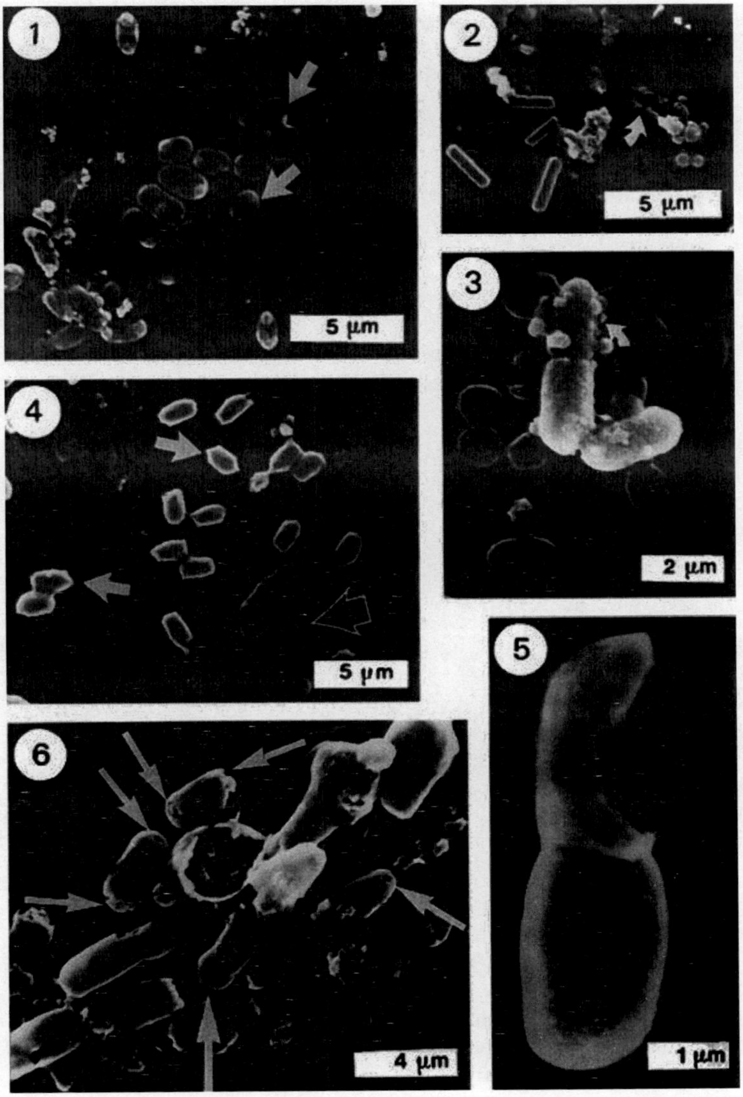

Fig. 5.1 First stages of bacterial Ca-carbonate precipitation in eutrophic conditions. (1) Bacterial cells with carbonate pimples (white arrows) and surrounding excretates (black arrow). Carbonate pimples cannot be confused with endospores because they appear on dividing cells (1). Carbonate pimples are sometimes located at one tip (2) or both tips (3) of the cells. Pimples form sometimes in the central zone of the cell against the inner part of the membrane (4). (2) Rods and cocci totally covered by carbonate (black arrows). The black triangle points at a rod which bears a tail made of excretates and a lateral pimple. The white arrow points at a ghost of bacterial rod which was destroyed by SEM preparation and which appears as a print into its produced carbonate material. (3) Cocooned rods (a couple of which is issued from a recent division) and surrounding excretates (white arrow) on a background of weakly calcified bacteria. (4) Carbonated bacterial bodies with hexagonal crystalline shapes (white arrows) on a background of poorly calcified bacterial bodies with excretates. The black arrow points at a calcified dense cluster of bacterial bodies. (5) Couple of young bacterial cells. The crack of the upper one (black arrow) shows the rigidity of the carbonate

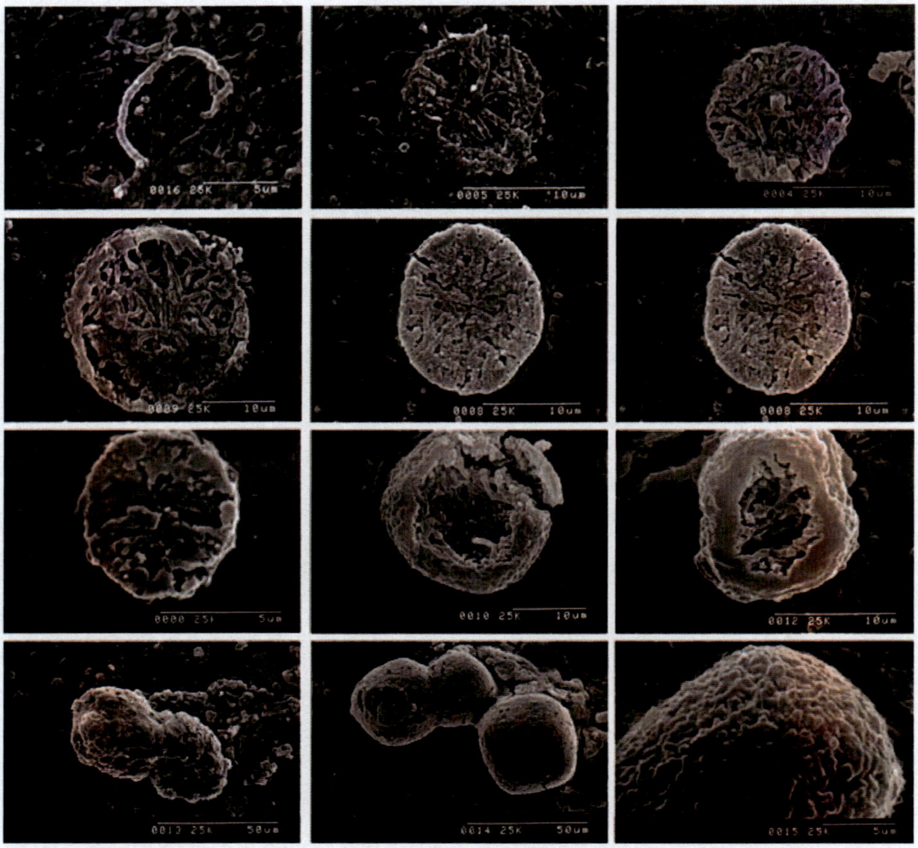

Fig. 5.2 SEM images of the sequence of biolith formation by *H. eurihalina* in liquid medium at 7.5% salts. Phot. 0016 = calcified bacterial chains or filaments; phot. 0005 = initiation of circular calcified forms; phot. 0004 = lightly calcified disc; phot. 0009 = lightly calcified disc with initiation of recrystallization; phot. 0008 = calcified disc, selectively recrystallized in external zones; phots. 0007 and 000 = recrystallized calcified discs; phots. 0010 and 0012 = progressive formation of three-dimensional bioliths from disc phase; phot. 0013 = roughly spherical three-dimensional forms; phot. 0014 = spherulites; phot. 0015 = detail of calcified bacterial bodies on biolith surface (with permission from Rivadeneyra et al. 1998)

can, in presence of calcium ions, foster CCP when a state of oversaturation develops (Fortin et al. 1997). Dupraz et al. (2009) talk of "alkaline engine" as a series of processes (microbial metabolism and environmental conditions impacting the CC SI) through which generation of carbonate ions occurs. Zavarzin (2002) talks of an alkaline barrier, formed as result of microbial activities, responsible for

---

Fig. 5.1 (continued) cocoon. (6) Part of a colony of cocooned cells showing the pores (arrows) which maintain nutritional and gaseous exchanges between cells in spite of the cocoons. A broken cocoon let appear the inside carbonate products (with permission from Castanier et al. 1999)

calcium-mediated precipitation. This alkaline barrier is primarily due to decomposition of anions rather than to production of alkali. Organisms which increase the medium pH by elimination of anions leave the scene ready for calcium precipitation. Rather than activities of single species or groups, activities of microbial communities should be considered in natural environments. Metabolic pathways involved in BCCP are described in Castanier et al. (1999); interactions of calcium cycle with carbon, nitrogen, and sulfur and the microbial groups involved are described by Zavarzin (2002).

Metabolic pathways involved in BCCP include autotrophic as well as heterotrophic pathways (both in aerobiosis and in anaerobiosis) with a different contribute.

Three main kinds of bacteria are involved in autotrophic production (Castanier et al. 1999): methanogenic archaebacteria, sulfurous or non-sulfurous, purple and green bacteria, and cyanobacteria. All obtain carbon from gaseous or dissolved CO_2, the origin of which is complex (atmosphere, eukaryotic and prokaryotic respiration, fermentation) and use it as carbon source to produce organic matter. These pathways induce local CO_2 depletion either of the medium or of the immediate environment of bacteria. When calcium ions are present in well-buffered alkaline or neutral media, such depletion favors CCP according to the overall reaction (5.1):

$$Ca^{2+} + 2HCO_3{}^- \leftrightarrow CaCO_3 + CO_2 + H_2O \tag{5.1}$$

Photosynthetic-induced calcification by cyanobacteria is regarded as the most common form of BCCP in aqueous environments such as marine and/or freshwater ones (Ehrlich 1998; McConnaughey and Whelan 1997). Cyanobacteria generate carbonate during consumption of bicarbonate in photosynthesis and thereby create alkaline surroundings (5.2–5.4), which favors the precipitation of carbonate by Ca^{2+} dissolved in water. According to Ehrlich (1998), the process is based on the metabolic utilization of dissolved CO_2, which exists in chemical equilibrium with $HCO_3{}^-$ and $CO_3{}^{2-}$ (5.2) in the medium surrounding the bacteria. This would induce a shift in the bicarbonate equilibrium and a subsequent pH rise in the bulk medium (5.3 and 5.4). Under such circumstances, precipitation could occur if soluble calcium ions are present.

$$2HCO_3{}^- \leftrightarrow CO_2 + CO_3{}^{2-} + H_2O \tag{5.2}$$

$$CO_2 + H_2O \rightarrow (CH_2O) + O_2 \tag{5.3}$$

$$CO_3{}^{2-} + H_2O \rightarrow HCO_3{}^- + OH^- \tag{5.4}$$

A well-known example of microorganism associated with this pathway is the cyanobacterium *Synechoccus* (Ehrlich 1998) which converts intracellular $HCO_3{}^-$ photosynthetically into reduced carbon (CH_2O) according to the reaction (5.5):

$$HCO_3{}^- + H_2O \rightarrow (CH_2O) + O_2 + OH^- \tag{5.5}$$

The intracellular OH^- is exchanged for extracellular HCO_3^- across the cell membrane. The now extracellular OH^- generates an alkaline pericellular region where CO_3^{2-} is generated from HCO_3^- (reaction 5.6):

$$HCO_3^- + OH^- \rightarrow CO_3^{2-} + H_2O \tag{5.6}$$

The resulting CO_3^{2-} immediately reacts with Ca^{2+} at the cell surface to form $CaCO_3$. The calcite deposits of *Synechoccus* can be in the form of marl sediment and massive bioherms. Other cyanobacteria lay down $CaCO_3$ as concretions around pebbles in streams.

In heterotrophy, Castanier et al. (1999) distinguished two bacterial carbonate precipitation-enhancing processes which may occur often concurrently: (1) ionic exchanges through the cell membrane (active); (2) metabolic pathways able to produce carbonate and bicarbonate ions and induce chemical modifications in the medium, mainly a pH increase (passive). According to Dupraz et al. (2009), all the processes derived from the activities of living cells are considered active, overcoming the previous distinction between active vs. passive processes. Process (1) is discussed in Sect. 5.3.2. Here we discuss process (2). Metabolic pathways belonging to the nitrogen and sulfur cycle are involved. In the nitrogen cycle, bacterial precipitation follows three different pathways: (1) ammonification of amino acids, (2) dissimilatory reduction of nitrate, (3) degradation of urea or uric acid. These pathways are more often associated with precipitation in soils and geological sediments, as well as precipitates linked to the urinary tract. These three pathways induce production of carbonate and bicarbonate ions and, as a metabolic end-product, ammonia, which induces a pH increase. When the H^+ concentration decreases, the carbonate–bicarbonate equilibria are shifted toward the production of CO_3^{2-} ions (reaction 5.7). If calcium ions are present, CCP occurs (reaction 5.8).

$$HCO_3^- \leftrightarrow CO_3^{2-} + H^+ \tag{5.7}$$

$$Ca^{2+} + CO_3^{2-} \leftrightarrow CaCO_3 \tag{5.8}$$

Urea or uric acid hydrolysis by the urease enzyme is a simple model and it has often been used in technological application of BCCP. The reaction takes place according to the following (5.9) and (5.10) (Wright 1999):

$$CO(NH_2)_2 + H_2O \rightarrow CO_2 + 2NH_3 \tag{5.9}$$

$$2NH_3 + CO_2 + H_2O \rightarrow 2NH_4^+ + CO_3^{2-} \tag{5.10}$$

In the sulfur cycle, the dissimilatory reduction of sulfate carried out by sulfate-reducing bacteria (SRB) is recognized to affect CCP. The environment must be anoxic and rich in organic matter, calcium, and sulfate. Using this pathway, bacteria produce carbonate, bicarbonate ions and hydrogen sulfide (H_2S). In presence of Ca^{2+} ions, CCP depends on the behavior of the hydrogen sulfide. Removal of the produced H_2S, and the resulting pH increase, is a prerequisite for carbonate precipitation to occur. In a natural environment, sulfide can be removed as gas or combined with iron to produce pyrite (Wright 1999), or it can be converted by anoxygenic sulfide phototrophic bacteria to elemental sulfur.

Precipitation via this pathway has been described in seawater, geological formations, during the biological treatment of acid mine drainage, in microbial mats (Hammes and Verstraete 2002). The reaction often starts with the dissolution of gypsum ($CaSO_4.2H_2O/CaSO_4$), which can be a pure physicochemical process (5.11). Under these circumstances, organic matter can be consumed by SRB, and sulfide and metabolic CO_2 is released (5.12).

$$CaSO_4.2H_2O \rightarrow Ca^{2+} + SO_4{}^{2-} + 2H_2O \tag{5.11}$$

$$2\ CH_2O + SO_4{}^{2-} \rightarrow H_2S + 2HCO_3{}^{-} \tag{5.12}$$

Interestingly, in several of the described examples for this pathway, dolomite and aragonite, instead of calcite, seem to be the most predominant forms of CC to precipitate (Hammes and Verstraete 2002).

The SRB *Desulfovibrio* can remove sulfates starting from gypsum in a process coupled to calcite production (Atlas and Rude 1988). Formation of crystals occurs by a combination of dissolution–precipitation and diffusion processes; calcium ions, released from gypsum when the bacteria reduce sulfates, react with carbon dioxide and result in the formation of calcite, according to the following proposed overall reaction (5.13):

$$6CaSO_4 + 4H_2O + 6\ CO_2 \rightarrow 6\ CaCO_3 + 4H_2S + 2S + 11O_2 \tag{5.13}$$

It is worth noting that these two simultaneous processes by *Desulfovibrio* have been exploited in bioremediation of monumental stones, both in removal of black gypsum crusts and, at the same time, in consolidating calcareous stone by calcite precipitation (Cappitelli et al. 2007).

In lithifying microbial communities, SRB have been recognized as key players in the precipitation of CC (Braissant et al. 2007). Seawater is supersaturated with respect to CC, but precipitation does not occur spontaneously due to various inhibiting factors. Metabolic activity of SRB, which are abundant in microbial mats, is believed to enhance CC precipitation. The activity of SRB affects the formation of carbonate minerals in several ways and constitutes a known example of how different mechanisms can concur to CCP by the same bacterial group. According to Braissant et al. (2007), first, sulfate reduction results in a pH increase, affecting the saturation index and therefore the precipitation of carbonate minerals. Second, when SRB use low-molecular-weight organic acids (e.g., lactate, acetate) as electron donors for growth, the availability of free calcium ions may increase due to the removal of carboxylic acids binding calcium. Third, by removing sulfate ions from solution, SRB alter the kinetic inhibition of dolomite formation. Through these processes, the metabolic activity of SRB can be considered as the environmental "engine" that sustains carbonate precipitation in lithifying microbial mats. In addition, the mere presence of SRB cells, even metabolically inactive, may favor CCP by providing heterogeneous nucleation sites. The role of SRB external polymers (EPS) in affecting precipitation and crystal growth is discussed in Sect. 5.4.

In complex natural environments, the different pathways from microbial communities above discussed can also combine to induce precipitation. A specific case of dolomite precipitation is described by Wright (1999): dead photosynthetic cyanobacteria, together with other organic matter, were degraded by SRB, which resulted in the production and release of DIC and ammonia. This was responsible for an increase in pH and carbonate levels, which eventually lead to precipitation.

Some authors consider metabolic activities of heterotrophic bacteria to be the most relevant mechanism in CCP. In heterotrophic bacterial communities, pathways of carbonate precipitation always appear to be a response to organic matter enrichment (Castanier et al. 1999).

According to Hammes and Verstraete (2002), the precipitation mechanisms mediated by metabolic pathways focus primarily on chemical changes induced in the macroenvironment surrounding the bacteria and are general in nature. This accounts for the common occurrence of BCCP and thereby, to some extent, neglects the relevance of the precipitation event in regard to the precipitating organism and its microenvironment. The latter aspect is discussed in Sect. 5.3.2.1.

5.3.2 Role of Calcium Carbonate Precipitation in Bacterial Metabolism

Although a lot of knowledge has been acquired on how metabolic pathways can influence calcium carbonate precipitation (Sect. 5.3.1), little is known on the possible role of calcium precipitation in bacterial metabolism. Among prokaryotes, there are no organisms with a specific function to precipitate CC, except for several cyanobacteria (Zavarzin 2002). Although there is no apparent known role for CC in bacterial metabolism (i.e., no calcite structures with a specific function seem to be built by prokaryotes), few exceptions of CCP established role are found in stromatolite formation (Erlich 1996; Barton et al. 2001; Sect. 5.2). Certain organisms precipitate calcite during their growth (Barton et al. 2001) and this suggests a defined role for CCP. Unique among bacteria, *Achromatium okaliferum* contains internal calcite inclusions during growth (Head et al. 1996).

This topic leads to a more general question on role of BCCP on bacterial physiology at the different scales of single bacterial cell, cell population, and microbial communities in a complex microenvironment. In this view, the question "how do bacteria precipitate CC" is strictly linked to "why do bacteria precipitate CC."

5.3.2.1 Role of Calcium Ions in the Cell and Bacterial Calcium Metabolism

Life evolved in an environment containing many different cations, including Ca^{2+}. Cations present during early evolution would also have participated in the random selection processes that eventually yielded the structural and functional components of cells. According to Smith (1995), the role that Ca^{2+} came to play in the early cells may have been conserved during subsequent evolution. The attributes of specific cations suited them to specific roles. Cations of related elements often are associated in pairs, with one needed for intracellular nutrition and the other not used intracellularly (Silver 1997). Unlike Mg^{2+}, which functions in many intracellular roles and must be transported inward and carefully regulated, Ca^{2+} frequently functions extracellularly, rather than intracellularly, in biological processes and is maintained at low intracellular levels by efflux transport pathways (Silver 1997).

Many of the functions that Ca^{2+} performs in eukaryotes may therefore be expected to be present in prokaryotes. The common evolutionary origin of the prokaryotes and eukaryotes and the many examples of evolutionary conservation of structure and function that have been shown to exist between them, support the concept that such evolutionary conservation should extend to the role of Ca^{2+} (Smith 1995).

Extensive investigations on Ca^{2+} in eukaryotic cells have shown its important role in signal transduction, as a signal transmitter in membrane depolarization events, as an intracellular second messenger and as an effector of actin–myosin contraction (for references, see Smith 1995). Relatively few systematic studies on the role of Ca^{2+} have been done in bacteria and its role is less well established. Recently, however, there has been an increased interest in the role of Ca^{2+} in prokaryotes and a number of investigations have reported data demonstrating a clear Ca^{2+} contribution to structure and regulatory functions of the prokaryotic cell (Norris et al. 1996; Smith 1995). The evidence in favor of Ca^{2+} as a mediator of regulatory phenomena in prokaryotes continues to accumulate. There is evidence that calcium is involved in a number of bacterial processes such as maintenance of cell structure, motility, cell division, gene expression, and cell differentiation processes such as sporulation, heterocyst formation, and fruiting body development (Dominguez 2004).

Since Ca^{2+} plays a pivotal role in numerous biological processes in both prokaryotes and eukaryotes, its intracellular concentration must be strictly regulated and maintained to constant values. The free intracellular Ca^{2+} concentration in bacteria is tightly regulated ranging from 100 to 300 nM, with similar values to those found in eukaryotic cells (Dominguez 2004), against large changes in external Ca^{2+}concentrations. This is also necessary to avoid toxic effects of free Ca^{2+} excess and irreversible damage to the cells, such as formation of calcium salts relatively insoluble. Thus the maintenance of low intracellular concentrations of calcium is essential both for the survival of an organism and for calcium to function as a secondary messenger. Since calcium was present during the early stages of evolution, there must have been some means to remove Ca^{2+} from the cell in order to prevent precipitation of cellular constituents. The ability to control Ca^{2+} content, or at least intracellular free Ca^{2+} concentration, may thus be a fundamental attribute of all cells. A wide array of mechanisms participate in attaining such a homeostatic condition. Calcium regulation in bacterial cells comprises influxes and effluxes dependent on active and passive transport mechanisms. Due to the fact that normal intracellular calcium concentrations are usually up to 10^3 times lower than extracellular concentrations, passive transport usually accounts for calcium influx (Norris et al. 1996). A net Ca^{2+} concentration gradient across the cell membrane is formed which by present-day values would be of the order of 10^{-3} to 10^{-4} M. This gradient has to be maintained by continuous exclusion of Ca^{2+} from the cell. The removal of Ca^{2+} by active extrusion requires energy to pump the Ca^{2+} against the electrochemical gradient. The metabolic apparatus that serves this function involves Ca^{2+} protein-based and non-proteinaceous channels, Ca^{2+} antiporters ($Ca^{2+}/2H^+$, Ca^{2+}/Na^+), and ATP-dependent Ca^{2+} pumps (Norris et al. 1996; Hammes and Verstraete 2002). ATP-driven calcium efflux appears to play a pivotal role in microbial calcium regulation (Naseem et al. 2009). Another way to remove Ca^{2+} is to bind it into a harmless form by Ca^{2+} chelating materials.

The possibility to form calcium minerals might be also seen as a tool to remove this cation, but a very few experimental studies are available to support it. Anderson et al. (1992) demonstrated that the ability of the soil bacterium *Pseudomonas fluorescens* to synthetize extracellular calcite in a calcium-stressed environment enabled the organism to regulate its cytosolic calcium content maintaining it at levels compatible with cell survival. The authors measured the calcium content in different cellular fractions of a *P. fluorescens* culture grown in a medium added with 10-mM calcium, during growth. As growth progressed, no significant increase in cytoplasmic calcium was recorded. After 16 h of growth, extracellular deposition of a calcium precipitate was observed which continued through to stationary phase and did not appear to have an inhibitory influence on cellular multiplication. Most of the calcium was sequestered in this precipitate, which was calcite with hexagonal crystal habits. No extracellular residue was evident if the culture medium was not inoculated, indicating that the metabolism of *P. fluorescens* was necessary for the formation of the calcite precipitate. While during the exponential phase of growth, the cytoplasmic calcium content appeared to be regulated by mechanism(s) other than biotransformation of the metal as an insoluble residue, calcite bioprecipitation could ensure a more efficient utilization of energy at the stationary phase of growth. The *P. fluorescens* crystalline $CaCO_3$ was then involved, according to the authors, in calcium homeostasis and allowed the organism to survive in the calcium-stressed medium by contributing to the maintenance of a low intracellular Ca^{2+} concentration. The authors concluded that the maintenance of innocuous levels of cytosolic calcium by the extracellular deposition of crystalline $CaCO_3$ is an uncommon occurrence, but we could suspect a larger occurrence and conservation of such a mechanism among bacteria. It is worth nothing that biomineralization as a metabolic detoxification process induced by increases in intracellular [Ca^{2+}] was proposed by Simkiss in 1977 (see Sect. 5.3.2.2).

As discussed in Sect. 5.2, some authors proposed a specific active involvement of bacteria in the CCP process. In particular, both Castanier et al. (1999) and McConnaughey and Whelan (1997) suggested that "active" precipitation could be linked to ion transport (specifically Ca^{2+}) across cellular membranes. Castanier et al. (1999) did not better precise the nature of these ionic exchanges, stating that they follow still poorly known mechanisms. McConnaughey and Whelan (1997) stated that calcification creates two types of products, minerals, and protons, according to (5.14):

$$Ca^{2+} + HCO_3^- \rightarrow CaCO_3 + H^+ \tag{5.14}$$

While production of calcareous skeletons by mineralization has an evident role in supporting and protecting the soft parts of many organisms, proton secretion is less obvious but plays major roles in carbon and nutrient assimilation by plants and photosynthetic symbioses. They postulated that such uses of calcification account for much of the massive carbonate accumulation in alkaline environments ranging from desert soils to coral reefs. The authors hypothesized that also photosynthetic microorganisms, such as algae and cyanobacteria, use active calcium metabolism coupled with photosynthetic CCP for the purpose of generating protons to assist the organisms in nutrient and bicarbonate uptake, according to (5.14) and (5.15).

$$H^+ + HCO_3^- \rightarrow CH_2O + O_2 \tag{5.15}$$

In alkaline waters, bicarbonate is the most abundant carbon source, but it should be inaccessible without a source of protons. By discharging the protons from calcification into their boundary layers, photosynthetic organisms can maintain or even elevate CO_2 concentrations despite CO_2 photosynthetic uptake, which gives to calcification a competitive advantage in light- and nutrient-deficient alkaline environments. In this view, the authors considered structural and defensive uses for calcareous skeletons sometimes overrated.

Some experimental support to this hypothesis can be found in the work of Yates and Robbins (1999), who found that extracellular precipitated calcium originated from inside the unicellular green alga *Nannochloris atomus*. The authors discussed that the incorporation of intracellular calcium into extracellular CC indicated that calcium expulsion from cells played a significant role in mineralization and raised the question of how Ca^{2+} cycling in cells could affect environments of CCP.

Based on McConnaughey and Whelan (1997), Hammes and Verstraete (2002) proposed an alternative view on the role of calcium metabolism in BCCP, based on the hypothesis that alkaline pH stress and subsequent bacterial calcium metabolism are key points in the precipitation process and on the relevance of precipitation in regard to the precipitating organism and its microenvironment. This microenvironment is constituted by the thin watery layer surrounding bacteria which forms an interface between the bacterial cell and the outside environment, where different concentrations of protons (pH), DIC, and Ca^{2+} can prevail. Coupled with the electronegative nature of the cell membranes (crystal nucleation sites, see Sect. 5.4), this can create a unique precipitation environment on microscale (Schultze-Lam et al. 1992). The proposed events are schematized in Fig. 5.3. A typical carbonate precipitation environment comprises high extracellular calcium concentrations (compared to intracellular) and low proton concentrations extracellular compared to intracellular (as a result of alkaline pH regimes). The combination of an extracellular alkaline pH and calcium ions is an inevitable stressful environment for bacteria: passive calcium influx as a result of the complementary $Ca^{2+}/2H^{+}$ electrochemical gradients will lead to intracellular calcium buildup and excessive proton expulsion (Fig. 5.3, Sect. A). At the cellular level, this event could be detrimental due to the (1) disruption of intracellular calcium-regulated signal processes, (2) alkalization of intracellular pH, and (3) depletion of the proton pool required for numerous other physiological processes (Norris et al. 1996). Survival under such conditions requires active export of intracellular calcium, e.g., via the ATP-dependent calcium pumps, which would reduce intracellular calcium ions and simultaneously compensate the proton loss (Fig. 5.3, Sect. B). The latter event could result in a localized increase in pH, due to proton uptake, in the same region as the calcium ions increase, which would form an ideal localized precipitation microenvironment (Fig. 5.3, Sect. C; 5.14). Even if spatial limitations restrict the formation of localized microenvironments, survival of the organism is dependent on active calcium metabolism. To accomplish this, energy is required, of which metabolic CO_2 is a natural by-product. Thus, survival and proliferation will lead to an increase in the extracellular [DIC] which would affect the solubility product of

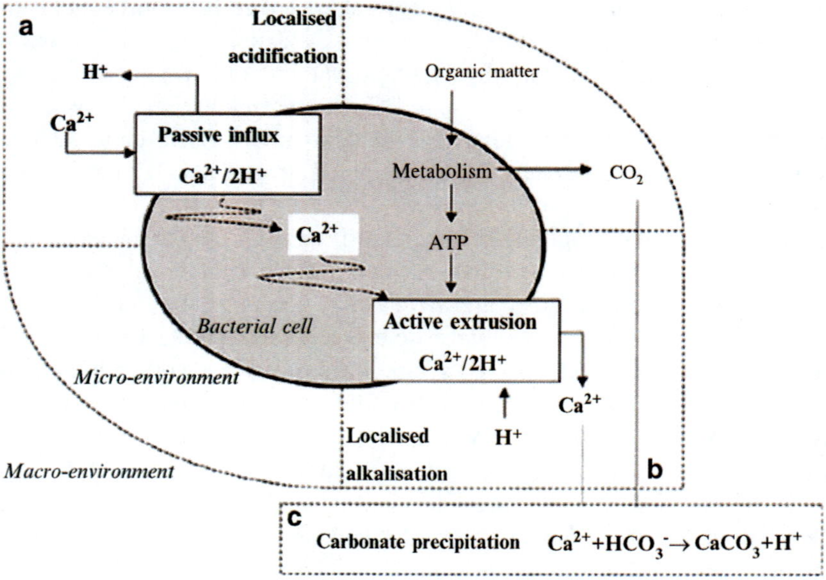

Fig. 5.3 Schematic presentation of suggested bacterial calcium metabolism and subsequent $CaCO_3$ precipitation under high-pH and high-Ca^{2+} extracellular conditions (with permission from Hammes and Verstraete 2002)

$CaCO_3$, eventually favoring precipitation. Moreover, carbonate precipitation will alter the extracellular environment according to 5.14 and Fig. 5.3, Sect. C (decrease soluble Ca^{2+} and increase acidity), so that it becomes more favorable for bacterial proliferation.

In this view, microbial calcium metabolism is an inevitable event under typical precipitation conditions, active calcium metabolism potentially creates unique precipitation conditions, and carbonate precipitation chemically favors bacterial survival and proliferation.

An active calcium metabolism controlling precipitation leads to the question if BCCP is under genetic control, i.e., specific molecules or mechanisms could be involved in controlling precipitation. In broad terms, all the metabolic pathways described in Sect. 5.3.1, which affect environmental conditions favoring precipitation, are under genetic control and manipulation of one of those pathways could enhance precipitation, for instance a higher urease activity increases calcite precipitation in *Sporosarcina pasteurii* (Achal et al. 2009).

To find genes which might specifically control calcite precipitation, the laboratory bacterium *Bacillus subtilis*, which can produce calcite precipitates, was used. The starting point was to look for mutant strains impaired in precipitation. Six mutants, impaired in calcite crystal formation, were isolated. In most cases, the putative function of the mutated genes was linked to fatty acid metabolism (Perito et al. 2000); moreover, mutations were found in *ysiB* and in the adjacent gene *ysiA*. Genes *ysiB* and *ysiA*, together with *lcfA*, *etfB*, and *etfA*, belong to a five-gene cluster, called *lcfA* operon. Further analysis of this cluster strongly suggested that the last gene, *etfA*, is essential for the precipitation phenotype (Barabesi et al. 2007). Putative functions assigned to the products of the *lcfA* operon genes are all involved in fatty acid metabolism.

The physiology behind the lack of precipitation in mutant FBC5 (*etfA* inactivated) was investigated, and compared to that of the wild-type strain *B. subtilis* 168, by Marvasi et al. (2010). Two main factors were considered as possibly affected by the *etfA* inactivation: EPS production (calcium-binding sites) and pH. EPS extracted from both strains revealed similar FT-IR patterns, with presence of bands typical of proteins and carbohydrates.

The pH decrease in FBC5 biofilm grown on solid medium was, on the contrary, the main process responsible for the absence of calcite formation. The phenomenon was reversible and crystal formation was restored when cells were incubated under alkaline conditions, suggesting that no specific inhibitors of precipitation were produced. The decrease of pH during FBC5 development resulted from an excess of H^+ production in the mutant; specifically, the H^+ extruded by the mutant was estimated to be 0.7 H^+ mole/L higher than in the wild-type strain. The excess of protons could result from a miss-regulation in some part of the EtfA pathway. Even though the function of the two EtfA-B proteins in *B. subtilis* is only putative, in *Clostridium acetobutylicum etfA* and *etfB* co-expression is essential for the butyryl-CoA dehydrogenase (BCD) activity, and *etfA-B* genes are involved in the oxidation of NADH in the butyryl-CoA synthesis pathway (Inui et al. 2008). If the reduction of β-hydroxybutyryl-CoA takes place in *B. subtilis* in the same way as in

C. acetobutylicum, an increase in NADH accumulation was expected. Experimental data showed cytoplasmic levels of NADH 37 times higher in the mutant than in the wild-type strain (Marvasi et al. 2010). Cytosolic accumulation of NADH could explain proton extrusion excess. If this deregulated proton extrusion might interfere with the active coupled $Ca^{2+}/2H^{+}$ transport through membrane involved in the precipitation model proposed by Hammes and Verstraete (2002; Fig. 5.3), it remains at the moment only a hypothesis.

5.3.2.2 Evolution and Meaning of CC Mineralization By Organisms

Since the early Precambrian, microbes have had an impact on the evolution of the Earth's surface, including the uppermost lithosphere and hydrosphere, as well as of the atmosphere (Ehrlich 1998). During the Precambrian, the calcium cycle (with both deposition and dissolution phenomena) was sustained by the prokaryotes, and played a key part in the emergence of the biosphere (Zavarzin 2002). The participation of bacteria in the geochemical calcium cycle is the most important factor maintaining neutral conditions on Earth. This cycle has profound influence on the fate of inorganic carbon, and, thereby, on the removal of CO_2 from the primitive atmosphere. Deposition of $CaCO_3$ minerals is one of the main processes of the biogeochemical calcium cycle. Most calcium deposits were formed in the Precambrian, when the prokaryotic biosphere predominated, as proved by the extensive development of stromatolite deposits. After that, calcium recycling based on biogenic deposition by skeletal organisms became the main process (Zavarzin 2002).

The ability of CC mineralization, as well as other biomineralization processes evolved then with prokaryotes (Ben Omar et al. 1997), even if apparently without a specific aim or function (Zavarzin 2002). The necessity to immobilize calcium outside the cell to avoid intracellular $[Ca^{2+}]$ rising to dangerous levels might have enabled bacteria to produce $CaCO_3$, as discussed above. Similarly to what might have occurred to Prokaryotes, calcium biomineralization could have occurred as an accident also to eukaryotes, which might have made a virtue of necessity. According to Brennan et al. (2004), marine shells would be a form of solid waste produced by marine organisms due to a geologic process which filled oceans with calcium. One of the most dramatic and confounding events in evolutionary and geologic history (an unresolved issue in Earth's history) was the sudden onset of biomineralization in the Early Cambrian during the "Cambrian explosion", i.e., the sudden onset of hard structures of insoluble calcium minerals built by organisms by biologically controlled calcification (see Sect. 5.2.1). Brennan et al. (2004) analyzed major ion composition of primary fluid inclusions (evaporated seawater) in Terminal Proterozoic (ca. 544 Ma) and Early Cambrian (ca. 515 Ma) marine halites and found that the major ion composition of seawater changed between 544 Ma and 515 Ma, highlighted by a large increase (threefold higher) in $[Ca^{2+}]$. The timing of this shift in seawater chemistry broadly coincides with the "Cambrian explosion", a brief drop in marine 87Sr/86Sr values, and an increase in tectonic activity, suggesting a link between the advent of biocalcification, hydrothermal mid-ocean-ridge brine production, and the composition of seawater. This increase in the seawater [Ca^{2+}] during the Early Cambrian may have created a chemical environment favorable for the initial development of CC and calcium phosphate hard parts, which have dominated marine biota ever since. Supporting this hypothesis, biocalcification in modern microorganisms can be affected by changes in seawater [Ca^{2+}] as well as the Mg^{2+}/Ca^{2+} ratio as observed in cyanobacteria (Sect. 5.2.3), coccolithophorids, and coralline algae. Calcifying cyanobacteria are abundant in the Phanerozoic rock record during times of high oceanic [Ca^{2+}] (Arp et al. 2001), and the first appearance of widespread calcifying cyanobacteria coincides with the proposed spike in the Early Cambrian.

During the Early Cambrian, the ambient oceanic [Ca^{2+}] may have risen to such a degree that cells could no longer effectively exclude or expel the Ca^{2+} ions, causing intracellular [Ca^{2+}] in marine organisms to reach toxic levels and triggering the metabolic changes that led to pervasive biocalcification, so spurring evolutionary changes and diversification in marine biota.

Even if hard shells and exoskeletons are considered defense mechanisms against predators, they would have occurred first as defense mechanisms against high levels of free Ca^{2+} (detoxification hypothesis according to Simkiss 1977).

5.4 Cell Surface Structures and Precipitation

Bacterial surfaces play an important role in CCP since they make exceptional interfaces for the precipitation of metal ions and the development of fine-grained minerals (Fortin et al. 1997). Bacterial surfaces, and consequently bacterial cells, can act as important sites for the absorption of cations and constitute particularly favorable templates for heterogeneous nucleation and crystal growth. Due to the presence of several negatively charged groups, at neutral pH, positively charged metal ions could be bound on bacterial surfaces, favoring heterogeneous nucleation (Fortin et al. 1997; Douglas and Beveridge 1998; Bäuerlein 2003).

Among surface structures, bacterial cell walls have been studied for their ability to complex metals, a process not well understood yet (Jiang et al. 2004). Negative charges result predominantly from deprotonation of carboxyl, phosphate, and hydroxyl functional groups exposed on the outer surface of the cell wall (Fein et al. 1997). The *B. subtilis* cell wall is one of the most studied for interactions with metals. Ca^{2+} ions are among the metals strongly bound to isolated walls of *B. subtilis*. Beveridge and Murray (1980) have predicted at least a two-step mechanism for the development of metal precipitation on the cell wall of *B. subtilis*. The first step is a stoichiometric interaction of metal with reactive chemical groups, which reside primarily in the peptidoglycan. After complexation, these sites nucleate the deposition of more metal as a chemical precipitate.

Surface reactivity of bacterial cells depends on the metabolic state of the cells (Jiang et al. 2004). In the absence of metabolic activity, passive interactions may occur in which microbial cells (inactive living or dead) behave as solid-phase sorbents of dissolved metals, and heterogeneous nucleation templates for authigenic mineral deposition (Beveridge 1989). Functional groups on the *B. subtilis* cell wall were considered able to bind dissolved Ca^{2+} in a calcite dissolution study with dead cells (Friis et al. 2003). *B. subtilis* dead cells, as well as a cell fraction comprising the cell wall, were demonstrated able to induce calcite formation in a laboratory test, acting as heterogeneous crystallization nuclei, with an applicative potential in the reinforcement of monumental calcareous stones (Barabesi et al. 2003). Precipitation occurring onto surface of inactive or dead cells as well as onto isolated cell fractions has to be considered as microbiologically influenced mineralization, according to Dupraz et al. (2009; Sect. 5.2.1).

Other bacterial cell surface components are known to favor precipitation and refer, in almost all cases, to extracellular polymeric substances or EPS.

EPS is a broad term that groups a large variety of organic polymers secreted by microbial cells in the environment. Polysaccharides are the major component of most EPS, commonly in combination with polypeptides, nucleic acids, phospholipids, and other polymeric compounds (Decho 2010). According to Costerton et al. (1995), the organic EPS matrix can be considered as an extension of the microbial cell. Microbial EPS can bind and accumulate ions. The interactions between metal ions and EPS are mediated by several functional groups (such as carboxylic acids or amino groups) present in the EPS matrix. Deprotonation of functional groups takes place when the pH increases, providing a negative charge to the polymer. In addition to sugar monomers, the EPS matrix may include noncarbohydrate acidic moieties such as sulfate or phosphate, which also contribute to the overall negative charge of the EPS (Braissant et al. 2007).

The role of EPS in CC nucleation and growth is well documented in laboratory experiments on bacterial cultures or isolated EPS (Braissant et al. 2007; Ercole et al. 2007; Tourney and Ngwenya 2009) as well as in natural environments such as microbial mats (Dupraz et al. 2009)

Bacterial EPS have been shown to be involved in the process of biocalcification by entrapping ions and serving as a nucleation site as well as by the action of specific proteins that influence precipitation and $CaCO_3$ polymorphism (Braissant et al. 2007; Kawaguchi and Decho 2002; Ercole et al. 2007; Tourney and Ngwenya 2009).

The S layer of the cyanobacterium *Synechoccus* acts as a template for fine-grain calcite formation by providing discrete, regularly arranged nucleation sites for the critical initial events in the mineralization process (Schultze-Lam et al. 1992). Pentecost and Bauld (1988) proposed that calcite deposition by cyanobacteria is initiated at sheath polymeric sites, on heteronuclei bound to sheath surfaces, or upon associated bacterial surfaces. The presence of cyanobacterial EPS also increases the viscosity of the medium, acting as a diffusion barrier, impacting calcium ions mobility and the kinetics of precipitation, and consequently the mineralogy of CC.

The role of EPS produced by SRB in $CaCO_3$ precipitation has been well documented in modern stromatolites and lithifying microbial mats, where SRB are closely associated with calcifying aragonitic micritic layers and high-Mg calcite micritic layers (Braissant et al. 2007).

According to Braissant et al. (2007), three main processes characteristic of EPS matrices have been shown to control the precipitation of carbonate minerals: (1) if calcium concentration exceeds the EPS-binding capacity under suitable pH conditions (i.e., pH >8.4), precipitation will occur inside the EPS matrix due to local super saturation; (2) self (re-)arrangement of acidic functional groups in the EPS matrix may create a template that favors the nucleation of carbonate minerals; (3) degradation of EPS by heterotrophic bacteria will contribute to the release of calcium, increasing the SI and enhancing carbonate precipitation.

Bontognali et al. (2008) have investigated mineralization within EPS in laboratory culture experiments carried out with *Desulfovibrio brasiliensis*, a SRB known to mediate dolomite formation under anoxic conditions. By combining CLSM and cryo-SEM imaging techniques, they showed that carbonate crystals nucleate as nanoforms within the EPS produced by the SRB, and not on bacterial cells, as previously proposed. Nanobacteria-like particles represent the early stage of carbonate nucleation within the EPS, which progressively evolve to larger globules displaying a grainy texture. In terms of microbial ecology, mineralization through excretions of EPS appears consistent with an expected self-preservation behavior of bacteria which remain prevalently mobile and are not entombed within the mineral. The authors suppose that mineralization through excretion of EPS could be a widespread process occurring both today and in the geological past.

According to Decho (2010), a common location for biopolymers, such as EPS, mediating biologically influenced precipitation (2.1), is the microbial "biofilm". Biofilms occur ubiquitously in a wide range of environments. Emerging evidence now suggests that the organic EPS matrix, which is an integral part of the microbial biofilms, plays a twofold role, either inhibiting or promoting carbonate formation, depending on the specific intrinsic (i.e., physicochemical) characteristics. The biofilm is also a microenvironment where precipitation can be facilitated or inhibited over a microspatial scale (i.e., μm to mm) and exhibit some spatial organization. Microorganisms, however, can accomplish a level of environmental control within the confines of the biofilm and EPS, e.g., in lithifying mats of marine stromatolites, may represent a primitive regulation of precipitation. The biofilm and its associated EPS matrix, therefore, serve as a useful starting point to investigate how organic molecules influence the precipitation process. According to Decho (2010), research on biopolymer-mediated precipitation is in its academic infancy, but it is poised to rapidly gain attention.

5.5 Conclusions

BCCP is a widespread process which bacteria carry out both in microbiologically induced (active) and/or in microbiologically influenced (passive) way (Dupraz et al. 2009).

Bacterial metabolic activities and cell surface structures and their interactions with environmental physicochemical parameters are commonly recognized as the key factors in BCCP. In complex natural environments and microbial communities, the different pathways and mechanisms can also combine to induce precipitation. The mechanisms of precipitation and the function of this process within both the cell physiology and the microbial ecology of the precipitating organism, however, still remain largely unresolved.

To understand how bacteria can affect the phenomenon at the geological scale, it is necessary to understand how BCCP occurs at the microscale, i.e., of the single cell. Understanding the molecular-scale events of BCCP is also necessary for applied purposes (Decho 2010).

A possible function of BCCP could be seen as a response of precipitating organisms to a stressful (micro)environment for calcium ions (Anderson et al. 1992). Intracellular calcium concentration must be maintained at low levels in the prokaryotic as well as in the eukaryotic cell, because of calcium's key role in the regulation of many fundamental processes and its potential harmfulness for cell structures (Smith 1995). The necessity to immobilize calcium outside the cell to avoid intracellular [Ca^{2+}] rising to dangerous levels might have enabled Prokaryotes, as well as Eukaryotes, to produce $CaCO_3$ (Brennan et al. 2004). Improved understanding of controlled, induced, and influenced types of CC mineral formation may reveal many common chemical and structural characteristics, even if the three processes are different (Dupraz et al. 2009).

BCCP as a detoxification mechanism for cell survival would require active mechanisms by the bacterial cells. A possible model of the role of active calcium metabolism in BCCP was proposed by Hammes and Verstraete (2002).

Further work has still to be done to better elucidate the genetic control of BCCP (Barabesi et al. 2007) as well as the molecular mechanisms acting at the cell microscale.

References

Achal V, Mukherjee A, Basu PC, Sudhakara Reddy M (2009) Strain improvement of Sporosarcina pasteurii for enhanced urease and calcite production. J Ind Microbiol Biotechnol 36:981–988

Anderson S, Appanna VD, Huang J, Viswanatha T (1992) A novel role for calcite in calcium homeostasis. FEBS Lett 308:94–96

Arp G, Reimer A, Reitner J (2001) Photosynthesis-induced biofilm calcification and calcium concentrations in Phanerozoic oceans. Science 292:1701–1704

Atlas RC, Rude PD (1988) Complete oxidation of solid phase sulfides by manganese and bacteria in anoxic marine sediment. Geochim Cosmochim Acta 52:751–766

Barabesi C, Galizzi A, Mastromei G, Rossi M, Tamburini E, Perito B (2007) Bacillus subtilis gene cluster involved in calcium carbonate biomineralization. J Bacteriol 189:228–235

Barabesi C, Salvianti F, Mastromei G, Perito B (2003) Microbial calcium carbonate precipitation for reinforcement of monumental stones. In: Saiz-Jimenez C (ed) Molecular biology and cultural heritage. AA Balkema Publishers, Lisse, The Netherlands, pp 209–212

Barton HA, Spear JR, Pace NR (2001) Microbial life in the underworld: biogenicity in secondary mineral formations. Geomicrobiol J 18:359–368

Bäuerlein E (2003) Biomineralization of unicellular organisms: an unusual membrane biochemistry for the production of inorganic nano- and microstructures. Angewandte Chemie International Edition 42:614–641

Bäuerlein E (2004) Biomineralization. Progress in biology, molecular biology and application. WILEY-VHC Verlag GmbH & Co KgaA, Weinheim

Ben Omar N, Arias JM, Gonzalez-Munoz MT (1997) Extracellular bacterial mineralization within the context of geomicrobiology. Microbiologia 13:161–172

Beveridge TJ (1989) Role of cellular design in bacterial metal accumulation and mineralization. Annu Rev Microbiol 43:147–171

Beveridge TJ, Murray RGE (1980) Sites of metal deposition in the cell wall of Bacillus subtilis. J Bacteriol 141:876–887

Bontognali TRR, Vasconcelos C, Warthmann RJ, Dupraz C, Bernasconi SM, McKenzie JA (2008) Microbes produce nanobacteria-like structures, avoiding cell entombment. Geology 36:663–666

Boquet E, Boronat A, Ramos-Cormenzana A (1973) Production of calcite (calcium carbonate) crystals by soil bacteria is a general phenomenon. Nature 246:527–529

Braissant O, Decho AW, Dupraz C, Glunk C, Przekop KM, Visscher PT (2007) Exopolymeric substances of sulfate-reducing bacteria: interactions with calcium at alkaline pH and implication for formation of carbonate minerals. Geobiology 5:401–411

Brennan ST, Lowenstein TK, Horita J (2004) Seawater chemistry and the advent of biocalcification. Geology 32:473–476

Cappitelli F, Toniolo L, Sansonetti A, Gulotta D, Ranalli G, Zanardini E, Sorlini C (2007) Advantages of using microbial technology over traditional chemical technology in removal of black crusts from stone surfaces of historical monuments. Appl Environ Microbiol 73:5671–5675

Castanier S, Métayer-Levrel L, Perthuisot J-P (1999) Ca-carbonates precipitation and limestone genesis-the microbiologist point of view. Sedimentary Geology 126:9–23

Costerton JW, Lewandowski Z, Caldwell DE, Korber DR, Lappin-Scott HM (1995) Microbial biofilms. Annu Rev Microbiol 49:711–745

De Muynck W, De Belie N, Verstraete W (2010) Microbial carbonate precipitation in construction materials: a review. Ecological Engineering 36:118–136

Decho AW (2010) Overview of biopolymer-induced mineralization: what goes on in biofilms? Ecological Engineering 36:137–144

Dominguez DC (2004) Calcium signalling in bacteria. Mol Microbiol 54:291–297

Douglas S, Beveridge TJ (1998) Mineral formation by bacteria in natural microbial communities. FEMS Microbiol Ecol 26:79–88

Dupraz C, Reid RP, Braissant O, Decho AW, Norman RS, Visscher PT (2009) Process of carbonate precipitation in modern microbial mats. Earth Sci Rev 96:141–162

Ehrlich HL (1996) Geomicrobiology, 3rd edn. Marcel Dekker, New York

Ehrlich HL (1998) Geomicrobiology: its significance for geology. Earth Sci Rev 45:45–60

Ercole C, Cacchio P, Botta AL, Centi V, Lepidi A (2007) Bacterially induced mineralization of calcium carbonate: the role of exopolysaccharides and capsular polysaccharydes. Microsc Microanal 13:42–50

Fein JB, Daughney CJ, Yee N, Davis TA (1997) A chemical equilibrium model for metal adsorption onto bacterial surfaces. Geochim Cosmochim Acta 61:3319–3328

Folk RL (1993) SEM imaging of bacteria and nanobacteria in carbonate sediments and rocks. J Sedim Petrol 63:990–999

Fortin D, Ferris FG, Beveridge TJ (1997) Surface-mediated mineral development by bacteria. Rev Mineral 35:161–180

Friis AK, Davis TA, Figueira MM, Paquette J, Mucci A (2003) Influence of bacillus subtilis cell walls and EDTA on calcite dissolution rates and crystal surface features. Environ Sci Technol 37:2376–2382

Hammes F, Verstraete W (2002) Key role of pH and calcium metabolism in microbial carbonate precipitation. Rev Environ Sci Biotechnol 1:3–7

Head IM, Gray ND, Clarke KJ, Pickup RW, Jones JG (1996) The phylogenetic position and ultrastructure of the uncultured bacterium *Achromatium okaliferum*. Microbiology 142: 2341–2354

Inui M, Suda M, Kimura S, Yasuda K, Suzuki H, Toda H, Yamamoto S, Okino S, Suzuki N, Yukawa H (2008) Expression of Clostridium acetobutylicum butanol synthetic genes in Escherichia coli. Appl Microbiol Biotechnol 77:1305–1316

Jiang W, Saxena A, Bongkeun S, Ward BB, Beveridge TJ, Myneni CB (2004) Elucidation of functional groups on Gram-positive and Gram-negative bacterial surfaces using infrared spectroscopy. Langmuir 20:11433–11442

Kawaguchi T, Decho AW (2002) A laboratory investigation of cyanobacterial extracellular polymeric secretions (EPS) in influencing $CaCO_3$ polymorphism. J Crystal Growth 240: 230–235

Little BJ, Wagner PA, Lewandowski Z (1997) Spatial relationship between bacteria and mineral surfaces. Rev Mineral 35:123–159

Lowenstam HA, Weiner S (1989) On biomineralization. Oxford University Press, Oxford

Mann S (2001) Biomineralization. Oxford University Press, New York

Marvasi M, Visscher PT, Perito B, Mastromei G, Casillas-Martinez L (2010) Physiological requirements for carbonate precipitation during biofilm development of Bacillus subtilis etfA mutant. FEMS Microbiol Ecol 71:341–350

McConnaughey TA, Whelan JF (1997) Calcification generates protons for nutrient and bicarbonate uptake. Earth Sci Rev 42:95–117

Murray J, Irvine R (1889–1890) On coral reefs and other carbonate of lime formations in modern seas. Proc Roy Soc Lond A 17:79–109

Naseem R, Wann KT, Holland IB, Campbell AK (2009) ATP regulates calcium efflux and growth in E. coli. J Mol Biol 391:42–56

Norris V, Grant S, Freestone P, Canvin J, Sheikh FN, Toth I, Trinei M, Modha K, Norman RI (1996) Calcium signalling in bacteria. J Bacteriol 178:3677–3682

Pentecost A, Bauld J (1988) Nucleation of calcite on the sheaths of cyanobacteria using a simple diffusion cell. Geomicrobiol J 6:129–135

Perito B, Biagiotti L, Daly S, Galizzi A, Tiano P, Mastromei G (2000) Bacterial genes involved in calcite crystal precipitation. In: Ciferri O, Tiano P, Mastromei G (eds) Of microbes and art: The role of microbial communities in the degradation and protection of cultural heritage. Plenum Publisher, New York, pp 219–230

Phoenix VR, Konhauser KO (2008) Benefits of bacterial biomineralization. Geobiology 6: 303–308

Rivadeneyra MA, Delgado R, del Moral A, Ferrer MR, Ramos-Cormenzana A (1994) Precipitation of calcium carbonate by *Vibrio* spp. from an inland saltern. FEMS Microbiol Ecol 13: 197–204

Rivadeneyra MA, Delgado G, Ramos-Cormenzana A, Delgado R (1998) Biomineralization of carbonates by Halomonas eurihalina in solid and liquid media with different salinities: crystal formation sequence. Res Microbiol 149:277–287

Rodriguez-Navarro C, Rodriguez-Gallego M, Ben Chekroun K, Gonzalez-Muñoz MT (2003) Conservation of ornamental stone by Myxococcus xanthus-induced carbonate biomineralization. Appl Environ Microbiol 69:2182–2193

Schultze-Lam S, Harauz G, Beveridge TJ (1992) Partecipation of a cyanobacterial S layer in fine-grain mineral formation. J Bacteriol 174:7971–7981

Silver S (1997) The bacterial view of the periodic table: specific functions for all elements. Rev Mineral 35:345–360

Simkiss K (1977) Biomineralization and detoxification. Calcif Tiss Res 24:199–200

Smith RJ (1995) Calcium and bacteria. Adv Microb Physiol 37:83–133

Tourney J, Ngwenya BT (2009) Bacterial extracellular polymeric substances (EPS) mediate $CaCO_3$ morphology and polymorphism. Chem Geol 262:138–146

Vasconcelos C, McKenzie JA, Bernasconi S, Grujic D, Tien AJ (1995) Microbial mediation as a possible mechanism for natural dolomite formation at low temperatures. Nature 377:220–222
von Knorre H, Krumbein WE (2000) Bacterial calcification. In: Riding RE, Awramik SM (eds) Microbial sediments. Springer, Berlin Heidelberg, pp 25–31
Weiner S, Dove PM (2003) An overview of biomineralization and the problem of the vital effect. Am Rev Mineral Geochem 54:1–31
Wright DT (1999) The role of sulphate-reducing bacteria and cyanobacteria in dolomite formation in distal ephemeral lakes of the Coorong region, South Australia. Sediment Geol 126 (1–4):147–157
Yates KK, Robbins LL (1999) Radioisotope tracer studies of inorganic carbon and Ca in microbiologically derived $CaCO_3$. Geochim Cosmochim Acta 63(1):129–136
Zamarreño DV, Inkpen R, May E (2009) Carbonate crystals precipitated by freshwater bacteria and their use as a limestone consolidant. Appl Environ Microbiol 75:5981–5990
Zavarzin GA (2002) Microbial geochemical calcium cycle. Microbiology: a translation of Mikrobiologiya 71:1–17

Chapter 6
Principles of Calcium-Based Biomineralization

Qingling Feng

Contents

Abstract The chapter provides some basic information on the formation principles of calcium carbonate in biological systems in marine environment in the point of view of materials science in order to provide strategies for biomimetic design and preparation of new functional materials. Many researchers try to explain the principles of biomineralization and get some valuable conclusions. This chapter introduces some calcium-based biominerals in aquatic organisms which mainly

Q. Feng (✉)
Department of Materials Science and Engineering, Tsinghua University, Beijing 100084, China
e-mail: biomater@mail.tsinghua.edu.cn

W.E.G. Müller (ed.), *Molecular Biomineralization*, Progress in Molecular and Subcellular Biology 52, DOI 10.1007/978-3-642-21230-7_6,

include calcium carbonate and calcium phosphate. Then it gives a presentation of the hierarchical structure of calcium carbonate-based and calcium phosphate-based biominerals, e.g., mollusc shell, pearl, carp otolith, tooth, and bone. Moreover, the chapter explains the principles of calcium carbonate mineralization from the aspects of the effects of additives and templates; it also gives some explanations to the principles of calcium phosphate mineralization.

Biomineralization involves the selective extraction and uptake of elements from the local environment and their incorporation into functional structures under strict biological control. The formation of hard bioinorganic materials such as bones and shells is univocally recorded in the fossil record; the biological processes were involved in inorganic mineralization dating as far back as 3,500 million years. Moreover, the fossils contain a record not only of the distant biology but also of the local climate and chemical conditions of the marine environment in the history (Mann 2001).

(Lowenstam and Weiner 1989) inaugurated modern studies on mineralization with the theory of "minerals formed by organisms," emphasized the important character of macromolecules in the mineralization process, also pointed out the differences between "biological controlled mineralization" and "biologically induced mineralization". Compared with an inorganic mineral, which is hard and stiff but brittle, biominerals are relatively soft, compliant but tough, so there is much to be gained in the mechanical design of life if the "organic toughening" is combined with the "inorganic strength." Biomineralization offers an organism more than just structural support and mechanical strength. As nature's master builder, it is involved in a wide variety of important biological functions such as: protection, motion, cutting and grinding, buoyancy, storage, optical, magnetic and gravity sensing. In summary, the big picture of biomineralization is one that contains many different subjects and perspectives, ranging from the global aspects of the earth sciences to the local niches of biology and the selection pressures on materials design, and to the anatomy of tissues and the microscopic world of cells.

The well-designed morphologies and hierarchical structure make biomaterials very attractive to the researchers. The architecture emerging from self-organization under ambient conditions provides a sophisticated model for materials science. The design of nanostructure materials with tailored morphologies, such as particles, rods, wires, tubes, and sheets, has attracted much interest because of their potential applications. Controlled assembly into a three-dimensional architecture is an important challenge in the broad application of the materials. Many biomaterials structure, such as nacre, abalone, ivory, human bone, tooth, etc., have been investigated. These materials, exhibiting amazing morphology and structure, have great potential as functional materials. The structural study of the fascinating biomaterials is important in biology, and may provide novel ideas for the design of synthetic materials. The structural investigation of the biomaterials can also consummate the biomineralization theory (Oaki & Imai 2005).

6.1 Calcium-Based Biominerals in Aquatic Organisms

There are over 60 different biological minerals in nature. H, C, O, Mg, Si, P, S, Ca, Mn, and Fe are common constituents of the 20–25 essential elements required by living organisms. Over half of the elements essential for life are incorporated in biomineral deposits. Of these, calcium is distinguished in being widespread and the common constituent of bones, teeth, and shells. It is interesting to note that bones are composed of calcium phosphate, while shells are built from calcium carbonate. The reasons for this significant difference are still not known. In both cases, however, the inorganic mineral is intimately associated with a complex assemblage of organic macromolecules, the organic matrix, is of fundamental importance.

Calcium carbonate and calcium phosphate minerals have high lattice energies and low solubilities, and are therefore thermodynamically stable within biological environments. In contrast, hydrated phases, such as calcium oxalate and calcium sulfate, are much more soluble and therefore less common. In general, the precipitation of calcium salts provides an effective means to control the Ca^{2+} ion concentration in biological fluids. This helps to maintain a steady-state condition corresponding to an intracellular calcium concentration of around 10^{-7} M (Mann 2001).

6.1.1 Calcium Carbonate

There are six calcium carbonate minerals with the same principal composition but different structure: calcite, aragonite, vaterite, calcium carbonate monohydrate, calcium carbonate hexahydrate, and amorphous calcium carbonate. But only the two most thermodynamically stable structures (calcite and aragonite) are deposited extensively as biominerals (Table 6.1). Magnesium ions are readily accommodated in the calcite lattice; many biological calcites also contain Mg^{2+} ions up to levels of 30 mol%.

6.1.1.1 Calcite and Aragonite

Mollusc Shell

Although molluscs build shells with all sorts of shapes and sizes that are generally conservative when it comes to the choice of mineral used, shells are made pure and simply of calcium carbonate usually in the form of calcite and aragonite. Interestingly, many types of seashell contain both calcite and aragonite but the minerals are spatially separated in distinct parts of the shells.

Table 6.1 Calcium carbonate biominerals (Mann 2001)

Mineral	Formula	Organism	Location	Function
Calcite	$CaCO_3$	Coccolithophores	Cell wall scales	Exoskeleton
		Foraminifera	Shell	Exoskeleton
		Trilobites	Eye lens	Optical imaging
		Molluscs	Shell	Exoskeleton
		Crustaceans	Crab cuticle	Mechanical strength
		Birds	Eggshells	Protection
		Mammals	Inner ear	Gravity receptor
Mg-calcite	$(Mg, Ca)CO_3$	Octocorals	Spicules	Mechanical strength
		Echinoderms	Shell/spines	Strength/protection
Aragonite	$CaCO_3$	Scleractinian corals	Cell wall	Exoskeleton
		Molluscs	Shell	Exoskeleton
		Gastropods	Love dart	Reproduction
		Cephalopods	Shell	Buoyancy device
		Fish	Head	Gravity receptor
Vaterite	$CaCO_3$	Gastropods	Shell	Exoskeleton
		Ascidians	Spicules	Protection
Amorphous	$CaCO_3 \cdot nH_2O$	Crustaceans	Crab cuticle	Mechanical strength
		Plants	Leaves	Calcium store

Usually, the outer layer (prismatic layer) of the shell, as well as the growing edge, consists of large crystals of calcite, whereas the inner region (nacre) is built from a "brick wall" of plate-like aragonite crystals (Fig. 6.1). The prismatic layer is first deposited and then the nacre is added with time as the shell grows in thickness. The remarkable switching between the calcium carbonate polymorphs is controlled by a layer of closely packed cells called the outer epithelium that is separated from the inner shell surface by a space filled with an aqueous solution – the extrapallial space and fluid, respectively. But the exact nature of how this process works is not known. The calcite–aragonite switch is of key functional importance throughout biomineralization and has been the research focus for decades.

The nacreous layer (mother-of-pearl) is a laminate of 0.5-μm thick aragonite polygonal tablets sandwiched between thin (approximately 30 nm) sheets of a protein–polysaccharide organic matrix. The matrix plays a key role in limiting the thickness of the crystals and is structurally important in the mechanical design of the shell. It reduces the number of voids in the shell wall and so inhibits crack propagation by dissipating the energy; associated with an expanding defect along the organic layers rather than through the inorganic crystals. This makes nacre approximately 3,000 times as tough as inorganic aragonite.

Many marine single-celled organisms produce elaborate externally mineralized structures that serve as all "exoskeleton" in which the organism lives. Many of the most fascinating structures are found in a group of marine algae commonly known as coccolithophores. The mineralized shells (coccospheres) consist of wonderfully sculpted calcite plates called coccoliths, with additional decorative features such as long trumpet-shaped spines.

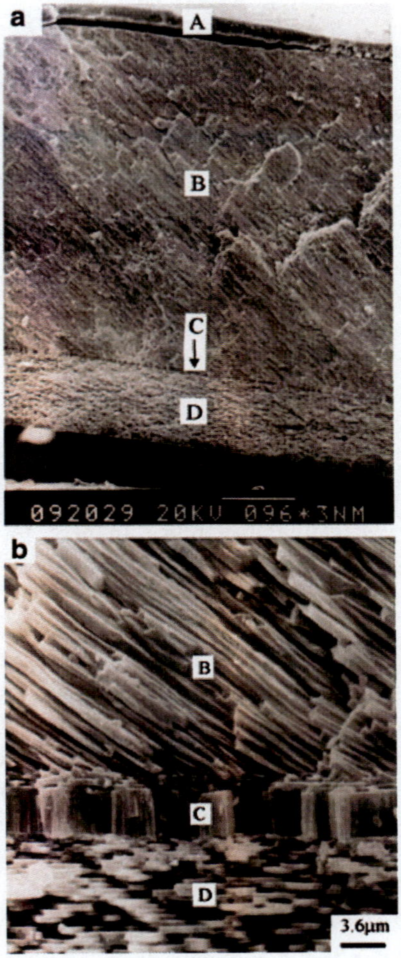

Fig. 6.1 (**a**) SEM morphology of the cross section of the *Mytilus edulis* shell (A: periostracum, B: oblique prismatic layer discussed in this paper, C: normal prismatic layer, D: nacreous layer); and (**b**) amplified image of (**a**) showing the detailed structure (Feng et al. 2000a)

Sensor

Calcite and aragonite are also used as gravity sensors in sea animals. These devices (generally referred to as statoliths, statoconia, otoliths, or otoconia) function in a similar way to the fluid in the semicircular canals (which detect changes in angular momentum). In the human ear, the crystals are made of calcite, spindle-shaped, and sited on a specialized membrane under which sensory cells are located. During a change in linear acceleration, the movement of the crystal mass relative to the delicate hair-like extensions of the cells results in the electrical signaling of the applied force to the brain.

The otoliths are calcium carbonate concretions in the inner ear of fish, which act as sound transducers and play an important role in the teleost fish balance system. Figure 6.2 shows the position of the otoliths in fish. Pannella discovered otolith microincrements (Pannella 1971), which led to an increasing number of studies

Fig. 6.2 (**a**) The digital camera graph of the wild carp and (**b**) the anatomy schematic illustrations of the otoliths

using microincrements as records of fishes' daily growth. The existence of daily increments in otolith is believed to be a widespread phenomenon among fish according to Campana and Neilson (1985). They considered that the otolith was composed of an incremental zone and a discontinuous zone. The former zone is wide and transparent, composed of calcium carbonate. The latter zone is narrow and opaque, composed of organic matrix. Current research works are mainly restricted to the field of the daily ring structure. Understanding microstructure of otolith is important for studying the larvae and juvenile fish. Campana and Neilson (1985) studied the ecological effect on otolith's microstructure, including the early life history, age, growth, recruitment, migration, mortality, stock structure, and recording the historical life information. Otolith microstructure examination and analysis could be used to discriminate populations, to show the feeding and growth history of fish experienced in wild or reared conditions in the elapsed days. The microstructure of otolith is related to many factors. First of all, the environmental conditions, such as water temperature, water enrichment elements, food supply, etc., could influence the increment width, contrast, transparence, and morphology. The otolith's growth and structure are also affected by proteins. Sollner et al. reported a gene, starmaker, which can control the crystal lattice structure and the shape of the otolith (Sollner et al. 2003).

Calcitic Microlenses in Brittlestar

Photosensitivity in most echinoderms has been attributed to "diffuse" dermal receptors. Aizenberg et al. (2001) reported that certain single calcite crystals used by brittlestars for skeletal construction are also a component of specialized photosensory organs, conceivably with the function of a compound eye. The analysis of arm ossicles in *Ophiocoma* showed that in light-sensitive species, the periphery of the labyrinthic calcitic skeleton extends into a regular array of spherical microstructures that have a characteristic double-lens design. These structures are absent in light-indifferent species. Photolithographic experiments in which a photoresist film was illuminated through the lens array showed selective exposure of the photoresist under the lens centers. These results provide experimental evidence that the microlenses are optical elements that guide and focus the light inside the tissue. The estimated focal distance (4–7 μm below the lenses) coincides with the location of nerve bundles, the presumed primary photoreceptors. The lens array is designed to minimize spherical aberration and birefringence and to detect light from a particular direction. The optical performance is further optimized by phototropic chromatophores that regulate the dose of illumination reaching the receptors. These structures represent an example of a multifunctional biomaterial that fulfills both mechanical and optical functions.

Echinoderms in general, and especially the brittlestars (Ophiuroidea), exhibit a wide range of responses to light intensity, from a largely light-indifferent behavior to pronounced color change and rapid escape behavior (Aizenberg et al. 2001). Figure 6.3 compares the appearance and the skeletal structure of two species of *Ophiocoma*, which represent the two extreme photosensitivity types. *Ophiocoma pumila* (Fig. 6.3a) shows no color change and little reaction to illumination. *Ophiocoma wendtii* is a highly photosensitive species, and it changes color

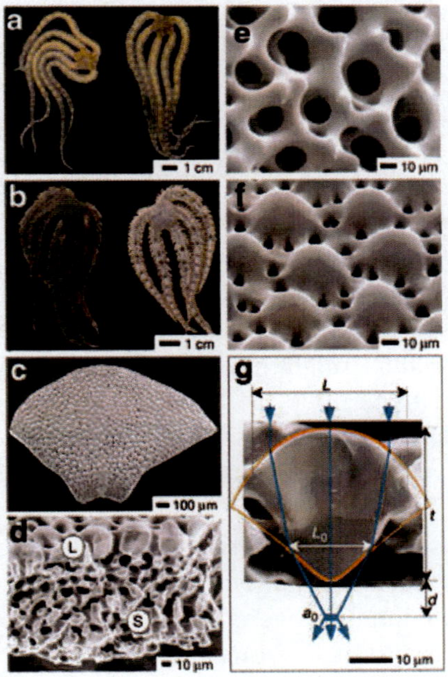

Fig. 6.3 Appearance and skeletal structure of ophiocomid brittlestars (Aizenberg et al. 2001). **(a)** Light-indifferent species *Ophiocoma pumila* shows no color change from day (*left*) to night (*right*). **(b)** Light-sensitive species *O. wendtii* changes color markedly from day (*left*) to night (*right*). **(c)** SEM of a dorsal arm plate (DAP) of *O. wendtii* cleansed of organic tissue. **(d)** SEM of the cross section of a fractured DAP from *O. wendtii* showing the typical calcitic stereom (S) and the enlarged lens structures (L) that constitute the peripheral layer. **(e)** SEM of the peripheral layer of a DAP of *O. pumila* showing that it lacks the enlarged lens structures. **(f)** SEM of the peripheral layer of a DAP from *O. wendtii* with the enlarged lens structures. **(g)** High-magnification SEM of the cross section of an individual lens in *O. wendtii*. Red lines represent the calculated profile of a lens compensated for spherical aberration. The operational part of the calcitic lens (L_0) closely matches the profile of the compensated lens (*bold red lines*). The light paths are shown in *blue*

markedly, from homogeneous dark brown during the day (Fig. 6.3b, *left*) to banded gray and black at night (Fig. 6.3b, *right*). Another conspicuous behavioral response to light is negative phototaxis: *O. wendtii* is able to detect shadows and quickly escape from predators into dark crevices, of which they are able to identify from several centimeters away. The latter reaction is particularly unexpected in these animals as the behavior is usually associated with the presence of discrete photosensory organs. No specialized eyes have, however, been documented in brittlestars and their reactions to light have been linked to diffuse dermal receptors.

The sensitivity to light seems to correlate with the specialized skeletal structure of the dorsal arm plates (DAPs). These ossicles protect the upper part of each joint in brittlestar arms (Fig. 6.3c). Skeletal elements of echinoderms are each composed of a single crystal of oriented calcite shaped into a unique, three-dimensional mesh (stereom). The diameter of the typical stereom in the DAPs of *Ophiocoma* is about 10–15 μm (Fig. 6.3d). In *O. wendtii* as well as in other photosensitive species, the outer surface of the DAP stereom bears a characteristic array of enlarged spherical structures 40–50 μm in diameter (Fig. 6.3d, f). In cross section, they have a remarkably regular double-lens shape (Fig. 6.3g). The optical axis of the constituent calcite is oriented parallel to the lens axis and perpendicular to the plate surface. The mean geometry of the lenses was inferred from the measurements of lens diameter (L) and thickness (t) in 20 random lenses sectioned through the center (Fig. 6.3g).

The absence of such structure on the DAPs in various relatively light-insensitive species, such as *O. pumila* (Fig. 6.3a, e) raises the possibility of the direct involvement of calcitic microlenses in photoreception. Calcitic microlenses were used by the trilobites. The presence of transparent regions of compact stereom has been reported also for sea stars and sea urchins. Aizenberg et al. (2001) proposed that these calcitic microstructures might have a function in directing and focusing the light on photosensitive tissues. On the basis of the results, they suggest that the array of calcitic microlenses with their unique focusing effect and underlying neural receptors may form a specialized photoreceptor system with a conceivable compound-eye capability.

The demonstrated use of calcite by brittlestars, both as an optical element and as a mechanical support, illustrates the remarkable ability of organisms, through the process of evolution, to optimize one material for several functions, and provides new ideas for the fabrication of smart materials.

6.1.1.2 Vaterite

Calcium carbonate, in its three non-hydrated crystalline polymorphs calcite, aragonite and vaterite, represents one of the most important inorganic materials with respect to the biomineralization processes in organisms. Of the three modifications, calcite is thermodynamically the most stable one, followed by aragonite and then vaterite, which is the least stable one. Vaterite is highly unstable, its crystal system is hexagonal, P63/mmc. When exposed to water, it can recrystallize to calcite

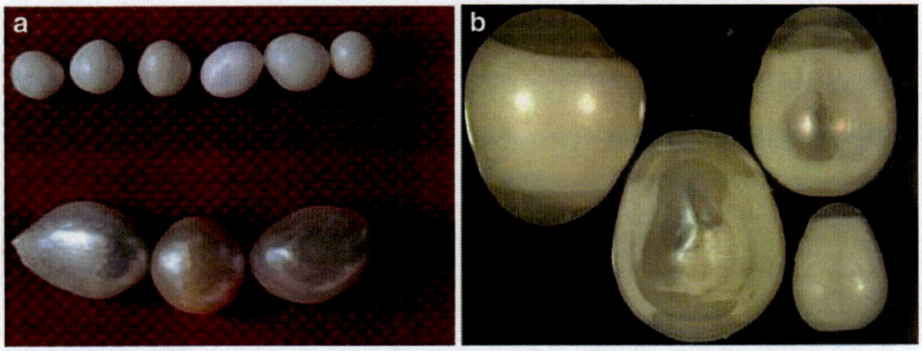

Fig. 6.4 Photos of freshwater pearls (**a**) vaterite pearls (*top*) and aragonite pearls (*bottom*) (**b**) semi-matt pearls with vaterite and aragonite

easily. There are a few reports of vaterite minerals precipitated in aquatic organisms. It occurs as elaborately shaped spicules in marine creatures called ascidians (the majority of calcareous sponges have magnesium-rich calcite spines), where it possibly acts as a structural support or as a deterrent against predators. Vaterite has also been observed in the inner ears of two species of fish. Some investigations showed that vaterite can exist in some extreme conditions with the precise control of pH, temperature, and pressure (Vecht & Ireland 2000). Even when vaterite is produced with organic additives or organic templates in lab, it is still difficult for vaterite to resist aqueous solutions or higher temperatures (Kanakis & Dalas 2000). Interestingly, living organisms can produce more vaterite in biomineralization processes, such as gallstones, urinary calculi, microbial biscuits, otolishs, and the eggshells of turtles. One of them, novel vaterite tablets are widely distributed among Chinese freshwater lackluster pearls. This vaterite generally coexists in a half-lackluster pearl with aragonite, the main component of the lustrous part of the pearl, or is present solely in a completely lackluster pearl, as shown in Fig. 6.4.

6.1.1.3 Calcium Carbonate – Amorphous Phase

Amorphous calcium carbonate is formed in the leaves of many plants as spindle-shaped deposits (cystoliths) that act as a store of calcium. Although this material is exceedingly unstable in inorganic systems due to rapid phase transformation in aqueous solution, the biomineral appears to be stabilized through the adsorption of biological macromolecules such as polysaccharides at the solid surface (Weiner et al. 2009).

It was assumed for a long time that organisms produce minerals directly from a saturated solution. A few exceptions were known, including the well-documented mineralized teeth of the chiton. In 1997, it was demonstrated that sea urchin larvae form their calcitic spicules by first depositing a highly unstable mineral phase called amorphous calcium carbonate. This strategy has since been shown to be used by animals from other phyla and for both aragonite and calcite. Recent evidence shows that vertebrate bone mineral may also be formed via a precursor phase of amorphous calcium carbonate. This strategy thus appears to be widespread. The challenge now is to understand the mechanisms by which these unstable phases are initially formed, how they are temporarily stabilized, and how they are destabilized and transform into a crystalline mature product.

The basic paradigms in biomineralization are that minerals from a saturated solution and that structured surfaces, as well as additives are intimately involved in controlling the mineral formation process. The first-formed mineral may be relatively disordered, and over time transforms into more stable phases (Ostwald's Rule of Stages). It may also be similar or identical to the mature mineral phase. The mineral formation process very much depends upon the mechanism of nucleation and the microenvironment in which this occurs. Studies of biomineralization processes in a variety of different organisms from various phyla show that transient minerals may first be formed and these subsequently crystallize upon structured substrates. Furthermore, their growth may be modified by ions and macromolecules. The recent studies focus on the development of the concept of an amorphous precursor phase strategy in biomineralization, and the distribution of this phenomenon, including the vertebrate phylum. This subject was briefly reviewed by Weiner et al. (2005).

6.1.2 Calcium Phosphate

Bone and teeth are made from calcium phosphate in the form of the mineral hydroxyapatite (HA), along with a large number of proteins. The structural chemistry of biological hydroxyapatite is very complex because the mineral is not compositionally pure (non-stoichiometric), often being calcium deficient and enriched in CO_3^{2-}, which replaces PO_4^{3-} ions in various lattice sites. In spite of that bone mineral is referred to as hydroxyapatite, it is often known as "carbonated apatite." The composition can be expressed as:

$$(Ca,Sr,Mg,Na,H_2O,[\,])_{10}(PO_4,HPO_4,CO_3,P_2O_7)_6(OH,F,Cl,H_2O,O,[\,])_2$$

where [] denotes the presence of lattice defects. For most purposes, it can be simplified as $Ca_{10}(PO_4)_6(OH)_2$.

Several other calcium phosphate phases have been identified as intermediates in the biomineralization of calcium phosphates (Table 6.2). In particular, there is evidence for an amorphous calcium phosphate phase in the early stage of bone and cartilage mineralization. Another phase, octacalcium phosphate $Ca_8H_2(PO_4)_6$, has also been identified in various tissues where it readily transforms to HA because of a close structural match between the unit cells of the two mineral phases.

Table 6.2 Calcium phosphate biominerals (Mann 2001)

Mineral	Formula	Organism	Location	Function
Hydroxyapatite	$Ca_{10}(PO_4)_6(OH)_2$	Vertebrates	Bone	Endoskeleton
		Mammals	Teeth	Cutting/grinding
		Fish	Scales	Protection
Octacalcium phosphate	$Ca_8H_2(PO_4)_6$	Vertebrates	Bone/teeth	Precursor phase
Amorphous	Variable	Chitons	Teeth	Precursor phase
		Gastropods	Gizzard plates	Crushing
		Bivalves	Gills	Ion store
		Mammals	Mitochondria	Ion store
		Mammals	Milk	Ion store

6.1.2.1 Bone

Bone comes in all sorts of shapes and sizes in order to serve the various functions of protection and mechanical support without compromising the requirement for mobility. More than any other biomineral, the nature of bone highlights the important distinction between the inorganic and bioinorganic material world. For example, bone is often thought of as a living mineral since it undergoes continual growth, dissolution, and remodeling in response to both internal signals, such as pregnancy and external force fields, such as gravity.

The mechanical properties of bone are derived from the organized mineralization of hydroxyapatite within a matrix of collagen fibrils, glycoproteins, and many other types of protein. The combination of inorganic and organic components provides an increased toughness compared with hydroxapatite alone. By sculpting these components into microanatomical structures-woven bone, cortical bone, etc., and controlling the amounts of mineral content, different levels of stiffness can be introduced into different bones according to their particular functions. A fast moving, highly agile animal such as a deer requires bones with high elasticity and relatively low mineral content (around 50 wt.%). By contrast, the bones of a large marine mammal like the whale are stiff, with a hydroxyapatite content greater than 80 wt.%. Bone is a kind of typical self-assembled biomaterial with hierarchical structure. Figure 6.5 shows the hierarchical structure of a long bone.

The non-stoichiometric nature of bone mineral may be responsible for the apparent piezoelectric response observed in this tissue. Although the precise mechanism is unknown, the application of pressure stimulates the growth of bone mineral. Bone contains a network of cells that live within the mineralized structure and are interconnected through small pores and channels. One possibility is that the osteocytes act as biological "strain gauges" that respond to changes in mechanical pressure and send chemical or electrochemical signals to the bone surface which then activate another type of cell called osteoblast to begin mineralization. The process of activation is further complicated because there is another type of cell called osteoclast whose job it is to degrade bone through acid and enzymes, and

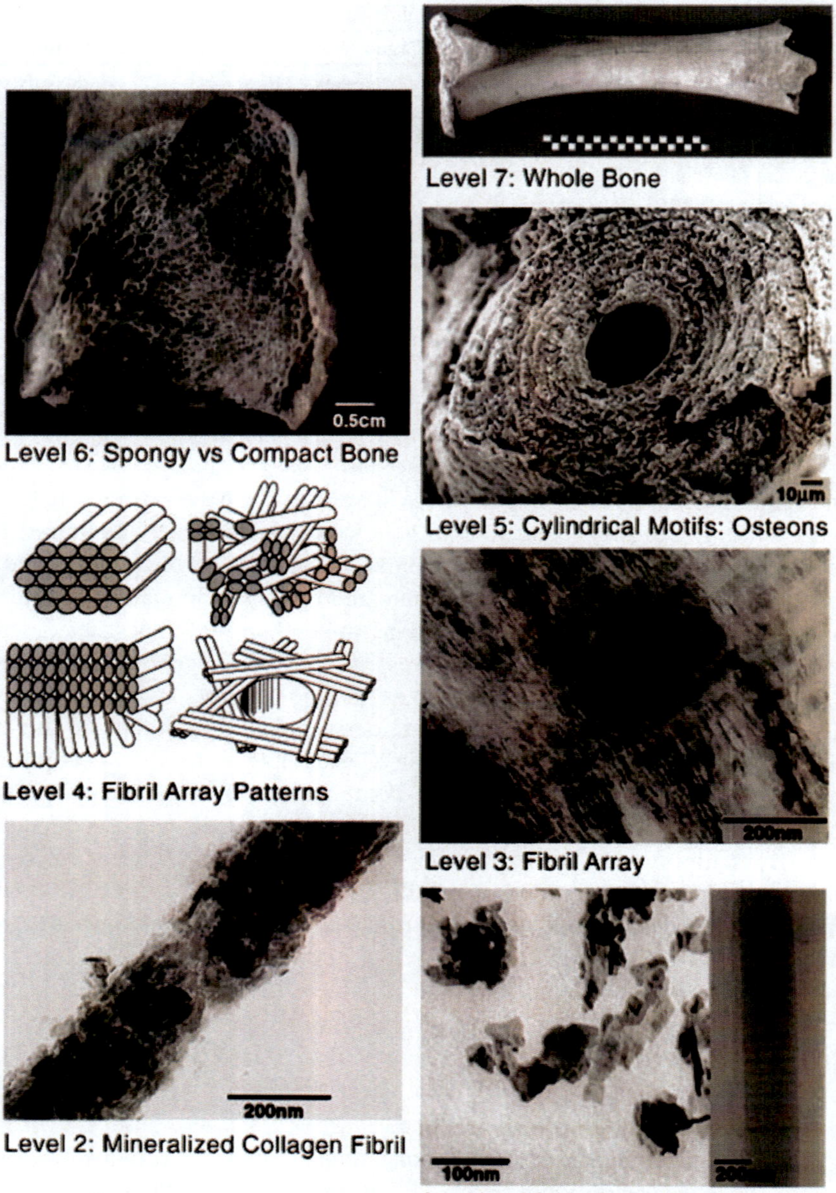

Fig. 6.5 Hierarchical structure of a long bone (Weiner & Wagner 1998). (**a**) Level 1: Major components, nanocrystals, (**b**) Level 2: Mineralized collagen fibril, (**c**) Level 3: Fibril array, (**d**) Level 4: Fibril array patterns, (**e**) Level 5: Cylindrical motifs: osteons, (**f**) Level 6: Spongy versus compact bone, and (**g**) Level 7: whole bone

who also responds to the signaling. Overall, the process is incredibly complex and subject to many forms of feedback controlled by a large number of biochemical triggers such as hormones that are circulating in the blood stream.

6.1.2.2 Tooth

The structure and organization of tooth enamel and dentine, like bone, derive from a highly complex system designed to withstand specific types of mechanical stress. Enamel, which is on the outside of the tooth, is much less tough than bone because it has close to 95 wt.% hydroxyapatite (human bone on average is around 65 wt.%) but gains some structural resistance by interweaving long ribbon-like crystals into an inorganic fabric. Interestingly, enamel starts out with a high proportion of proteins (mainly amelogenin and enamelin), which are progressively removed as the biomineral matures to produce the high mineral volume fraction of the erupted tooth. Dentine, on the other hand, which resides within the central regions of the tooth, contains collagen and is more similar in structure and composition to bone.

A principal cause of the general increase in dental health in many societies is the use of fluoride in drinking water and in numerous toothpastes. The F ion is readily incorporated into the hydroxyapatite lattice where it stabilizes the lattice and reduces the solubility of the mineral phase. Interestingly, fish teeth consist of a structure very similar to enamel (called enameloid) but contains high levels of natural fluoride. For example, the fluoride concentration in shark enameloid is over 1,000 times compared with that in human enamel.

6.2 Hierarchical Structure of Calcium Carbonate-Based Biomineral in Aquatic Organisms

Many natural biominerals have been found to have hierarchical structure, such as human bone, human enamel, nacre, ivory, etc.

Living organisms make up hierarchically organized materials through self-organization from precursors in aqueous solution, and scientists have developed various biomimetic techniques to prepare and organize building blocks. It is believed that an exquisite association of organic and inorganic compounds is required for the construction of bioinorganic superstructures; therefore, understanding the roles of macromolecules in the biomineralization process is a significant challenge in biomimetic materials. For example, the discovery of calcitic microlenses in brittlestars, chiral morphologies with stereochemical recognition, and the handedness of a snail shell bring out hidden elaborate structures and properties of biominerals. The nacreous layer has attracted the interest of researchers in a broad range of disciplines, especially in terms of its detailed structure, defects in different scales, incorporated macromolecules, mechanical strength, formation mechanisms, and mimetics. An understanding of the real hierarchically organized structure in biominerals is required to advance to the next stage of chemistry, biology, and materials science.

6.2.1 *Carp Otolith*

Fish otolith is a kind of typical natural calcium carbonate biomineral, which is composed of a pair of lapillus, sagitta, and asteriscus, respectively, as shown in Fig. 6.2. Otoliths are involved in the perception of sound and the maintenance of postural equilibrium in fishes. Although the relationship between microstructure and the growth history of fish otolith is widely known, the hierarchical structure of the otolith, especially at the nanometer level is not well understood. Li et al. (2009) observed the nanometer scale morphology of the otoliths and investigated the hierarchical levels of organization in wild carps otoliths in order to have a better understanding of the otolith structure–ecology relations. Investigation on the otolith hierarchical structure plays a key role in the study of the hydro environmental chemistry and calcium carbonate biomineralization.

6.2.1.1 Hierarchical Structure of Lapillus

Lapillus, which is composed of aragonite, is a type of mineralized material with highly complex hierarchical structure. The other major component in the lapillus is protein which accounts for 4–5% in weight. The structure of lapillus can be separated into seven levels.

Level 1: the aragonite nanocrystals

The basic mineral structure unit of lapillus is nanometer-scale aragonite crystal (Fig. 6.6a), the (111) and (002) planes are labeled. The aragonite nanocrystals build the first level of the lapillus minerals.

Level 2: the aragonite fibrils

The aragonite nanocrystals grow along the *c*-axis (the black arrow in Fig. 6.6b) and form a nanofibril structure with a diameter of 60 nm. The aragonite fibrils are covered by the proteins.

Level 3: fibril arrays

In the lapillus, mineral fibrils are present in arrays aligned along their diameter. The fibrils are intimately associated and the banding patterns in neighboring fibrils are in connection with each other (Fig. 6.6c). The fibril array is clearly highly ordered in two dimensions and forms the layered structure. All the different layers accumulate compactly and parallel to each other. The black arrow shown in Fig. 6.6c shows the array direction of the fibrils which is perpendicular to the long axes of the fibrils.

Level 4: three-dimensional stick

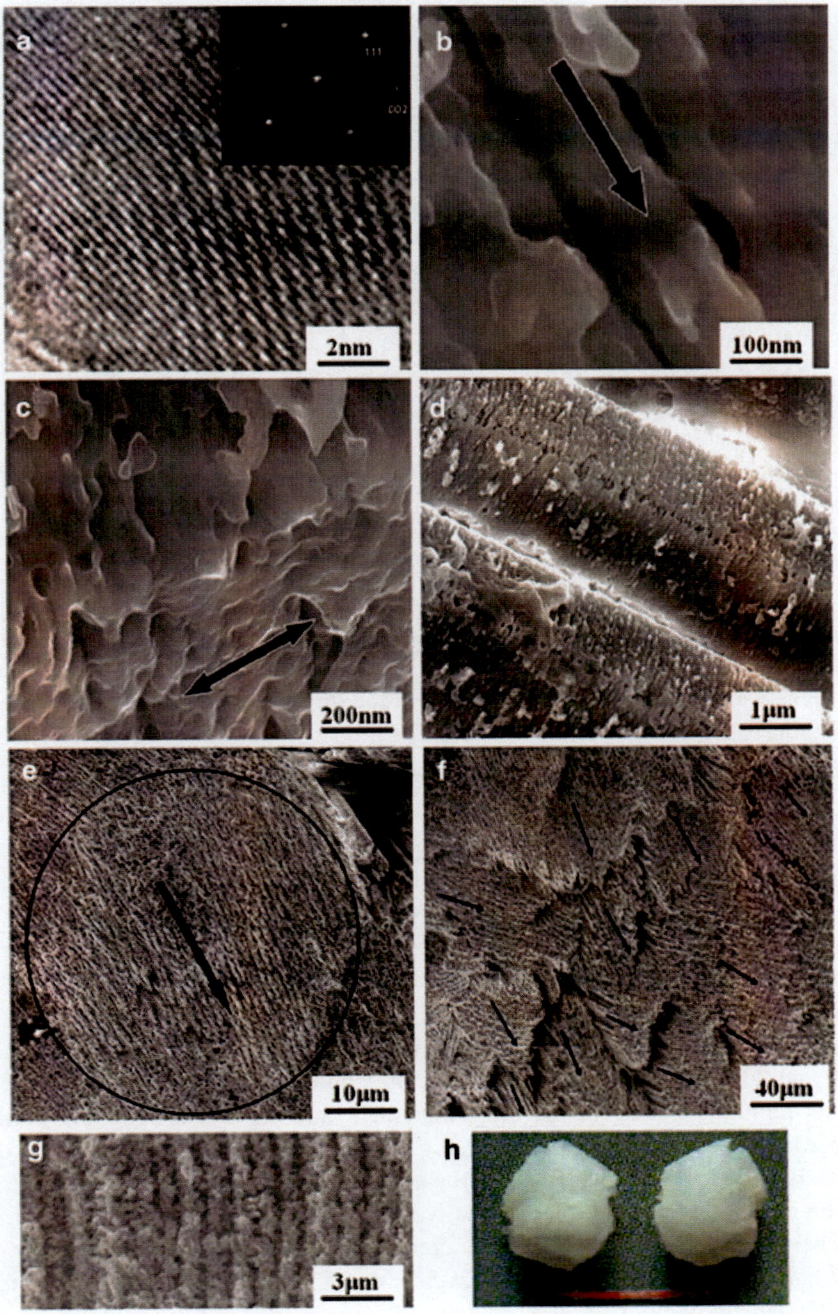

Fig. 6.6 The seven hierarchical levels of organization of the carp lapillus (Li et al. 2009). (**a**) Level 1: Isolated crystals from lapillus observed by HRTEM and SEAD, (**b**) Level 2: SEM micrograph of an aragonite fibril, (**c**) Level 3: SEM micrograph of an aragonite layer and the extension direction of the layers, bar = 200 nm, (**d**) Level 4: SEM micrograph of the sticklike

The long axes of aragonite fibrils (level 2), which are about 1.5 μm in length (Fig. 6.6d), are arranged along x-direction; the layers (level 3) accumulate along z-direction and extend along y-direction. The three directions are vertical to each other and form the three-dimensional sticklike structure. The long axis of the stick which is tens of microns in length is along y-direction.

Level 5: the domain structure

Based on the investigations of the morphology in a single ring structure, it is proposed that there exists a domain structure of the stick (level 4) orientation in the lapillus. Within micron range, hundreds of sticks appear to be arranged at the same orientation and make up the domain structure. The black circle whose diameter is about 100 μm in Fig. 6.6e is the area of a domain and the arrow shows the direction of the sticks' arrangement.

Level 6: the domain arrays

Level 6 pattern is essentially the combination of Level 5 structure into hundreds of microns and even millimeter scale. All these domains appear in the same ring. Level 6 structure represents one daily increment. According to Fig. 6.6f, the long axis direction of the stick is not inside the daily increment layer and most of the neighboring domains have the similar direction.

Level 7: the daily increments

The daily ring structure is shown in Fig. 6.6g. The width of each daily increment (the transparent area in the pattern) is between 1 and 2 μm and the absent area is the proteins which were etched. Figure 6.6h is the digital camera photograph of a pair of lapilli.

6.2.1.2 Hierarchical Structure of Sagitta

The aragonite crystals of a sagitta have the same structure of level 1 as a lapillus (Fig. 6.7a). The hierarchical structure of a sagitta is much more simple than that of a lapillus. Figure 6.7b shows the transect of single increment where we can distinguish clearly the aragonite fibril, which is about 1.5 μm in length and 250 nm in diameter (level 2). The aragonite fibrils in sagitta are larger than those in lapillus, but sagitta has not as many multilevels as lapillus. The aragonite fibrils are the direct building blocks for the daily increment and the direction of the fibrils is vertical to the daily increment layers (Fig. 6.7c). Figure 6.7d shows clearly the daily growth structure of sagitta. Because the core is not in the center of the transect, the width in different parts of the same daily increment is quite different. Figure 6.7e is a schematic illustration of the daily increment structure of a sagitta which has only one core (sometimes there are two or three cores in the otolith). Figure 6.7f is the digital camera photograph of a whole sagitta.

Fig. 6.6 (continued) structure, **(e)** Level 5: SEM micrograph of an orientation domain, **(f)** Level 6: SEM micrograph of interior structure of a daily increment, **(g)** Level 7: SEM micrograph of daily increments, bar = 10 μm, and **(h)** digital camera graph of the lapillus, bar = 5 mm

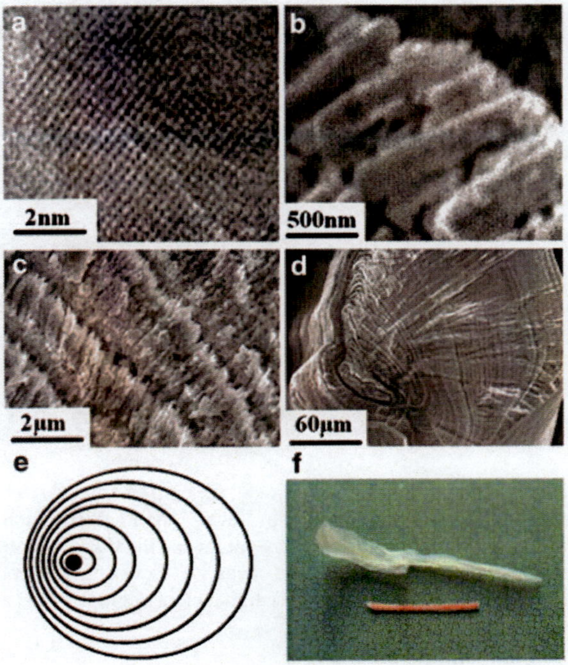

Fig. 6.7 The four levels of organization of the carp sagitta (Li et al. 2009). **(a)** Level 1: HRTEM micrograph of isolated crystals from sagitta, **(b)** Level 2: SEM micrograph of the aragonite fibril which is about 1.5 μm in length and 250 nm in diameter; **(c)** Level 3: SEM micrograph of the single daily increment; **(d)** Level 3: the daily increments structure of sagitta; **(e)** Schematic illustrations of the daily increments structure in the sagitta; and **(f)** digital camera graph of the sagitta, bar = 5 mm

6.2.1.3 Hierarchical Structure of Asteriscus

Level 1: the vaterite nanocrystals

Asteriscus, composed of vaterite, is a typical biomineral with hierarchical structure. Scanning electron microscopy (SEM) and transmission electron microscopy (TEM) tests show that the structure could be divided into 5 levels. According to high-resolution transmission electron microscopy (HETEM) and selected area electron diffraction (SEAD) results, the first level is vaterite nanocrystals, the size is about 20 nm (Fig. 6.8a).

Level 2: vaterite crystal stick

The vaterite nanocrystals stack together could form a vaterite crystal stick with a diameter of about 300 nm, as shown in Fig. 6.8b.

Level 3: the vaterite crystal layer

The vaterite crystal sticks arrange parallel in the two-dimensional direction, and form the third hierarchical level: the vaterite crystal layer. As shown in Fig. 6.8c, each layer is extended along the direction of which the arrows show, and the width of each layer is about 2 μm. On the perpendicular direction, adjacent layers pile up

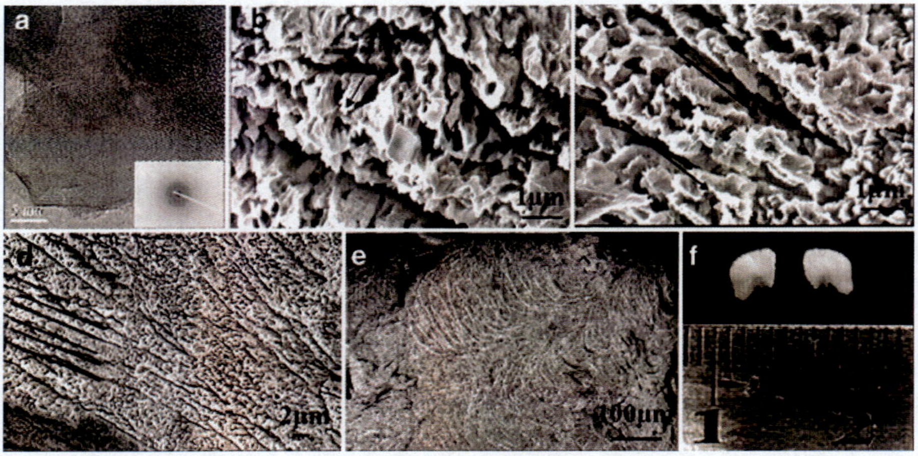

Fig. 6.8 The five hierarchical levels of organization of the carp asteriscus. (**a**) Level 1: the vaterite nanocrystals observed by HRTEM and SEAD, (**b**) Level 2: SEM micrograph of vaterite crystal sticks; (**c**) Level 3: SEM micrograph of vaterite crystal layers, the thickness of each layer is about 2 μm; (**d**) Level 4: SEM micrograph of domain areas; (**e**) Level 5: SEM micrograph of daily increments, the width is about 10 – 20 μm; and (**f**) digital camera graph of the asteriscus, diameter is about 4 mm

to each other, and the vaterite crystal sticks in different layers are parallel one another.

Level 4: daily growth ring

Figure 6.8e shows the daily ring structure of asteriscus. The width of each growth increment is about 10–20 μm. A daily ring comprises an increase belt and interval belt, the increase belt is wide and bright, mainly composed of calcium carbonate, and the interval belt is narrow and opaque, mainly composed of organic matrices.

Level 5: the whole asteriscus

Different growth rings pile up in certain ways and form the whole asteriscus, whose diameter is about 4 mm.

From the hierarchical structures of lapilli, sagittae, and asterisci, it can be seen that each kind of otolith has its particular structure and morphology. Both materials sciences and biology can benefit from this basic study which will be also relevant to the fish ecology and environmental chemistry. The self-assembled structure of natural biomaterials is always intimately associated with the versatile properties. The orientation domain structure, which is found in the lapillus, is not unique in the otolith. Feng and her group had found a domain structure in nacre of the *Mytilus edulis* shell (Feng et al. 1999). Those domains consisted of 3–10 continuous flat platelets perpendicular to the nacre plane. The aragonite crystals of the neighboring domains had the same *c*-axes but different *a*-axes. The directions of the assembled aragonite sticks of lapilli are more complicated than those of nacre platelets, because the long axes of the sticks do not point to the same directions; furthermore, the aragonite sticks, which are bundled of parallel fibrils, are not the basic units of the lapillus hierarchical structure. The aragonite fibrils in the sagittae are all perpendicular to the daily increment layer so that the section of sagitta seems to be radial. Figure 6.7d shows clearly the asymmetrical location of the nucleus, which is close to the sulcate side (the left side in Fig. 6.7d). The unusual shape of the otolith is partly explained by the distribution of the epithelium cells, and particularly the sensory epithelium.

The discovery of the hierarchical structure of the carp otolith can help us to investigate the biologic self-assembling mechanism and the relationship between the microstructure and the physiological function of the otolith. The hierarchical structure of the otoliths can give us more information for investigating the fish ecology and the hydrochemistry. The particular crystal array of otoliths can also provide the inspiration on biomimetic and synthetic materials.

6.2.2 Hierarchical Structure of Nacreous Layer in Mollusc Shells

Oaki and Imai (2005) investigated the nacreous layer of Japanese pearl oyster: *Pinctada fucata*. As shown in Fig. 6.9, mother-of-pearl has a hierarchical structure, with tiers 1–3 mediating the oriented assembly in two ways. The layered structure (tier 1, Fig. 6.9a, d) consists of aragonite plates (tier 2, Fig. 6.9b, e) that are about 1–5 mm wide and 200–700 nm thick. The magnified FESEM (field emission SEM) image (Fig. 6.9b) clearly indicates the presence of smaller components in each aragonite plate. An FETEM image of such a nanobuilding block clearly demonstrates the pseudohexagonal habit of aragonite (tier 3, Fig. 6.9c, f). The appearance of the nanobuilding blocks was neither attributable to the sample preparation process nor to radiation damage arising from FETEM. The high-resolution image on the fringe of the nanobuilding block shows a lattice spacing of 0.423 nm, which corresponds to the (110) plane of aragonite. The lengths of the nanobuilding blocks were in the range of 20–180 nm, whereas the nanobuilding blocks in the mother-of-pearl shell of the oyster were smaller.

Oriented assembly mediated the formation of the hierarchical architecture. The layered structure of mother-of-pearl (tier 1) is an oriented assembly of the aragonite

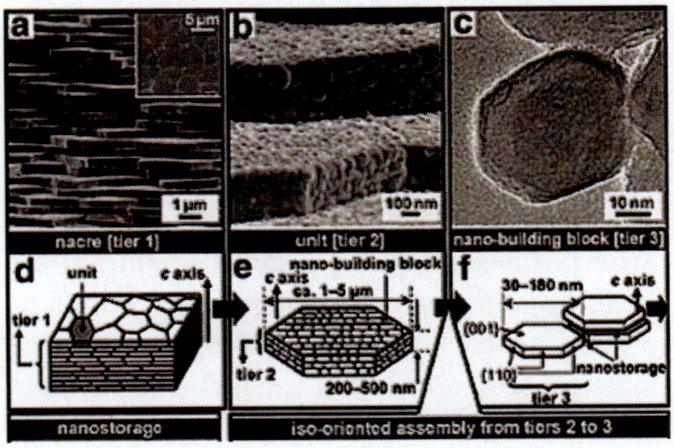

Fig. 6.9 Hierarchically organized structure of the nacreous layer. (**a**, **b**) FESEM and (**c**) FETEM images of tiers 1–3; (**d–f**) schematic representations of tiers 1–3 (Oaki & Imai 2005)

units (tier 2) along the *c* axis (Fig. 6.9d), although the orientation of the *a* and *b* axes in a layer remains unclear. The mineral-bridges model of Schaeffer et al. implies that the *a* and *b* axes should be perfectly oriented in all the layers in tier 1. However, Dai and Sarikaya reported dark-field TEM images showing that the *a* and *b* axes were not perfectly aligned in all the layers. XRD analysis indicates that the layers are perpendicular to the *c* axis. Consequently, the oriented assembly of the plates (tier 2) in the *c* axis make up the layered structure (tier 1). A single unit (tier 2) is also an oriented assembly of nanobuilding blocks (tier 3). Aggregations of the nanobuilding blocks showed that the nacreous layer is a three-level hierarchical architecture associated with two forms of oriented assembly (Fig. 6.9). Electron microscope analysis determined the morphology of the nanobuilding blocks and their oriented assembly into platy units.

These results indicate that the aragonite–biopolymer composite behaves as a host for organic molecules. Oaki and Imai (2005) have called this property "nanostorage." To gain a more comprehensive understanding of the superstructures, Oaki and Imai (2005) showed the detailed structures of two hierarchical architectures that have "nanostorage" properties, and they discussed the mutual growth process associated with the two respective roles of each polymer. The results imply that the manipulation of crystals and polymers could lead to a novel type of excellent inorganic–organic hybrid composites under ambient conditions. They concluded that the unit, like the nacreous layer, is an oriented assembly of nanobuilding blocks. They have elucidated the hierarchical architecture in nacre and identified its ability to host organic molecules. This model case suggests that a hierarchy similar to that of nacre can be induced through an appropriate combination of inorganic crystals and organic polymers. The specific interaction of the two components generates the nanoscopic architecture, and the switching between the modes of growth leads to the formation of macroscopic structure. Furthermore, an improved understanding of real and mimetic biominerals holds promise for the further development of chemical, biological, and materials sciences.

Storage, an additional nanoscopic function leading to the incorporation of versatile organic dye molecules, is attributed to the aragonite–biopolymer nanohybrid. The results are beneficial to the understanding of the overall architecture in the nacreous layer from the nanoscopic to the macroscopic scale.

6.2.3 Lackluster Pearl

Normal pearl is formed by nacre, which has attracted much attention because of its complex architectures, superior mechanism properties, and applications in materials design. Nacre inspires bright luster due to its regular structure layers of uniformly thick tablets of aragonite and high mechanical performance thanks to the organic matrix lying between neighboring tablets and lamellae to form a "brick and mortar" structure. Much research has revealed that the aragonite tablets in nacre have a strong texture basically with their *c*-axis perpendicular to the tablet plane, and mineral bridges keep neighboring crystals maintaining the same crystal orientation in three dimensions as a domain structure. In recent years, some research results seem to be different from classical theories, for instance, independent nucleation sites for each crystal tablet, nanostructure in single aragonite tablets, and a continuous layer of disordered amorphous $CaCO_3$ covering aragonite platelets. Therefore, understanding how nacre is formed has represented a major challenge in the field of nacre biomineralization.

In south China, freshwater cultured pearls are generally gestated in *Hyriopsis cumingii Lea*. Vaterite is found in these freshwater cultured lackluster pearls. This kind of vaterite generally coexists in a half-lackluster pearl with aragonite, the main component of the lustrous part of the pearl, or presents individually in a complete lackluster pearl. Vaterite often acts as a precursor in the formation of aragonite or calcite and is a very unstable phase of calcium carbonate rarely occurring in nature. Different from normal biomineralization, it is difficult to obtain simple and pure vaterite phase in abnormal mineralization, often presented with calcite or aragonite. Although vaterite in lackluster pearl is deposited abnormally under the control of environment, this deposit is often pure, and its dimension can extend to the macroscopic scale. Its aggregation is also one of the biggest minerals precipitated with vaterite completely in a biological system.

The existence of vaterite is one of the key factors influencing the quality of freshwater cultured pearls. Compared to the number of elegant studies on the calcite–aragonite switch, little is known about the aragonite–vaterite interface in biomineralization. The phase, morphology, structure, and orientation of vaterite crystals deposited in lackluster pearls were studied to control the quality of freshwater pearls.

Pearl powder is very similar to the patterns of vaterite samples referring to the standard PDF card 72–0506, without any peak of other calcium carbonate polymorphs. The only mineral phase in a complete lackluster pearl is vaterite. Powder *X-ray diffraction* (XRD) patterns of the half-lackluster pearls revealed that they are composed of both aragonite and vaterite. Figure 6.10 shows the microarchitecture of lackluster pearls, which reveals that the deposition of vaterite is in the form of a "brick and mortar" hierarchy similar to nacre. Figure 6.10a shows that the vaterite tablet is an elongated rectangle with dimensions of approximately $8 \times 2\ \mu m^2$; they align tightly with parallel lengths in two dimensions and assemble layer by layer. The side elevation of vaterite tablets in Fig. 6.10b shows that the vaterite layer is a lamella of 0.4 μm thick tablets, thinner and more irregular than nacre, which may result in the lackluster characteristic in the pearls. Figure 6.10c depicts that another section showing width and thickness is a rhombus. Interestingly, the crystals between successive sheets appear to nucleate close to the center of preexisting tablets located in the layer below, as shown in Fig. 6.10c, which is a remarkable "stack-of-coins" structure of gastropod shells and quite different from that observed in nacre. Here we give a three-dimensional structure sketch of both vaterite tablets and layers in lackluster pearls, as shown in Fig. 6.10d. Planes *x*, *y*, and *z* stand for the observed surfaces in Fig. 6.10a, b, c, respectively (Qiao et al. 2007).

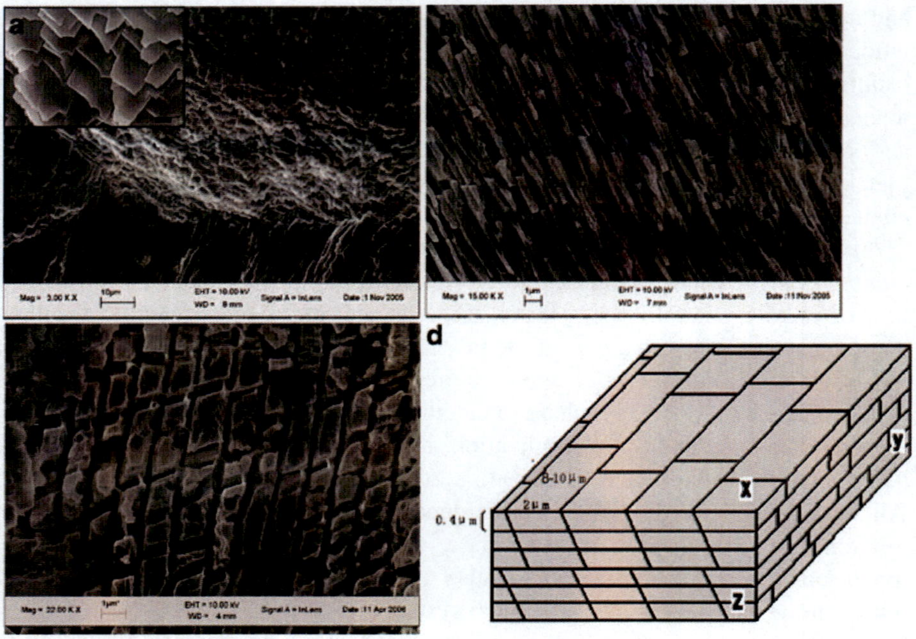

Fig. 6.10 SEM images of vaterite tablets and layers in lackluster pearls after 10 wt.% EDTA-2Na treatment: (**a**) etched surface of lackluster pearl (scale of amplificatory image, 1 μm), (**b**) etched section of length by thickness of vaterite tablets, (**c**) etched section of width by thickness of vaterite tablets, and (**d**) three-dimensional structure sketch of vaterite crystal in lackluster pearls (Qiao et al. 2007)

It is well known that protein can modulate calcium carbonates not only in polymorphs but also in morphology. The batten-like vaterite tablet found in lackluster pearls is a new kind of morphology, the reason for which may be considered to contribute to changeable crystal face-specific growth rates under biological control. Protein modifies the surface energies via preferential adsorption on crystal surfaces; the final form of crystals would show the slow-growing surfaces. On the other hand, another function of a matrix is a structure-matching template. The matrix acts as an organic template to only align crystals perpendicular to the matrix or array crystal iso-oriented with three-dimension crystallographic alignment. Due to the uniform arrangement in two dimensions, vaterite tablets should be an iso-oriented arrayed. Therefore, the structure of vaterite crystals in lackluster pearls performs a superior biomineralization process and may undergo a biologic control. This "brick and mortar" structure reveals that the growth mechanism of vaterite crystals in lackluster pearls is the same as that of nacre. About the growth mechanism of nacre, mineral bridge theory for the gastropod shell, and template theory for the bivalve shell are the most important hypotheses to explain the "brick and mortar" structure and uniform crystal orientation. For nacre of bivalve shells and pearl, each successive layer of crystals is offset, and the lateral growth of the aragonite tablets occurs to a much greater extent before the next sheet is added.

At present, nanostructured and amorphous $CaCO_3$ are believed to play an important role in the forming of nacre, which seems to contradict the epitaxial match between the structural organic matrix and the formed mineral. The growth mechanism of vaterite crystals in lackluster pearls is in demand for further study.

The investigation provided a similar hierarchy to nacre. It is well worth to mention that two kinds of contact modes between aragonite and vaterite tablets (side-side and front-back) were observed, which highlights the coexisting state of aragonite and vaterite in one pearl. It is reported that strong texture of [010], [101], and [102] was found in different scale with various methods. Vaterite tablets have a high degree of oriented arrangement in three dimensions from several neighboring tablets to macroscopic scale. The distribution of misorientation angles showed the domain structure and the cluster character in vaterite tablets.

In conclusion, the formation of vaterite crystals in lackluster pearls has typical biomineralization characteristics: (1) the size and morphology of inorganic crystals are regular; (2) the crystals are oriented in arrays; (3) the transition from vaterite to aragonite tablets is abrupt. Thus, both aragonite and vaterite in pearls have the same growth mechanism according to the semblable morphology and structure. The study of the aragonite–vaterite switch is noteworthy as the remarkable calcite–aragonite switch in shell. The (010) plane in vaterite layers is a significant crystalline surface, just as the (001) plane in aragonite tablets of nacre.

6.2.4 Crab (Meyers et al. 2008)

Arthropods are the largest animal phylum. They include the trilobites, chelicerates, myriapods, hexapods, and crustaceans. All arthropods are covered by an exoskeleton, which is periodically shed as the animal grows. The exoskeleton of arthropods consists mainly of chitin. In the case of crustaceans, there is a high degree of mineralization, typically calcium carbonate, which gives mechanical rigidity.

The arthropod exoskeleton is multifunctional: it supports the body, resists mechanical loads, and provides environmental protection and resistance to desiccation. The outermost region is the epicuticle, a thin, waxy layer which is the main waterproofing barrier. Beneath the epicuticle is the procuticle, the main structural part, which is primarily designed to resist mechanical loads. The procuticle is further divided into two parts, the exocuticle (outer) and the endocuticle (inner), which have similar composition and structure. The endocuticle makes up around 90 vol.% of the exoskeleton. The exocuticle is stacked more densely than the endocuticle. The spacing between layers varies from species to species. Generally, the layer spacing in the endocuticle is about three times larger than that in the exocuticle. The exoskeleton is highly anisotropic, both in structure and mechanical properties.

A striking feature of arthropod exoskeletons is their well-defined hierarchical organization, which reveals different structural levels. At the molecular level, there are long-chain polysaccharide chitins that form fibrils, 3 nm in diameter and 300 nm in length. The fibrils are wrapped with proteins and assemble into fibers of about 60 nm in diameter. These fibers further assemble into bundles. The bundles then arrange themselves parallel to each other and form horizontal planes. These planes are stacked in a helicoid fashion, creating a twisted plywood structure. A stack of layers that have completed a 180° rotation is referred to as a Bouligand structure. These structures repeat to form the exocuticle and endocuticle. The same Bouligand structure is also characteristic of collagen networks in compact bone, cellulose fibers in plant cell walls, and other fibrous materials. In crab exoskeletons, the minerals are in the form of calcite or amorphous calcium carbonate, deposited within the chitin–protein matrix.

In the direction normal to the surface (the *z*-direction), there are well-developed, high-density pore canals containing long, flexible tubules penetrating through the exoskeleton. These tubules play an important role in the transport of ions and nutrition during the formation of the new exoskeleton after the animals molt.

The crab exoskeleton is a three-dimensional composite comprising brittle chitin–protein bundles arranged in a Bouligand pattern (the *x*–*y* plane) and ductile pore canal tubules in the direction normal to the surface (the *z*-direction). The pore canal tubules possess ductile mechanical properties even in a dry condition.

The structure and mechanical properties of arthropod exoskeletons have been much studied. It is important for materials scientists to understand the design of natural composites, in order to develop novel composite materials with enhanced properties.

6.3 Hierarchical Structure of Calcium Phosphate-Based Biomineral

6.3.1 Zebrafish Bone

The zebrafish (*Danio rerio*) has proved to be an important system for studying vertebrate-specific problems of development because it is a powerful model organism for embryology, developmental biology, and genetics. Related researches utilizing the zebrafish system have been involved in many areas, such as developmental neurobiology, cardiovasology, etc.

Bone is a type of mineralized material with highly complex hierarchical structure. Weiner has described the seven levels of hierarchical organization of human long bone. The basic building block of the bone materials is the mineralized collagen fibril (level 2), which is composed of very hard material, the mineral and much softer material, the collagen fibrils (level 1). Mineralized collagen fibrils are always present in bundles or arrays aligned along their length (level 3). These fibril arrays organize into four common patterns: arrays of parallel fibrils, woven fiber structure, plywood-like structure, and radial fibril arrays (level 4). At a higher level of organization, the initially deposited primary bone undergoes internal remodeling and forms the secondary bone with a central canal for blood vessels

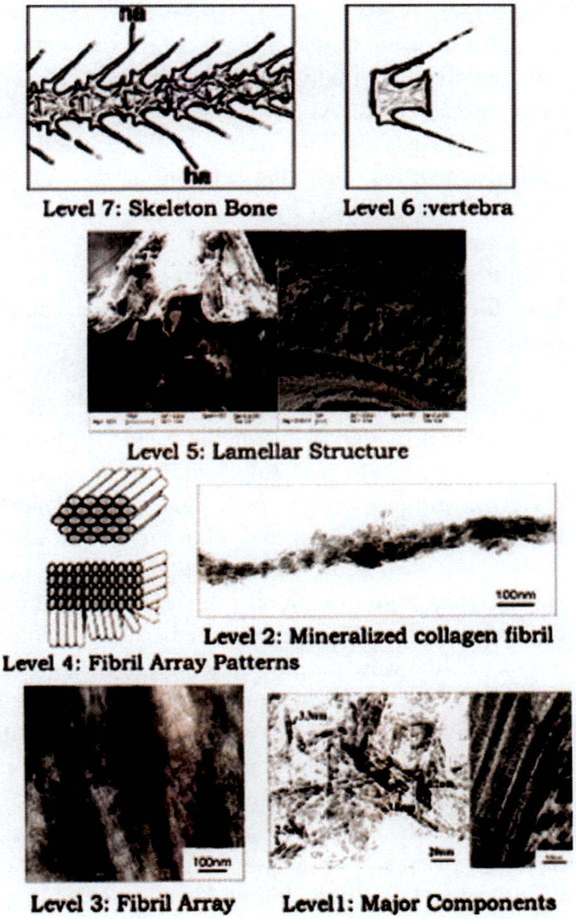

Fig. 6.11 The seven hierarchical levels of organization of the zebrafish skeleton bone. Level 1: Isolated crystals (*left side*) and part of stained collagen fibril (*right side*). Level 2: TEM micrograph of a mineralized collagen fibril. Level 3: TEM micrograph of a thin section of mineralized bone. Level 4: Two fibril array patterns of organization found in the zebrafish skeleton bone. Level 5: SEM micrograph of lamellar structure in one vertebra. Level 6: Schematic drawing of a vertebra. Level 7: Schematic drawing of part of the skeleton bone (Wang et al. 2004)

and nerves, which is called "haversian system" (level 5). The levels 6 and 7 refer to solid versus spongy bone and whole bones, respectively.

Wang et al. (2004) investigated the hierarchical levels of organization in wild-type in order to have a better understanding to the bone structure– function relations and the bone brittleness in bone diseases. They observed the similar hierarchical structure in zebrafish skeleton bone (Fig. 6.11). It is deserved to point out that only two of the most common fibril array patterns of organization are observed, which is parallel fibrils arrays and plywood-like structure, respectively. Beyond that, the "haversian system" does not exist in zebrafish bone.

6.3.2 Tooth

The microstructure in terms of hierarchical assembly exists widely in both natural and synthesized biomaterials, in which various bone tissues have been reported

(Weiner & Wagner 1998). However, similar investigations on dental enamel, another important family of hard connective tissues, have not been well developed, although there are many structural, mechanical, and functional similarities between these biomaterials. As the hardest connective tissue in the human body, dental enamel consists of 96% mineral, 4% organic material, and water. Enamel has unique properties, including extraordinarily high hardness, outstanding resistibility to wear, and stability over a lifetime of use within the physically demanding environment of the oral cavity (Ge et al. 2006). The enamel structure and its biomineralization have therefore been of interest in order to provide theoretical basis for both the treatment of enamel disease and the biomimic synthesis of novel biomaterials.

Recently, the hierarchical assembly of enamel structure was investigated and depicted via various microscopic explorations from nanoscale to microscale, which supposedly is related to its functions and the physical requirements placed upon it in the oral cavity. On the other hand, the mechanical diversity within enamel was reviewed and proposed to have close corresponding relationships with the microstructure of enamel in terms of hierarchy. Human enamel taken from mature third molars was explored (Cui & Ge 2007). Integrating the microscopic observations revealed the high complexity of the well-organized enamel structure in terms of hierarchical assembly. Based on these observations, seven hierarchical levels of the microstructure were proposed and described, using a scheme representing a complete spectrum of the organization in detail, covering a range from microscale to nanoscale: hydroxyapatite crystals (Level 1) at first form mineral nanofibrils (Level 2); the nanofibrils always align lengthways, aggregating into fibrils (Level 3) and further thicker fibers (Level 4); prism/interprism continua (Level 5) are then composed of them. At the microscale, prisms assemble into prism bands (Level 6), which present different arrangements across the thickness of the enamel layer (Level 7). Analysis of the enamel and bone hierarchical structure suggests similarities of scale distribution at each level. The study also aimed to understand further the structural–mechanical relations at each hierarchical level.

The prevailing concept of enamel structure is that the basic elements of enamel are nanosized fibril-like hexagonal hydroxyapatite crystals, which are further attached into groups. The most readily apparent structural blocks of enamel, termed prisms and interprisms, are imposed on this arrangement. It has been proved that in both sites the composition is identical; the only substantial difference is the orientation of the crystals. The crystals in the prisms, particularly along the center, tend to align lengthways and lie parallel to the prism axis, but deviate more and more as their distance from the center increases.

In the latter location, the crystals tend to orientate perpendicular to the incremental lines. Where the crystals within these two locations meet, the structure is discontinuous, leaving a gap, the prism sheath. The prisms and interprisms further construct a distinct structural pattern in the enamel. Generally, human molar enamel contains a superficial layer of enamel with parallel prisms, so-called radial enamel, within which the prisms are orientated radially and intercept the occlusal surface perpendicularly. In the inner two-thirds of the enamel, the varying complexity often consists of orientation differences, which are generally referred to as "decussation."

Figure 6.12 presents a schematic diagram of the hierarchically assembled structure of enamel, which has been depicted as a spectrum covering the range from nanometers to millimeters. As revealed by HRTEM and SAD analysis, the major component of enamel is hexagonal hydroxyapatite crystals (Level 1). At the nanoscale, the crystals at first form mineral nanofibrils (Level 2), the most unique structural units of enamel, which always align lengthways and aggregate into fibrils (Level 3) and further thicker fibers (Level 4). Then, the fibrils and fibers cluster parallel to each other in two different preferential orientations, assembling into prism/interprism continua (Level 5) at the mesoscale. At the microscale, prisms assemble into prism bands (Level 6), which present differing arrangements across the thickness of the enamel layer (Level 7) to match the mechanical and physical requirements of enamel in the oral environment.

The hierarchical structure of enamel may imply that the formation of enamel crystals undergoes two stages: first, the crystals rapidly elongate along their *c*-axes, parallel to each other, and later grow in width and thickness into the nanofibrils with their special aspect ratio. Some of the nanofibrils have the possibility to attach into small groups as they grow thicker. They are always separated by organic material and have very well-defined shapes. Then, concurrent with the massive efflux of the organic components and water, the crystals undergo the second growth spurt and are eventually bound into aggregations, mainly by enamelin cleavage products. However, until now it was not fully understood whether the fibrils and fibers are the original structural components of enamel tissues or are created during acid erosion. The facts found in this study, that the patterns are invisible in the intact samples, as presented in Fig. 6.13, but become more and more evidently revealed as the surfaces are eroded, may suggest that, even though it is partly because of the enhancement of the surface relief, a combined presumption is preferred. The patterns observed are created during acid erosion; however, they reflect the

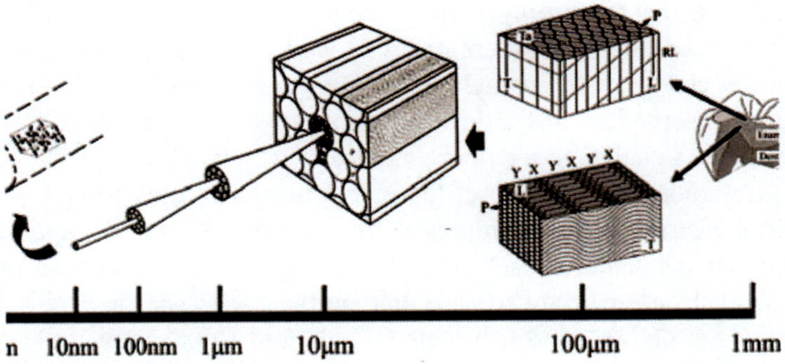

Fig. 6.12 Schematic illustration (not drawn to scale) of the hierarchical assembly of enamel structure, from the millimeter to the nanometer scale. The ruler below the diagram demonstrates the typical scale distribution of each assembly level. *IP* interprism, *L* longitudinal plane, *P* prism, *RL* Retzius line, *T* transverse plane, *ta* tangential plane; X, Prisms appear as bands of approximately cross-sectioned; Y, Prisms are relatively longitudinally arrayed (Cui & Ge 2007)

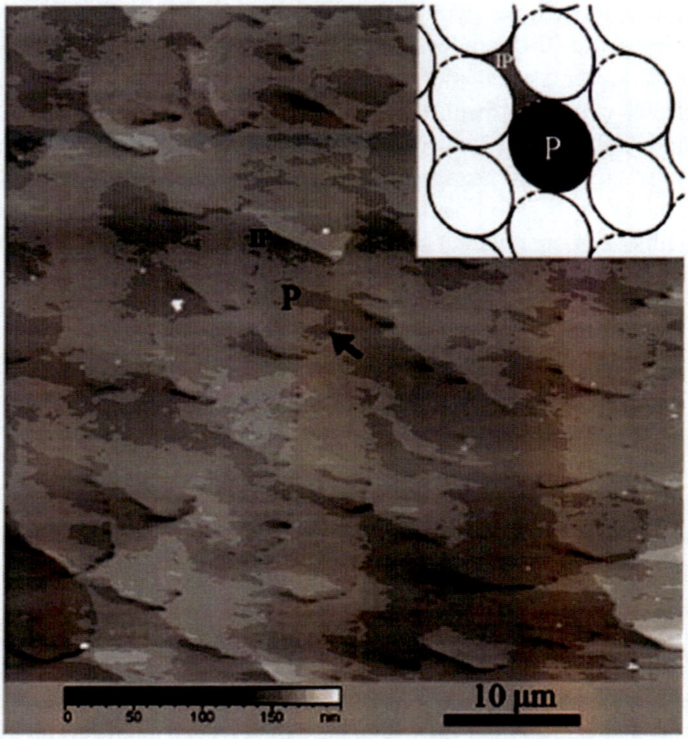

Fig. 6.13 AFM image of the enamel surface without etching treatment. The sheaths are indicated by the black arrows. The insert illustrates the area definition. The fiber and fibril patterns are invisible in the image. *IP* interprism, *P* prism (Cui & Ge 2007)

divisions inside the enamel crystals, determined by the two-stage mineralization process. It can be presumed that the minerals in the fusion interfaces between the nanofibrils or their small groups are vulnerable in acidic environments; thus, the divisions of the crystals are eventually sculptured in the eroded surface. This is because of the unique maturation stage, as well as the significant structural and componential changes that enamel undergoes. An explanation of the formation of the patterns observed in this study is complicated, and providing further direct evidence is still a challenge for future studies.

One interesting characteristic of the hierarchical assembly to be highlighted is that the scale distribution of the representative patterns at all levels appears regular, indicating that the scale of each level is around ten times than that of the next primary level. It is important to note that similar characteristics are found in the structures of both human and fish bone (Wang et al. 2004). Taken together, these results may imply an inherent regulation of mineralized hard connective tissues. Moreover, such a regulation may exist not only in the natural biomaterials but also in some self-assembly synthesized materials (Zhang et al. 2003a). Therefore, such structural motifs presumably have the advantage of stabilizing systems that consist of a large amount of small subunits. Although giving theoretical explanations for this statement is a challenge, we can speculate that the stability of the system with hierarchical structures would be fairly adjusted.

The hierarchical distribution of structures within biomaterials reflects adaptation to function, among which the mechanical function is the first to be highlighted. Recently, as advanced techniques such as nanoindentation are being more and more widely applied in dental enamel research, the diversity of the mechanical properties within enamel at various scales has been studied. Inspecting current achievements of measuring the nanomechanical distribution in human dental enamel implies that it fine-turns the hierarchical assembly of enamel structure from the microscale to the mesoscale. Cuy et al. (2002) have measured the nanomechanical property mapping of enamel by nanoindentation, which represents the mechanical distribution throughout the thickness of enamel layer. From the viewpoint of the hierarchical assembled structure, such mapping is determined to some extent by the differing arrangements of the prism bands, indicating that the orientations and alignments of prisms influence the mechanical diversity at the microscale. At a higher resolution of mesoscale, the nanomechanical distributions within the prism/interprism continuum, including the nanohardness and elastic modulus of the prisms, interprisms and sheaths and the anisotropy of a single prism, have been explored (Ge et al. 2006). Merging the mechanical analysis at this level could result in an integrated system that well reflects the properties of the prism/interprism continuum. However, using currently available technologies, it is still a challenge to measure the mechanical property distribution within smaller areas of enamel accurately. We can look forward to such expecting measurements of enamel as the measuring facilities are further refined.

The clarification of the hierarchical assembly of enamel structure provides a novel insight into the microstructure of dental enamel, which may be important for understanding the unique mechanical properties of enamel and its stable chemical properties. In addition, it may have potential value in developing therapeutic strategies for dentists. Furthermore, from a materials science viewpoint, the insights gained from the study of these fascinating materials are not only important biologically, but may be helpful in developing our understanding of the relationship between the structures and mechanical properties of materials, as well as well providing new schemata that can be applied to the design and synthesis of advanced materials by biomimic methods.

6.4 Study on the Principles of Calcium Carbonate Mineralization

Many leading scientists over the world like Williams, Mann and Weiner et al. are devoted to the theories that illustrate the process of biomineralization as organic–inorganic interfacial recognition, molecule recognition and molecular geometric match, etc. The researches developed from microscale into nanoscale, from structure conformation into biomimic preparation, from theory analysis into biomimetic synthesis, from cell mediation to gene mediation ((Davis 2004); (Ameye et al. 2001); (Hunter 1996); (Choi & Kim 2000); (Ogasawara et al. 2000); (Sarikaya et al. 1999); (Mann 1996)).

Calcium-based materials are the most important inorganic phase of biomineralization systems. Calcium carbonate is one of the most important biominerals that have been a research focus for decades. There are three crystal forms of anhydrous calcium carbonate: calcite, aragonite and vaterite, and their crystal systems are rhombohedral, orthorhombic, and hexagonal, respectively (de Leeuw & Parker 1998). In aqueous system at 25°C, they have decreasing stabilities and increasing solubility limits. Their solubility constant (K_{sp}) values are $10^{-8.48}$, $10^{-8.34}$, and $10^{-7.91}$ (Plummer & Busenberg 1982), respectively. Geological minerals of calcium carbonate are almost all calcite with a few aragonites, since its Gibbs free energy is the lowest of the three. In biominerals, calcite and aragonite are the most common forms of calcium carbonate crystals, which exist mainly in mollusk shells and bird eggs (de Leeuw & Parker 1998). Vaterite is metastable, the most unstable crystal form of calcium carbonate, and would automatically transform into calcite or aragonite in aqueous solution. But in natural systems like carp asteriscus (Li & Feng 2007) and lackluster pearls (Hang 1994) in fresh water, it was found that vaterite could exist stably. Based on their unique characteristics, vaterite and amorphous calcium carbonate (ACC) have become hot research spots in recent years. In order to understand how protein matrices mediate calcite, aragonite, and vaterite in living organisms, and how complicated microstructures are formed, many experiments were processed in vitro to simulate biomineralization.

6.4.1 *Effects of Additives on Calcium Carbonate Mineralization*

6.4.1.1 Soluble Matrices (SM) in the Solution as Additives

In most biomaterials, inorganic components are dominant with over 95% of the mass or volume. Protein matrices, microelements, polysaccharides, and others share the rest 5%. Studies for years have proved that proteins are the most important influence factors in the formation of calcium carbonate crystals. One of the most commonly used method to study the effect of proteins on mineralization of calcium carbonate crystals is to extract, separate, add them into simulation systems, and to study their functions under different conditions in vitro.

In previous studies, organic matrices were extracted by their solubility in sodium ethylenediaminetetraacetic acid (EDTA), and were classified as insoluble matrix (IM) proteins and soluble matrix (SM) proteins. IM and SM are found to have different roles in controlling the polymorph and morphology of calcium carbonate. IM proteins are normally structure molecules, and provide SM joint sites as substrates. In the mineralization process, IM could control crystal growth, which influences the size and orientation of the crystals. SM proteins are mainly composed of acidic macromolecules, distributing on the surface of IM, could either directly touch with crystals, or distribute in crystals. The SM matching the holes of IM could combine calcium, and provide nucleation sites. On the contrast, when SM in solutions, they could inhibit crystal formation. SM proteins mainly determine the crystal forms in organism, may possibly control crystal growth, and also act importantly in biominerals concern with cell activities, such as ion transportation, enzyme regulation, and hormone (Qiao et al. 2008a). In studies of Feng et al. (2000a), Falini et al. (1996), Samata et al. (1999), and Kono et al. (2000), it showed that in the mineralization process, IM proteins mainly act as structure frames, providing nucleation sites for calcium carbonate crystals, while SM proteins control the polymorphs of calcium carbonate crystals.

Many in vitro mineralization experiments have proofed that SM extracted from calcite biominerals could induce calcite growth, and SM extracted from aragonite biominerals could induce aragonite growth. For example, the protein matrices extracted from nacreous layer (aragonite crystals) in mollusk shell *Mytilus edulis* induced aragonite formation, while those extracted from prismatic layer (calcite crystals) induced calcite formation (Feng et al. 2000b). SM extracted by (Belcher et al. 1996) from shells of *Haliotis refescens* abalone could well induce aragonite, so they believed that SM alone could control the polymorph and morphology of calcium carbonate crystals, IM was not necessary. Feng et al. (2000b) proved that IM from nacre of shell *Mytilus edulis* could influence the size and density of the crystals (Fig. 6.14). While Falini et al. (1996) pointed out that besides SM, IM could also control calcium carbonate crystal polymorphs. They chose SM from biominerals with different calcium carbonate crystals, while IM from β-chitin and silk-fibroins of other animals, adding them into the system with different combinations. The results showed that SM and IM together could induce the same $CaCO_3$ crystal form with the biominerals of which SM were extracted, but β-chitin and SM together without silk fibroin could only induce calcite crystals, no matter whether the SMs were extracted from nacreous layer (aragonite crystals)or prismatic layer (calcite crystals). On the other hand, IM alone did not have any influence on the polymorph. The results show that SM is one of the most important factors for crystal control, but not the only one. From the above results, we can see that different results with different conditions could explain a same natural phenomenon, which means that we are still not holding all the information of protein matrices-mediated biomineralization.

With the development and fulfillment of theories and technologies, new fields of protein-mediated biomineralization emerged. New protein matrices extraction methods were developed by Lopez group of National Museum of Natural Science in Paris (Pereira-Mouriès et al. 2002). Acetic acid and Milli-Q water, replacing EDTA, were used in experiments in order to avoid the potential damage of protein structures. In this way, water-soluble matrix (WSM), acid-soluble matrix (ASM), and acid-insoluble matrix (AIM) can be extracted from pearls. Pearls produced by the freshwater mussel *Hyriopsis cumingii* crystallize sometimes in the form of vaterite instead of aragonite. Using this new extraction protocol, the so-called water-soluble organic matrix, WSM, was extracted and used as additive in controlled calcium carbonate growth experiments. Vaterite crystals were grown for the

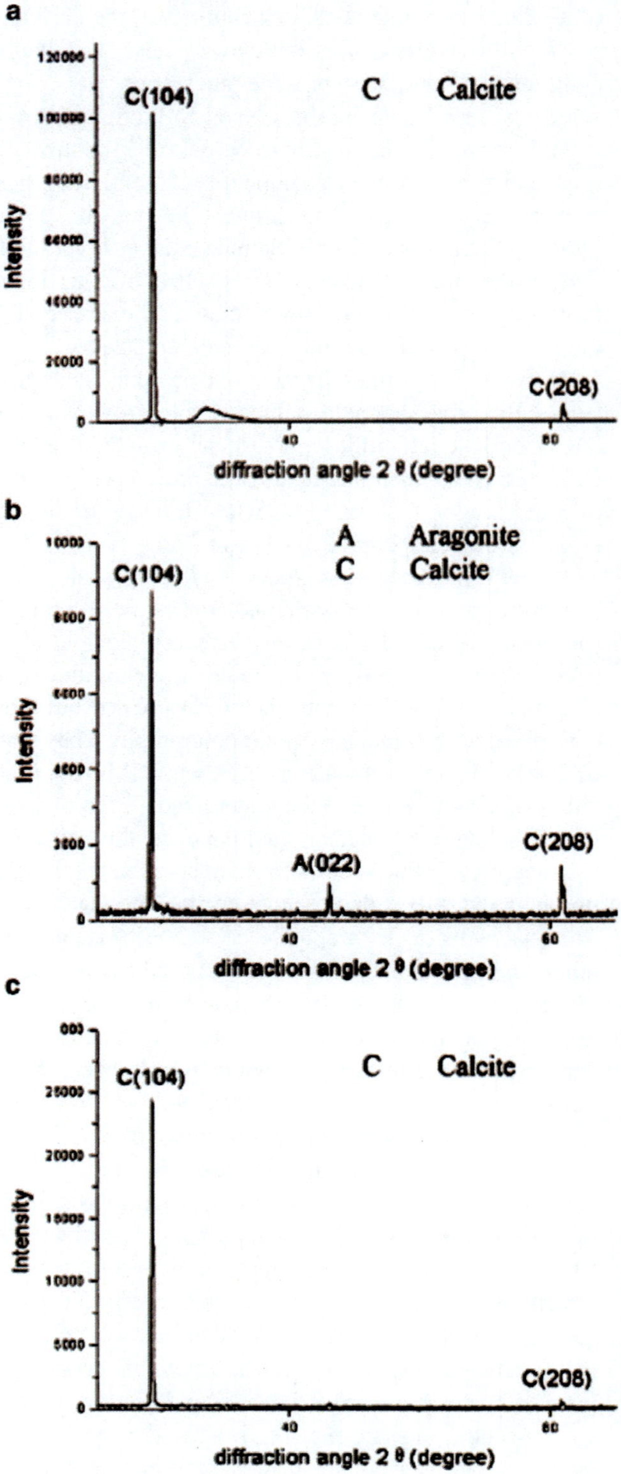

Fig. 6.14 XRD patterns and SEM images of the calcium carbonate crystals grown on single-crystal silicon (0 0 1), (**a**) without protein, (**b**) with soluble proteins from nacreous layer, and (**c**) with soluble proteins from prismatic layer (Feng et al. 2000b)

greater part when WSM from vaterite pearls existed in the solution. The formation of vaterite appears to be clearly connected to the water-soluble matrices. Such kind of extraction method could also define different effects of protein matrices in a more detailed way. Using biochemical methods like reverse-phase high-performance liquid chromatography and SDS-PAGE, scientists have successfully obtained purified proteins, such as nacrein (Miyamoto et al. 1996), lustrin A (Shen et al. 1997), MSI60 (Sudo et al. 1997), N16 (Sarikaya et al. 1999), pearlin (Miyashita et al. 2000), mucoperlin (Marin et al. 2000), N14 (Kono et al. 2000), N66 (Kono et al. 2000), perlucin (Weiss et al. 2000), perlustrin (Weiss et al. 2000), p20 (Bedouet et al. 2001), MSI7 (Zhang et al. 2003b), AP7 (Michenfelder et al. 2003), AP24 (Michenfelder et al. 2003), p10 (Zhang et al. 2006), and AP8 (Fu et al. 2005). Some of their effects on mineralization under in vitro conditions were studied. Almost all these protein matrices were extracted by EDTA or weak acids (acetic acid normally) dissolving the calcium carbonate crystals, and were classified by their dissolubility. But, the formation of natural biominerals is a very complicated process, in which a series of proteins were involved, even though initial researches about how proteins work in the mineralization process in organisms, a lot of issues were still not figured out clearly, and those proteins with even tiny content might be extremely in such processes.

6.4.1.2 Amino Acids in the Solution as Additives

Amino acid analysis of SM indicated that the three most important amino acids are glycine, aspartic acid, and glutamic acid (Lavi et al. 1998).

Hou and Feng (2006b) studied the function of glycine in the mineralization system; they selected two methods to precipitate $CaCO_3$ particles: titration method and diffusion method. In titration method, under stirring condition, with short depositing time, both calcite and vaterite particles were formed, and the proportion of vaterite increased with the increase of the glycine concentration. The shape of calcite and vaterite are normally rhombic and spherical, respectively (Fig. 6.15). With the extending of deposition time, all calcium carbonate crystals are vaterite particles with spherical and spindly morphologies, regardless the concentration of glycine. The results showed that stirring condition itself was sufficient to induce formation of spindly vaterite when the dripping velocity was slow enough. The possible reason is that stirring changed the activation energy of nucleation (ΔG_n) of calcite and vaterite in the aqueous system and upset their nucleation priorities (Mann et al. 1993). They also found that rapid titration produced local instantaneous high supersaturation to overcome the nucleation energy barrier of both calcite and vaterite. On the other hand, slow titration provided a longer time for nucleation and limited the supersaturation value to induce vaterite. In the diffusion method with glycine in the solution without stirring, depositions were almost all spherical vaterite particles without any spindly ones, which implied that the spindly shape occurring in the titration experiments is directly caused by stirring. And they supposed that glycine may change the activation energy of nucleation in

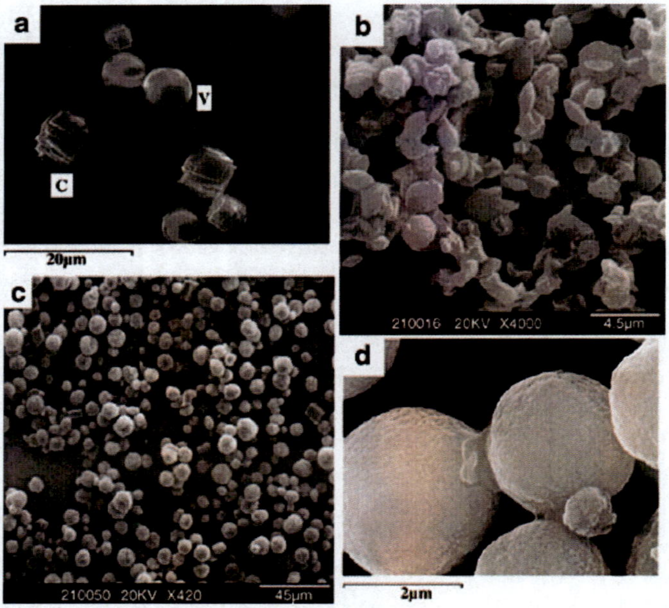

Fig. 6.15 SEM morphologies of vaterite crystals obtained by the titration and diffusion method. (**a**) [Gly] = 10^{-3} M, dripping velocity 15 ml/min, *C* calcite, *V* vaterite; (**b**) [Gly] = 10^{-3} M, dripping velocity 2 ml/min; sample without glycine showed similar result; (**c**) [Gly] = 10^{-3} M, diffusion method, almost all vaterite; and (**d**) Detailed structure of vaterite, composed of oviods with a size about 0.1 nm (Hou & Feng 2006b)

aqueous solutions, like the function of stirring. In the presence of glycine, ΔG_n ($\Delta G_n = \Delta G_{\text{surface}} + \Delta G_{\text{bulk}}$) of calcite became relatively higher than that of vaterite, which would make vaterite to deposit in the first place. The possible reason is that, in solution, carboxyl group of glycine may affect hydrated surface energies (de Leeuw & Parker 1997) by interacting with calcium ions like the behavior of surfactant (Donners et al. 2002a). The solution height may also be an important factor since the determinant factor in the Gibbs formula is no more surface energy but bulk lattice energy, which was in accordance with the result of all aragonite formation. When the height of the solution was lowered, the bulk lattice energy of aragonite became the lowest.

Glutamic acid and aspartic acid are both acidic amino acid that exist in biominerals, and their influences on calcium carbonate mineralization have been studied intensively. It is illustrated that their functions may lay in the negative electricity of *R-radical*, which would absorb dissociate Ca^{2+} ions or Ca^{2+} ions on crystal surfaces, so as to change the crystallization process of calcium carbonate. It is found that both glutamic acid and aspartic acid have the ability to induce vaterite growth in solution. Tong et al. (2004) succeeded in mediating porous vaterite crystals, considering it was due to the absorption between aspartic acid and the surface of calcium carbonate crystals, and inhibited formation of calcite. (Manoli and Dalas 2001) found that glutamic acid could stabilize vaterite crystals.

In mineralization process, not only crystal forms, but also morphologies are influenced by amino acids. In very low solution height systems (Hou & Feng 2006a), which is more likely to happen in biominerals, only calcite crystals were formed, vaterite and aragonite were avoided. The calcite morphologies changed with different amino acids in the solution. Aggregates were found in both glycine and aspartic acid systems, and the difference between glycine and aspartic acid is that aspartic acid is an acidic amino acid that has two $–COO^-$ groups in one molecule. This is a possible reason for which aspartic acid has the ability of agglutinating two or more nuclei, glycine does not seem to have this ability.

6.4.1.3 Magnesium Ions and Collagen in Solution as Additives

Since sea water contains 0.13 wt.% of Mg, people have been interested in how Mg^{2+} influences calcium carbonate mineralization in marine animals for a long time. It is found that Mg^{2+} can replace Ca^{2+} in calcite, but cannot inter crystal lattice in aragonite. So, in the condition of high Mg/Ca concentration, the formation of calcite crystal nucleus would be inhibited, which promoted aragonite nucleus formation; in this way, calcium carbonate crystal form was mediated (Mann 2001). People also discovered that Mg^{2+} could promote amorphous calcium carbonate (ACC) formation. (Raz et al. 2003) studied ACC in the spicule of *Juvenile* sponge, they found that only calcite was formed without Mg^{2+} in the solution, and the spicule protein could not mediate ACC formation, which showed that Mg^{2+} was extremely important in ACC stabilization. They also pointed that calcite form was transformed from ACC in high Mg^{2+} concentration (Fig. 6.16). Loste et al. (2003)

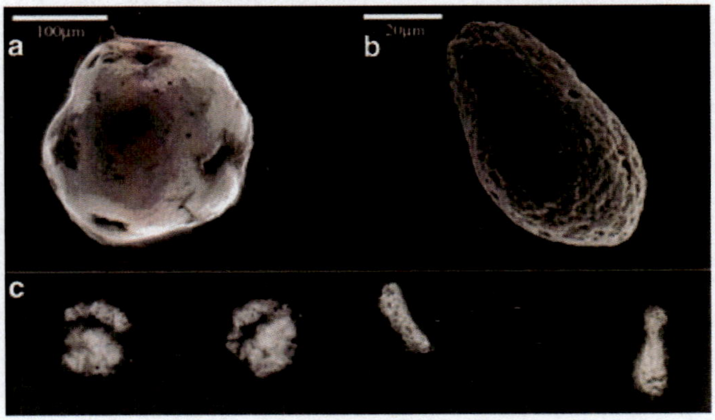

Fig. 6.16 Calcium carbonate precipitates grown in the presence of Mg/Ca = 2:1 and the addition of 35 nmol ml^{-1} of (**a**, **c**) macromolecules extracted from 48 h spicules; (**b**, **d**) macromolecules extracted from 72-h spicules. (**a**, **b**) SEM images; (**c**, **d**) images taken under cross-polarized light. The particles in (**d**) are single crystals, which extinguish each in a different position of the polarizer, while the particle in (**c**) has homogeneous extinction in all positions (It is found that extracts of spicules from 48-h prism stage embryos induce the formation of a transient ACC phase, whereas extracts from spicules of 72-h plutei induce the formation of single calcite crystals) (Raz et al. 2003)

also found that Mg^{2+} combination could defer ACC transformation for the most part, and such effect improved with the increase of Mg^{2+} amount.

In biomineralization processes, the formation of inorganic crystals is controlled by organic macromolecules such as proteins ((Mann 1996); (Addadi & Weiner 1985)). Collagen is the most important water-insoluble fibrin which represents the framework of extracellular matrix. The basic structure of collagen is tropocollagen, and its primary structure has a repetitive sequence of $(Gly\text{-}x\text{-}y)_n$, of which x usually represent Pro, and y of Hy-Pro of Hy-Lys. Shen et al. (2002) studied collagen-mediated calcium carbonate crystal in vitro in order to reveal the principle of protein-mediated mineralization process. The results of XRD and SEM of calcium carbonate crystal deposition showed that only calcite was formed, and the calcite growth is more and more inhibited as collagen is increased in concentration (Fig. 6.17). It means that collagen does not change the polymorph of calcium carbonates. With the increasing concentration of collagen, the morphology of calcite changed from perfect rhombohedral to well-faceted rhombohedral crystals with little disfigurement, and then to overgrown calcite crystals with new planes. The thickness of new layers became thinner with the increasing of collagen, some of calcite crystal planes would develop into a flowerlike form, and they become spherulites as the concentration further increases (>10 g/l). They provided an explanation, why collagen would absorb on the edges of {1 0 4} planes of calcite, which are parallel to the *c*-axis, and inhibited its growth, thus new planes appeared. The new planes started to appear at the edges of rhombohedral calcite crystals when collagen was present in the solution, since crystal growth starts at edges and corners, which provided good attachment sites; it led to the decrease of growth rate and new plane growth on such positions. New planes formed have better attraction for collagen, which would eventually alternate the morphologies. Collagen additives not only influenced morphology of the crystals, but also the quantity. Number of crystals increased, while size diminished with the increase of the protein concentration. The adsorption of protein from solution onto a solid plane is determined by the stability of its structure. One unique characteristic of collagen is its structure stability, and behaves as colloidal particles and only adsorbs onto ionic planes.

Jiao et al. (2006) studied the joint influence of Mg^{2+} and collagen. It was confirmed that Mg^{2+} could stabilize amorphous calcium carbonate and control calcite morphologies. They also found that Mg^{2+} could induce spherical aragonite and vaterite (small amount), and such action was amplified by collagen. But collagen alone had no significant influence on calcium carbonate crystals, and almost all calcite crystals were formed, which showed that collagen has a promotional effect on magnesium ions in controlling the polymorph of $CaCO_3$ crystals (Fig. 6.18). They also provided an explanation; magnesium is likely to react with collagen to change the stereochemical structure of collagen molecules, and thus induced aragonite or vaterite.

As a summary of the above discussion, we think that polymorphs of the formed crystals essentially lie on their nucleation energies in the system. There are three ways to change nucleation priority: influence on $\Delta G_{\mathrm{surface}}$ like glycine, influence on ΔG_{bulk} like Mg ions, and influence on the relationship between $\Delta G_{\mathrm{surface}}$ and ΔG_{bulk} (influence on the nucleation type) like choosing low solution height.

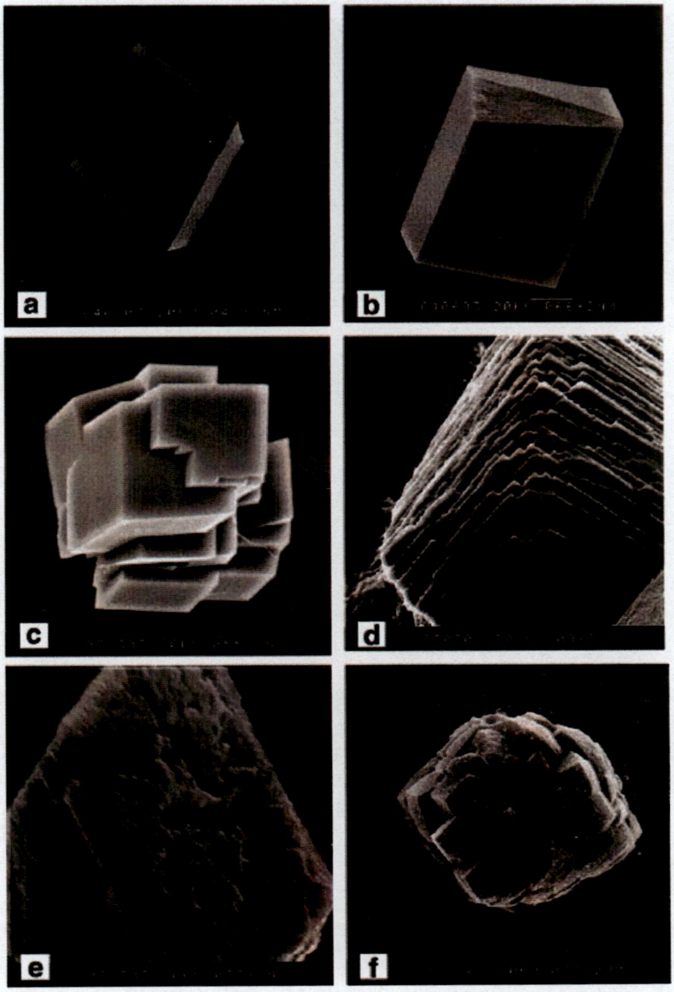

Fig. 6.17 SEM morphologies of the calcite crystals precipitated in the solution. (**a**) Rhombohedral calcite crystal grown in the solution without collagen. (**b**) Rhombohedral calcite crystal with little disfigurement, collagen concentration: 0.1 g/l. (**c**) Overgrown calcite crystal with new planes, collagen concentration: 0.1–5 g/l. (**d**) Multiple layer calcite crystal with thinner layer thickness, collagen concentration: 5–10 g/l. (**e**) Some planes with flowerlike pattern, collagen concentration: 5–10 g/l. (**f**) Spherulitic calcite aggregates at higher collagen concentration: >10 g/l (Shen et al. 2002)

6.4.2 Effects of Templates on Calcium Carbonate Mineralization

The formation of biominerals is also controlled by organic template molecules resulting in materials with unique shapes and properties. Templates used in in vitro mineralization refer in particular to those substrates with particular sequence fragments that could mediate special mineral crystals. It is generally believed that

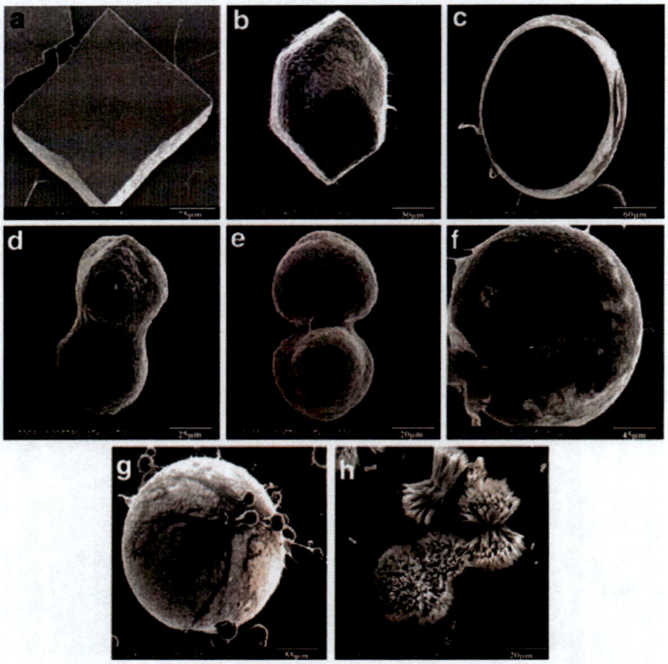

Fig. 6.18 SEM morphologies of the calcium carbonate crystals precipitated in the solution, collagen concentration: 0.1 g/l. (**a**) Irregular rhombohedral calcite crystal grown in the solution without magnesium. (**b**) Irregular lumpish crystals with lamellar growth structure (Mg/Ca:1). (**c, d, e**) Discoid and dumbbell calcium carbonate crystals. (**f**) Spherical aragonite crystals at higher magnesium concentration (Mg/Ca:5). (**g**) Spherical aragonite crystals with more regular shape (Mg/Ca:5, collagen concentration:0.4 g/l). (**h**) Aragonite crystals with needlelike shape without collagen (Mg/Ca:5) (Jiao et al. 2006)

the template molecules act as nucleators for the inorganic material and that the surface chemistry of the template induces oriented nucleation of the complementary crystal face. The history of using templates in in vitro mineralization experiments could date back to 1988, when Mann et al. (1988) chose *Langmuir* monomolecular membrane, took advantage of its ordered arrangement of radicals to mediate calcium carbonate crystal growth.

6.4.2.1 Natural Biomineral Template

Qiao et al. (2008a) chose the fresh cross section of nacre surface of freshwater lustrous pearls (aragonite) as templates without any additives in solutions. XPS results showed that the surface is mainly composed of organic matrix particles, AFM micrograph showed its size of 70 nm. The deposition proved that the formation process was a complex and multistep one, from an ACC layer (Fig. 6.19), iso-oriented nano-stacks to hexagonal aragonite tablets (Fig. 6.20). The result not only proved the existence of ACC in aragonite formation process,

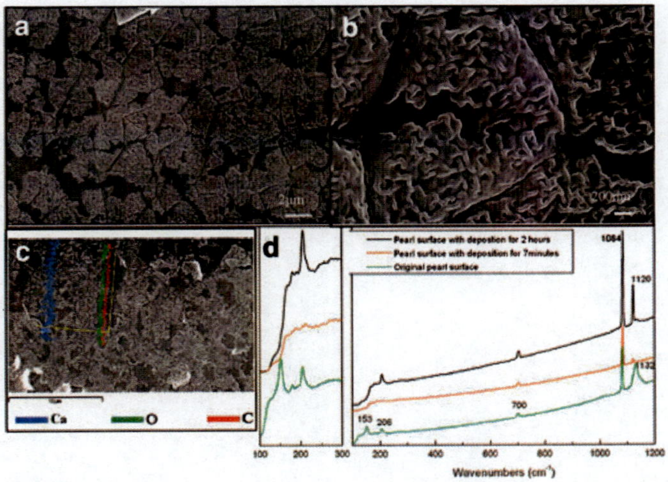

Fig. 6.19 ACC grown on the surface of aragonite pearl in the first 5 min (**a**) SEM image, (**b**) expansion of (**a**), (**c**) EDS result, and (**d**) Raman result (Qiao et al. 2008a)

but also verified that organic matrices could induce the same calcium carbonate crystal polymorph in vitro as in vivo.

In order to further study different effects of WSM and ASM, only WSM and ASM of asteriscus were added into the solution without templates. SEM results showed that almost all the crystals were vaterite; there was a little difference in crystal morphology between the vaterites induced by WSM and ASM respectively, as showed in Fig. 6.21, the possible reason lies in different effects on the growth rate of *a*-axis and *c*-axis.

Similar experiments were carried out by Li (2008), the templates were the fresh cross section of lapillus and asteriscus. SEM and XRD showed that perfect aragonite formed on the surface of lapillus (aragonite), and perfect vaterite crystals formed on the surface of asteriscus (vaterite) without any additive in solutions, similar as the crystal forms in natural carp otolith. Calcium carbonate crystal depositions on lapillus templates with lapillus WSM or ASM in solutions were all tested to be aragonite ones, and their morphologies were both similar to that without any additives (Fig. 6.22). While, when adding asteriscus WSM into the system of asteriscus template, vaterite layers were viewed by SEM, and we deduce that WSM was absorbed from solution onto vaterite layer, and then induced vaterite formation on another exposed surface. An organic–inorganic complex structure was then formed accordingly. When adding asteriscus ASM into the system of asteriscus template, crystal particles formed were more intensive, and the morphology was not the same as those deposited in systems without protein, or with WSM in the solution, it tended to grow a platy form particles in solution (Fig. 6.23), rather than grow a spherical form particles on the surface of asteriscus.

Another important natural biomineral template used in in vitro mineralization is IM. Falini et al. (1996) showed that soluble shell protein matrix can determine the polymorph of crystals grown on a substrate of squid chitin and silkworm fibroin, in

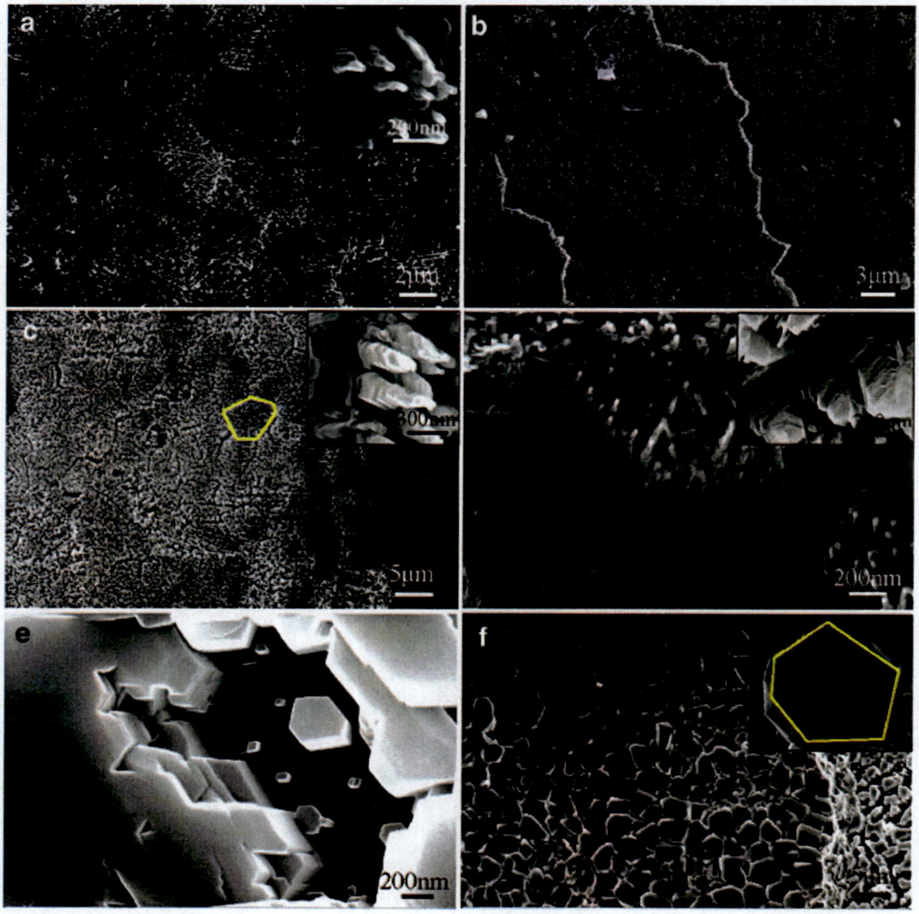

Fig. 6.20 In vitro growth process of nacre-like aragonite tablets on the surface of aragonite pearl (**a**) 10 min, (**b**) 30 min, (**c**) 1 h, (**d**) 2 h, (**e**) 3 h, (**f**) 10 h (Qiao et al. 2008a)

contrast to earlier work that polymorph controlled by networks of insoluble protein matrix extracted from shells. AIM of aragonite and vaterite pearls were used as templates in in vitro experiments, and results showed that perfect calcite crystals were formed on both two templates. The crystal sizes were all 30 μm, which showed that AIM alone has no influences on crystal polymorphs or morphologies. The crystals deposited on AIM templates were considered to be formed in solution and then fell onto the template.

6.4.2.2 Amino Acid Template

Construction of organic–inorganic hybrid materials with controlled mineralization analogous to those produced by nature is now of current interest for both organic

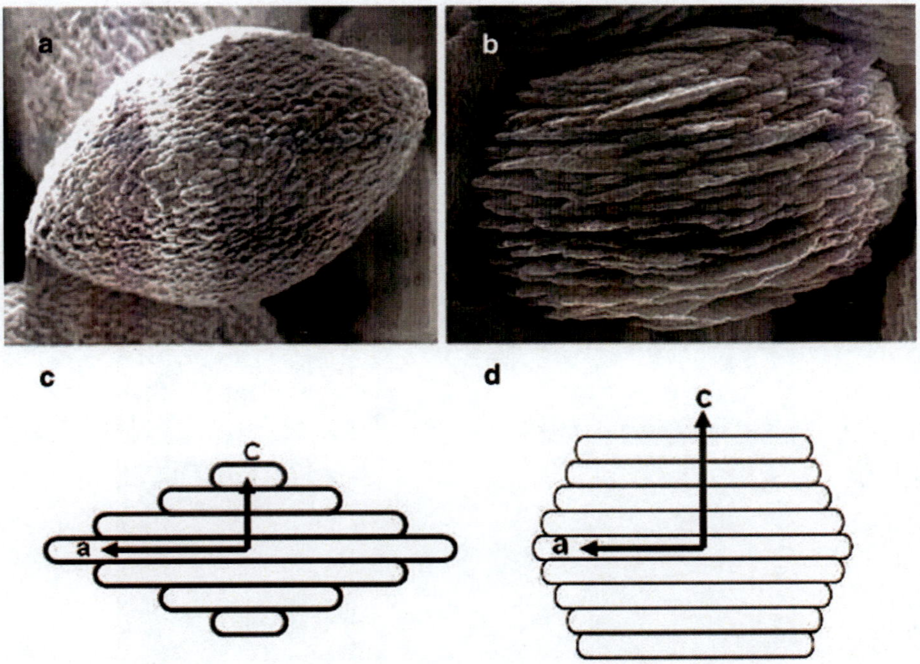

Fig. 6.21 Morphologies and sketch map of vaterite crystal growth under different protein matrix function (**a**, **c**) WSM in solution, (**b**, **d**) ASM in solution (Li 2008)

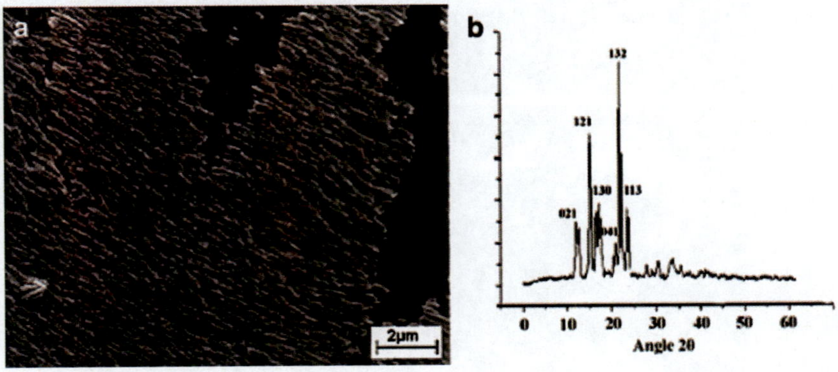

Fig. 6.22 SEM (**a**) and XRD (**b**) result of aragonite crystals on lapillus templates without additive in the solution (Li 2008)

and inorganic chemists. In order to seek out industrial and technological applications, it is necessary to understand the mechanism of the natural biomineralization process. Many templates with different characteristics were used in experiments, and we only introduce some examples so as to expand visual fields.

Amino acid modified calcite substrate was used as templates by Qiao et al. (2008b), and the characteristic of amino acid was able to deposit oriented calcite

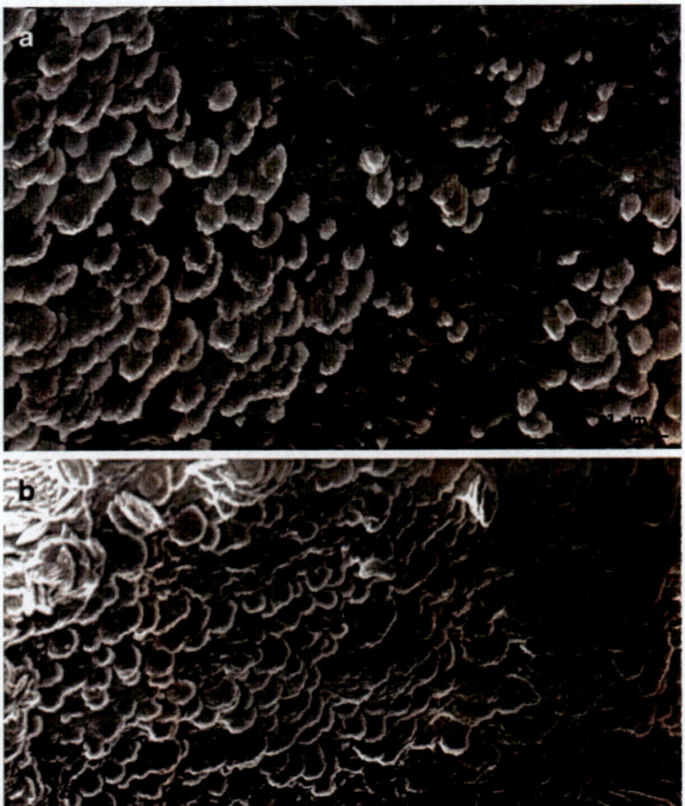

Fig. 6.23 SEM results of vaterite crystals on asteriscus templates without additive (**a**), with WSM (**b**), and with ASM (**c**) in the solution (Li 2008)

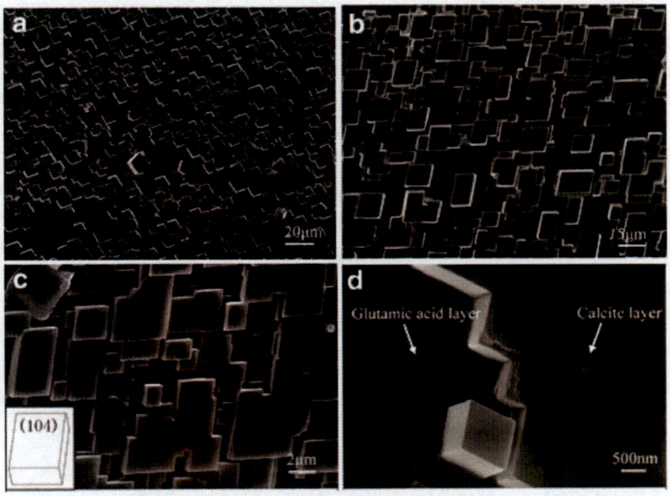

Fig. 6.24 SEM micrographs of calcite tablets and calcite layers deposited on amino acid layers. (**a**) Calcite tablets grown on glutamic acid layer for 1 h, (**b**) calcite tablets grown on arginine layer for 1 h, (**c**) the bigger and conjunctive calcite tablets grown on glutamic acid layer for 1.5 h, and (**d**) the calcite layer grown on glutamic acid layer for 2 h (Qiao et al. 2008b)

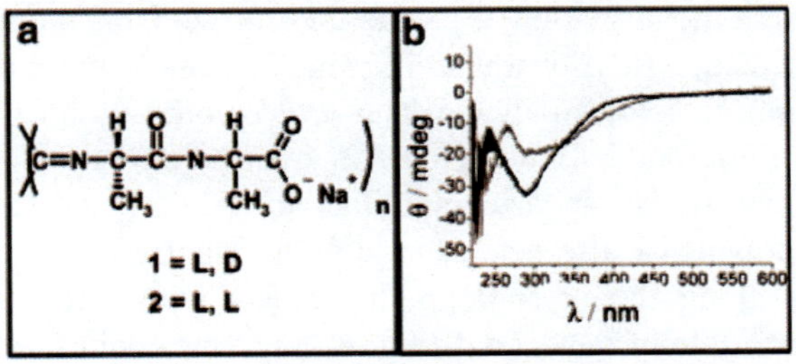

Fig. 6.25 (**a**) Chemical structures of polymers 1, and 2. (**b**) CD spectra of aqueous solutions of $1/Ca^{2+}$ (*black*) and $2/Ca^{2+}$ (*gray*); Ca^{2+}/repeat unit = 1:1, [polymer] = 1.3 mM (in repeat units) (Donners et al. 2002a)

crystals as showed in Fig. 6.24. The size, distribution, and orientation of the calcite crystals deposited on template were uniform, and the crystal plane exposed was (104). The size and density of the crystals changed in accordance with amino acids.

Donners et al. (2002b) reported a shape-persistent polymeric template composed of alanyl-alanine-derived poly (isocyanide)s 1 and 2, of which the rigid macromolecules possessed a regular distribution of carboxylic acid-terminated side chains, and its molecule formula was given as Fig. 6.25.

SAXS and CD spectrum indicated the persistence of the rigid helical structure of the macromolecules under the conditions used for crystallization experiments. It also showed that the presence of Ca^{2+} ions even stabilized the polymer architecture,

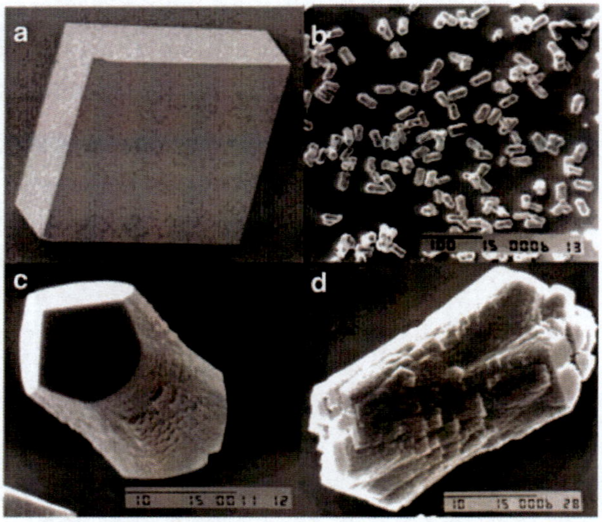

Fig. 6.26 SEM of (**a**) calcite grown in the absence of polymer, (**b**, **c**) calcite grown in the presence of polymer 1, and (**d**) crystals grown in the presence of polymer 2 (Donners et al. 2002b)

probably because complex calcium ions screened the peptide bonds in the polymer side chains from water molecules, thereby prohibiting the slow, but gradual disruption of the hydrogen bonds. The introduction of such polymer into a crystallization solution resulted in the formation of calcite crystals with apple core-type morphology, and the nucleation density was in the range normally observed for *Langmuir* monolayer templates (Heywood & Mann 1994), but significantly higher than nucleation densities found for templates in bulk solution. The high nucleation densities, the low spread in size, and the absence of rhombohedral crystals suggested that polymer 1 also acts as an efficient nucleator. The formed crystals were elongated along the crystallographic *c*-axis with three {104} end faces expressed on each side of the crystal (Fig. 6.26c). A model of which the polymer absorbed on to (011) calcite showing the orientational match between the carboxylate groups of the template and the carbonate ions in the nucleated crystal face, thus inhibited growth in these directions and allowing growth only along the *c*-axis.

Calcium carbonate is one of the most common biomineral; its growth occurs according to a variety of phases and of morphologies under different conditions: rhombohedra for calcite, needles for aragonite and spherical polycrystalline aggregates for vaterite. The above results showed that the mechanism of growth is affected by ions and molecules as amino acids or proteins. We also showed that they can be influenced by the natural or synthetic (or even inorganic) templates used as substrates. By means of additive proteins or organic templates as support, biological tissues can control the allotropic shape, the morphology but can also stabilize amorphous forms as ACC. Thus, chemistry at ambient temperature plays with additional use of organic molecules to open the door to a much wider range of possible forms. It is not surprising that when evolution requires precipitating large single crystals of calcium carbonate structure, it selects calcite; for transportation or growth, it selects ACC or vaterite.

6.4.2.3 Modified Single Crystal Silicon Template

In in vitro mineralization experiments, glass substrate is often used because of its good characters like extensive resources, convenient usage, cheap price, etc., but glass surface may induce aragonite crystals. Single crystal silicon is another commonly used substrate. The single crystal silicon substrate should be treated by HF acid to remove surface oxide layer, then cleaned by acetone, absolute ethyl alcohol, and deionized water to expose the (100) crystal plane. (Li 2008) tentatively studied radical influence on polymorph of calcium carbonate crystals by surface modification with –OH, $-NH_2$ and –COOH on single crystal silicon substrate. WSM and ASM were absorbed on such substrate to prepare specific templates, and then in vitro mineralization experiments were carried out. SEM and XRD results showed that calcite aggregates were formed on –OH surface, surface with $-NH_2$ had no influence on crystal polymorph, only calcite was precipitated, a small amount of vaterite was formed on –COOH surface due to the acidic electrostatic adsorption (Fig. 6.27).

Calcium carbonate depositions on templates in vitro, of which carp lapillus WSM and ASM absorbed on modified single crystal silicon substrate, were characterized to have different crystal morphologies. WSM of lapillus (aragonite) on $-NH_2$ and –COOH templates could induce aragonite crystals, while amorphous calcium carbonate was also observed in the first 15 min. ASM of lapillus (aragonite) on $-NH_2$ and –COOH templates could induce perfect needlelike aragonite particles, which were similar to the aragonite sticks in natural lapillus, but ACC was not found in the process. These results revealed that WSM and ASM have different effects in the mineralization process: ASM could mediate perfect crystals compared with WSM, but it did not have the size-mediation effect as WSM did. All the results above explained that the formation of calcium carbonate crystals in lapillus is a very complex process in which different kinds on protein matrices combined and functioned. While, when ASM and WSM of carp asteriscus (vaterite) adsorbed on the modified silicon substrate, the deposited crystal particles were similar to those on template without any protein, they are calcite particles with only a few vaterite ones (Fig. 6.28). Why WSM and ASM did not show the mediating effect, the possible reason might lie in the absence of AIM during the extraction process, on which WSM and ASM would absorb, so as to affect the mineralization process of calcium carbonate.

6.5 Principles of Calcium Phosphate Mineralization

Mineralized calcium phosphate by collagen in vertebrates can be considered as a material produced by self-assembly at ambient temperature (Cui et al. 2007). The concept of self-assembling was defined by (Whitesides & Grzybowski 2002), i.e., self-assembling is the autonomous organization of components into patterns or

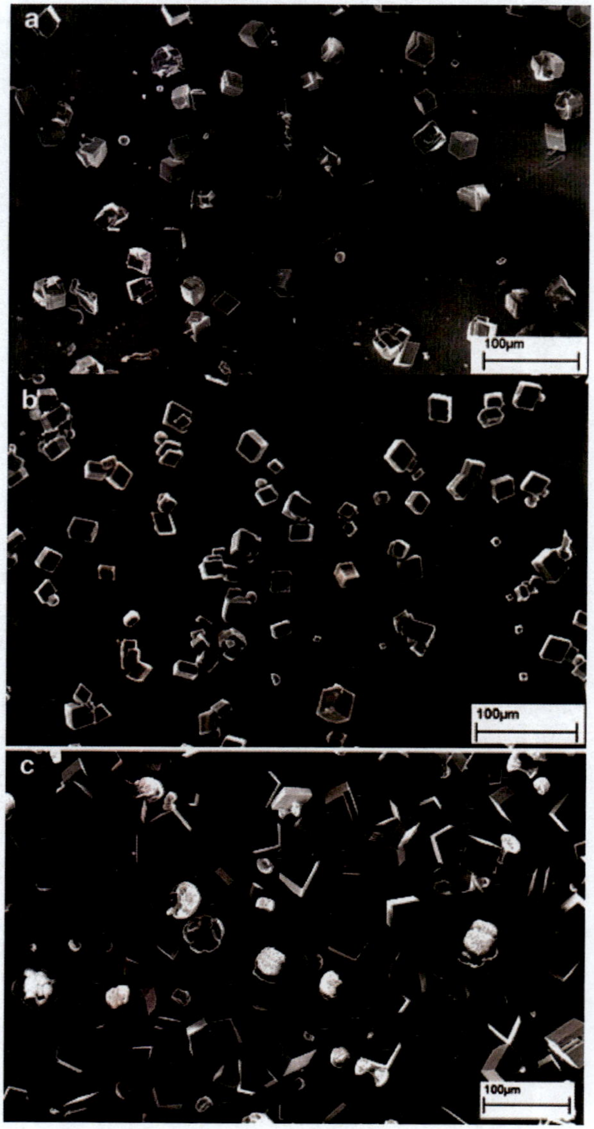

Fig. 6.27 SEM result of calcium carbonate precipitates on the –OH, –NH_2, and –COOH modified surface of single silicon: (**a**) –OH modified surface, (**b**) –NH2 modified surface, and (**c**) –COOH modified surface (Li 2008)

structures without human intervention. It is now considered that self-assembling processes are common throughout natural biomineralization. There are many interesting researches on the structure, formation, and properties of mineralized calcium phosphate, as well as biomimetic synthesis of new materials with the structure of mineralized collagen. The focus is mainly on type I collagen, of which at least 20 collagens have so far been discovered. By mineralized collagen, it refers mostly to calcium phosphate–based crystals, which in bone are found to consist primarily of calcium and phosphate ions, with traces of magnesium, carbonate, hydroxyl, chloride, fluoride, and citrate ions.

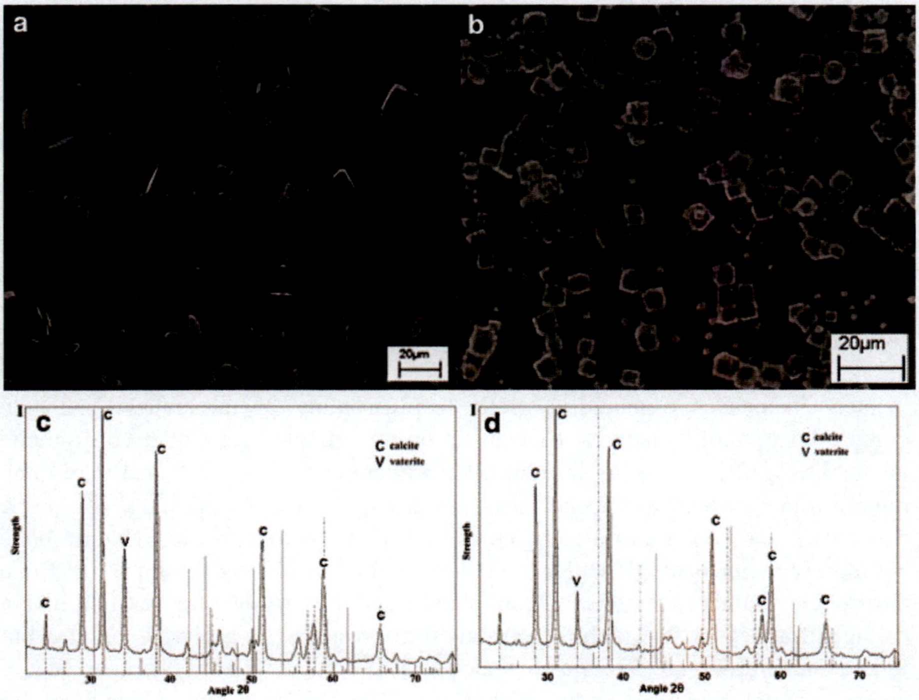

Fig. 6.28 SEM and XRD results of soluble protein matrices–mediated $CaCO_3$ crystals (**a**, **c**) $CaCO_3$ crystals formed on modified silicon with WSM in solution; (**b**, **d**) $CaCO_3$ crystals formed on modified silicon with ASM in solution (Li 2008)

Calcium phosphates are among the most advanced structural composite materials known to be made of macromolecular building blocks. A wide range of tissues, each possessing very different properties, are successfully synthesized in natural environments with only the same basic macromolecular design. These tissues show some common features – they are assembled in numerous assembly ways that allow control of the formation of varying hierarchical structures, from the nanometer scale to macroscale. The concept of hierarchical assembly has been recognized and emphasized by more and more scientists over the last decades, as exemplified by the investigations of numerous biological materials. The hierarchical levels of organization with highly specific interconnectivity and with unique architectures are designed to give the required spectrum of properties for each oriented composite system. Based on these lessons in biology, the laws for the formation of complex composite systems for functional macromolecular assemblies have been probed ((Mann 2001); (Dove et al. 2003)). In addition to gaining knowledge of the fundamental mechanisms for assembly of such materials, the ability to build architectures as a direct consequence of the precision in assembly would certainly open the gate to some new areas of materials science. Examples could be the design and construction of inorganic materials with specified structure, size, shape, crystal orientation, and number of defects and the integration of these architectures into functioning devices for anticipated electrical, optical, magnetic, and chemical outputs. Only with an appreciation and understanding of the principles of calcium phosphate mineralization in such biosystems can such proposals be achieved.

Collagens comprise a family of extracellular matrix molecules responsible for the integrity and mechanical properties of both soft and hard connective tissues, including cornea, skin, tendon, cartilage, and bone. Almost all of the connective tissues with collagen fibrils as the basic building blocks have remarkably similar chemistry at the macromolecular and fibrillar levels of structure. However, differentiation in the hierarchical structure takes place as these fibrils are arranged in the specific architecture required from the construction of special tissues each with unique functions.

Here it focuses principally on the self-assembly of mineralized collagen composites in hard connective tissues and on the relative involvement of mimetic insoluble organic structures in controlled mineralization. In most cases of such mineralization where collagen fibrils are involved, the matrix is a polymeric framework that consists of a complex assembly of macromolecules. Natural bone is a representative example with a typical hierarchically ordered organization. Bone tissues are mainly constructed from nano-sized hydroxyapatite crystals and a collagen framework in which the crystals form, resulting in a highly complex but ordered mineral–organic composite material. This composite itself is organized into layers or lamellae, each with the thickness of a few microns, and in turn these are arranged into higher order structures in a variety of ways depending on the specific bone types (Cui et al. 2007).

6.5.1 Collagen-Induced Calcium Phosphate Mineralization

The hierarchical structure of bone formed by assembly of an orderly deposition of HA minerals within the type I collagen matrix. The crystallographic *c*-axis of the crystals is oriented preferentially parallel along the longitudinal axis of the collagen fibrils. Both investigations and simulations of the hierarchical nanofibril structure in naturally occurring materials can offer some new ideas in the design and fabrication of new functional materials, such as tissue engineering scaffolds and biomimetic engineering materials. Scientists have attempted to mimic the collagen-mineralization process in vitro in order to achieve a better understanding of the organizational structure in naturally occurring tissues in which the major organic matrix is collagen. Numerous studies about mineralized collagen have been reported.

By combining the collagen fibril assembly and the calcium phosphate formation in one process, Bradt et al. (1999) obtained a homogeneously mineralized collagen gel, consisting of a three-dimensional network of collagen fibrils covered with calcium phosphate. The initial precipitate, along with the fibril assembly, was amorphous calcium phosphate. This was then transformed into a crystalline apatite-like phase. The addition of polyaspartate to the reaction mixture was found to

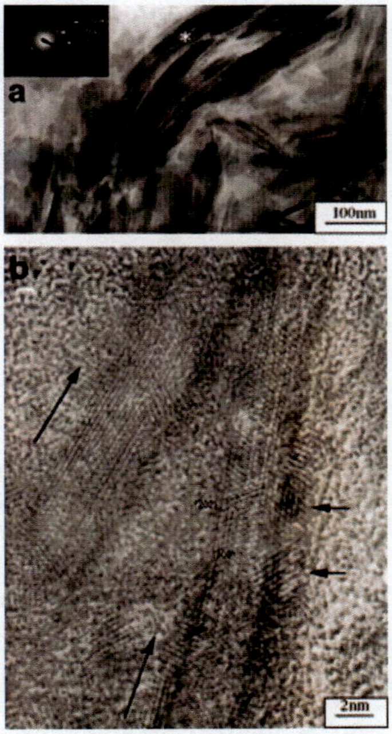

Fig. 6.29 (**a**) TEM image of mineralized collagen fibrils. The insert is a selected area electron diffraction pattern of the mineralized collagen fibrils. (**b**) HRTEM image of mineralized collagen fibrils. The two long arrows indicate the longitudinal direction of the collagen fibrils. The two short arrows indicate two HA crystals (Zhang et al. 2003)

improve the attachment between the collagen fibrils and the calcium phosphate crystals. Rhee et al. (2000) investigated the nucleation of Ca–P crystal through chemical interaction with collagen by soaking a collagen membrane in a supersaturated simulated body fluid solution. Goissis et al. (2003) reported both in vitro and in vivo studies of biomimetic mineralization of charged collagen matrices. Their results showed that the calcium phosphate deposited in close resemblance to the D-periodicity seen for assembly of collagen fibrils in vivo. Additionally, the in vitro results suggested that amide hydrolysis may have introduced into the matrix, signs for the controlled mineralization of collagen fiber. Amide hydrolysis was found from TEM investigations to occur near the overlap and gap zones. Pederson et al. (2003) reported a strategy for exploiting temperature-driven self-assembly of collagen and thermally triggered liposome mineralization to form a mineralized collagen composite. Cui et al. have now synthetically prepared nanofibrils of mineralized collagen as a self-assembly model system, with the objective of evaluating the possibility of synthesizing materials with hierarchical structures similar to those found in nature (Ge et al. 2006). They used different compositions of monomeric collagen and solutions containing calcium and phosphate ions, and then used either pH or temperature to induce the formation of collagen fibrils. TEM investigations (Fig. 6.29a) of unstained samples at low magnification revealed that the composites formed consist of an intertwined assembly of collagen fibrils bundles more than 1 μm long. Each collagen fibril is surrounded by a layer of HA nanocrystals. Each mineralized bundle of collagen fibrils is much thicker than the self-assembled collagen fibrils, implying that the self-assembled collagen nanofibrils act as the template for HA precipitation. Additionally, in order to discern

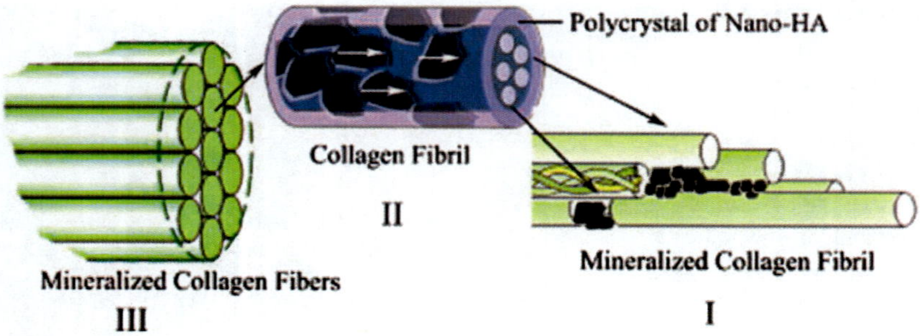

Fig. 6.30 The hierarchical structure of a self-assembled HA–collagen composite (Cui et al. 2007). (I) The first level of the hierarchy is the organization of collagen molecules with the nano-HA crystals formed initially in the gap zones between the collagen fibrils. (II) The second level of the hierarchy is the organization of collagen fibrils with respect to HA crystals. The HA crystals are sheetlike and grow on the surface of these fibrils in such a way that their *c*-axes are oriented along the longitudinal axes of the fibrils, as indicated by the white arrows in the figure. (III) The third level of the hierarchy is the organization of the mineralized collagen fibrils. A number of mineralized collagen fibrils align parallel to each other to form mineralized collagen fibers

the relative orientation of the HA crystals with respect to collagen fibrils, electron diffraction investigation have been carried out. The results demonstrate the preferential alignment of the HA crystallographic *c*-axis with the collagen fibril longitudinal axis. To study the relationship between the newly formed crystals and the collagen fibrils directly, high-resolution TEM (HRTEM) analysis of the lattice plane arrangements was performed (Fig. 6.29b). The HRTEM analysis of the parallel-aligned mineralized collagen fibrils revealed that crystal lattice is seen not only on the side area of the collagen fibrils, but also in the middle area, and that the electron density on the surface of the collagen fibrils is higher than in the interior area. These findings indicated that HA crystals grew on the surface of the collagen fibrils, giving the first direct evidence to support previous theories (Cui et al. 2007).

A schematic picture of self-assembled HA/collagen composites comprising multiple levels of hierarchical organization has been depicted, as shown in Fig. 6.30. The first level of this hierarchy is the organization of collagen molecules with some particulates of nano-HA crystals. The collagen fibrils are formed by self-assembly of collagen triple helices, and the HA crystals are formed initially in the gap zones between the collagens. Considering that the diameter of the collagen molecule is 1.5 nm, the diameter of the collagen fibrils in the five-stranded packing model should be approximately 4.0 nm. The second level of the hierarchy is the organization of these collagen fibrils with respect to the growth of the HA crystals. HA crystals, which have a sheetlike shape, grow on the surface of these fibrils in such a way that their *c*-axes are oriented along the longitudinal axes of the fibrils and surround the fibrils. This arrangement implies that the nucleation and growth of HA crystals are not random but rather are controlled by the fibrils themselves. The third level of the hierarchy is the organization of these mineralized collagen fibrils, which are aligned parallel to one another to form mineralized collagen fibers. The epitaxial growth mechanism may possibly be used to explain the assembly. As mentioned previously, it has been shown that the negatively charged groups of the collagen molecules are the nucleation sites for HA crystals. The positions of the oxygen atoms in the hydroxyl groups of HA crystals have epitaxial relationships with those in the carboxylate groups of collagen fibrils (Cui et al. 2007).

The nanoscale organization of the composites resembles that of HA crystals in mineralized tissue in which the HA crystals also align their *c*-axes with the longitudinal axes of the collagen fibrils. Such alignment is the most impressive characteristic of bone minerals. Development of novel self-assembled structures should therefore improve the understanding of collagen-mediated mineralization in other calcified tissues, and point the way to the development of new functional materials for biomimetic engineering. Moreover, these fundamental studies provide the basic theoretical support for the fabrication of HA/collagen composites and their application in bone regeneration.

6.5.2 Peptide–Amphiphilic Nanofibers-Induced Calcium Phosphate Mineralization

It is a challenge that the preparation of any material with multiple levels of hierarchical organization based on nano-building blocks. Fabrication of materials that resemble bone, even at the lowest level of hierarchical organization, is even more difficult because it involves two dissimilar organic and inorganic nanophases each of which have a specific spatial relation with respect to each another. One way to accomplish this in an artificial system is to prepare an organic nanophase designed to exert control over crystal nucleation and growth of the inorganic component, as has been widely investigated for a long time. Studies on such template crystal growth methods have suggested that nucleation occurs on surfaces which expose repetitive patterns of anionic groups. These anionic groups tend to concentrate the inorganic cations creating a local supersaturation followed by oriented nucleation of the inorganic crystal phase. At present, there is increasing interest in the fabrication of HA/peptide composites using designed self-assembling systems (Cui et al. 2007).

In this context, Stupp et al. (1997) have reported several studies on the use of self-assembly and mineralization to prepare a nanostructured composite material ((Zubarev et al. 1999b); (Silva et al. 2004)); the structural assembly of collagen and HA in bone has been recreated (Hartgerink et al. 2001). In their work, the composites are prepared by self-assembly, covalent capture, and mineralization of a peptide–amphiphile, which is synthesized by standard solid phase chemistry ending with alkylation of the NH_2 terminus of the peptide. The peptide–amphiphile consists of five key structural segments. The first region is a long alkyl tail that conveys hydrophobic character and makes the molecules amphiphilic. Four consecutive cysteine residues are also incorporated in the sequence following the alkyl tail in order to give covalent capture ability for polymerization of the self-assembled structure. The third region is a group of three glycine residues used to make the hydrophilic head of the peptide flexible. The fourth segment is a single phosphorylated serine residue that makes the assembled molecules interact strongly with calcium ions and thus able to nucleate the formation of HA. Finally, since it is beneficial for biomedical applications if the fibers can promote the adhesion and growth of cells on their surfaces; the designed peptide also contains a ligand Arg-Gly-Asp (RGD) at the end of the sequence. According to existing knowledge of amphiphile assembly, such molecules with an alkyl tail coupled to an ionic peptide should assemble in water into cylindrical micelles, due to the overall conical shape of the amphiphiles. TEM examination has indicated that the peptide–amphiphile assembles into nanofibers, which are stable in alkaline solutions. Moreover, HRTEM observations have a shown donut-shaped pattern in the cross section of the fibers, indicating that the hydrophobic alkyl tails pack on the inside of the fiber micelle and leave the acidic moieties of the peptide exposed to the aqueous environment. The chemistry of the peptide region is thus repetitively displayed on the surface. To investigate the mineralization properties of these nanofibers, mineralization experiments were designed that could be performed directly on holey, carbon-coated TEM grids. It was thereby shown that the fibers are able to nucleate HA on their surfaces. The negatively charged surfaces promote mineralization by establishing local ion supersaturation. More significance is the observation that the *c*-axes of the HA crystals are aligned with long axes of the fibers. This fact implies that the orientation of the crystalline nuclei and the subsequent crystal growth are not random but are controlled by the micelles.

The recent development of recombinant protein expression technology provides a reliable, predictable, and chemically defined source of purified humanlike collagen multi-peptide that is free of animal components. These triple-helical multi-peptides have the same amino acid sequence as human tissue-derived collagen and are free of the concerns related to the use of animal-derived collagen, such as the risk of causing immunogenic reactions and transmission of infection.

Purified recombinant collagens are capable of undergoing spontaneous alignment to form collagen fibrils and defined features that are characteristic of collagen. Recent studies (Wang & Cui 2006; Zhai & Cui 2006) have revealed that recombinant humanlike collagen has the same characteristics in the initial minerlization stage as natural collagen. Additionally, it also can induce the deposition and direct the growth of HA nanocrystals in vitro, in the form of self-assembly of nanofibrils of mineralized collagen resembling an extracellular matrix.

Molecular self-assembly is a powerful approach for the synthesis of novel supramolecular architectures. Zhang et al. (2003a) have focused on the fabrication of several self-assembling peptides and proteins for a variety of studies of biomaterials. Their studies have shown that a broad range of peptides and proteins have the ability to produce very stable nanofibers, which are very well ordered and possess remarkable regularity and helical periodicity. Moreover, these nanofibers are similar in scale to the extracellular matrices that are crucial in manufacturing artificial functional tissues. Furthermore, work in their group has demonstrated that a variety of cells, including neuronal cells, encapsulated and grown in three-dimensional peptide scaffolds, show interesting functional cellular behaviors, including proliferation, functional differentiation, active migration, and extensive production of their own extracellular matrices.

The chapter provides some basic information on the formation principles of calcium carbonate in biological systems in marine environment in the point of view of materials science in order to provide strategy for biomimetic design and preparation of new functional materials. In spite of intensive studies on the structure–function relation of the biological materials, and simulations on the biological processes of biomineralization, there are still so many questions that remain unanswered. As a consequence, we have to study carefully the biological systems and understand further their formation mechanisms.

Acknowledgments The author is grateful to the financial support of the National Basic Research Program of China (No. 2007CB815604) and the National Natural Science Foundation of China (51072090, 51061130554).

References

Addadi L, Weiner S (1985) Interactions between acidic proteins and crystals: stereochemical requirements in biomineralization. Proc Natl Acad Sci USA 82:4110–4114

Aizenberg J, Tkachenko A, Weiner S et al (2001) Calcitic microlenses as part of the photoreceptor system in brittlestars. Nature 412:819–822

Ameye L, De Becker G, Killian C, Wilt F, Kemps R, Kuypers S, Dubois P (2001) Proteins and saccharides of the sea urchin organic matrix of mineralization: characterization and localization in the spine skeleton. J Struct Biol 134:56–66

Bedouet L, Schuller MJ, Marin F, Milet C, Lopez E, Giraud M (2001) Soluble proteins of the nacre of the giant oyster *Pinctada maxima* and of the abalone *Haliotis tuberculata*: extraction and partial analysis of nacre proteins. Comp Biochem Physiol B Biochem Mol Biol 128:389–400

Belcher AM, Wu XH, Christensen RJ, Hansma PK, Stucky GD, Morse DE (1996) Control of crystal phase switching and orientation by soluble mollusc-shell proteins. Nature 381:56–58

Bradt JH, Mertig M, Teresiak A, Pompe W (1999) Biomimetic mineralization of collagen by combined fibril assembly and calcium phosphate formation. Chem Mater 11:2694–2701

Campana SE, Neilson JD (1985) Microstructure of fish otoliths. Can J Fish Aquat Sci 42:1014–1032

Choi CS, Kim YW (2000) A study of the correlation between organic matrices and nanocomposite materials in oyster shell formation. Biomaterials 21:213–222

Cui FZ, Ge J (2007) New observations of the hierarchical structure of human enamel, from nanoscale to microscale. J Tissue Eng Regen Med 1:185–191

Cui FZ, Li Y, Ge J (2007) Self-assembly of mineralized collagen composites. Mater Sci Eng R 57 (1–6):1–27

Cuy JL, Mann AB, Livi KJ, Teaford MF, Weihs TP (2002) Nanoindentation mapping of the mechanical properties of human molar tooth enamel. Arch Oral Biol 47:281–291

Davis ME (2004) How life makes hard stuff. Science 305:480–481

de Leeuw NH, Parker SC (1997) Atomistic simulation of the effect of molecular adsorption of water on the surface structure and energies of calcite surfaces. J Chem Soc Faraday Trans 93:467–475 de Leeuw NH, Parker SC (1998) Surface structure and morphology of calcium carbonate polymorphs calcite, aragonite, and vaterite: an atomistic approach. J Phys Chem B 102:2914–2922

Donners JJJM, Heywood BR, Meijer EW, Nolte RJM, Sommerdijk NAJM (2002a) Control over calcium carbonate phase formation by dendrimer/surfactant templates. Chem Eur J 8:2561–2567

Donners JJJM, Nolte RJM, Sommerdijk NAJM (2002b) A shape-persistent polymeric crystallization template for $CaCO_3$. J Am Chem Soc 124:9700–9701

Dove PM, Yoreo DJJ, Weiner S (2003) Biomineralizationed. Mineralogical Society of America and Geochemical Society, Washington, DC

Falini G, Albeck S, Weiner S, Addadi L (1996) Control of aragonite or calcite polymorphism by mollusk shell macromolecules. Science 271:67–69

Feng QL, Li HB, Cui FZ, Li HD, Kim TN (1999) Crystal orientation domains found in the single lamina in nacre of the *Mytilus edulis* shell. J Mater Sci Lett 18:1547–1549

Feng QL, Li HB, Pu G, Zhang DM, Cui FZ, Li HD (2000a) Crystallographic alignment of calcite prisms in the oblique prismatic layer of *Mytilus edulis* shell. J Mater Sci 35:3337–3340

Feng QL, Pu G, Pei Y, Cui FZ, Li HD, Kim TN (2000b) Polymorph and morphology of calcium carbonate crystals induced by proteins extracted from mollusk shell. J Cryst Growth 216:459–465

Fu G, Valiyaveettil S, Wopenka B, Morse DE (2005) CaCO3 Biomineralization: acidic 8-kDa proteins isolated from aragonitic abalone shell nacre can specifically modify calcite crystal morphology. Biomacromolecules 6:1289–1298

Ge J, Cui FZ, Wang XM, Feng HL (2006) Property variations in the prism and the organic sheath within enamel by nanoindentation. Biomaterials 26:3333–3339

Goissis G, Silva-Maginador SV, Conceicao-Amaro-Martins V (2003) Biomimetic mineralization of charged collagen matrices: in vitro and in vivo study. Artif Org 27(5):437–443

Hang YM (1994) The geological characteristics and the processing technology of fresh water pearl in Ezbou, Hubei province. Master dissertation, Guilin University of Technology (GUT), Guilin (now Guangxi University)

Hartgerink JD, Beniash E, Stupp SI (2001) Self-assembly and mineralization of peptide-amphiphile nanofibers. Science 294:1684–1688

Heywood BR, Mann S (1994) Molecular construction of oriented inorganic materials: controlled nucleation of calcite and aragonite under compressed *Langmuir* monolayers. Chem Mater 6:311–318

Hou WT, Feng QL (2006a) Morphologies and growth model of biomimetic fabricated calcite crystals using amino acids and insoluble matrix membranes of *Mytilus edulis*. Cryst Growth Des 6:1086–1090

Hou WT, Feng QL (2006b) Morphology and formation mechanism of vaterite particles grown in glycine-containing aqueous solutions. Mater Sci Eng C 26:644–647

Hunter GK (1996) Interfacial aspects of biomineralization. Curr Opin Solid State Mater Sci 1:430–435

Jiao YF, Feng QL, Li XM (2006) The co-effect of collagen and magnesium ions on calcium carbonate biomineralization. Mater Sci Eng C 26:648–652

Kanakis J, Dalas E (2000) The crystallization of vaterite on fibrin. J Cryst Growth 219:277–282

Kono M, Hayashi N, Samata T (2000) Molecular mechanism of the nacreous layer formation in *Pinctada maxima*. Biochem Biophys Res Commun 269:213–218

Lavi Y, Albeck S, Brack A, Weiner S, Addadi L (1998) Control over aragonite crystal nucleation and growth: an in vitro study of biomineralization. Chem Eur J 4:389–396

Li Z (2008) Studies on hierarchical structure of otolith and biomineralization mechanism of calcium carbonate controlled by otolith's proteins. Doctoral dissertation, Tsinghua University, Beijing

Li Z, Feng QL (2007) Analysis of polymorphs of Carp's otoliths. Rare Met Mater Eng 36:47–49

Li Z, Gao YH, Feng QL (2009) Hierarchical structure of the otolith of adult wild carp. Mater Sci Eng C 29:919–924

Loste E, Wilson RM, Seshadri R, Meldrum FC (2003) The role of magnesium in stabilising amorphous calcium carbonate and controlling calcite morphologies. J Cryst Growth 254:206–218

Lowenstam HA, Weiner S (1989) On biomineralization. Oxford University Press, New York

Mann S (1996) Inorganic materials, 2nd edn. Wiley, Chichester, p 255

Mann S (2001) Biomineralization. Oxford University Press, New York

Mann S, Heywood BR, Rajam S, Birchall JD (1988) Controlled crystallization of $CaCO_3$ under stearic acid monolayers. Nature 334:692–695

Mann S, Archibald DD, Didymus JM, Douglas T, Heywood BR, Meldrum FC, Reeves NJ (1993) Crystallization at inorganic-organic interfaces: biominerals and biomimetic synthesis. Science 261:1286–1292

Manoli F, Dalas E (2001) Calcium carbonate crystallization in the presence of glutamic acid. J Cryst Growth 222:293–297

Marin F, Corstjens P, de Gaulejac B, Vrind-De Jong ED, Westbroek P (2000) Mucins and molluscan calcification – molecular characterization of mucoperlin, a novel mucin-like protein from the nacreous shell layer of the fan mussel *Pinna nobilis* (Bivalvia, Pteriomorphia). J Biol Chem 275:20667–20675

Meyers MA, Chen PY, Lin AYM, Seki Y (2008) Biological materials: structure and mechanical properties. Progress in Materials Science 53:1–206

Michenfelder M, Fu G, Lawrence C, Weaver JC, Wustman BA, Taranto L, Evans JS, Morsel DE (2003) Characterization of two molluscan crystal-modulating biomineralization proteins and identification of putative mineral binding domains. Biopolymers 70:522–533

Miyamoto H, Miyashita T, Okushima M, Nakano S, Morita T, Matsushiro A (1996) A carbonic anhydrase from the nacreous layer in oyster pearls. Proc Natl Acad Sci USA 93:9657–9660

Miyashita T, Takagi R, Okushima M, Nakano S, Miyamoto H, Nishikawa E, Matsushiro A (2000) Complementary DNA cloning and characterization of pearlin, a new class of matrix protein in the nacreous layer of oyster pearls. Mar Biotechnol 2:409–418

Oaki Y, Imai H (2005) The hierarchical architecture of nacre and its mimetic material. Angew Chem Int Ed Engl 44:6571–6575

Ogasawara W, Shenton W, Davis SA, Mann S (2000) Template mineralization of ordered macroporous chitin-silica composites using a cuttlebone-derived organic matrix. Chem Mater 12:2835–2837

Pannella G (1971) Fish otoliths: daily growth layers and periodical patterns. Science 173:1124–1127

Pederson AW, Ruberti JW, Messersmith PB (2003) Thermal assembly of a biomimetic mineral/collagen composite. Biomaterials 24:4881–4890

Pereira-Mouriès L, Almeida MJ, Ribeiro C, Peduzzi J, Barthélemy M, Milet C, Lopez E (2002) Soluble silk-like organic matrix in the nacreous layer of the bivalve *Pinctada maxima*: a new insight in the biomineralization field. Eur J Biochem 269:4994–5003

Plummer LN, Busenberg E (1982) The solubilities of calcite, aragonite and vaterite in CO_2-H_2O solutions between 0 and 90°C, and an evaluation of the aqueous model for the system $CaCO_3$-CO_2-H_2O. Geochim Cosmochim Acta 46:1011–1040

Qiao L, Feng QL, Lu SS (2008a) In vitro growth of nacre-like tablet forming: from amorphous calcium carbonate, nanostacks to hexagonal tablets. Cryst Growth Des 8(5):1509–1514

Qiao L, Feng QL, Li Z (2007) Special vaterite found in freshwater lackluster pearls. Cryst Growth Des 2:275–279

Qiao L, Feng QL, Li Z, Lu SS (2008b) Alternate deposition of oriented calcite and amino acid layer on calcite substrates. J Phys Chem B 112:13635–13640

Raz S, Hamilton PC, Wilt FH, Weiner S, Addadi L (2003) The transient phase of amorphous calcium carbonate in sea urchin larval spicules: the involvement of proteins and magnesium ions in its formation and stabilization. Adv Funct Mater 13:480–486

Rhee SH, Lee JD, Tanaka J (2000) Nucleation of hydroxyapatite crystal through chemical interaction with collagen. J Am Chem Soc 83:2890–2892

Samata T, Hayashi N, Kono M, Hasegawa K, Horita C, Akera S (1999) A new matrix protein family related to the nacreous layer formation of *Pinctada fucata*. FEBS Lett 462:225–229

Sarikaya M, Fong H, Frech DW, Humbert R (1999) Biomimetic assembly of nanostructured materials. Bioceramics 293:83–97

Shen XY, Belcher AM, Hansma PK, Stucky GD, Morse DE (1997) Molecular cloning and characterization of lustrin A, a matrix protein from shell and pearl nacre of *Haliotis rufescens*. J Biol Chem 272:32472–32481

Shen FH, Feng QL, Wang CM (2002) The modulation of collagen on crystal morphology of calcium carbonate. J Cryst Growth 242:239–244

Silva GA, Czeisler C, Niece KL, Beniash E, Harrington DA, Kessler JA, Stupp SI (2004) Selective differentiation of neural progenitor cells by high-epitope density nanofibers. Science 303:1352–1355

Sollner C, Burghammer M, Nentwich EB, Berger J, Schwarz H, Riekel C, Nicolson T (2003) Control of crystal size and lattice formation by starmaker in otolith biomineralization. Science 302:282–286

Stupp SI, LeBonheur VV, Walker V, Li LS, Huggins KE, Keser M, Amstutz A (1997) Supramolecular materials: self-organized nanostructures. Science 276:384–389

Sudo S, Fujikawa T, Nagakura T, Ohkubo T, Sakaguchi K, Tanaka M, Nakashima K, Takahashi T (1997) Structures of mollusc shell framework proteins. Nature 387:563–564

Tong H, Ma WT, Wang LL, Wan P, Hu JM, Cao LX (2004) Control over the crystal phase, shape, size and aggregation of calcium carbonate via a l-aspartic acid inducing process. Biomaterials 25:3923–3929

Vecht A, Ireland TG (2000) The role of vaterite and aragonite in the formation of pseudo-biogenic carbonate structures: implications for Martian exobiology. Geochim Cosmochim Acta 64:2719–2725

Wang Y, Cui FZ (2006) Research on an affective model. Mater Sci Eng C 26(4):635–637

Wang XM, Cui FZ, Ge J, Wang Y (2004) Hierarchical structural comparisons of bones from wild-type and *liliput*^{dtc232} gene-mutated Zebrafish. J Struct Biol 145:236–245

Weiner S, Wagner HD (1998) The material bone: structure–mechanical function relations. Annu Rev Mater Sci 28:271–298

Weiner S, Sagi I, Addadi L (2005) Choosing the path less traveled. Science 309:1027–1028

Weiner S, Mahamid J, Politi Y, Ma Y, Addadi L (2009) Overview of the amorphous precursor phase strategy in biomineralization. Front Mater Sci 3(2):104–108

Weiss IM, Kaufmann S, Mann K, Fritz M (2000) Purification and characterization of perlucin and perlustrin, two new proteins from the shell of the mollusc *Haliotis laevigata*. Biochem Biophys Res Commun 267:17–21

Whitesides GM, Grzybowski B (2002) Self-assembly at all scales. Science 295:2418–2421

Zhai Y, Cui FZ (2006) Recombinant human-like collagen directed growth of hydroxyapatite nanocrystals. J Cryst Growth 291(1):202–208

Zhang S (2003) Fabrication of novel biomaterials through molecular self-assembly. Nat Biotechnol 21:1171–1178

Zhang W, Liao SS, Cui FZ (2003a) Hierarchical self-assembly of nanofibrils in mineralized collagen. Chem Mater 15:3221–3226

Zhang Y, Xie LP, Meng QX, Jiang TM, Pu RL, Chen L, Zhang RQ (2003b) A novel matrix protein participating in the nacre framework formation of pearl oyster, Pinctada fucata. Comp Biochem Physiol B Biochem Mol Biol 135:565–573

Zubarev ER, Pralle MU, Li L, Stupp SI (1999a) Conversion of supramolecular clusters to macromolecular objects. Science 283:523–526

Zhang C, Li S, Ma ZJ, Xie LP, Zhang RQ (2006) A novel matrix protein p10 from the nacre of pearl oyster (*Pinctada fucata*) and its effects on both $CaCO_3$ crystal formation and mineralogenic cells. Mar Biotechnol 8:624–633

Zubarev ER, Pralle MU, Li L, Stupp SI (1999b) Conversion of supramolecular clusters to macromolecular objects. Science 283:523–526

Chapter 7
Molecular Aspects of Biomineralization of the Echinoderm Endoskeleton

P.U.P.A. Gilbert and Fred H. Wilt

Contents

Abstract Echinoderms possess a rigid endoskeleton composed of calcite and small amounts of occluded organic matrix proteins. The test (i.e., the shell-like structure of adults), spines, pedicellariae, tube feet, and teeth of adults, as well as delicate endoskeletal spicules found in larvae of some classes, are the main skeletal structures. They have been intensively studied for insight into the mechanisms of biomineralization. Recent work on characterization of the mineral phase and occluded proteins in embryonic skeletal spicules shows that these simple-looking structures contain scores of different proteins, and that the mineral phase is composed of amorphous calcium carbonate (ACC), which then transforms to an

---

P.U.P.A. Gilbert (✉)
Department of Physics, University of Wisconsin-Madison, 1150 University Ave, Madison, WI 53706, USA
e-mail: pupa@physics.wisc.edu

W.E.G. Müller (ed.), *Molecular Biomineralization*, Progress in Molecular and Subcellular Biology 52, DOI 10.1007/978-3-642-21230-7_7,

anhydrous ACC and eventually to calcite. Likewise, the adult tooth shows a similar transition from hydrated ACC to anhydrous ACC to calcite during its formation, and a similar transition is likely occurring during adult spine regeneration. We speculate that: (1) the ACC precursor is a general strategy employed in biomineralization in echinoderms, (2) the numerous occluded proteins play a role in post-secretion formation of the mature biomineralized structure, and (3) proteins with "multi-valent" intrinsically disordered domains are important for formation of occluded matrix structures, and regulation of crucial matrix–mineral interactions, such as ACC to calcite transitions and polymorph selection.

7.1 Introduction

There has been substantial interest in the echinoderm skeleton, especially that of sea urchins, for centuries. Skeletal elements of echinoderms are abundant in the fossil record and, hence, important in paleontology and evolutionary studies. The enveloping shell-like surface, the test, is often admired for its decorative beauty, and for its toughness and strength, which are very different from pure calcite.

The phylum Echinodermata is comprised of five extant classes: sea urchins and sand dollars (echinoidea), sea lilies (crinoidea), sea stars (asteroidea), brittle stars (ophiuroidea), and sea cucumbers (holothuroidea). This phylum and the phylum Hemichordata are the closest relatives of the chordates, which include the vertebrates. Members of these three phyla are the only animals with deuterostome mode of embryonic development, i.e., the site of gastrulation becomes the anal pore of the larva or adult. The echinoderms have a hard, mineralized endoskeleton, which is a composite of calcite and an organic matrix, while the vertebrate endoskeleton uses calcium phosphate.

The sea urchin embryo is well suited for biochemical and molecular studies, and identification and characterization of biomineralization proteins have steadily increased in recent years. There is also intense interest in the mineral phase of skeletal elements in both larva and adult, as well as in the relationships of organic matrix and mineral, and biomechanical properties of the skeletal elements. We shall concentrate in this review on a discussion of recent work on matrix proteins, on relationships of matrix to mineral, and on recent progress in understanding the crystal orientation in skeletal elements, and the mineral precursor phases.

It is not feasible to cite or discuss all the important and pioneering studies on these subjects. Much of the older literature is thoughtfully discussed in books by Simkiss and Wilbur (1989), and by Lowenstam and Weiner (1989). The chapter by Raup (1966) on the echinoderm skeleton is also useful. Work on biomineralization in classes other than echinoids is sparse, and is discussed in a review by Wilt et al. (2003). Decker and Lennarz (1988) have surveyed earlier studies of spicule formation, and Wilt and Ettensohn (2007), and Wilt and Killian (2008) have published reviews of the recent findings on the genes that encode proteins associated with biomineralization in sea urchins.

7.2 Formation of the Endoskeleton in the Embryo

7.2.1 Spicule Formation

The development of the larval endoskeleton in indirectly developing sea urchins (i.e., those that develop through a larval stage) has been studied extensively by embryologists since the late 1800s. The formation of the larval endoskeleton can be followed in real time under the microscope. At the dawning of modern experimental biology, the endoskeletal spicules found in sea urchin larvae were the morphological characters Boveri used in his discovery that chromosomes were the carriers of heredity (Laubichler and Davidson 2008).

The embryology of spicule development has been described in detail by Okazaki (1975a, b), and more recently by Wilt and Ettensohn (2007). At the fourth cell division, four of the resultant 16 cells are clustered at one pole, and are termed micromeres, since they are smaller than the other blastomeres (see Fig. 7.1). At the fifth division, each micromere gives rise to two cells, the larger of which is uniquely dedicated to formation of a skeleton. The so-called large micromeres continue to divide, forming a cohort of 32–64 cells (depending on the species) located in the wall of the vegetal hemisphere of the blastocoel. Just before the invagination movements of gastrulation begin, these micromere descendants, which are epithelial cells, transform into motile mesenchymal cells and burrow through the basement membrane, entering the blastocoel, where they wander, exploring the wall of the blastula for several hours, concomitant with the invagination of the archenteron (primitive gut) during the early phases of gastrulation. These large micromere

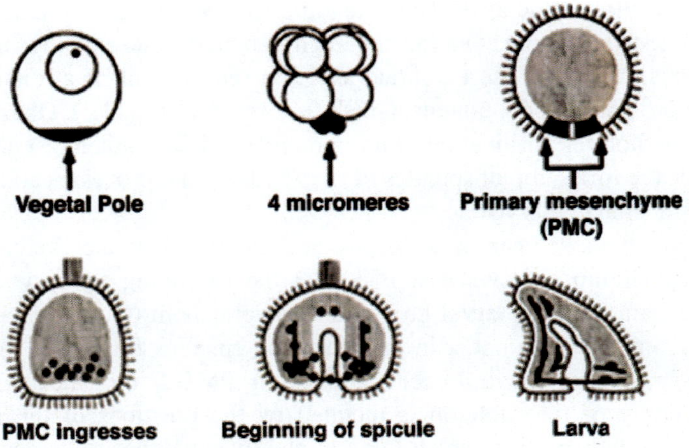

Fig. 7.1 The embryonic development of endoskeletal spicules. Diagrams of several stages are shown, with emphasis on the origin of the spicules. The primary mesenchyme cells and the cytoplasm of the egg from which they derive are black. Spicules are indicated by thick solid lines in the mid-gastrula and pluteus larval stages. The egg is about 100 μm minimum diameter and the larva is about twice that size. Reprinted by permission of Elsevier Publishing from Wilt (1999)

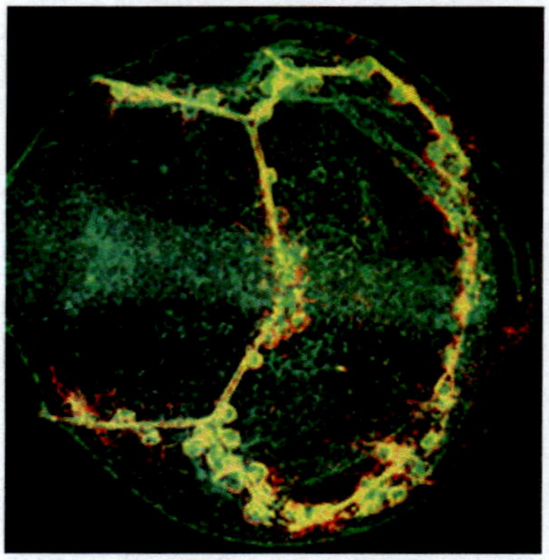

Fig. 7.2 A stained prism-stage embryo. A late prism-stage embryo of *Lytechinus pictus* was stained with an antibody to the spicule matrix protein, LpSM30 (*green*) and with an antibody to a PMC-specific cell surface antigen (*red*). The doubly stained PMC cellular syncytium and spicule are yellow. There is sufficient background staining with the anti-LpSM30 antibody to outline the larva and the developing gut, which runs approximately on the horizontal in this micrograph. The prism-stage embryo is about 180 μm in diameter. Photo courtesy of C.E. Killian and F.H. Wilt

descendants, which are now called primary mesenchyme cells (PMC), then form a stereotypical array in the vegetal portion of the blastocoel, and adjacent PMCs fuse to form a multicellular syncytium, as shown in Fig. 7.2.

The fused PMCs form long cables between cell bodies, and soon after, syncytium formation granules of calcite can be detected in two ventrolateral locations where PMCs are congregated. The embryonic endoskeleton forms by extension of three rays of $CaCO_3$ from each of two calcite rhombohedra, first in a plane defined by the *a* crystallographic axes. The extending spicules, adding mineral principally at the tips and to some extent increasing in girth, then bend to extend in the direction of the *c* axis, and form the elaborate skeleton reminiscent of a Victorian easel; hence, the larva is called a pluteus, Greek for easel (see Fig. 7.2). Okazaki (1975a) devised a method for purification of micromeres, which can then be cultured. They recapitulate the formation of spicules *in vitro* by the same processes and at the same tempo as the intact embryo.

Secondary branches can arise to produce a more elaborate skeleton in many species, and clumps of syncytial PMCs at the extending tip serve as sites of further elongation during larval growth and development (Gustafson and Wolpert 1967). It is important to underline the fact that spicules form only in very close association with the syncytial cables connecting PMC bodies; hence, the macroscopic anatomy of the skeleton is dictated by the positions of the PMCs. The membrane-limited space in which mineral and matrix deposition occurs seems to be entirely enclosed by the cell membrane of the syncytial cables, a seeming vacuole, though more recent studies support the idea that the membrane surrounding the spicule is actually surface plasmalemma that enrobes the spicule; hence, the spicule is a result of vectorial (i.e., directionally secreted) secretion into a privileged space inside the syncytium that is topologically outside the cell, although it is enrobed by the cell membranes of many PMC cells fused together. A more detailed discussion of this issue can be found in Wilt and Ettensohn (2007) and Wilt (2002).

The morphology of the spicules formed is dictated primarily by PMCs. But the gross anatomical placement of the spicules within the blastocoel depends on the properties of the overlying ectodermal layer of the embryo (Ettensohn and Malinda 1993; Guss and Ettensohn 1997). It has recently been shown that the initiation of calcite formation depends on signaling by the ligand VEGF (Duloquin et al. 2007), which emanates from a small number of ectodermal cells that immediately overlie the congregations of PMCs in the ventrolateral aspects of the blastocoel of the early gastrula stage. A small number of cells that express the homeobox transcription factor, *Otp*, produce the VEGF signal (Dibernardo et al. 1999). PMCs express VEGF receptors for transduction of this ectodermal signal, and initiation and maintenance of calcification depend on this signaling.

7.2.2 Calcium

The calcium of the spicule (and presumably the Mg as well, though it has not been studied) comes from the sea water, as shown by use of isotope studies. (Nakano et al. 1963) Reduction of the Ca concentration of sea water (normal is ~10 mM) below 2–4 mM causes spicules that develop to be malformed. Even lower Ca levels result in irregular, small, mineralized masses, and a generalized deterioration of normal embryonic development (Okazaki 1956). The PMCs must possess very active Ca transporters, presumably with high capacity and low affinity, but they have not been identified or characterized. Beniash et al. (1999) visualized intracellular deposits by electron microscopy that were identified as Ca precipitates, presumably $CaCO_3$. These did not display x-ray diffraction patterns and were believed to be amorphous; after heating of the sample, these deposits diffracted as calcite. After the initiation of triradiate spicule formation, birefringent granules have not been observed in PMCs in living embryos or cultures (see Okazaki 1960); so, either the imported Ca is segregated in intracellular vesicles containing supersaturated Ca solutions, or the precipitated material must be below the resolution of the light microscope, or in a state that is not birefringent.

This role of intracellular calcium deposits was recently addressed by using the Ca fluorophore calcein, a fluorescein derivative that is intensely fluorescent when incorporated into calcium-containing precipitates. Wilt et al. (2008b) used calcein pulse labeling to follow calcium delivery to spicules. They found that brief pulses of calcein-containing sea water, followed by rinsing and culture in normal sea water, resulted in an initial labeling of intracellular submicron sized granules, followed by their clearance from the cell and subsequent appearance of label in developing spicules, particularly near the tips of the extending spicules. They interpreted this as evidence for an intracellular precipitated, noncrystalline Ca precursor that is shuttled along the syncytial cytoplasmic strand and secreted into the syncytium in which the spicule forms.

7.2.3 Occluded Proteins

Spicule matrix proteins are also synthesized and secreted by PMCs; thereafter, the proteins can be found occluded within the mineral phase. The earlier literature on this subject can be found in the review of Decker and Lennarz (1988). There has been very little work done on the synthesis, delivery, and occlusion of these proteins during spicule formation. Benson et al. (1989) demonstrated, using polyclonal antibodies against an undefined mixture of matrix proteins from the spicule, that intracellular vesicles, which contained putative matrix proteins, could be identified in PMCs. The first matrix proteins studied by modern methods, SM50 and SM30, have been the subject of several studies. Ingersoll et al. (2003) using immunoelectron microscopy and affinity-purified antibodies showed that both of these proteins are found in the Golgi apparatus and in transport vesicles (approximately 50 nm in size), that deliver their contents to the space in which the spicule is formed. Wilt et al. (2008b) followed GFP-tagged versions of SM30 and SM50; in both instances, the PMCs harboring the transgene secreted GFP-tagged SM30/50 into the spicule space near the PMC of origin. This finding contrasts with the delivery of calcium, which shuttles primarily to the extending tip. The kinetics of secretion of ^{35}S-labeled SM30 were followed in PMCs by immunoprecipitation of the protein in the cell and spicule, formed in culture. It took about 20 min after the introduction of label for secreted proteins to begin accumulation in the forming spicule, and the SM30 delivered to the spicule had a somewhat lower molecular mass, indicating that processing of the protein had occurred during the secretion and/or spicule assembly steps.

There are a variety of studies that characterize the cell biology of spicule formation. Inhibitors of metalloproteases stop spicule elongation (Ingersoll and Wilt 1998; Huggins and Lennarz 2001), though this inhibition does not interfere with initial formation of the calcite rhombohedra in the PMCs. It is no surprise that inhibitors of ion transport have a deleterious effect (e.g., Mitsunaga et al. 1986). The mechanism (s) by which vectorial secretion of matrix proteins is accomplished is unknown. We do know that PMCs secrete many other molecules, including collagens and proteoglycans, into the blastocoel (Benson et al. 1990), but such molecules are not found occluded in the spicule, nor are spicule proteins found in the blastocoel.

Finally, it is important to remember that spicule formation may only represent one mode of biomineralization. The biomineral is not assembled on a scaffold at some distance from the cell, as is the case in mollusks, or vertebrate bone, but in intimate contact with the cell. The spicule is completely enrobed by syncytial membrane and a cytoplasmic sheath without any intervening space, as was shown by Beniash et al. (1997).

7.2.4 Formation of Postembryonic Skeletal Elements

Larvae that survive the rigors of life in the plankton will eventually settle on a suitable substrate, and undergo metamorphosis. The larval structures subsequently wither and disappear, leaving a juvenile sea urchin (Smith et al. 2008). Drawings of advanced larvae show the relationships of the echinus rudiment to the larva more clearly than do microphotographs (see Fig. 7.3). The juvenile urchin forms within the body space, the former blastocoel of the embryo while the growing larva is still part of the plankton. A portion of the larval foregut, in association with cellular descendants of the "small micromeres," which are generated at the fifth cleavage division, gives rise to a rudimentary structure (called the echinus rudiment) that gradually forms a miniature version of the mature sea urchin. The rudiment gradually assumes a more familiar morphology possessing the pentaradial form characteristic of echinoderms. Early in development of the rudiment, small test plates form and become calcified. Tube feet and spines will also become apparent. By the time the juvenile is a few millimeters in diameter, it will have calcified test plate elements, spines, and teeth.

The morphological details of appearance and growth of the adult endoskeleton have been little studied. Some recent work by Yajima and Kiyomoto (2006); (Yajima 2007) has established that cells responsible for juvenile calcified endoskeleton are not PMCs, but rather a related yet distinct embryonic lineage called secondary mesenchyme cells. Smith et al. (2008) worked out a detailed atlas of developmental stages of metamorphosis of *Strongylocentrotus purpuratus* and were able to use specific antibodies directed against endoskeleton-specific proteins to chart the early development of the endoskeleton. Calcified structures in the juvenile are often found arising in close conjunction with larval spicules. Continued development of spines, test plates, and pedicellariae has been studied by electron microscopy (Ameye et al. 1999, 2001). It is generally believed that cells closely associated with biomineralized structures in the adult are responsible for their deposition, a reasonable supposition, though detailed evidence is often lacking, except for the cases of the spine and tooth, which we shall consider in due course.

Classical descriptions of the gross and microanatomy of the juvenile endoskeleton – the test plates, teeth, and spines – can be found in Hyman (1955). The presence in the adult of a few of the characterized matrix proteins of sea urchins has been verified by Western blotting of extracted matrix proteins (Killian and Wilt 1996) or identification of the cognate mRNAs (George et al. 1991; Livingston et al. 2006). Immuno-labelling electron microscopic localization of SM30 and SM50 in pedicellaria and spines of the adult was done by Ameye et al. (1999). We shall restrict subsequent discussion to more recent work on the fine structure of the spicules, the tooth, and the spine.

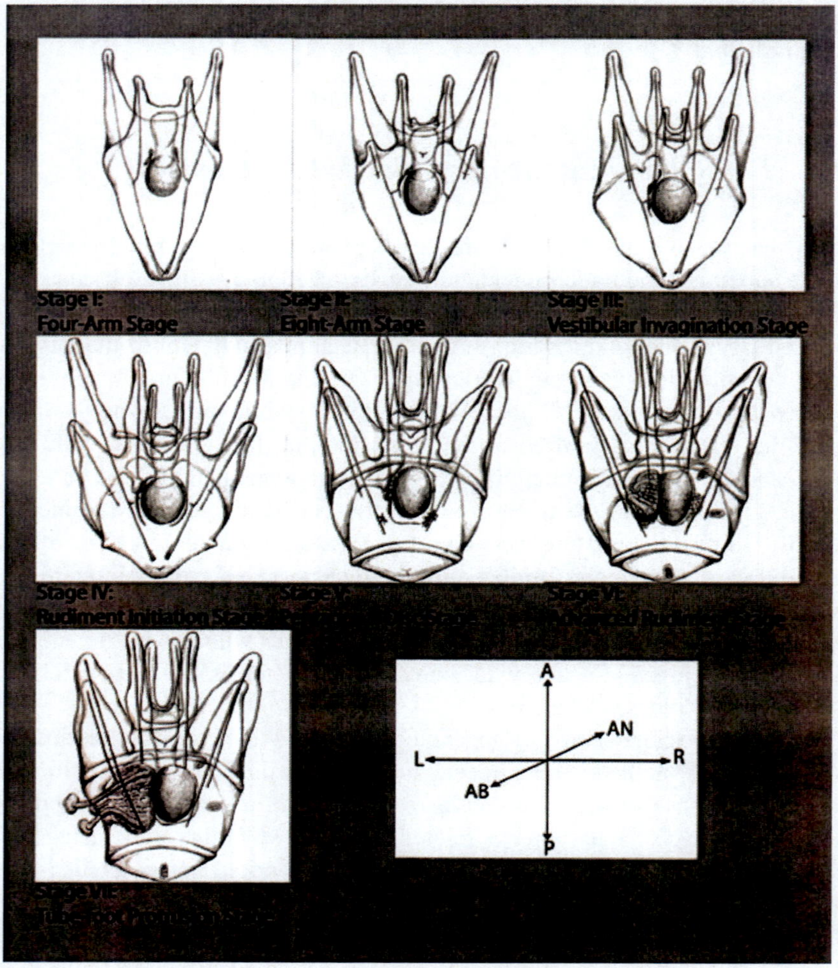

Fig. 7.3 Development of the larva of *S. purpuratus*. Seven stages of development of the pluteus larva during its residence in the plankton. The development of the arms, the skeletal rods, and the gut are emphasized. The echinus rudiment, which will form the juvenile sea urchin after metamorphosis, can be seen emerging from the left side of the midgut in stages V–VII. The body axes are: *L-R* = left-right; *A-P* = anterior-posterior; *AB-AN* = Aboral–Abanal. Reprinted by permission of John Wiley and Sons from Smith et al. (2008)

7.2.5 Recent Work on the Structure and Composition of the Embryonic Spicule

The gross morphology of the endoskeletal spicule in the pluteus stage is a species-specific character. The location of branches, small spurs, and fenestrations is reproducible and specific. Okazaki and Inoue (1976) carried out pioneering work on the orientation of the crystal axes, and showed that the founding mineral granule appeared as a rhombohedron of calcite. Various workers (Okazaki 1960; Beniash et al. 1997) demonstrated a putative surface "envelope" composed of organic material, and Benson et al. (1983) gave clear morphological evidence for a network of occluded organic molecules in the demineralized spicule. Gentle etching of fractured surfaces revealed a lamellar organization of the mineral, much like growth rings of a section of a tree trunk (Seto et al. 2004). It was proposed that these lamellae could be formed from the periodic deposition of additional layers of mineral that are responsible for increase in girth of the spicule during its development, but there is no direct evidence for that proposal (Seto et al. 2004).

Using affinity-purified antibodies against SM30 and SM50 Seto et al. (2004) showed by immunoelectron microscopy that fractured, etched surfaces were specifically labeled, thereby providing direct evidence for occlusion of these matrix proteins. Both these proteins, especially SM50, are also found on the external surface of the spicule. The picture that emerged is one of well (but not perfectly) aligned domains of calcite (Berman et al. 1993) in which a fibrous network of occluded matrix proteins traverse boundaries between domains.

One goal of research on biomineralization is the identification, enumeration, and function of organic molecules found on, or occluded within, the mineralized structures. Polysaccharides and proteins dominate the lists in the various biomineralized tissues that have been closely examined. Application of the methods of molecular biology has been especially helpful since occluded matrix proteins often resist routine methods of protein purification and characterization. An extended discussion of the approaches and results can be found in reviews by Wilt and Ettensohn (2007) and Killian and Wilt (2008). Suffice it to say that even though the protein content of the sea urchin spicule is very low (~0.1% by mass, Wilt 1999), there are apparently over 40 different proteins of which most, but not all, are acidic and glycosylated[1]. Only one of these 40 has been subjected to a rigorous test of its functional role: synthesis of SM50 during embryonic development is essential for spicule formation (Peled-Kamar et al. 2002; Wilt et al. 2008a). While SM50 is necessary for spicule formation, the exact nature of the role SM50 plays has not been elucidated. It might be instructive to engineer and express counterfeit versions of SM50 to see if particular portions of SM50 can act as a "dominant negative".

The genome of *S. purpuratus* has been sequenced and annotated, and an enumeration of genes known to be involved in spicule formation and/or structure compiled (Livingston et al. 2006). These proteins are mostly acidic, glycosylated, secreted and contain a C-lectin domain. The necessity and/or function of any of them, except SM50, is unknown. We should also remember that proteins important for spicule formation may not necessarily end up occluded in the composite, e.g., the apparent important role of metalloproteases.

[1]Recent proteomic work demonstrated over 200 proteins in the spicule (Mann et al. 2010).

PMC-specific gene expression has also been analyzed by analysis of ESTs of a PMC cDNA library (Zhu et al. 2001). They identified a transmembrane protein, P16, that is essential for spicule formation, as judged by gene knock-down experiments (Cheers and Ettensohn 2005). It is a straightforward matter to use new, powerful methods of protein identification to obtain a complete list of occluded proteins in the spicule, which has been done for the sea urchin tooth and spine (Mann et al. 2008a, b; Mann et al. 2010).

7.3 ACC: Discovery, Importance, and Implications in Other Systems

A resurgence of interest in amorphous minerals was motivated by the discovery by Beniash et al. (1997) that spicules isolated from sea urchin embryos, especially those from earlier stages (e.g., prism) prior to the mature pluteus larva, have substantial amounts of amorphous calcium carbonate (ACC) as identified by Fourier Transform InfraRed (FTIR) spectroscopy. This has been confirmed by a variety of other physical techniques, including visible light polarization and X-ray absorption near-edge structure (XANES) spectroscopy (Politi et al. 2006, 2008). Though stable forms of ACC containing equimolar amounts of hydration water and $CaCO_3$ are known in ascidians, crustaceans, and other animals and plants, (Lowenstam and Weiner 1989), the ACC of the sea urchin embryo slowly transforms to calcite, so that the developing larva (a day or two after attaining the pluteus form) has little ACC. Furthermore, isolated spicules that are stored at −20°C still slowly transform ACC to calcite (Beniash et al. 1997). Synthetic ACC prepared in the laboratory is unstable and quickly transforms to the most stable polymorph, calcite, in minutes, not hours or days.

The ACC found in spicules is apparently not an isolated example. The presence of amorphous precursor minerals was also observed in regenerating spines of sea urchins (Politi et al. 2004), in the forming end of the sea urchin tooth (Killian et al. 2009), in continuously growing fin rays of fish (Mahamid et al., 2008), and in forming tooth enamel from mouse incisors (Beniash et al. 2009). ACC has also been implicated in the formation of mollusk shells (Weiss et al. 2002; Nassif et al. 2005), although a recent report did not find ACC in newly deposited nacre (Kudo et al. 2010). An obvious implication of these diverse findings is that formation of an amorphous phase of the mineral, as a metastable precursor to calcite, aragonite, or carbonated apatite, might be a general tactic used in biomineralization. We shall consider that proposition near the end of this review.

Possible atomic structures of synthetically produced ACC have been investigated using x-ray scattering (Michel et al. 2008; Goodwin et al. 2010), but heretofore the ACC structure has not been investigated in biomineral amorphous precursors. The mode of crystal formation and propagation has been analyzed by Politi et al. (2008) and by Killian et al. (2009). Politi et al. found that there are two amorphous precursor phases in the sea urchin spicule, one of which is hydrated, and shows spectra exactly equivalent to those of synthetic hydrated ACC. Another phase may be anhydrous ACC, but at the time that work was published, there was no synthetic equivalent to compare the spectra to. The final phase is calcite, as expected, and its spectra are similar to those of geologic and synthetic calcite (Politi et al. 2008). A recent study by Radha et al. (2010) showed that the synthetic anhydrous ACC and spicule ACC have the same enthalpy of transformation into calcite, thus providing strong evidence that the intermediate phase in spicules is indeed anhydrous ACC.

Although echinoderm biominerals behave as single crystals of calcite in polarized light and X-ray diffraction, they do not cleave as crystals. Their fracture surfaces are conchoidal, with curved surfaces and curved edges reminiscent of amorphous glasses. Elegant work by Berman et al. gives compelling evidence that the conchoidal fracture is due to proteins. Calcite rhombohedra were grown in vitro, in the presence of acidic glycoproteins extracted from sea urchin skeletal elements. Not only did the synthetic calcite rhombohedra occlude the proteins, but they also fractured conchoidally (Berman et al. 1988).

This pioneering experiment, and many others that followed it, raise an interesting question: how can a very small amount of organic molecules, e.g., 0.1w% in spicules, make the fracture surface so macroscopically different from a cleavage plane of a crystal? We propose that layers (atomic layers in a crystal or sheets of paper in a sheaf) do not need to be "glued" together extensively. A few spots of "organic glue" between each pair of subsequent layers are sufficient to keep the sheaf or the crystal together, thereby preventing cleavage into separate layers. Hopefully, this hypothesis can be evaluated by obtaining more detailed information about the exact locations of occluded proteins in the skeletal structures discussed here.

7.4 Recent Work on the Adult Spine

The iconic feature of sea urchins is their adornment of spines, which can range from millimeters to scores of centimeters, depending on their location on the test, and on the species. Spines are long and tapered columns of calcite, ending in a sharp point, well designed for defense and abrasion. Their spongy texture provides spaces for several different kinds of dermal cells, and coelomocytes. The surface is covered with an epithelium, meaning that these hard conspicuously external elements are truly endoskeletal.

Figure 7.4 shows the mineralized portion of a spine after bleaching and transverse fracture. The young spine is trabecular and fenestrated, as is the forming test plate. This morphological structure is termed "a stereom." As the spine matures, it becomes more and more heavily mineralized, displaying radially arranged sectors connected by transverse bridges. Sometimes the center is hollow and occupied by cells, sometimes occupied by both cells and extracellular matrix.

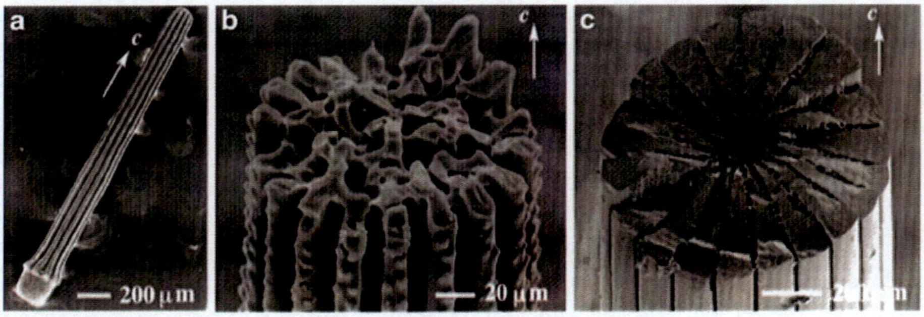

Fig. 7.4 Scanning electron micrographs of secondary spines from the sea urchin *Paracentrotus lividus*. The direction of the c axis of the calcite crystal is indicated by the *arrows*. (**a**) Intact spine. (**b**) Fracture surface of a young spine, showing the spongy structure of the stereom. (**c**) Fracture surface of the mature spine, showing the development of the sectors that filled the stereom. Note the difference in the sizes of the young and mature spines. Data from Aizenberg et al. (1997). Reprinted by permission American Chemical Society

Examination by X-ray diffraction or polarized light demonstrates that the spine is a remarkably co-oriented single crystal, with the *c* crystallographic axis parallel to the long axis of the spine. Figure 7.5 shows a recent result demonstrating that the spine co-orientation is accurate even when observed at much higher resolution with electron diffraction.

Synchrotron X-ray diffraction measurements show that there are domains of perfect crystallinity about 210–235 nm along the long axis (160 nm along the orthogonal transverse axis), and these domains are very well aligned, displaying about 0.130° of variation in alignment of neighboring domains. Geological calcite shows perfect domains of about 800 nm with only 0.003° of misalignment (Berman et al. 1993; Magdans and Gies 2004). The difference between the spine and pure calcite is thought to arise because of the presence of about 0.1w% of occluded organic material (mainly protein) and variable amounts of Mg^{+2} ions (~2–12%) in the calcite of the spine. The Mg content of the calcite is believed to confer additional hardness, and is somewhat higher toward the base of the spine. Mg content can also vary with ocean temperature as well as species.

The structure of the spine at the light and electron microscope level was examined in detail by Heatfield and Travis (1975), looking at the cells in the stereom and epithelial covering as well as the mineral (also, Märkel 1983a, b). The spine elongates by deposition of fenestrated columns of mineral by dermal sclerocytes (sometimes called calcoblasts).

Not much is known about the proteins occluded in the calcite of the spine. Earlier work on occluded proteins of embryonic spicules showed, in passing, that both SM50, and some forms of SM30 are present in the spine Killian and Wilt (1996), and Ameye et al. (1999, 2001) used immunohistochemical methods to demonstrate the presence of both these proteins in forming pedicellariae and spines. More recently, Killian et al. (2009) have used polymerase chain reaction (PCR) to analyze which isoforms of SM30 mRNA are present in spine tissues, and found that

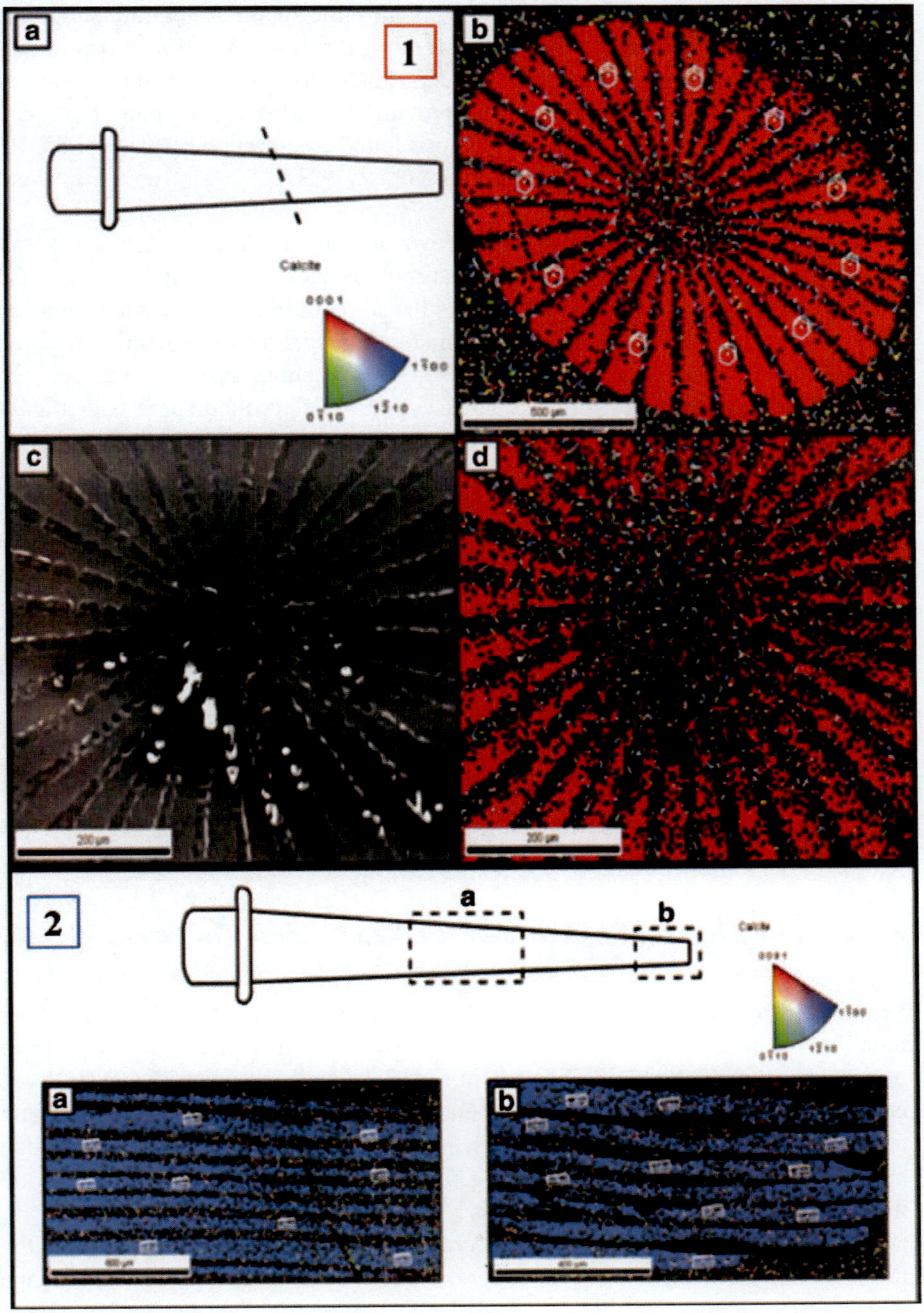

Fig. 7.5 Electron back-scattered diffraction (EBSD) analyses of *P. lividus* spine. (**1**): Transverse and (**2**): Longitudinal cuts. (**1a**): Cartoon of spine with position of section indicated. (**1b**) and (**1d**): EBSD crystallographic orientation map, according to color key in (**1a**), of region presented in (**1c**) as the secondary electron image. Wire frames in (**1c**) indicate that the c-axis is parallel to the spine long axis. (**2**). Cartoon of spine with positions of sections in (**2a**) and (**2b**) indicated. (**2a**) and (**2b**)

SM30 D is dominant, with only low levels of some of the other forms. This is unusual since SM30 D is completely missing in the embryo and is a very minor component in test, teeth, and tube feet. Mann et al. (2008a) have recently published a very thorough proteomic study of the proteins occluded in the spine; there is a very large number, including many of those found in embryonic spicules. SM50, C-lectin-containing proteins, carbonic anhydrase, MSP 130, and proteases were also well represented.

Most of the characterization of spine growth and maturation has been carried out by studying the process of spine regeneration, a process known at least since the mid-nineteenth century. Figure 7.6 depicts the stereom of the tip of a regenerating spine. Elongation of the spine occurs first in the center of existing truncated material by deposition of thin trabeculae; subsequently, lateral "branches" of mineral are laid down and girth is increased. Newly deposited spine is gradually filled in with additional mineral so that the spongy nature of the stereom decreases and prominent columns of calcite dominate. A review of studies of spine regeneration by Dubois and Ameye (2001) provides a good overview of the subject.

The nature of the newly deposited mineral at the tip of a regenerating spine has been studied by Politi et al. (2004). They provided evidence using etching, FTIR, and electron microscopy that the newly deposited mineral is ACC, initially probably in a hydrated form, which is later gradually transformed into anhydrous ACC state, and finally into calcite. This is reminiscent of the findings on the growing tip of the sea urchin embryo spicule, and corroborates the idea that deposition of ACC as a precursor is a general mode of biomineralization of calcite in echinoderms.

7.5 Recent Work on the Adult Tooth

7.5.1 The Mineral Structure of the Sea Urchin Tooth

Sea urchins use their teeth to bite food, but also to burrow into rocks and shelter their bodies from predators (Moore 1966; Nelson and Vance 1979) and from pounding waves (Otter 1932), as seen in Fig. 7.7. In all sea urchins, five teeth are continuously forming at their proximal end (the plumula) and are worn by grinding at their distal end (the tooth tip). They are arranged and supported in a jaw-like apparatus called Aristotle's lantern (Fig. 7.7), as it was first described by Aristotle in his *Historia Animalium*, in 343 BCE.

The tooth structures in several sea urchin species are remarkably similar: the tooth is elongated, slightly curved, approximately 2-cm long, and it has a T-shape cross section (Kniprath 1974; Ma et al. 2008; Märkel and Titschack 1969; Wang

Fig. 7.5 (continued) are EBSD crystallographic orientation maps according to the same color key These maps also indicate that the c-axis is parallel to spine long axis. Data from Moureaux et al. (2010). Reprinted by permission from Elsevier Publishers

Fig. 7.6 The tip of a regenerating spine. This early stage of a regenerating spine tip shows the fenestrated, reticular stereom of the spine. *Arrows* indicate the presence of lateral bridges between developing columns of calcite sectors. The spine was prepared by chemical debridement of tissue using NaOCl and examined by SEM. Reprinted by permission of John Wiley and Sons from Dubois and Ameye (2001)

Fig. 7.7 (**a**) Purple sea urchins (*S. purpuratus*) in the intertidal zone at Pt. Arena, California. Notice that each urchin is sheltered into a hole, which it dug into the rock substrate using its teeth. Photo courtesy of C. E. Killian. (**b**) The five tooth tips slowly open and close radially. Photo courtesy of P. Gilbert. (**c**) The Aristotle lantern extracted from the animal, with the tooth tips visible at the bottom, and the side of one tooth indicated by the *arrow*. Photo courtesy of P. Gilbert

et al. 1997; Wang 1998; Ma et al. 2007). The top part of the T is commonly termed the flange; the vertical part of the T is the keel. The flange contains curved calcitic plates, spaced at a few microns from each other. Calcitic fibers with varying

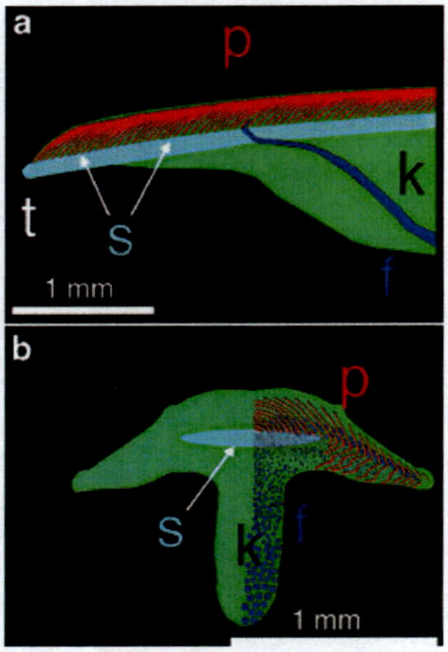

Fig. 7.8 Schematic of a *Strongylocentrotus purpuratus* tooth tip in longitudinal section (**a**) and cross section (**b**). The plates (*p*) are highlighted in *red*, the fibers (*f*) in *blue*. Plates and fibers are cemented together by a polycrystalline matrix (*green*). Notice that the fibers change diameter across the keel (*k*), becoming thicker as they grow away from the plates. The stone part (*s*), highlighted in cyan, is an elliptical region at the center of the tooth cross section, in which the nanoparticles of the polycrystalline matrix reach their highest Mg concentration. At the grinding tip (*t*), the stone part is exposed, after plates and fibers are shed off. Images courtesy of P. Gilbert

diameter start at the ends of the plates, and extend with an S-shape morphology across the keel. Figure 7.8 provides a schematic view of the tooth and its elements.

The five calcitic teeth of an adult sea urchin are continuously growing at a rate of approximately 10 μm per hour at the forming proximal end (Holland 1965; Orme et al. 2001), while the grinding distal end wears off and self-sharpens Killian et al. (2011). The larger structural components of the tooth – plates and fibers – are formed by syncytia of odontoblasts in the plumula at the proximal end of the tooth (Kniprath 1974; Ma et al. 2008), as shown in Fig. 7.9.

As for all other echinoderm biominerals (Brusca and Brusca 1990), the mineralized structures in the sea urchin tooth are highly co-oriented (Berman et al. 1993; Killian et al. 2009). Killian et al. (2009) recently showed that all plates are topologically connected by mineral bridges, thus a single crystal orientation propagates through all plates with spatial continuity. The bridges are shown in Fig. 7.10. Sea urchin species from different oceans possess these same bridges, demonstrating that plate co-orientation via mineral bridges is a highly conserved strategy in biominerals found in sea urchins.

A polycrystalline matrix of ~10 nm particles Yang et al. (2011) of Mg-rich calcite ($Ca_{1-x}Mg_xCO_3$), with x varying between 0.3 and 0.45 in different species (Wang et al. 1997; Killian et al. 2009; Robach et al. 2009), subsequently fills the space between the plates and the fibers, effectively cementing all components together. This matrix was initially believed to be softer (Ma et al. 2007), but

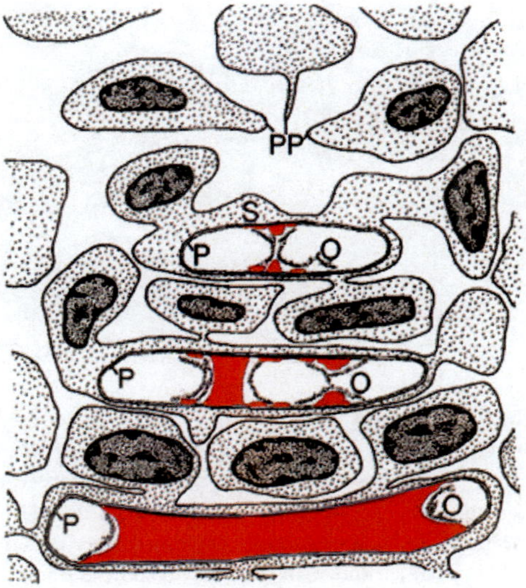

Fig. 7.9 Schematic diagram of the formation of plates in the plumula of the tooth of *P. lividus*. Free odontoblasts (above and at micrograph edges) send the pseudopods (*PP*) to meet and fuse with each other. The plate sheath (*P*) becomes a syncytium, and then calcification (*red*) begins at the two opposite surfaces of the plate sheath. (*O*) is the organic material displaced as the mineral grows and fills the syncytium. Adapted from Kniprath (1974) by permission of Springer Verlag

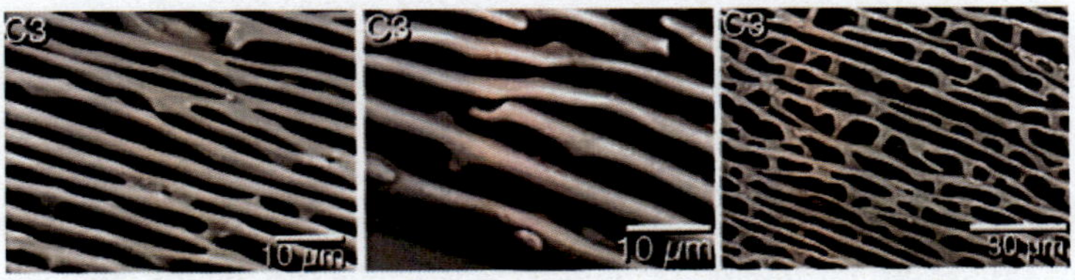

Fig. 7.10 The mineral bridges connecting the plates in the Pacific *S. purpuratus* (*left*), the Mediterranean *P. lividus* (*center*), and Atlantic *L. variegatus* (*right*). Data from Killian et al. (2009). See the original publication and its supporting information to visualize the location of the bridges on the sides of the flange. By permission of American Chemical Society

recently found to be harder (Ma et al. 2008) than plates and fibers. It is clear from Fig. 7.8a that it is the polycrystalline matrix in the stone part that becomes the sharp, hard tip of the tooth that does the grinding.

Recently, Ma et al. (2009) reported that the polycrystalline matrix is highly co-oriented in *P. lividus*. Furthermore, Robach et al. (2009) demonstrated that polycrystalline matrix and fibers share the same orientation in *L. variegatus*, whereas Killian et al. (2009) and Yang et al. (2011) made the same observation in *S. purpuratus*. These concurring observations raise an important question: how does the polycrystalline matrix form and co-orient its crystal nanoparticles? Are the nanoparticles aggregating before or after crystallizing? In other words, is this the result of oriented attachment of crystalline nanoparticles, as first observed by Penn

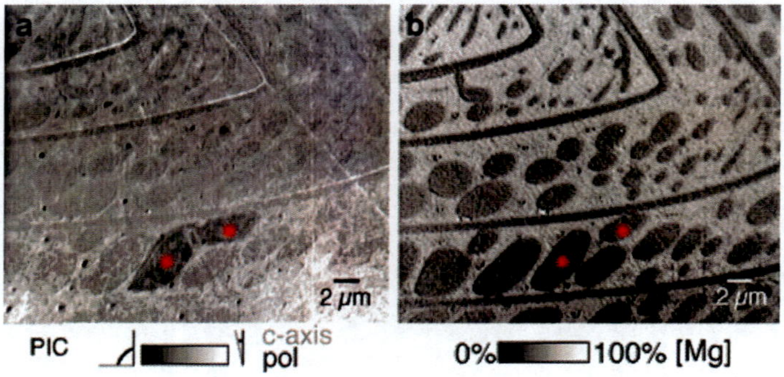

Fig. 7.11 Two stray fibers reveal how co-orientation arises. Spectromicroscopy results from a region of a cross section at the mature end of an *S. purpuratus* tooth. (**a**) Map in which gray level indicates crystal orientation. This map shows two strongly misoriented fibers (red asterisks). (**b**) Mg distribution map from the same region in A, showing well-defined and sharp elliptical fiber edges. Such strongly misaligned fibers are extremely rare in the sea urchin tooth, in which all other fibers are co-oriented with each other and with the polycrystalline matrix. Despite the sharp edges of the fibers in the Mg map, the map in A shows that the polycrystalline matrix surrounding and between the two stray fibers is as misoriented as the fibers themselves. This indicates that the nanoparticles in the polycrystalline matrix get their orientation from the fibers. Data from Killian et al. (2009). By permission of American Chemical Society

and Banfield in TiO_2 (Penn and Banfield 1999), and FeOOH (Banfield et al. 2000), and later in many other synthetic mesocrystals (Cölfen and Antonietti 2008)? Or, is this the result of an amorphous precursor phase forming first, with crystallinity propagating through it subsequently (Aizenberg et al. 2003; Politi et al. 2008)?

This question was addressed in detail by Killian et al. (2009), who showed that, indeed, amorphous precursor phases are aggregated first, and then crystallinity propagates through them via a mechanism of secondary nucleation. Figure 7.11 shows the propagation front of crystallinity and crystal orientation, starting from the fibers and expanding into the surrounding polycrystalline matrix.

Interestingly, spectroscopic analysis at the forming end of the *S. purpuratus* tooth revealed for the first time that there are not one but two amorphous precursor minerals (Killian et al. 2009). These precursor phases are identical to those reported by Politi et al. (2008) in *S. purpuratus* larval spicules. The phases are hydrated ACC, crystalline calcite, and another phase, which is presumably intermediate, and presumably anhydrous ACC, although the order in which the phases occur during tooth formation is unclear and difficult to detect. The aforementioned microcalorimetry results by the Navrotsky group indicate that the enthalpy of transformation from dehydrated synthetic ACC and calcite is very similar to that of forming spicules. This similarity provides the first evidence, to the best of our knowledge, that in sea urchin spicules the temporal sequence of phases is hydrated ACC $\rightarrow$ anhydrous ACC $\rightarrow$ calcite. In addition, the microcalorimetry data confirm that this sequence is thermodynamically exothermic, and thus energetically downhill (Radha et al. 2010). We predict that the same sequence of transformations takes place in the tooth.

7.5.2 Matrix Proteins of the Tooth

The tooth elements are also known to contain organic material occluded in the calcite. The entire mineralized portion of the tooth, ground and cleaned of adherent material with NaOCl, has been subjected to proteomic analysis by Mann et al. (2008b). Some 138 proteins were identified, 56 of which had been previously identified by this group in test and spine. Major components were identical to abundant proteins in test and spine, while some apparent tooth-specific proteins were especially rich in alanine and proline.

Killian et al. (2010) using PCR methods demonstrated that mRNA encoding SM30 E was very prominent both in mineralized portions and the plumula, and earlier work showed that SM50 is present. Recent work by Veis and his collaborators (Alvares et al., 2009) has identified a number of occluded phosphoproteins. Two of them are identical to proteins identified in an EST library of PMCs of the embryo, notably P16 and P19. P16 had been shown by Cheers and Ettensohn (2005) to be a transmembrane protein whose function was essential for embryonic spicule deposition. Alvares et al. (2009) showed using specific antibody staining that P16, dubbed UTMP16 in the tooth, is found in syncytial membranes in contact with mineral. Mann et al. (2010) have recently analyzed the tooth proteome for the presence of phosphoproteins; they found 15 phosphorylated proteins, 13 of which are unique to tooth tissue.

One of these, named phosphodontin, is rather prominent and contains 35 repeats of an acidic 11–12 amino acid motif that is phosphorylated. Though the amino acid sequence is not orthologous to any known vertebrate tooth protein, the sequence and charge of the peptides indicate it is probably an intrinsically disordered protein.

7.6 Generalizations

Biomineralization processes in Echinoderms display certain general features. Mineralization proceeds in privileged spaces closely apposed by cellular processes, i.e., surrounded by phospholipid membranes. Large numbers of proteins, many of which (but not all) are acidic glycoproteins, are secreted into this space, and ACC, perhaps in the hydrated form, is also deposited into the same places. The ACC gradually transforms to calcite, and some of the proteins are occluded within the forming skeletal element. The details of initial assembly of ACC and protein are not clear, but the slow conversion of ACC to calcite probably occurs by some atomic level reorientation of anhydrous ACC to calcite via a secondary nucleation type of propagation, presenting, at least temporarily, a patchwork of ACC and calcite.

In some respects, this is not so different from proposals made by others (Veis 2008): in this view, cells construct a structural matrix in a defined space, other molecules then assist in orderly nucleation and regulation of crystal growth, habit, shape, and size. One could argue that formation of shells in the extrapallial space adjoining the mantle of mollusks utilizes similar processes. These generalizations, while providing a useful viewpoint, do not easily lead to experimentally testable models. Just what are the matrix molecules? How do they participate? What molecules preside over regulation of the transition from ACC to calcite or aragonite? Of amorphous calcium phosphates to carbonated apatite? How is the secretion of molecules linked to attaining particular crystalline polymorphs?

New facts have emerged that may help inform some models. In our judgment, these are:

1. Amorphous forms of the mineral can and do serve as precursors,
2. Very large numbers of protein-modifying enzymes are occluded in the amorphous mineral and could regulate transitions to crystalline state, polymorph selection, resistance to fracture, and other properties of the crystal,
3. While there are few, if any, orthologs of occluded "matrix" proteins present in shells, stereoms, or vertebrate teeth and bones, proteins with intrinsically disordered domains play a crucial role in regulation of nascent mineralization.

We have already touched on the first point, and while the presence of an amorphous precursor has not been shown to be near-universal, the examples have been reported from three different phyla: echinoderms (Beniash et al. 1997; Politi et al. 2004, 2008; Killian et al. 2009), mollusks (Weiss et al. 2002; Nassif et al. 2005), and chordates (Mahamid et al. 2008; Beniash et al. 2009). Thus the idea of generalizing amorphous precursors is not unreasonable. Indeed, it helps explain how crystalline biominerals can attain the wide variety of shapes that they do. It should be noted, however, that Kudo et al. (2010) did not observe ACC in nacre deposition in the Japanese oyster, *Crassostrea nippona*.

The second point is hinted at in the recent proteomic studies of echinoderm teeth, test, and spines. An extraordinary number of different proteins are occluded; many of them are present in amounts that are unlikely to be due to contamination, and indeed, were identified earlier by methods with less resolution (Killian and Wilt, 1996). Of course, the presence of some organic components could be due to contamination. Mann et al. (2008b) did compare protein content of ground tooth fragments with and without treatment with NaOCl. The intentional "contamination" indicated that many of the proteins provisionally classified as "occluded" (such as carbonic anhydrase, metalloproteases, cyclophilins) are probably authentic occluded proteins, although some very rare ones (e.g., histones) are more likely contaminants. Therefore, the putative presence of proteases, e.g., indicates that post-secretory modifications of matrix proteins can and probably do occur, and thus, could be expected to participate in the regulation of changes in the mineral phase.

Third, and finally, there is now enough genomic information from vertebrates, from the sea urchin, and from some mollusks and invertebrates, to prudently conclude that major matrix proteins from one clade are not present in others. For instance, the SM30 and SM50 protein families of echinoderms are not found in mollusks or vertebrates. Dentinal phosphoproteins of vertebrate teeth are not found in echinoderms or mollusks, and so on. What does emerge, however, is the presence of proteins in mollusks, (AP24, Pif, n16, etc.), echinoderms, and vertebrates

(Sibling family proteins) that have extended domains termed "intrinsically disordered" and can adopt different conformations when interacting with different "targets", such as crystal surfaces. Such proteins have been shown to participate in polymorph selection (aragonite/calcite) in mollusks (Evans 2008; McMahon et al. 2005; Metzler et al. 2010). In addition to IDP domains, domains with dense clusters of negative charge, such as phosphates found in phosphophoryn of dentin, could conceivably play similar roles in different biomineralizing systems. Perhaps we should look for analogous function of certain domains, rather than specific proteins, as a common thread in biomineralization.

There has been a radical shift in the kinds of models considered to explain the formation of biominerals. Just a few decades ago, it would have been difficult to foresee amorphous precursors and proteins with domains of poorly defined conformation playing roles in biomineralization, but these are now well established. There are exciting developments before us. Current research is so robust that we can foresee considerable progress in the near future.

Acknowledgments The authors gratefully acknowledge support for research in their respective laboratories from DOE grant DE-FG02-07ER15899, NSF grant DMR&CHE-0613972, and UW-Madison Hamel Award to PUPAG, and NSF grant 0444724 and Committee on Research of UC Berkeley to FW.

References

Aizenberg J, Hanson J, Koetzle TF, Weiner S, Addadi L (1997) Control of macromolecule distribution within synthetic and biogenic single calcite crystals. J Am Chem Soc 119:881–886

Aizenberg J, Muller DA, Grazul JL, Hamann DR (2003) Direct fabrication of large micropatterned single crystals. Science 299:1205–1208

Alvares K, Dixit SE, Lux E, Veis A (2009) Echinoderm phosphorylated matrix proteins UTMP16 and UTMP19 have different functions in sea urchin tooth mineralization. J Biol Chem 284:26149–26160

Ameye L, Hermann R, Wilt F, Dubois P (1999) Ultrastructural localization of proteins involved in sea urchin biomineralization. J Histochem Cytochem 47:1189–1200

Ameye L, Becker G, Killian C, Wilt F, Kemps R, Kuypers S, DuBois P (2001) Proteins and saccharides of the sea urchin organic matrix of mineralization: characterization and localization in the spine skeleton. J Struct Biol 134:56–66

Banfield JF, Welch SA, Zhang HZ, Ebert TT, Penn RL (2000) Aggregation-based crystal growth and microstructure development in natural iron oxyhydroxide biomineralization products. Science 289:751–754

Beniash E, Aizenberg J, Addadi L, Weiner S (1997) Amorphous calcium carbonate transforms into calcite during sea urchin larval spicule growth. Proc R Soc Lond Biol 264:461–465

Beniash E, Addadi L, Weiner S (1999) Cellular control over spicule formation in sea urchin embryos: a structural approach. J Struct Biol 125:50–62

Beniash E, Metzler R, Lam RSK, Gilbert PUPA (2009) Transient amorphous calcium phosphate in forming enamel. J Struct Biol 166:133–143

Benson S, Jones EME, Benson N, Wilt F (1983) Morphology of the organic matrix of the spicule of the sea urchin larva. Exp Cell Res 148:249–253

Benson NC, Benson SC, Wilt F (1989) Immunogold detection of glycoprotein antigens in sea urchin embryos. Am J Anat 185:177–182

Benson S, Smith L, Wilt F, Shaw R (1990) Synthesis and secretion of collagen by cultured sea urchin micromeres. Exp Cell Res 188:141–146

Berman A, Addadi L, Weiner S (1988) Interactions of sea-urchin skeleton macromolecules with growing calcite crystals-a study of intracrystalline proteins. Nature 331:546–548

Berman A, Hanson J, Leiserowitz L, Koetzle TF, Weiner S, Addadi L (1993) Biological control of crystal texture: a widespread strategy for adapting crystal properties to function. Science 259:776–779

Brusca RC, Brusca GJ (1990) Invertebrates. Sinauer Associates, Sunderland, MA

Cheers MS, Ettensohn CA (2005) P16 is an essential regulator of skeletogenesis in the sea urchin embryo. Dev Biol 283:384–396

Cölfen H, Antonietti M (2008) Mesocrystals and nonclassical crystallization. John Wiley & Sons, Chichester, UK

Decker GL, Lennarz WJ (1988) Skeletogenesis in the sea urchin embryo. Development 103:231–247

Dibernardo M, Castagnetti S, Bellomonte D, Oliveri P, Melfi R, Palla F, Spinelli G (1999) Spatially restricted expression of *Pl OTP* of *P. lividus* orthopedia related homeobox gene, is correlated with oral ectoderm patterning and skeletal morphogenesis in late cleavage sea urchin embryos. Development 126:2171–2179

DuBois P, Ameye L (2001) Regeneration of spines and pedicellariae in Echinoderms: a review. Micros Res Tech 55:427–437

Duloquin L, Lhomond G, Gache C (2007) Localized VEGF signaling from the ectoderm to mesenchyme cell controls morphogenesis of the sea urchin embryo skeleton. Development 134:2293–2302

Ettensohn CE, Malinda KM (1993) Size regulation and morphogenesis: a cellular analysis of skeletogenesis in the sea urchin embryo. Development 119:155–167

Evans JS (2008) "Tuning in" to mollusk shell nacre- and prismatic-associated protein terminal sequences: Implications for biomineralization and the construction of high performance inorganic-organic composites. Chem Rev 108:4455–4462

George NC, Killian CE, Wilt FH (1991) Characterization and expression of a gene encoding a 30.6 kD *Strongylocentrotus purpuratus* spicule matrix protein. Dev Biol 147:334–342

Goodwin AL, Michel FM, Phillips BL, Keen DA, Dove MT, Reeder RJ (2010) Nanoporous Structure and Medium-Range Order in Synthetic Amorphous Calcium Carbonate. Chemistry of Materials 22:3197–3205

Guss KA, Ettensohn CA (1997) Skeletal morphogenesis in the sea urchin embryo: regulation of primary mesenchyme gene expression and skeletal rod growth by ectoderm-derived cues. Development 124:1899–1908

Gustafson T, Wolpert LM (1967) Cellular movement and contact in sea urchin morphogenesis. Biol Rev 42:441–498

Heatfield BM, Travis DF (1975) Ultrastructural studies of regenerating spines of the sea urchin *Strongylocentrotus purpuratus* I. Cell types without spherules. J Morphol 145:3–50

Holland ND (1965) An autoradiographic investigation of tooth renewal in purple sea urchin (*Strongylocentrotus purpuratus*). J Exp Zool 158:275–282

Huggins L, Lennarz WJ (2001) Inhibition of procollagen C-terminal proteinase blocks gastrulation and spicule elongation in the sea urchin embryo. Dev Growth Differ 43:415–424

Hyman LH (1955) The invertebrates: echinodermata. McGraw-Hill, New York

Ingersoll EP, Wilt FH (1998) Matrix metalloproteinase inhibitors disrupt spicule formation by primary mesenchyme cells in the sea urchin embryo. Dev Biol 196:95–106

Ingersoll E, Wilt F, MacDonald K (2003) The ultrastructural localization of SM30 and SM50 in the developing sea urchin embryo. J Exp Zool 300:101–112

Killian CE, Wilt FH (1996) Characterization of the proteins comprising the integral marix of embryonic spicules of *Strongylocentrotus purpuratus*. J Biol Chem 271:9150–9155

Killian CE, Wilt FH (2008) Molecular aspects of biomineralizatiionn of the echinoderm endoskeleton. Chem Rev 108:4463–4474

Killian CE, Metzler RA, Gong YT, Olson IC, Aizenberg J, Politi Y, Addadi L, Weiner S, Wilt FH, Scholl A, Young A, Doran A, Kunz M, Tamura N, Coppersmith SN, Gilbert PUPA (2009) The mechanism of calcite co-orientation in the sea urchin tooth. J Am Chem Soc 131:18404–18409

Killian CE, Croker L, Wilt FH (2010) SpSM30 gene family expression patterns in embryonic and adult biomineralized tissues of the sea urchin, *Strongylocentrotus purpuratus*. Gene Expr Patterns 10(2–3):135–139

Killian CE, Metzler RA, Gong YUT, Churchill TH, Olson IC, Trubetskoy V, Christensen MB, Fournelle JH, De Carlo F, Cohen S, Mahamid J, Wilt FH, Scholl A, Young A, Doran A, Coppersmith SN, Gilbert PUPA (2011) Self-sharpening mechanism of the sea urchin tooth. Adv Funct Mater 21:682–690

Kniprath E (1974) Ultrastructure and growth of the sea urchin tooth. Calc Tiss Res 14:211–228

Kudo M, Kameda J, Saruwatari K, Ozaki N, Okano K, Nagasawa H, Kogure T (2010) Microtexture of larval shell of oyster, *Crassostrea nippona*: a FIB-TEM study. J Struct Biol 169:1–5

Laubichler MD, Davidson EH (2008) Boveri's long experiment: sea urchin merogones and the establishment of the role of nuclear chromosomes in development. Dev Biol 314:1–11

Livingston BT, Killian C, Wilt FH, Cameron AC, Landrum MJ, Ermolaeva O, Sapojnikov V, Maglott DR, Ettensohn C (2006) A genome-wide analysis of biomineralization-related proteins in the sea urchin, *Strongylocentrotus purpuratus*. Dev Biol 300:335–348

Lowenstam H, Weiner S (1989) On biomineralization. Oxford University Press, New York

Ma Y, Weiner S, Addadi L (2007) Mineral deposition and crystal growth in the continuously forming teeth of sea urchins. Adv Funct Mater 17:2693–2700

Ma Y, Cohen SR, Addadi L, Weiner S (2008) Sea urchin tooth design: an "all-calcite" polycrystalline reinforced fiber composite for grinding rocks. Adv Mater 20:1555–1559

Ma Y, Aichmayer B, Paris O, Fratzl P, Meibom A, Metzler RA, Politi Y, Addadi L, Gilbert PUPA, Weiner S (2009) The grinding tip of the sea urchin tooth exhibits exquisite control over calcite crystal orientation and Mg distribution. Proc Natl Acad Sci USA 106:6048–6053

Magdans U, Gies H (2004) Single crystal structure analysis of sea urchin spine calcites. Eur J Miner 16:261–268

Mahamid J, Sharir A, Addadi L, Weiner S (2008) Amorphous calcium phosphate is a major component of the forming fin bones of zebrafish: indications for an amorphous precursor phase. Proc Natl Acad Sci USA 105:12748–12753

Mann K, Poustka AJ, Mann M (2008a) The sea urchin (*Strongylocentrotus purpuratus*) test and spine proteomes. Proteome Sci 6:22–32

Mann K, Proustka AJ, Mann M (2008b) In-depth, high-accuracy proteomics of sea urchin tooth organic matrix. Proteome Sci 6:33–44

Mann K, Poustka AJ, Wilt FH (2010) The sea urchin (*Strongylocentrotus purpuratus*) spicule proteome. Proteome Sci 8:33

Mann K, Wilt FH, Poustka AJ (2010) Proteomic analysis of sea urchin (*Strongylocentrotus purpuratus*) spicule matrix. Proteome Science 8

Märkel K, Titschack H (1969) Morphology of sea-urchin teeth (in German). Z Morph Tiere 64:179–200

Märkel K, Röser U (1983a) The spine tissues in the Echinoid *Eucidaris tribuloides*. Zoomorphology 103:25–41

Märkel K, Röser U (1983b) Calcite-resorption in the spine of the Echinoid *Eucidaris tribuloides*. Zoomorphology 103:43–58

McMahon SA, Miller JI, Lawton JA, Kerkow DE, Hodes A, Marti-Renom MKA, Doulatov S, Narayaanan E, Sali A, Miller JF, Ghosh P (2005) The C-type lectin fold as an evolutionary solution for massive sequence variation. Nat Struct Mol Biol 12:886–892

Metzler RA, Evans JS, Killian CE, Zhou D, Churchill TH, Appathurai NP, Coppersmith SN, Gilbert PUPA (2010) Nacre protein fragment templates lamellar aragonite growth. J Am Chem Soc 132:6329–6334

Michel FM, MacDonald J, Feng J, Phillips BL, Ehm L, Tarabrella C, Parise JB, Reeder RJ (2008) Structural characteristics of synthetic amorphous calcium carbonate. Chemistry of Materials 20:4720–4728

Mitsunaga K, Makihara R, Fujino Y, Yasumasu I (1986) Inhibitory effects of ethacrynic acid, furosemide and ifedipne on the calcification of spicules in cultures of micromeres from *H. pulcherrimus*. Differentiation 30:197–205

Moore HB (1966) Ecology of echinoids. In: Boolootian RA (ed) Physiology of echinodermata. John Wiley and Sons, New York, pp. 75–86

Moureaux C, Pérez-Huerta A, Compère P, Zhu W, Leloup T, Cusack M, Dubois P (2010) Structure, composition and mechanical relations to function in sea urchin spine. J Struct Biol 170(1):41–49

Nakano E, Okazaki K, Iwamatsu T (1963) Accumulation of radioactive calcium in larvae of the sea urchin *Pseudocentrotus depressus*. Biol Bull 125:125–133

Nassif N, PInna N, Gehrke N, antoniettei M, Jager C, Colfen H (2005) Amorphous layer around aragonite platelets in nacre. Proc Natl Acad Sci USA 102:12653–12655

Nelson BV, Vance RR (1979) Diel foraging patterns of the sea urchin *Centrostephanus coronatus* as a predator avoidance strategy. Mar Biol 51:251–258

Okazaki K (1956) Skeleton formation of the sea urchin larvae. I. Effect of Ca concentration of the medium. Biol Bull 110:320–333

Okazaki K (1960) Skeleton formation of sea urchin larvae. II. Organic matrix of the spicule. Embryologia 5:283–320

Okazaki K (1975a) Spicule formation by isolated micromeres of the sea urchin embryo. Amer Zool 15:567–581

Okazaki K (1975b) Normal development to metamorphosis. In: Czihak G (ed) The sea urchin embryo. Springer, Berlin, pp 177–216

Okazaki K, Inoue S (1976) Crystal property of the larval sea urchin spicule. Dev Growth Differ 188:567–581

Orme CA, Noy A, Wierzbicki A, McBride MT, Grantham M, Teng HH, Dove PM, DeYoreo JJ (2001) Formation of chiral morphologies through selective binding of amino acids to calcite surface steps. Nature 411:775–779

Otter GW (1932) Rock burrowing echinoids. Biol Rev Camb Philos Soc 7:89–107

Peled-Kamar M, Hamilton P, Wilt FH (2002) The Spicule matrix protein LSM34 is essential for biomineralization of the sea urchin spicule. Exp Cell Res 272:56–61

Penn RL, Banfield JF (1999) Morphology development and crystal growth in nanocrystalline aggregates under hydrothermal conditions: Insights from titania. Geochim Cosmochim Acta 63:154915–154957

Politi Y, Arad T, Klein E, Weiner S, Addadi L (2004) Sea urchin spine calcite forms via a transient amorphous calcium carbonate phase. Science 306:1161–1164

Politi Y, Levi-Kalisman Y, Raz S, Wilt F, Addadi L, Weiner S, Sagi I (2006) Structural characterization of the transient amorphous calcium carbonate precursor phase in sea urchin embryos. Adv Funct Mater 16:1289–1298

Politi Y, Metzler RA, Abrecht M, Gilbert B, Wilt FH, Sagi I, Addadi L, Weiner S, Gilbert PUPA (2008) Transformation mechanism of amorphous calcium carbonate into calcite in the sea urchin larval spicule. Proc Natl Acad Sci USA 105:17362–17366

Radha AV, Forbes TZ, Killian CE, Gilbert PUPA, Navrotsky A (2010) Transformation and crystallization energetics of synthetic and biogenic amorphous calcium carbonate. Procs Natl Acad Sci USA 107:16438–16443

Raup DM (1966) The endoskeleton. In: Boolootian RA (ed) Physiology of echinodermata. John Wiley and Sons, New York, pp 379–395

Robach JS, Stock SR, Veis A (2009) Structure of first-and second-stage mineralized elements in teeth of the sea urchin *Lytechinus variegatus*. J Struct Biol 168:452–466

Seto J, Zhang Y, Hamilton P, Wilt F (2004) The localization of occluded matrix proteins in calcareous spicules of sea urchin larvae. J Struct Biol 148:123–130

Simkiss K, Wilbur KM (1989) Biomineralization. Academic Press, San Diego

Smith MM, Smith LC, Cameron RA, Urry L (2008) A larval staging scheme for *Strongylocentrotus purpuratus*. J Morphol 269:713–733

Veis A (2008) Crystals and life: an introduction. In: Sigel A, Sigel H, Sigel RKO (eds) Biomineralization. From nature to application, vol. 4 of metal ions in life sciences. John Wiley and Sons, Ltd, Chichester, pp 2–35

Wang RZ (1998) Fracture toughness and interfacial design of a biological fiber-matrix ceramic composite in sea urchin teeth. J Am Ceram Soc 81:1037–1040

Wang RZ, Addadi L, Weiner S (1997) Design strategies of sea urchin teeth: structure, composition and micromechanical relations to function. Phil Trans R Soc Lond B 352:469–480

Weiss I, Tuross N, Addadi L, Weiner S (2002) Mollusc larval shell formation: amorphous calcium carbonate is a precursor for aragonite. J Exp Zool A 293:478–491

Wilt FH (1999) Matrix and mineral in the sea urchin larval skeleton. J Struct Biol 126:216–226

Wilt FH (2002) Biomineralization of the spicules of sea urchin embryos. Zool Sci 19:253–261

Wilt FH, Ettensohn CE (2007) Morphogenesis and biomineralization of the sea urchin larval endoskeleton. In: Baeuerlein E (ed) Handbook of biomineralization. Wiley-VCH, Weinheim, pp 183–210

Wilt FH, Killian CE (2008) What genes and genomes tell us about calcium carbonate biomineralization. In: Sigel A, Sigel H, Sigel RKO (eds) Biomineralization. From nature to application, vol. 4 of metal ions in life sciences. John Wiley and Sons, Ltd, Chichester, pp 36–69

Wilt FH, Killian CE, Livingstson B (2003) Development of calcareous skeletal elements in invertebrates. Differentiation 71:237–250

Wilt FH, Killian CE, Hamilton P, Croker L (2008a) The dynamics of secretion during sea urchin embryonic skeleton formation. Exp Cell Res 314:1744–1752

Wilt FH, Croker L, Killian CE, McDonald K (2008b) Role of LSM34/SpSM50 proteins in endoskeletal spicule formation in sea urchin embryos. Invert Biol 127:452–459

Yajima M (2007) A switch in the cellular basis of skeletogenesis in late-stage sea urchin larvae. Dev Biol 307:272–281

Yajima M, Kiyomoto M (2006) Study of larval and adult skeletogenic cells in developing sea urchin larvae. Biol Bull 211:183–192

Yang L, Killian CE, Kunz M, Tamura N, Gilbert PUPA (2011) Biomineral nanoparticles are space-filling. RSC-Nanoscale 3:603–609

Zhu X, Mahairas G, Illies M, Cameron RA, Davidson EH, Ettensohn CA (2001) A large scale analysis of the mRNAs expressed by primary mesenchyme cells of the sea urchin embryo. Development 128:2615–2627

Chapter 8
Echinoderms as Blueprints for Biocalcification: Regulation of Skeletogenic Genes and Matrices

Valeria Matranga, Rosa Bonaventura, Caterina Costa, Konstantinos Karakostis, Annalisa Pinsino, Roberta Russo, and Francesca Zito

Contents

Abstract Echinoderms have an extensive endoskeleton composed of magnesian calcite, a form of calcium carbonate that contains small amounts of magnesium carbonate and occluded matrix proteins. Adult sea urchins have several calcified structures, including test, teeth, and spines, composed of numerous ossicles which form a three-dimensional meshwork of mineral trabeculae, the stereom. The biomineral development begins in 24-hour-old embryos within the primary mesenchyme cells (PMCs), the only cells producing a set of necessary matrix proteins. The deposition of the biomineral occurs in a privileged extracellular space produced by the fused filopodial processes of the PMCs. We showed for the first time that signals from ectoderm cells overlying PMCs play an important role in the regulation of biomineralization-related genes. It is believed that growth factors

V. Matranga (✉)
Consiglio Nazionale delle Ricerche, Istituto di Biomedicina e Immunologia Molecolare "Alberto Monroy", Via Ugo La Malfa 153, 90146 Palermo, Italy
e-mail: matranga@ibim.cnr.it

W.E.G. Müller (ed.), *Molecular Biomineralization*, Progress in Molecular and Subcellular Biology 52, DOI 10.1007/978-3-642-21230-7_8,

are produced by ectoderm cells and released into the blastocoel where they interact with cognate receptor tyrosine kinases restricted to PMCs, which activate signaling cascades regulating the expression of biomineralization-related genes. We demonstrated the implication of a TGF-beta family factor by a perturbation model in which skeleton elongation was indirectly blocked by monoclonal antibodies to an extracellular matrix (ECM) protein located on the apical surface of ectoderm. Thus, it was inferred that interfering with the binding of the ECM ligand, a member of the discoidin family, to its cell surface receptor, a βC integrin, disrupts the ectodermal cell signaling cascade, resulting in reduced or aberrant skeletons. During the last few years, we analyzed the expression of biomineralization-related genes in other examples of experimentally induced skeleton malformations, produced by the exposure to toxic metals, such as Cd and Mn or ionizing radiations, such as UV-B and X-rays. Besides the obvious toxicological implication, since the mis-expression of spicule matrix genes paralleled skeleton defects, we believe that by means of these studies we can dissect the molecular steps taking place and possibly understand the physiological events regulating embryonic biomineralization.

8.1 The Basis of Biomineral Formation

Biomineralization refers to the biological processes employed by living organisms to form minerals as a result of regulated processes. A biomineral represents a complex material which incorporates both mineral and organic components exhibiting advantageous properties compared to its inorganically formed counterpart. Compared to abiotic minerals, biominerals possess additional physical and chemical characteristics which offer increased flexibility and duration. They vary in morphology, shape, and size as well as in element composition. The structure of a biomineral involves a mosaic of crystalline domains separated by occluded proteinaceous material forming a framework (Wilt 1999). The structure exhibits single crystal diffraction properties as shown by X-ray diffraction studies (Simkiss 1986).

Classically, according to the degree of biological control over the precipitated mineral, biomineralization processes can be categorized into two groups: the "biologically induced" (Lowenstam 1981) and the "biologically controlled" mineralization (Mann 1983).

In biologically induced mineralization, cell surfaces may act as causative nucleation agents which lead to crystal growth. Mineral growth is indirectly affected, but not controlled, by the biological system. The adopted mineral form is favored by metabolic processes which define the chemical conditions of the microenvironment (i.e., pH, pCO_2, concentration of products resulting from secretion) (Frankel and Bazylinski 2003). As environmental conditions play a potential role in the formation of the biologically induced minerals, these biominerals exhibit heterogeneity in elemental composition, in water content, and in particle size, resulting in various external morphologies.

In biologically controlled mineralization, cellular activities direct all the stages for the precipitation of the mineral: from the determination of the initial deposition site, to nucleation, growth, and formation of the final crystalline morphology. Resulting biominerals acquire reproducible, species-specific, genetically determined structures and properties. Controlled biomineralization requires an isolated environment which serves as the mineralization site. This environment can be extracellular, intercellular or intracellular, with respect to the cells which control the process. In general, biologically induced mineralization is found in bacteria and lichens, whereas biologically controlled mineralization is found in foraminifera, cephalopod statoliths, mollusks shells, bryozoan exoskeletons, scleractinian corals, echinoderms, human bones, and teeth.

In nature, almost 50% of biominerals are calcium-bearing minerals (Lowenstam and Weiner 1989). Calcium content in living organisms is highly regulated, as it has key roles in metabolic processes at concentrations varying from 0.01 μM to 10 μM. The most abundant calcium-bearing biominerals precipitate acquiring one of the eight known calcium carbonate polymorph forms. These include seven crystalline forms: calcite, Mg-calcite, aragonite, vaterite, monohydrocalcite, protodolomite, hydrocerussite, and one amorphous calcium carbonate (ACC) form (Addadi et al. 2003). Most models of biomineralization invoke the involvement of membrane ion transporters (channels and pumps) in the delivery of Ca^{2+} and other ions to the calcification site (Simkiss and Wilbur 1989).

An important parameter for the formation of a biomineral is the regulation of the "isotopic composition" (Weber and Raup 1966) which establishes a physicochemical equilibrium with the microenvironment. The medium from which the mineral forms is a saturated solution, which is required at the site of mineralization only. Supersaturation can be achieved by the presence of additives which prevent the deposition of crystalline phases, such as magnesium in concentrations similar to those of the seawater (Raz et al. 2000). Another inhibitory factor of crystallization is the presence of proteins which, serving as substrates, influence the solubility of the mineral phase and stabilize different polymorphs, either ACC (Aizenberg et al. 1996) or crystalline calcium carbonate forms. In conclusion, crystal shape depends on inorganic and organic factors: pI, temperature, microenvironment growth, solubility of the mineral phase, concentration of occluded macromolecules and ions. As a result, calcium carbonate growth undergoes a series of phases and morphologies to form the final calcite structure. On the contrary, the effect of whole extracts of proteinaceous matrixes on the in vitro precipitation of calcium carbonate was shown to selectively induce the precipitation of particular polymorphs. Some authors have pointed out that the soluble or insoluble matrix, extracted from nacre, controls calcium carbonate crystal polymorphs toward aragonite or calcite formation (Falini et al. 1996). Others used chitin as a substrate and Mg^{2+} as an additive to induce aragonite double-layered composite film formation (Kato 2000). Studies such as those mentioned above indicate that template molecules can act as nucleators for the precipitation of the inorganic material. The surface chemistry of each template molecule and the arrangement of the amino acid groups guide the oriented nucleation of the cognate crystal face.

8.2 Biomineral Contents and Shapes

Examples of calcium-bearing biominerals are found in the echinoderms, a phylum of marine invertebrates. Biomineral formation has an important role in the phylum, which includes Echinoids (sea urchins and sand dollars), Asteroids (sea stars), Ophiuroids (brittle stars), Crinoids (feather stars) and Holothuroids (sea cucumbers). It supports the construction of the skeleton of the living organisms, offering support and protection. Echinoderms employ a genetically regulated process to form their calcareous skeleton by the precipitation of magnesian calcite, a form of calcium carbonate that contains small amounts of magnesium carbonate in a ratio given by the formula: $(Mg_xCa_{1-x})CO_3$. In a high Mg/Ca ratio, amorphous calcium carbonate or aragonitic nucleation is favored over calcite (Raz et al. 2000; Mann 2001).

The magnesium carbonate content can vary from 2.5% to 39% depending on the classes and species. In general, the hardness of the biomineral is directly proportional to the concentration of magnesium. Low magnesium biominerals are found in the softer Ophiuroids and high magnesium biominerals are found in the harder Asteroids (Dubois and Chen 1989).

The skeletons of all echinoderms are made of microscopic bony plates, also called ossicles, deposited as a three-dimensional meshwork called a stereom, made of magnesian calcite and a network of interconnected holes filled with living tissue. Different types of stereom are indicative of the type of living tissue that penetrates the plates (e.g., cells, tube feet, others). The skeleton is covered by an epidermis and contains a network of internal water-filled canals or encloses a coelomic cavity bathed in coelomic fluid. The structure of the biomineral reflects complexity. Skeletal plates may remain simple or may fuse to form composite plates. They can also form tubercles, granules, fixed, or movable spines. Examples of the diversity of stereoms within the one species, *Paracentrotus lividus (P. lividus)*, can be observed in Fig. 8.1. Among the classes, the most diverse skeletal organizations are evident in Echinoids and Holothuroids. In the former, organisms are completely surrounded by a calcified test embedded in the mesodermal stroma tissue, with only a thin layer of muscular tissue. On the contrary, Holothuroids possess an endoskeleton reduced to microscopic ossicles dispersed throughout the dermis, with a highly muscularized body wall (Smith et al. 2010). In the last 40 years, advanced high-resolution imaging techniques, such as SEM and TEM, have helped researchers to provide information about the fine structure and composition of the stereom, as well as the cellular basis of echinoderm calcification.

Various studies have been aiming to identify the proteins involved in the mineralization of the adult sea urchin test, teeth, and spines (Berman et al. 1993). The proteome of the mineralized parts of the adult sea urchin was recently described using mass spectrometry-based methods. In the *Strongylocentotus purpuratus* (*S. purpuratus*) sea urchin, 138 (Mann et al. 2008a) and 110 (Mann et al. 2008b) proteins were identified in the tooth and test/spines organic matrix respectively.

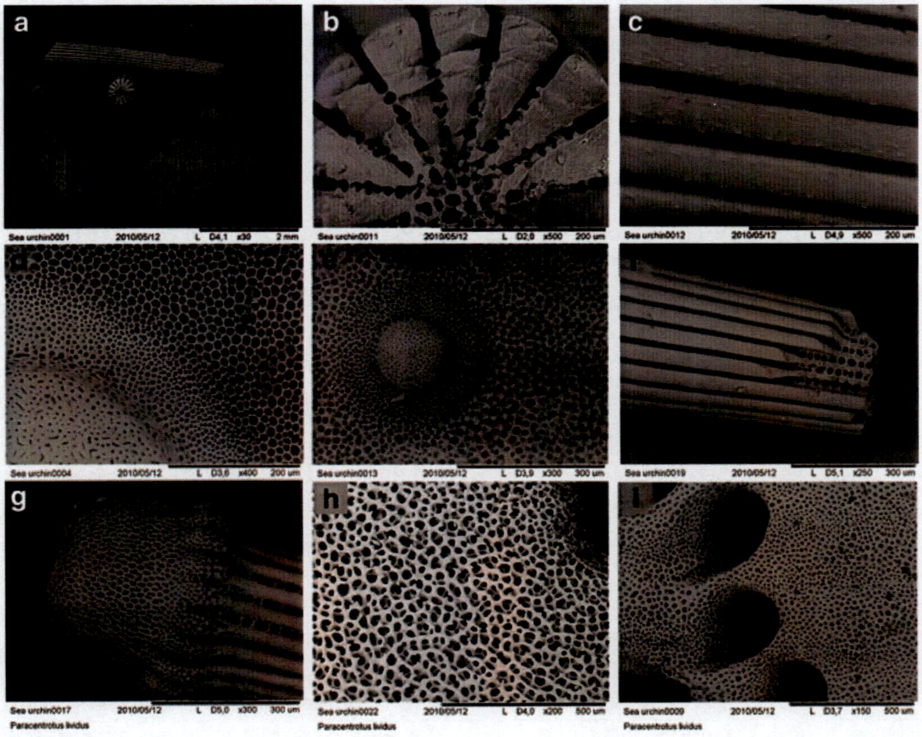

Fig. 8.1 *Paracentrotus lividus* stereoms. Scanning electron micrographs of different parts of tests and spines. (**a**) Low magnification of all samples observed in B-I. (**b**) cross section of a spine fragment, (**c**) longitudinal surface of the spine, (**d**) high and (**e**) low magnification of external test portions,(**f**) fractured spine tip, (**g**) indented base which fits like a ball-and-socket joint over a tubercle of the test, and (**h**, **i**) integument plate. This plate shows rectilinear (box-like) stereom characteristic of the ectoderm

In sea urchin embryos, 231 proteins were identified in spicule matrix extracts (Mann et al. 2010). Among the most abundant proteins found are SM30 and SM50, which were originally purified biochemically and belong to the C-type lectin family. Various other lectins include: SM29, SM32, SM37, PM27 and Sp-Clect_13, as well as metalloproteases and carbonic anhydrase. Some of these proteins are expressed both in embryonic spicules and adult mineralized parts, e.g., the phosphoproteins P16 and P19 (Alvares et al. 2009) and isoforms of the C-type lectin SM30 (Killian et al. 2010). The great number of proteins identified in the above-mentioned studies has been possible thanks to the genome-wide analysis of biomineralization-related proteins (Livingston et al. 2006). Most of the proteins were found to be sea urchin specific, meaning they have no apparent homologues in other invertebrate deuterostomes or vertebrates. They include several families: the spicule-matrix proteins, the msp-130 family, cyclophilins, collagens, carbonic anhydrase, P-16, P-19, secreted Ca-binding phosphoproteins, transcription factors, ECM molecules, and proteins involved in cell-ECM interaction, such as secreted proteases. In conclusion, although for some of those proteins the biological functions have been proved, the future challenge will be to characterize and demonstrate the specific role of each protein in the growth and patterning of the skeleton.

8.3 Cells Involved in Adult Echinoderms Biomineralization

Teeth, spines, and the test of adult echinoderms are all formed by a terminally differentiated type of cells of mesodermal origin, called sclerocytes, specializing in the deposition of skeletal parts. Sclerocytes are the only cell population to contact the stereom and occupy the mesodermal stroma tissue of an intact mineral structure. They produce thin cell processes that closely surround the trabeculae, forming the so-called skeletal cytoplasmic sheath (Heatfield and Travis 1975). The thin processes show no organelles, while the cell body includes a well-developed and dilated Golgi complex, with associated vesicles and other organelles (Stricker 1985; Märkel et al. 1986; Dubois and Ameye 2001). Calcifying vacuoles are formed within the cytoplasmic layers, which are separated from the calcite by a vascular space. All developing ossicles appear to be surrounded by a syncytial network of sclerocytes, though it is reported that the ossicle rudiment appears either in a single cell (Crinoids, Echinoids, and Ophiuroids) or in a syncytium (Holothuroids) (Dubois and Chen 1989). The mineral skeleton is produced intracellularly or intrasyncytially (Märkel and Röser 1985). During the calcification process, the vacuoles increase in size and ramify, producing the characteristic projection of the primary plates, and new sclerocytes are incorporated in the pseudopodial syncytium (Kniprath 1974; Märkel et al. 1986). This syncytium may either envelop preexisting calcite surface on which the biomineral grows, or form independently from the preexisting stereom (Heatfield and Travis 1975).

8.3.1 Biomineral Formation and Regenerative Events

Echinoderms show a remarkable self-repairing ability as adults and a latent potential for regeneration of large parts of their bodies. This potential is common to all classes of the phylum as an adaptive mechanism for survival and dispersion. Lines of cells preserve the ability to differentiate into somatic and germ tissues, but little is known concerning the nature of cells and the molecular pathways involved in the different phases of echinoderm regeneration such as wound healing, growth, morphogenesis, and cell differentiation. Regeneration is in fact a characteristic type of developmental process that can involve cell turnover and tissue repair, reconstruction of external and internal organs, and regrowth of new complete adults from detached body fragments (Candia Carnevali et al. 2009). It has been reported that the dermis of regenerating spines is characterized by active sclerocytes, the principal skeleton-forming cells also known as calcoblasts or odontoblast (Märkel et al. 1986; Dubois and Ameye 2001).

Signals, activated in response to initial inflammatory events, trigger a series of intracellular processes associated with survival, cell proliferation, migration, attachment, differentiation, and eventually matrix deposition. All these steps are required to accelerate tissue repair and reconstruction, including neural, muscular, and skeletogenic tissues. Any regenerative process implies the existence of stem cells present in the circulating fluids or in the tissues in the form of resident cells, ready to be recruited after trauma (Pinsino et al. 2007). Scarce information is available on the biomineralization process following regeneration. Pioneering studies at the molecular level demonstrated the overexpression of three mRNA specific of the primary mesenchyme cells (PMCs) in regenerating spines. By in situ hybridization, it was shown that gene products were localized primarily in calcoblasts that accumulated at the regeneration sites (Drager et al. 1989). More studies are awaited in tissue calcification events, which are of fundamental importance, as they ensure that biomineralization proceeds normally in the proper sites, while ectopic mineralization is prevented elsewhere.

8.4 Cellular Signaling and Biomineral Formation in the Sea Urchin Embryo

Among echinoderms, the sea urchin embryo has been known for its versatility and suitability since the end of the nineteenth century, when classical embryologists performed the earliest studies on the basic mechanisms of embryo development, facilitated by the optical transparency of the embryo, along with its simplicity in shape and organization (Hörstadius 1939). Since then, the sea urchin has been utilized to a great extent for studies in several scientific fields, ranging from basic developmental biology to ecotoxicology and applied research. In addition, the ability to form a larval endoskeleton encourages the use of the sea urchin embryo for detailed studies on the mechanisms underlying biomineralization, an extremely interesting topic for nano(bio)technology. It is becoming increasingly clear that the biomineralization process is genetically controlled and a great number of steps in such biological control could exist.

Biomineralization in the sea urchin initiates from the early embryonic developmental stages and is mediated by the PMCs which express genes under a complex signaling control of transcription and growth factors (for a review see Ettensohn 2009). Schematic drawings of the initial phases of embryonic skeleton deposition and PMCs arrangement are presented in Fig. 8.2. Accumulating evidence demonstrates that the PMCs express genes which control the formation and remodeling of carbonate crystals. In sea urchin embryos, ACC is a transient phase leading to the formation of calcitic spicules (Beniash et al. 1997; Weiss et al. 2002). The formation of ACC results from a supersaturated solution produced by ions and proteins (Raz et al. 2003).

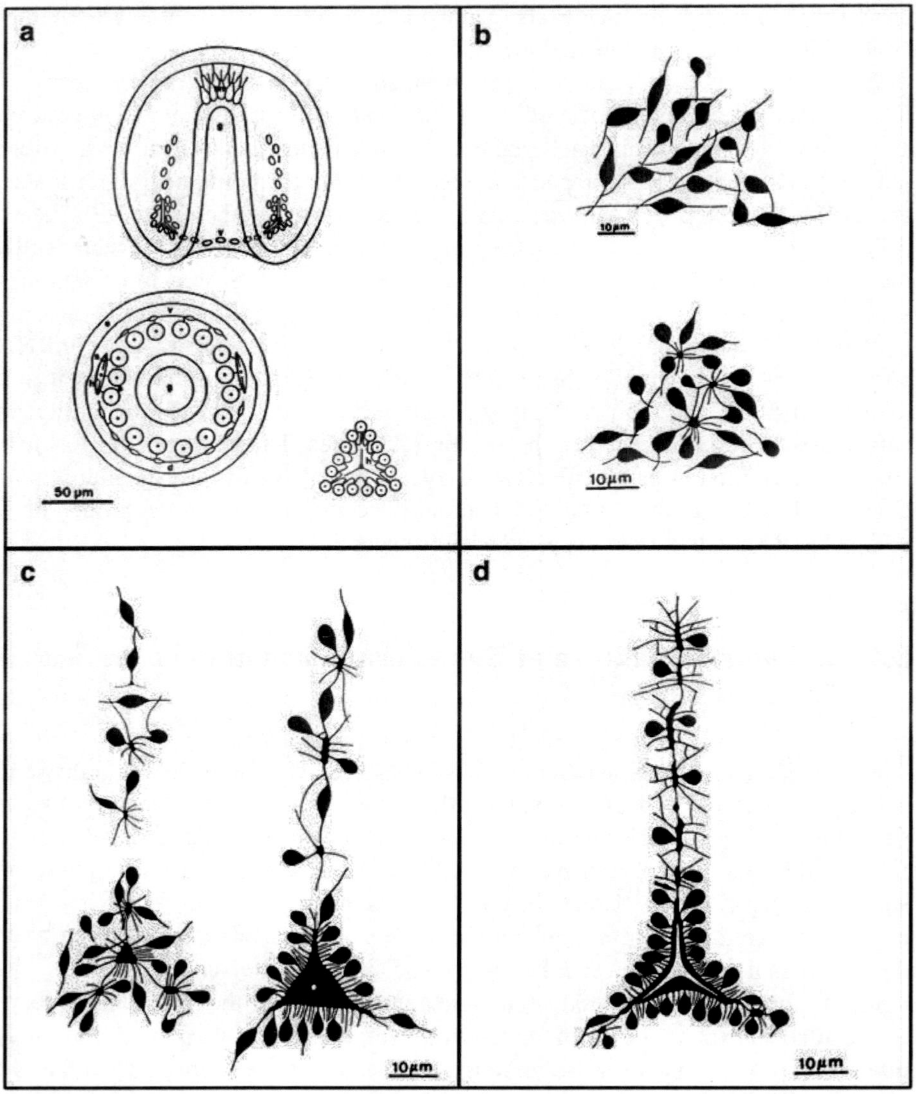

Fig. 8.2 Schematic drawing the location and arrangements of PMCs. (**a**) Gastrula frontal and vegetal views showing the PMCs aggregates (clusters) at opposite sides of the archenteron and one of the clusters enclosing the spicule triradiate rudiment; (**b**) formation of PMCs aggregates and initial pseudopodial fusion; (**c**) alignment of PMCs and enlargement of the fused cytoplamic space, note the first sign (white dot) of biomineral deposition; and (**d**) development of spicule rudiment inside the cell cluster (white area), assuming the shape of a triradiate crystal (modified from Dubois and Chen 1989)

The formation of the sea urchin calcareous embryonic skeleton occurs in specialized vesicles/vacuoles within the PMCs. Cells tightly control the concentration of the ions precursors and nucleation is initiated by deposition on an organic template. The pH, pCO_2, and the concentration of trace elements are regulated by membrane trafficking. After the initial precipitation occurring inside the cells, crystals are secreted to the extracellular blastocoelic environment, by fusion of vesicles with the plasma membrane, where they interact with matrix proteins to acquire the final crystal form (Wilt et al. 2008; Yang et al. 2011). Thus, the mineral is believed to be enrobed by the syncytial membrane and the cytoplasm.

The larval skeleton displays considerable morphological diversity among sea urchin species, including variations in the number, structure/shape, and size of rods. Thus, when describing the single parts of the skeleton, one should refer to the species taken into consideration. Here, we want to briefly describe for the first time the development of the skeleton of *P. lividus* sea urchin embryo, identifying the chains of PMCs which will give rise to the different rods. In particular, as shown in Fig. 8.3, the ventral chain will give rise to the ventral transverse rod, the longitudinal chain will form the anterolateral rod, and the dorsal chain will give rise to the body and postoral rods. The larval skeleton morphology and its differences among species is interesting for evolutionary studies, both from a developmental viewpoint, to understand how these differences are produced, and from an ecological viewpoint, to understand why such differences have been generated (Zito and Matranga 2009; Ettensohn 2009).

As already mentioned, the only cells competent to produce a skeleton in sea urchin embryos are the PMCs. Their morphological features and behavior, as well as the main cellular events leading to the formation of the embryonic skeleton, have

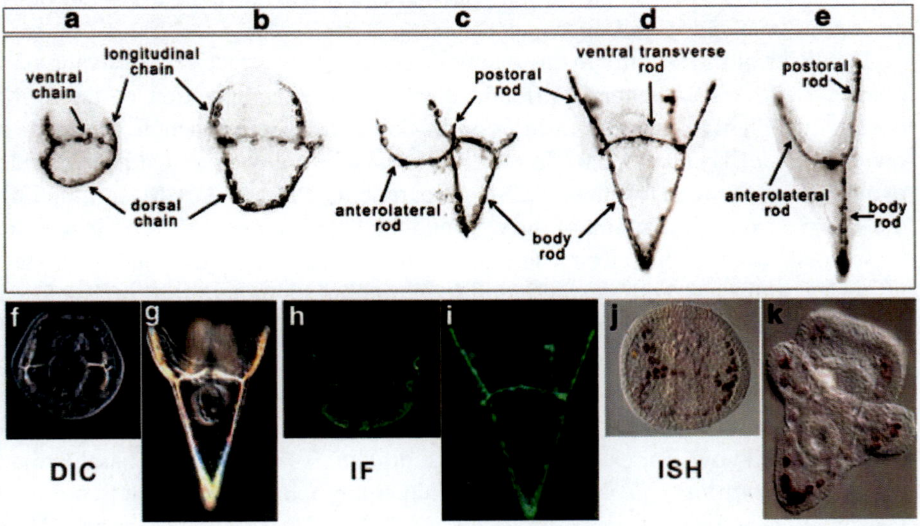

Fig. 8.3 Development of *Paracentrotus lividus* skeleton. Schematic drawings of skeleton development observed at (**a**) late gastrula; (**b**) prism; (**c**) early pluteus; (**d**) pluteus, ventral view; (**e**) pluteus, lateral view. Ventral chain, longitudinal chain, and dorsal chain indicate set of PMCs which will give raise to the ventral transverse rod, anterolateral rod and body and postoral rod, respectively. Tools to study skeleton formation in late gastrula (**f, h, j**) and pluteus (**g, i, k**) stage embryos. (**f, g**) differential interference contrast; (**h, i**) immunofluorescence with antibody to msp130; (**j, k**) in situ hybridization with msp130 probe

been extensively described (see also this book, Chapter by Goldberg and Wilt). Considerable progress has been made in elucidating the molecular basis of PMC specification by the analysis of a gene regulatory network that operates in the large micromere descendants (Ettensohn 2009). Upstream components include several maternal and zygotic transcription factors; downstream are gene products that directly control the PMCs morphogenetic behaviors, such as ingression, migration, fusion, and deposition of the biomineralized endoskeleton. In the following, we will focus on selected aspects of skeletogenesis concerning the external cues stimulating PMCs to produce the right patterned skeleton. In particular, we will discuss on the involvement of extracellular matrix (ECM) molecules and growth factors in the process of skeletogenesis.

8.4.1 Extracellular Matrix

A great number of in vitro and in vivo studies highlight the great importance of the ECM during morphogenesis. In fact, in addition to its function as an extracellular space-filling scaffold, it plays a fundamental role in cell–substrate interactions, providing both spatial and temporal information to adherent cells, thus influencing cellular proliferation and viability, differentiation, and gene expression. Accordingly, interest has increased in recent years toward the identification, purification, and functional studies on ECM components, along with their ligands, and other molecules involved in cell–ECM adhesion.

The ECM of the sea urchin embryo is a very complex structure, consisting of a number of layers with many different elements. The organization of the ECM occurs in a highly regulated fashion during different developmental steps (for a review see McClay et al. 1990). In addition, it must be pointed out that it is possible to distinguish between two main ECM compartments: the apical ECM, surrounding the embryo from fertilization of the egg, and the basal lamina, localized inside the blastocoelic cavity, which forms during the late cleavage stage (see Fig. 8.4). So far, no ECM molecule with an active direct role in biomineralization has been documented in the sea urchin embryo. Nevertheless, there are a number of evidences of indirect roles played by ECM molecules during the formation of the skeleton. Among the ECM components present in the blastocoel, only ECM molecules that function in supporting or directing PMCs movements have been described until now. For example, Pamlin, a protein isolated from the basal lamina of the *Hemicentrotus pulcherrimus* sea urchin embryo, has been shown to promote PMC binding and migration in vitro (Katow 1995). In *Lytechinus variegatus*, PMCs have been observed by light microscopy to directly interact with a class of ECM fibers, localized on the basal surface of the ectoderm via their filopodia (Hodor et al. 2000). A component of such fibers is ECM3, a high-molecular-weight protein accumulated selectively in the basal lamina adjacent to the ectoderm in all regions except for the animal pole (Wessel and Berg 1995). ECM3 has been suggested as a strong candidate as a PMCs substrate molecule, with a role in providing guidance

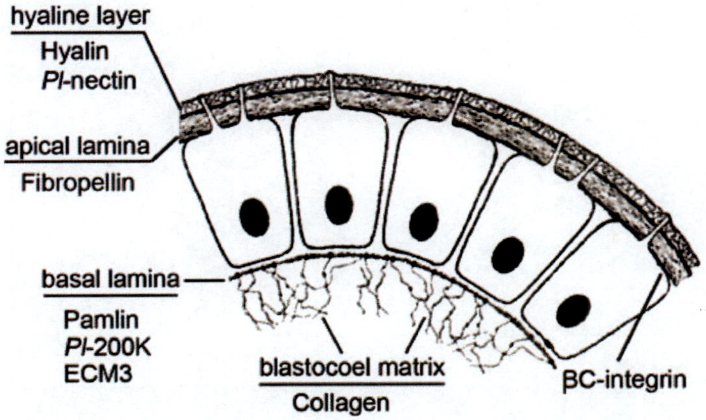

Fig. 8.4 Schematic diagram of the sea urchin blastula cell monolayer. Compartments of the extra-embryonic (hyaline layer, apical lamina) and blastocoelic (basal lamina, blastocoel matrix) extracellular matrixes are indicated. The location of ECM proteins of interest is shown

information to migrating PMCs. Interestingly, ECM3 contains an N-terminal domain similar to the mammalian chondroitin sulfate proteoglycan core protein NG2, required for the responsiveness of some cell types to PDGF, supporting the idea that it may play a role in growth factor presentation (Hodor et al. 2000). The central region of ECM3 contains five tandem repeats of a motif similar to the domain contained within the regulatory Ca^{2+}-binding loop of the Na^{+}-Ca^{2+} exchange protein, which is similar in several respects to MAFp4, a component of the aggregation factor molecular complex, mediating the species-specific sorting of sponge cells (Hodor et al. 2000).

Other ECM proteins located on the basal lamina have been described to have a role in spicule formation. For example, by microinjection of specific antibodies into the blastocoelic cavity, it has been shown that collagen in *S. purpuratus* (Wessel et al. 1991) and *Pl*-200 K protein in *P. lividus* (Tesoro et al. 1998) are essential for skeleton elongation and patterning.

Sea urchin metalloproteinases can modify collagen and other ECM components, as well as glycoproteins that associate with collagens and carbohydrates in the ECM. It has been shown that several inhibitors of metalloproteinases inhibit the continuation of skeletogenesis in both *S. purpuratus* and *L. pictus embryos* (Ingersoll and Wilt 1998).

The functional role played by ECM molecules in vivo has often been investigated by means of monoclonal antibodies (McAb) which interfere with their functions. Among the ECM proteins already known, *Pl*-nectin, isolated from *P. lividus* (Matranga et al. 1992) as a collagen-binding molecule, is the first described as an "indirect actor" in the ecto-mesoderm signaling (Zito et al. 1998). The protein, a discoidin family member whose complete sequence and domain architecture has been recently characterized (Fig. 8.5a) (Costa et al. 2010), is localized on the apical surface of ectoderm and endoderm cells from the blastula and gastrula stage onwards (Fig. 8.5b). By in vitro assays, it has been shown to

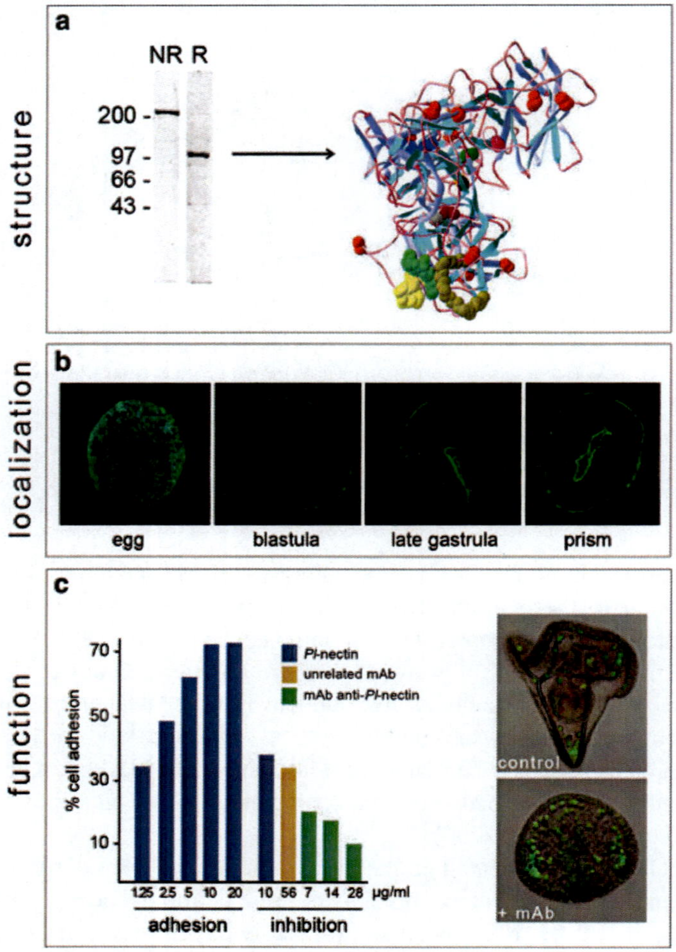

Fig. 8.5 *Pl*-nectin structure, localization and function. (**a**) Structure: on the left SDS-PAGE of the purified molecule showing a 210 kDa homodimer under non-reducing conditions (NR), consisting of two 105 kDa monomers under reducing conditions (R), which are held together by S-S bridges (modified from Matranga et al. 1992). On the right, three-dimensional structure of the 105 kDa C-shaped monomer, based on in silico homology modeling (modified from Costa et al. 2010). (**b**) Localization: indirect immunofluorescence on section with antibody to *Pl*-nectin. In the unfertilized egg, the protein is found within granules uniformly distributed throughout the cytoplasm. After fertilization, the protein is polarized to the apical surface of ectoderm and endoderm cells during all the developmental stages examined, until the prism stage (modified from Matranga et al. 1992). (**c**) Function: on the left cell adhesion assays. The protein promotes the adhesion of blastula cells to the substrate in a dose-dependent manner (histogram) (modified from Matranga et al. 1992). A monoclonal antibody (mAb) to *Pl*-nectin inhibits the adhesion of blastula cells to *Pl*-nectin-coated substrates in "in vitro" functional assays (histogram). On the right, in vivo perturbation experiments with mAb to *Pl*-nectin show embryos with major defects in skeleton elongation and patterning (compare control embryo with treated embryo). Primary mesenchyme cells are stained with an antibody to Msp130 showing that PMCs enter the blastocoel and position correctly (modified from Zito et al. 1998)

mediate cell–substrate adhesion in a dose-dependent fashion (Fig. 8.5c), suggesting a functional role during development (Matranga et al. 1992). A homologous protein has been purified from *Temnopleurus hardwickii*, and named *Th*-nectin (Yokota et al. 1994). In the outer ECM, two other proteins called fibropellins are present after fertilization and throughout early development. At the blastula stage, fibropellins become organized into distinct fibers which form a mesh-like network over the surface of the embryo (Bisgrove et al. 1991).

The function of *Pl*-nectin in vivo has been studied by morphogenetic assays in which embryos were cultured in the presence of a McAb to *Pl*-nectin (Fig. 8.5c). A high number of embryos with serious skeleton defects, but with normally developed ectoderm and endoderm structures, were found (Zito et al. 1998, 2003). It was demonstrated that skeleton deficiency was not due to a reduction in the number of PMCs ingressing the blastocoel, which were found as normal in number (usually 32 cells in *P. lividus*) and regularly arranged in a subequatorial ring (Zito et al. 2003). On the other hand, skeleton-deficient embryos showed a decrease in the expression levels of SM30 (Zito et al. 2003), while SM50 and MSP130 expression were not affected by the treatment with a McAb to *Pl*-nectin (Zito, personal communication). Concurrently, a strong reduction in the expression levels of univin, a TGF-β family growth factor, was found, while its mis-expression obtained by the injection of its mRNA was sufficient to rescue defects in skeleton elongation and SM30 expression (Zito et al. 2003). Based on these results, we proposed for the first time, a working model in which the secretion of univin, or other growth factor(s), into the blastocoel by ectodermal cells drives PMCs to synthesize SM30 and other spicule matrix proteins required for spicule growth. The competence of these ectodermal cells to produce the signal depends on their binding to *Pl*-nectin (see Fig. 8.6).

Some parts of the proposed model have been later confirmed by our laboratory and a few others. Recently, it has been shown by in vitro immunoprecipitation and affinity chromatography experiments that *Pl*-nectin binds to a βC integrin subunit, suggesting that the interaction of *Pl*-nectin with ectoderm cells is mediated by a βC-containing integrin receptor (Zito et al. 2010). Interestingly, an integrin recognition motif (LDT) has been found in the *Pl*-nectin protein sequence (Costa et al. 2010). An LDT motif has been originally identified in the human mucosal addressin cell adhesion molecule (MAdCAM-1) as the binding site of the α4/β7 integrin receptor. Thus, the binding of *Pl*-nectin with ectoderm cells surface could result from the interaction of the LDT motif with the identified integrin receptor. Adhesion assays using competing synthetic peptides (LDT, and others) are in progress in our laboratory to confirm this hypothesis. In addition, the structure of *Pl*-nectin, consisting of six tandemly repeated discoidin domains, provides it with the ability to: (1) bind to ECM molecules bearing galactose and N-acetylglucosamine carbohydrate moieties (including collagen, or cell membrane surface glycoproteins), (2) bind to cell surface proteins, such as tyrosine kinase receptors and G protein-coupled receptors, and (3) form multimeric structures by the discoidin domains self-binding (Costa et al. 2010). The last property would explain the nectosome structures observed by TEM described in *T. hardwikii* (Kato et al. 2004).

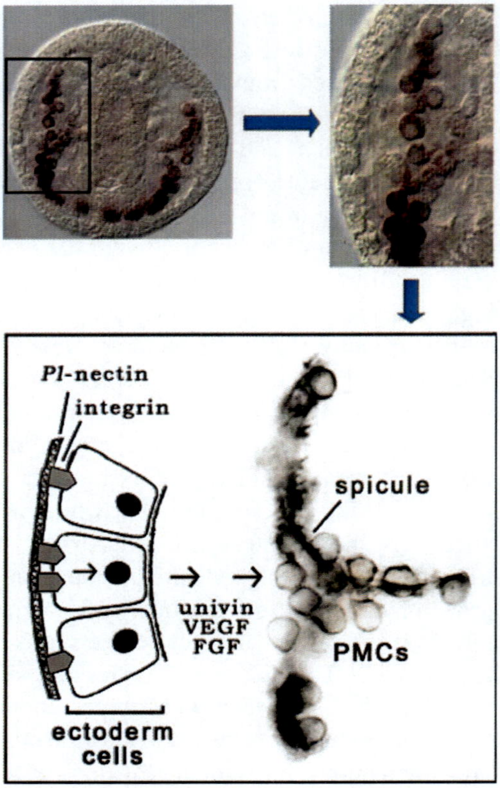

Fig. 8.6 The interaction of ectoderm cells with *Pl*-nectin conditions ecto-mesoderm induction and skeleton growth. In the upper panels: low and high magnification of late gastrula embryos in which PMCs are expressing skeletogenic genes, as shown by in situ hybridization with msp130 probe. In the lower panel: schematic representation of ectoderm cells, properly interacting with the *Pl*-nectin contained in the ECM (hyaline layer), secrete into the blastocoel growth factors, i.e., univin, VEGF, and/or FGF, which signal PMCs to synthesize the spicule. The interaction of ectoderm cells with *Pl*-nectin is mediated by an integrin receptor and activates a yet unknown signaling pathway

8.4.2 Growth Factors

Besides the synthesis and secretion of spicule matrix proteins and the regulation of matrix–mineral interactions, PMCs require several types of cues from the surrounding embryo, including axial, temporal, and scalar signals, in order to synthesize a normal sized and patterned skeleton. Such cues have been shown to originate from the overlying ectoderm and the apical ECM (Guss and Ettensohn 1997; Zito et al. 2003; Kiyomoto et al. 2004; Duloquin et al. 2007; Röttinger et al. 2008). The attractive idea that skeleton formation is regulated by the ectodermal cues was first proposed more than 70 years ago (von Ubish, 1937), although the molecular cues implicated in such interactions are being identified only in recent years. So far, essential signals released by the ectoderm have been identified among growth factors and include univin, VEGF, and FGF (Zito et al. 2003; Duloquin et al. 2007; Röttinger et al. 2008). It seems that each of these growth factors is required for controlling skeleton morphogenesis, probably using independent pathways that are not functionally redundant.

Univin is the first gene encoding a member of the TGF-β superfamily to be identified in the sea urchin embryo (Stenzel et al. 1994). The univin amino acid sequence is closely related to zDVR-1 (zebrafish), BMPs-2 and 4 (human), decapentaplegic protein (dpp) (Drosophila), and Vg-1 (Xenopus) (Stenzel et al. 1994; Lapraz et al. 2006). At the blastula stage, univin is expressed in a circum-equatorial ring of ectodermal cells and during gastrulation is expressed in the archenteron. Univin transcripts are confined to bilateral regions of the ectoderm between the arms of the young pluteus larva (Stenzel et al. 1994; Range et al. 2007). We have shown that embryos with skeletal defects caused by ECM–ectoderm cell-binding inhibition (see above, Fig. 8.6 proposed model), and expressing low levels of univin and SM30, could be rescued by univin mRNA microinjection with the contemporaneous partial resumption of SM30 mRNA levels (Zito et al. 2003). This was the first evidence showing that a growth factor produced by ectoderm cells is the inductive signal responsible for skeleton growth. Later, Duloquin et al. (2007) demonstrated the guidance role of VEGF/VEGFR signaling for the positioning and differentiation of PMCs during gastrulation. At the early gastrula stage, the VEGFR is expressed only in the PMCs, while VEGF is expressed in two restricted areas of the ventrolateral ectoderm overlying the PMCs clusters. At the pluteus stage, VEGF transcripts are detected only in a few ectodermal cells, located at the tip of the arm buds, which face VEGFR-expressing PMCs. The impairment of VEGF/VEGFR expression, obtained by morpholinos injection, causes the inhibition of SM50 and SM30 and leads to skeleton-lacking embryos. MSP130 expression is not significantly affected, suggesting that it is not under the control of the VEGF/VEGFR signaling machinery. In the same way, Röttinger et al. (2008) have identified and characterized the FGFA gene, and its putative receptor genes FGFR1 and FGFR2. The co-localization of their transcripts during development, observed by in situ hybridization, as well as loss-of-function experiments, involving morpholinos injection, have shown that they regulate PMCs migration and dramatically affect skeleton elongation. Again, a block of SM30 and SM50 expression is demonstrated in the PMCs.

8.5 Ecotoxicological Approaches to the Study of Skeletogenesis

Toxicology is one of the oldest sciences. In the 1500s, Paracelsus found a way to overcome the difficulties in defining which substances are toxic and which are not, by his famous aphorism: "only dose makes the difference". Modern ecotoxicology, can be defined as the integration of toxicology and ecology, and aims to quantify the toxic effects of stressors upon changes in the ecosystem composition (Truhaut 1977; Chapman 2002). The most relevant chemical/physical stressors for all organisms, including the marine ones, are: (1) physical agents such as ionizing radiation (UV-B and X-rays); (2) inorganic chemicals, such as heavy metals, nitrates, and nitrites; (3) organic chemicals, such as pesticides, oil, haloorganics, hydrocarbons, phenols, synthetic materials; and (4) emerging stressors, such as metallic and engineered nanoparticles, CO_2, pH, temperature. Here, we focused on reports about the negative influences of chemical/physical stressors on biomineralization in echinoderms.

8.5.1 Metals Affecting Biomineralization

In general, although metallic elements have certain properties in common, each is a distinct element with its own physicochemical characteristics which determine its biological or toxicological properties (Duffus 2002). Exposure of cells/organisms to high levels of nonessential or essential metals causes toxicity which gained attention of many research groups (ATSDR 2008; Gerber et al. 2002; CICAD 2004; Lima et al. 2008). As an example, manganese (Mn), which at physiological concentrations is involved in cellular protection, replication and bone mineralization (ATSDR 2008; Santamaria 2008; Daly 2009), can cause toxicity when its concentration are in excess (Satyanarayana and Saraf 2007).

In sea urchins, toxicity tests on embryos and larvae have been proposed for the characterization of the effects causes by a variety of toxicants, including essential and heavy metals. Metals have been classically used as inducers of skeleton malformations in sea urchin and sea star embryos. These include cadmium (Cd), zinc (Zn), copper (Cu), chromium (Cr), lead (Pb), nickel (Ni), lithium (Li), manganese (Mn), gadolinium (Gd), whose exposure causes: reduced or absent skeleton, abnormal branching patterns, extranumerary origin sites were observed (Hardin et al. 1992; Radenac et al. 2001; Coteur et al. 2003; Russo et al. 2003; Roccheri et al. 2004; Kobayashi and Okamura 2004; Agca et al. 2009; Kiyomoto et al. 2010; Pinsino et al. 2010; Saitoh et al. 2010). In some cases, (Cd and Mn exposures) defects in biomineralization have been correlated to the expression of stress proteins, such as hsp60 and hsp70 (Roccheri et al. 2004; Pinsino et al. 2010), apoptosis (Filosto et al. 2008), and Metallothionein mRNA overexpression (Russo et al. 2003). The expression of biomineralization-related genes has been investigated only in lithium-treated and in zinc-treated embryos, by a high throughput analysis based on gene array screening and whole mount in situ hybridization, providing evidence of the differential expression of more than 4,000 and 250 genes, respectively (Poustka et al. 2007). Four genes expressed in PMCs, namely, Pmar, SMAD2, P19, SM50, seemed to be lightly upregulated upon treatment. Further investigation is needed in this direction.

8.5.2 Ionizing Radiations

An increase in UV-B radiation due to the thinning of the ozone hole can reach the Earth, including marine ecosystems. Although the penetration of UV-B into natural waters can vary, depending on the concentration and optical qualities of dissolved organic matter, phytoplankton, or other suspended particles (Dunne and Brown 1996), its consequences can have serious effects on organisms. Some planktonic embryos, larvae, and adults of many species are more prone to UV-B as they dwell in the top layers of the water column (Hader 2000). Harmful effects of water-penetrating UV-B radiation cause strong impairment of development as well as modifications in cellular protein composition, especially in the modulation of stress

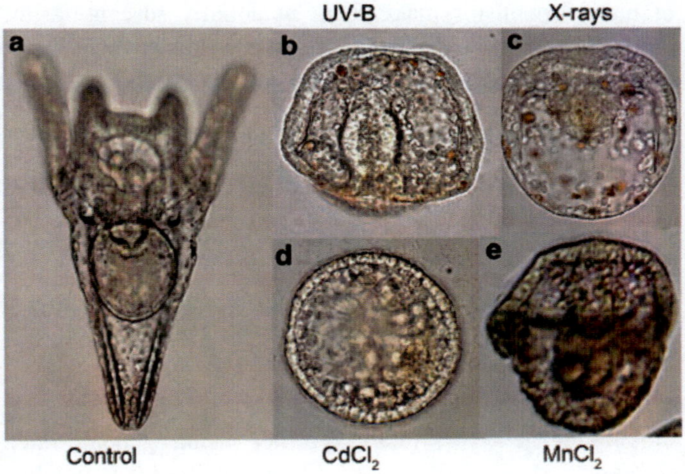

Fig. 8.7 Effects of physical and chemical impacts on skeleton growth. Morphologies of *P. lividus* sea urchin embryos observed 48 h after fertilization. (**a**) control embryo at the pluteus stage, (**b**) UV-B embryo pulse-irradiated at the mesenchyme blastula stage (modified from Bonaventura et al. 2005), (**c**) X-rays embryo pulse-irradiated at the cleavage stage (modified from Matranga et al. 2010), (**d**) embryo continuously exposed to $CdCl_2$ from fertilization (modified from Russo et al. 2003), and (**e**) embryo continuously exposed to $MnCl_2$ from fertilization (modified from Pinsino et al. 2010)

proteins levels, mis-regulation in gene expression and DNA damage (Batel et al. 1998; Schröder et al. 2005; Bonaventura et al. 2005, 2006; Holzinger and Lütz 2006; Tedetti and Sempéré 2007; Banaszak and Lesser 2009). In the following, we will briefly review the effects of ionizing radiation on the formation of the skeleton.

The effects of UV-B radiation on skeletons of echinoderms were studied at early and late stages of development on *P. lividus*. Embryos experimentally exposed to UV-B were lacking the skeleton or had severe skeleton abnormalities (Fig. 8.7). In addition, increased levels of the heat shock proteins 70 (hsp70) and the activation of p38 MapK were found (Bonaventura et al. 2005, 2006). Russo et al. (2010) demonstrating a direct dose-dependent relationship between UV-B exposure of sea urchin embryos and the mRNA levels of *Pl*-14-3-3ε, a gene involved in stress, survival, and apoptosis, suggesting its implication in the regulative cascade activated in response to UV-B irradiation. Recent evidence demonstrated that embryos exposed to high doses of X-rays lack spicules and their PMCs were delocalized inside the blastocoel. These results, together with a reduced expression of SM30 by RT-PCR and msp130 by ISH, indicate that X-rays strongly affect sea urchin embryo biomineralization (Matranga et al. 2010).

8.5.3 Impacts of Ocean Acidification on Biocalcification

Calcification is one of the primary targets for many studies on the impact of CO_2-driven climate change in the oceans since the calcium carbonate shells or skeletons of many organisms make them potentially susceptible to dissolution in acidic waters (Orr et al. 2005). The pH shift consequent to CO_2 dissolution in seawater changes the equilibrium between bicarbonate and carbonate, depleting the available carbonate pool (Gattuso et al. 2010). It has been demonstrated that CO_2-induced seawater acidification alters skeletogenesis of different developing sea urchin species: *Hemicentrotus pulcherrimus* and *Echinometra mathaei* (Kurihara and Shirayama 2004), *L. pictus* (O'Donnell et al. 2010), *P. lividus* (Dupon et al. 2010), and *Tripneusteus gratilla* (Sheppard-Brennand et al. 2010). Embryos showed significant perturbation on both size and shape, which were enhanced upon simultaneous seawater warming and acidification, suggesting that although much emphasis has been placed on ocean acidification, embryos may not reach the skeletogenic stage because of warm waters (Sheppard-Brennand et al. 2010).

When looking at gene expression profiles using genome microarrays of *L. pictus*, O'Donnell et al. (2010) showed that, among genes involved in energy metabolism and biomineralization, only a few were downregulated, including Suclg-1succinyl-CoA synthetase, SM30-like, osteonectin, whereas only the Atp2a1-Ca^{2+} ATPase, involved in ion regulation and acid–base balance pathways was upregulated. All together, these results suggest that, although larvae are able to form an endoskeleton, increased CO_2 levels have consequences on larval transcriptome. Contrary to expectations, no visible differences in developmental timing or obvious developmental skeleton abnormalities were associated with CO_2 treatments in *S. purpuratus*. However, the trascriptomic approach has evidenced the downregulation of a few biomineralization genes, namely: cyclophylins, MSP130, MSP130-related, collagens (COLP3α, COLP4α, etc.), P19, P16, P16-like, osteonectin (Todgham and Hofmann 2009).

8.6 Concluding Remarks

The formation of the sea urchin skeleton offers a good model for the in vivo study of biocalcification both in adults and embryos, the latter where most of the studies were performed. In summary, to build the magnesian calcitic embryonic skeleton, cellular, microenvironmental conditions and a biomolecular toolkit are needed. PMCs, the only cells in the embryo controlling skeleton origin and growth, provide a set of specific spicule matrix proteins and supersaturated calcium carbonate niches. The process initiates with the formation of intermediate transient aggregating nanoparticles of ACC, which then crystallizes into macroscopic single rhombohedral crystals of calcite (Yang et al. 2011). Radial growth of the crystals and its branching is probably dictated by more than 200 PMCs-specific proteins, the most famous being SM30 and SM50, the first to be fully described (reviewed by Wilt 1999), which take part in the above-mentioned mechanism, possessing different roles. Within the gene repertoire of PMCs, among a great number of annotated

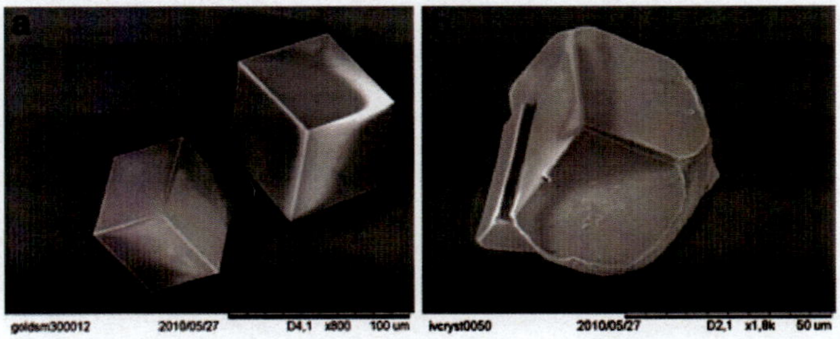

Fig. 8.8 In vitro crystallization assay of calcite precipitation. Scanning electron photographs of: (**a**) negative control and (**b**) acetic acid-soluble extract from *Paracentrotus lividus* tests. Partial inhibition of crystal formation is induced by proteins contained in the preparation

biomineralization-related proteins, potential spicule matrix proteins are categorized as such when having at least one of the following characteristics: a lectin domain (or a C-type lectin domain), a signal sequence, pro/gly-rich or asp-rich regions (Livingston et al. 2006; Mann et al. 2010).

Despite extensive information at the gene level, there is still very little information about the cellular and/or molecular mechanisms behind the biomineralization occurring in sea urchin embryos. More studies are required for a thorough description of the process which might proceed following a few main experimental directions. The first, aimed at the characterization of the role of each protein could involve in vitro crystallization tests showing the inhibition of calcium carbonate precipitation, as shown for the calprismin and caspartin, two small acidic proteins, purified from the mollusc *Pinna nobilis* (Marin et al. 2005). Preliminary experiments utilizing a crude extract obtained from acetic acid extraction of *P. lividus* tests powder (see Fig. 8.8) encourage the use of such an assay with purified recombinant spicule matrix proteins. A second experimental approach relates to studies on protein–mineral interactions where proteins might be challenged as template molecules having an effect on the patterning of the crystal.

In conclusion, despite the difference in composite "design" between vertebrates and echinoderms, the proteins related to mineralization of calcium-based skeletal tissue have remarkable similarities. Therefore, studying differences in the mineral end-products of one group will continue to illuminate processes in the other.

Acknowledgments This work was supported by the BIOMINTEC Project (European Commission N° PITN-GA-2008-215507 grant). Authors are indebted to Prof. Frederic Marin for access to and help in the use of the SEM at the UMR 5561 CNRS Biogéosciences, Université de Bourgogne, 21000 Dijon, France. Authors thank Mr. Mauro Biondo for his valuable technical assistance.

References

Addadi L, Raz S, Weiner S (2003) Taking advantage of disorder: amorphous calcium carbonate and its roles in biomineralization. Adv Mat 15:959–970

Agca C, Klein WH, Venutia JM (2009) Respecification of ectoderm and altered Nodal expression in sea urchin embryos after cobalt and nickel treatment. Mech Dev 126:430–442

Aizenberg J, Lambert G, Addadi L, Weiner S (1996) Stabilization of amorphous calcium carbonate by specialized macromolecules in biological and synthetic precipitates. Adv Mat 8:222–226

Alvares K, Dixit SN, Lux E, Veis A (2009) Echinoderm Phosphorylated Matrix Proteins UTMP16 and UTMP19 Have Different Functions in Sea Urchin Tooth Mineralization. J Biol Chem 284:26149–26160

ATSDR (2008) Draft toxicological profile for manganese. Agency for toxic substances and disease registry. Division of toxicology and environmental medicine/applied toxicology branch, Atlanta, Georgia. Available via DIALOG. http://www.atsdr.cdc.gov/tox profiles/tp151-p.pdf

Banaszak AT, Lesser MP (2009) Effects of solar ultraviolet radiation on coral reef organisms. Photochem Photobiol Sci 8:1276–1294

Batel R, Fafandjel M, Blumbach B, Schröder HC, Hassanein HM, Müller IM, Müller WE (1998) Expression of the human XPB/ERCC-3 excision repair gene-homolog in the sponge Geodia cydonium after exposure to ultraviolet radiation. Mutat Res 409:123–133

Beniash E, Aizenberg J, Addadi L et al (1997) Amorphous calcium carbonate transforms into calcite durino sea urchin larval spicule growth. Proc R Soc Lond B 264:461–465

Berman A, Hanson J, Leiserowitz L, Koetzle TF, Weiner S, Addadi L (1993) Biological control of crystal texture: A widespread strategy for adapting crystal properties to function. Science 259:776–779

Bisgrove BW, Andrews ME, Raff RA (1991) Fibropellins, products of an EGF repeat-containing gene, form a unique extracellular matrix structure that surrounds the sea urchin embryo. Dev Biol 146:89–99

Bonaventura R, Poma V, Costa C, Matranga V (2005) UVB radiation prevents skeleton growth and stimulates the expression of stress markers in sea urchin embryos. Biochem Bioph Res Co 328:150–157

Bonaventura R, Poma V, Russo R, Zito F, Matranga V (2006) Effects of UV-B radiation on the development and hsp 70 expression in sea urchin cleavage embryos. Mar Biol 149:79–86

Candia Carnevali MD, Thorndyke MC, Matranga V (2009) Regenerating echinoderms: a promise to understand stem cells potential. In: Stem cells in marine organisms (eds: Rinkevich B, Matranga V) Springer, New York, pp 165–186

Chapman PM (2002) Integrating toxicology and ecology: putting the "eco" into ecotoxicology. Mar Poll Bull 44:7–15

CICAD (2004) Manganese and its compounds: environmental aspects. Concise international chemical assessment document63. WHO, Geneva, Switzerland Available via DIALOG. http://www.who.int/ipcs/publications/cicad/cicad63_rev_1.pdf

Costa C, Cavalcante C, Zito F, Yokota Y, Matranga V (2010) Phylogenetic analysis an homology modelling of Paracentrotus lividus nectin. Mol Divers 14:653–665

Coteur G, Gosselin P, Wantier P, Chambost-Manciet Y, Danis B, Pernet P, Warnau M, Dubois P (2003) Echinoderms as bioindicators, bioassays, and impact assessment tools of sediment-associated metals and PCBs in the North Sea. Arch Environ Contam Toxicol 45:190–202

Daly MJ (2009) A new perspective on radiation resistance based on Deinococcus radiodurans. Nat Rev Microbiol 7:237–244

Drager BJ, Harkey MA, Iwata M, Whitele AH (1989) The expression of embryonic primary mesenchyme genes of the sea urchin, Strongylocentrotus purpuratus, in the adult skeletogenic tissues of this and other species of echinoderms. Dev Biol 133:14–23

Duffus JH (2002) Effect of Cr(VI) exposure on sperm quality. Ann Occup HygMar 46:269–270

Dunne RP, Brown BE (1996) Penetration of solar UVB radiation in shallow tropical waters and its potential biological effects on coral reefs; results from the central Indian Ocean and Andaman Sea. Mar Ecol Prog Ser 144:109–118

Dupon S, Ortega-Martinez O, Thorndyke M (2010) Impact of near-future ocean acidification on echinoderms. Ecotoxicology 19:449–462

Ettensohn CA (2009) Lessons from a gene regulatory network: echinoderm skeletogenesis provides insights into evolution, plasticity and morphogenesis. Development 136:11–21

Falini G, Albeck S, Weiner S, Addadi L (1996) Control of aragonite or calcite polymorphism by mollusk shell macromolecules. Science 271:67–69

Filosto S, Roccheri MC, Bonaventura R, Matranga V (2008) Environmentally relevant cadmium concentrations affect development and induce apoptosis of Paracentrotus lividus larvae cultured in vitro. Cell Biol Toxicol 24:603–610

Frankel RB, Bazylinski DA (2003) Biologically induced mineralization by bacteria. Rev Mineral Geochem 54:95–114

Gattuso JP, Gao K, Lee K, Rost B, Schulz KG (2010) Approaches and tools to manipulate the carbonate chemistry. In: Riebesell U, Fabry VJ, Hansson L, Gattuso J-P (eds) Guide to best practices for ocean acidification research and data reporting. Publications Office of the European Union, Luxembourg, pp 41–52

Gerber GB, Leonard A, Hantson Ph (2002) Carcinogenicity, muta- genicity and teratogenicity of manganese compounds. Crit Rev Oncol Hematol 42:25–34

Guss KA, Ettensohn CA (1997) Skeletal morphogenesis in the sea urchin embryo: regulation of primary mesenchyme gene expression and skeletal rod growth by ectoderm-derived cues. Development 124:1899–1908

Hader DP (2000) Effects of solar UV-B radiation on aquatic ecosystems. Adv Space Res 26:2029–2040

Hardin J, Coffman JA, Black SD, McClay DR (1992) Commitment along the dorsoventral axis of the sea urchin embryo is altered in response to NiCl2. Development 116:671–685

Heatfield BM, Travis DF (1975) Ultrastructural studies of regenerating spines of the sea urchin Strongylocentrotus purpuratus. II. Cells with spherules. J Morphol 145:51–72

Hodor PG, Illies MR, Broadley S, Ettensohn CA (2000) Cell-substrate interactions during sea urchin gastrulation: migrating primary mesenchyme cells interact with and align extracellular matrix fibers that contain ECM3, a molecule with NG2-like and multiple calcium-binding domains. Dev Biol 222:181–194

Holzinger A, Lütz C (2006) Algae and UV irradiation: effects on ultrastructure and related metabolic functions. Micron 37(190–606):207

Hörstadius S (1939) The mechanics of sea urchin development, studied by operative methods. Biol Rev 14:132–179

Ingersoll EP, Wilt FH (1998) Matrix metalloproteinase inhibitors disrupt spicule formation by primary mesenchyme cells in the sea urchin embryo. Dev Biol 196:95–106

Kato T (2000) Polymer/calcium carbonate layered thin-film composites. Adv Mater 12:1543–1546

Kato KH, Abe T, Nakashima S, Matranga V, Zito F, Yokota Y (2004) 'Nectosome': a novel cytoplasmic vesicle containing nectin in the egg of the sea urchin, Temnopleurus hardwickii. Develop Growth Differ 46:239–247

Katow H (1995) Pamlin, a primary mesenchyme cell adhesion protein, in the basal lamina of the sea urchin embryo. Exp Cell Res 218:469–478

Killian CE, Croker L, Wilt FH (2010) SpSM30 gene family expression patterns in embryonic and adult biomineralized tissues of the sea urchin, Strongylocentrotus purpuratus. Gene Expr Patterns 10:135–139

Kiyomoto M, Zito F, Sciarrino S (2004) Commitment and response to inductive signals of primary mesenchyme cells of the sea urchin embryo. Dev Growth Differ 46:107–114

Kiyomoto M, Morinaga S, Ooi N (2010) Distinct embryotoxic effects of lithium appeared in a new assessment model of the sea urchin: the whole embryo assay and the blastomere culture assay. Ecotoxicology 19:563–770

Kniprath E (1974) Ultrastructure and growth of the sea urchin tooth. Calc Tiss Res 14:211–228

Kobayashi N, Okamura H (2004) Effects of heavy metals on sea urchin embryo development. Chemosphere 55:1403–1412

Kurihara H, Shirayama Y (2004) Effects of increased atmospheric CO2 on sea urchin early development. Mar Ecol Progr Series 274:161–196

Lapraz F, Röttinger E, Duboc V et al (2006) RTK and TGF-β signaling pathways genes in the sea urchin genome. Dev Biol 300:132–152

Lima PDL, Vasconcellos MC, Bahia MO, Montenegro RC, Pessoa CO, Costa-Lotufo LV, Moraes MO, Burbano RR (2008) Genotoxic and cytotoxic effects of manganese chloride in cultured human lymphocytes treated in different phases of cell cycle. Toxicol In Vitro 22:1032–1037

Livingston BT, Killian CE, Wilt F et al (2006) A genome-wide analysis of biomineralization-related proteins in the sea urchin Strongylocentrotus purpuratus. Dev Biol 300:335–348

Lowenstam HA (1981) Minerals formed by organisms. Science 211:1126–1131

Lowenstam HA, Weiner S (1989) On Biomineralization. Oxford University Press, New York

Mann S (1983) Mineralization in biological systems. Struct Bonding 54:125–174

Mann S (2001) Biomineralization: principles and concepts in bioinorganic materials chemistry. Oxford University Press, New York

Mann K, Poustka AJ, Mann M (2008a) In-depth, high-accuracy proteomics of sea urchin tooth organic matrix. Proteome Sci 6:33

Mann K, Poustka AJ, Mann M (2008b) The sea urchin (Strongylocentrotus purpuratus) test and spine proteomes. Proteome Sci 6:22

Mann K, Wilt FH, Poustka AJ (2010) Proteomic analysis of sea urchin (Strongylocentrotus purpuratus) spicule matrix. Proteome Science 8:33

Marin F, Amons R, Guichard N, Stigter M, Hecker A, Luquet G, Layrolle P, Alcaraz G, Riondet C, Westbroek P (2005) Caspartin and calprismin, two proteins of the shell calcitic prisms of the Mediterranean fan mussel Pinna nobilis. J Biol Chem 280:33895–33908

Märkel K, Röser U (1985) Comparative morphology of echinoderm calcified tissues: Histology and ultrastructure of ophiuroid scales (Echinodermata, Ophiuroida). Zoomorphology 105:197–207

Märkel K, Röser U, Mackenstedt K (1986) Ultrastructural investigations of matrix-mediated biomineralization in echinoids (Echinodermata, Echinoidea). Zoomorphology 106:232–243

Matranga V, Di Ferro D, Zito F, Cervello M, Nakano E (1992) A new extracellular matrix protein of the sea urchin embryo with properties of a substrate adhesion molecule. Roux's Arch Dev Biol 201:173–178

Matranga V, Zito F, Costa C, Bonaventura R, Giarrusso S, Celi F (2010) Embryonic development and skeletogenic gene expression affected by X-rays in the Mediterranean sea urchin Paracentrotus lividus. Ecotoxicology 19:530–537

McClay DR, Alliegro MC, Black SD (1990) The ontogenetic appearance of extracellular matrix during sea urchin development. In Organization and assembly of plant and animal extracellular matrix (eds: Adair WS, Mecham R). pp 1–13 Academic Press, San Diego, CA

O'Donnell MJ, Todgham AE, Sewell MA, Hammond LM, Ruggiero K, Fangue NA, Zippay ML, Hofmann GE (2010) Ocean acidification alters skeletogenesis and gene expression in larval sea urchins. Mar Ecol Progr Series 398:157–171

Orr JC, Fabry VJ, Aumont O, Bopp L, Doney SC, Feely RA, Gnanadesikan A, Gruber N, Ishida A, Joos F, Key RM, Lindsay K, Maier-Reimer E, Matear R, Monfray P, Mouchet A, Najjar RG, Plattner GK, Rodgers KB, Sabine CL, Sarmiento JL, Schlitzer R, Slater RD, Totterdell IJ, Weirig MF, Yamanaka Y, Yool A (2005) Anthropogenic ocean acidification over the twenty-first century and its impact on calcifying organisms. Nature 437:681–686

Pinsino A, Thorndyke MC, Matranga V (2007) Coelomocytes and post-traumatic response in the common sea star Asterias rubens. Cell Stress Chap 12:332–342

Pinsino A, Matranga V, Trinchella F, Roccheri MC (2010) Sea urchin embryos as an in vivo model for the assessment of manganese toxicity: developmental and stress response effects. Ecotoxicology 19:555–562

Poustka AJ, Kühn A, Groth D, Weise V, Yaguchi S, Burke RD, Herwig R, Lehrach H, Panopoulou G (2007) A global view of gene expression in lithium and zinc treated sea urchin embryos: new components of gene regulatory networks. Genome Biol 8:R85

Radenac G, Fichet D, Miramand P (2001) Bioaccumulation and toxicity of four dissolved metals in Paracentrotus lividus sea-urchin embryo. Mar Environ Res 51:151–166

Range R, Lapraz F, Quirin M, Marro S, Besnardeau L, Lepage T (2007) Cis-regulatory analysis of nodal and maternal control of dorsal–ventral axis formation by Univin, a TGF-β related to Vg1. Development 134:3649–3664

Raz S, Weiner S, Addadi L (2000) The formation of high magnesium calcite via a transient amorphous colloid phase. Adv Mater 12:38–42

Raz S, Hamilton P, Wilt F, Weiner S, Addadi L (2003) The transient phase of amorphous calcium carbonate in sea urchin larval spicules: the involvement of proteins and magnesium ions in its formation and stabilization. Adv Funct Mater 13:480–486

Roccheri MC, Agnello M, Bonaventura R, Matranga V (2004) Cadmium induces the expression of specific stress proteins in sea urchin embryos. Biochem Biophys Res Commun 321:80–87

Röttinger E, Saudemont A, Duboc V et al (2008) FGF signals guide migration of mesenchymal cells, control skeletal morphogenesis and regulate gastrulation during sea urchin development. Development 135:354–365

Russo R, Bonaventura R, Zito F, Schroder HC, Muller I, Muller WEG, Matranga V (2003) Stress to cadmium monitored by metallothionein gene induction in Paracentrotus lividus embryos. Cell Stress Chaperones 8:232–241

Russo R, Zito F, Costa C, Bonaventura R, Matranga V (2010) Transcriptional increase and misexpression of 14-3-3 epsilon in sea urchin embryos exposed to UV-B. Cell Stress Chaperones 15:993–1001

Saitoh M, Kuroda R, Muranaka Y, Uto N, Murai J, Kuroda H (2010) Asymmetric inhibition of spicule formation in sea urchin embryos with low concentrations of gadolinium ion. Dev Growth Differ 52:735–746

Santamaria AB (2008) Manganese exposure, essentiality & toxicity. Indian J Med Res 128:484–500

Satyanarayana YV, Saraf R (2007) Iron and manganese contamination: sources, adverse effects and control methods. J Environ Sci Eng 49:333–336

Schröder HC, Di Bella G, Janipour N, Bonaventura R, Russo R, Müller WE, Matranga V (2005) DNA damage and developmental defects after exposure to UV and heavy metals in sea urchin cells and embryos compared to other invertebrates. Prog Mol Subcell Biol 39:111–137

Sheppard-Brennand H, Soars N, Dworjanyn SA, Davis AR, Byrne M (2010) Impact of ocean warming and ocean acidification on larval development and calcification in the sea urchin tripneustes gratilla. PLoS One 5:e11372

Simkiss K (1986) The processes of biomineralization in lower plants and animals-an overview. In: Leadbeater BSC, Riding R (eds) Biomineralization in lower plants and animals, vol 30. Oxford University Press, New York, pp 19–37

Simkiss K, Wilbur K (1989) Biomineralization. Cell Biology and Mineral Deposition. Academic Press, Inc., San Diego

Smith LC, Ghosh J, Buckley MK, Clow AL, Dheilly MN, Haug T et al (2010) Echinoderm immunity. In: Soderhall K (ed) Invertebrate immunology. Landes Bioscience, Inc

Stenzel P, Angerer LM, Smith BJ, Angerer RC, Vale WW (1994) The univin gene encodes a member of the transforming growth factor-beta superfamily with restricted expression in the sea urchin embryo. Dev Biol 166:149–158

Stricker SA (1985) The ultrastructure and formation of the calcareous ossicles in the body wall of the sea cucumber Leptosynapta clarki (Echinodermata, Holothuroida). Zoomorphology 105:209–222

Tedetti M, Sempéré R (2007) Penetration of ultraviolet radiation in the marine environment. A review Photochem Photobiol 82:389–397

Tesoro V, Zito F, Yokota Y, Nakano E, Sciarrino S, Matranga V (1998) A protein of the basal lamina of the sea urchin embryo. Dev Growth Differ 40:527–535

Todgham AE, Hofmann GE (2009) Transcriptomic response of sea urchin larvae Strongylocentrotus purpuratus to CO2-driven seawater acidification. J Exp Biol 212:2579–2594

Truhaut R (1977) Eco-toxicology – objectives, principles and perspectives. Ecotoxicology and Environm Safety 2:151–173

Weber JN, Raup DM (1966) Fractionation of the stable isotopes of carbon and oxygen in marine calcareous organisms—the Echinoidea. Part II. Environmental and genetic factors. Geochim Cosmochim Acta 30:705–736

Weiss IM, Tuross N, Addadi L, Weiner S (2002) Mollusk larval shell formation: amorphous calcium carbonate is a precursor for aragonite. J Exp Zool 293:478–491

Wessel G, Berg L (1995) A spatially restricted molecule of the extracellular matrix is contributed both maternally and zygotically in the sea urchin embryo. Dev Growth Diff 37:517–527

Wessel GM, Etkin M, Benson S (1991) Primary mesenchyme cells of the sea urchin embryo require an autonomously produced, nonfibrillar collagen for spiculogenesis. Dev Biol 148:261–272

Wilt F (1999) Matrix and mineral in the sea urchin larval skeleton. J Struct Biol 126:216–226

Wilt FH, Killian CE, Hamilton P, Croker L (2008) The dynamics of secretion during sea urchin embryonic skeleton formation. Exp Cell Res 314:1744–1752

Yang L, Killian CE, Kunz M, Tamura N, Gilbert PUPA (2011) Biomineral nanoparticles are space-filling. Nanoscale 3:603–609

Yokota Y, Matranga V, Zito F, Cervello M, Nakano E (1994) Nectins in sea urchin eggs and embryos. J Mar Biol Ass UK 74:27–34

Zito F, Matranga V (2009) Secondary mesenchyme cells as potential stem cells of the sea urchin embryo. In Stem cells in marine organisms (eds: Rinkevich B, Matranga V). Springer, New York, pp 187–213

Zito F, TesoroV McClay DR, Nakano E, Matranga V (1998) Ectoderm cell–ECM interaction is essential for sea urchin embryo skeletogenesis. Dev Biol 196:184–192

Zito F, Costa C, Sciarrino S, Poma V, Russo R, Angerer LM, Matranga V (2003) Expression of univin, a TGF-beta growth factor, requires ectoderm–ECM interaction and promotes skeletal growth in the sea urchin embryo. Dev Biol 264:217–227

Zito F, Burke RD, Matranga V (2010) Pl-nectin, a discoidin family member, is a ligand for betaC integrins in the sea urchin embryo. Matrix Biol 29:341–345

Part III
Biosilica – and its Application

Chapter 9
The Unique Invention of the Siliceous Sponges: Their Enzymatically Made Bio-Silica Skeleton

Werner E.G. Müller, Xiaohong Wang, Ailin Chen, Shixue Hu, Lu Gan, Heinz C. Schröder, Ute Schloßmacher, and Matthias Wiens

Contents

Abstract Sponges are sessile filter feeders that, among the metazoans, evolved first on Earth. In the two classes of the siliceous sponges (the Demospongiae and the Hexactinellida), the complex filigreed body is stabilized by an inorganic skeleton composed of amorphous silica providing them a distinct body shape and plan. It is proposed that the key innovation that allowed the earliest metazoans to form larger specimens was the enzyme silicatein. This enzyme is crucial for the formation of the siliceous skeleton. The first sponge fossils with body preservation were dated

W.E.G. Müller (✉)
ERC Advanced Grant Research Group at the Institute for Physiological Chemistry, University Medical Center of the Johannes Gutenberg University Mainz, Duesbergweg 6, D-55128 Mainz, Germany
e-mail: wmueller@uni-mainz.de; http://www.biotecmarin.de

W.E.G. Müller (ed.), *Molecular Biomineralization*, Progress in Molecular and Subcellular Biology 52, DOI 10.1007/978-3-642-21230-7_9,

back prior to the "Precambrian-Cambrian" boundary [Vendian (610–545 Ma)/Ediacaran (542–580 Ma)]. A further molecule required for the formation of a hard skeleton was collagen, fibrous organic filaments that need oxygen for their formation. Silicatein forming the spicules and collagen shaping their morphology are the two organic components that control the appositional growth of these skeletal elements. This process starts in both demosponges and hexactinellids intracellularly and is completed extracellularly where the spicules may reach sizes of up to 3 m. While the basic strategy of their formation is identical in both sponge classes, it differs on a substructural level. In Hexactinellida, the initial silica layers remain separated, those layers bio-fuse (bio-sinter) together in demosponges. In some sponge taxa, e.g., the freshwater sponges from the Lake Baikal, the individual spicules are embedded in an organic matrix that is composed of the DUF protein. This protein comprises clustered stretches of amino acid sequences composed of pronounced hydrophobic segments, each spanning around 35 aa. We concluded with the remark of Thompson (1942) highlighting that "the sponge-spicule is a typical illustration of the theory of 'bio-crystallisation' to form 'biocrystals' ein Mittelding between an inorganic crystal and an organic secretion." Moreover, the understanding of the enzymatic formation of the spicules conferred sponge biosilica a considerable economical actuality as a prime raw material of this millennium.

9.1 Introduction

Sponges are sessile filter-feeding organisms with an extremely effective and complex network of water-conducting channels and choanocyte chambers lined with flagellated choanocyte cells. Until not too long ago the "ground" material, the mesohyl, between the external pinacoderm and the internal choanoderm (endopinacoderm), the two cell layers that seal sponges against the environment, was thought to consist of mostly functionally independent cells (Pechenik 2000). Such a setup would result in the formation of amorphous, unorganized creatures (Pechenik 2000). However, during the last few years, the existence of cell surface-bound receptors and their extracellular as well as intracellular segments could be verified: this led to the conclusion that also sponges possess molecules that allow the establishment of a distinct body plan. The discovery of the metazoan novelties first developed during evolution in sponges, comprising cell-cell/-matrix, signal transduction-, immune-, neuronal-, and morphogenetic molecules, helped to overcome the long-standing debate whether sponges are specialized protists or true Metazoa (Hyman 1940). The phylum Porifera is subdivided into three classes, Hexactinellida, Demospongiae, and Calcarea. Until very recently, the phylogenetic positions of these classes remained unresolved. Like any other metazoan, also sponges have a defined Bauplan; this has most artistically been illustrated by Haeckel (1872a). But unlike other Metazoa, adult sponges are considered to have no pronounced anterior/posterior polarity; surely they do not have a dorsal ventral axis. In higher metazoans, the patterning along the anterior–posterior axis is regulated among others by the "famous" family of Homeobox genes. Homeobox-related genes that have been identified in sponges display, however, a more general function as transcription factors active in all sponge cells (Seimiya et al. 1998).

The body plan of sponges is defined and its orientation is fixed by an inorganic skeleton. In most sponges, this solid support, the spicules, is composed of hydrated, amorphous, noncrystalline silica (SiO_2/H_2O) as in the classes of Demospongiae and Hexactinellida, or of calcium carbonate ($CaCO_3$) as in the class of Calcarea. The secretion of spicules occurs in specialized cells, the sclerocytes. While in demosponges and hexactinellids silica is deposited around an organic filament, no organic axial structure is found in the spicules from Calcarea. Exciting new data on spicule formation and the enzymatic silicatein-mediated biosynthesis of silica have been summarized recently (Müller et al. 2005 and 2006b). It had been mainly the siliceous skeleton that put the sponges into the position to survive adverse climatic periods and to enable them as the first metazoans to occupy new aquatic biotopes.

9.2 The Key Innovation During the Proterozoic: The Skeleton of Metazoa

9.2.1 Proterozoic, Silica-Rich Ocean

The "Precambrian-Cambrian" boundary interval was the crucial time for the metazoan evolution; it dates back to the Vendian [610–545 million years ago (Ma) (Gaucher et al. 2004)]/Ediacaran [542–580 Ma (Gradstein et al. 2005)] period, preceded by the Marinoan (635–660 Ma) and Sturtian glaciations (700–750 Ma); Fig. 9.1. During this time, the Earth changed drastically with respect to (1) biological events, (2) geochemical processes, and (3) environmental conditions. Prior to the two ice periods the metazoans evolved, and the Vendian mega-continent was formed (the early Gondwana) through the amalgamation of the micro-continents West Gondwana (North and South America, Siberia, North China, Baltica) and East Gondwana (India, Australia, East Antarctica, South China, and East Africa); Shanker et al. (2001). The ocean of the Precambrian-Cambrian period had a higher concentration of silica than the present-day marine environment (Simonson 1985; Siever 1992). Two components are pertinent for the understanding of the ancient ocean; (1) silica weathering and (2) sulfate reduction. These two parameters controlled alkalinity, total CO_2 (TCO_2), and in turn also the saturation index of the mineralic components.

The major proportion of dissolved silica in the Pre-Cambrian ocean came from silica weathering processes that also influenced decisively the ocean's alkalinity. Decomposition of the continental crust, primarily of the tectosilicate minerals from the feldspar family (the plagioclase) $NaAlSi_3O_8$–$CaAl_2Si_2O_8$, occurred. During weathering and dissolution of those minerals, two alkaline conditions are formed, the Ca-bound alkalinity and the Na-bound alkalinity (Kazmierczak et al. 2004).

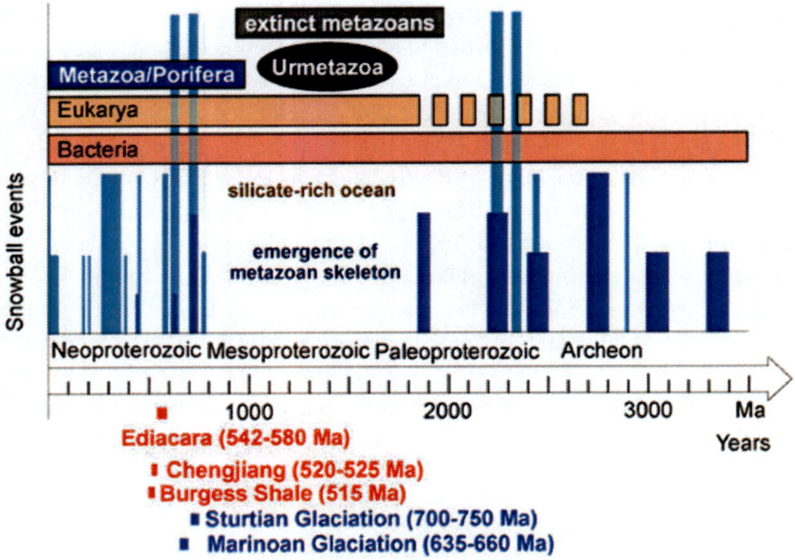

Fig. 9.1 The evolution of sponges (phylum Porifera) and the appearance of the Metazoa, with the Urmetazoa as the first hypothetical ancestor, preceded the Sturtian glaciation (700–750 Ma) and the Marinoan/Varanger glaciation (635 Ma). The frequency of major glacial periods is shown in *blue*; the *light blue bars* indicate the major glacial periods, whereas the *dark blue bars* mark the occurrence of banded iron formation (modified after Hoffman and Schrag 2002). Soon after the "Snowball Earth" period, the "Cambrian Explosion" occurred as can be monitored by the fossils collected from the Ediacara era such as the Chengjiang biota and the Burgess Shale fossils

While the Ca-bound alkalinity is removed in the course of $CaCO_3$ deposition, the Na-bound alkalinity requires long-term processes and reverse weathering in deep floor hydrothermal reactions during which soluble OH^- is removed. Worth mentioning is that in contrast to weathering of carbonate rocks, followed by the precipitation of an equivalent amount of carbonate minerals by marine organisms, the dissolution process of silicate minerals, e.g., plagioclase, is followed by a net consumption of atmospheric CO_2 (reviewed in Street-Perrott and Barker 2008). The absorption/sink of atmospheric CO_2 during weathering of Ca- and Mg silicates is followed by rinsing of the weathered minerals into rivers and groundwater. Besides of atmospheric CO_2, also organic carbonic acids contribute decisively to the weathering processes (Berner and Berner 1996).

In the present oceans, sulfate $[(SO_4)^{2-}]$ reduction processes are insignificant, with the exception of the deep anaerobic basins and the Black Sea (Kazmierczak et al. 2004). However, in the Precambrian oceans, the sulfate ions consumption contributed considerably to the increase in alkalinity due to the production of anion equivalents, which is $(HCO_3)^-$. Noteworthy is also the fact that during the transition from the Precambrian (Sturtian and Marinoan glaciations) to the Phanerozoic, the level of sulfate increased strongly from 1 to 10 mM (Canfield and Farquhar 2009). Simultaneous and in consequence of silicate weathering, calcium was liberated. However, the level of free Ca^{2+} remained relatively low, in comparison to present-day ocean, very likely due to the pH in the archaic "soda ocean" and the likewise high level of phosphate. Phosphate, if available, resulted in a rapid formation of Ca-phosphate [$Ca_3(PO_4)_2$] that is insoluble at $<100\ \mu M$ (Einsele 2000).

As generally accepted, the Precambrian oceans were richer in dissolved silica in comparison to the present-day marine environment (Simonson 1985; Siever 1992; Kazmierczak et al. 2004). However, in order to judge/assess the level of silica, an estimation of the pH in Precambrian oceans must be given. The solubility of silica depends strongly on the pH of the aqueous system; at pH 8, the solubility is lowest (2 mM) and a strong increase is seen above pH 9 (>10 mM); Iler (1979), Morey and Rowe (1964). The pH of the present-day ocean varies around 7.5–8.4 and the level of silica is the range of 10–180 μM in the deep sea and in the coastal regions only less than 3 μM (Maldonado et al. 1999). The higher concentration of silica (1 mM) in the Precambrian oceans had been enhanced due to the increased pH (Kempe and Degens 1985; Siever 1992). These authors proposed that during the Proterozoic, the pH shifted from 10 to 8.5 allowing a "soda ocean" to occur. Hence the dominant components in the "soda ocean" were Na_2CO_3 and $NaHCO_3$ as well as high concentrations of Fe^{2+}. While only a portion of that silica was immobilized and disappeared from the ocean through pore water diffusion, the biogenetic consumption of silica was negligible. The consequence was an increase of dissolved silica in the ocean close to (super)saturation levels (around 2 mM). Because silica can persist for months at pH 7–8 (Morey and Rowe 1964) and silica is nontoxic as amorphous material at those levels (Bramm et al. 1980), it provided the starting material for the formation of a poly(silicate)-based skeleton and hence could be fixed by biogenic processes (Simonson 1985), especially by sponges. As can be deduced from present-day sponges, some of which live in a high pH/alkaline milieu as the Lake Chagytai (Siberia; Wiens et al. 2009), these animals could have coped with those extreme physical conditions. There is no solid evidence that the diatoms that are the dominant silica consuming and precipitation organisms in the present-day oceans, existed already in the Proterozoic (Sims et al. 2006), most likely they evolved in the Mesozoic oceans (230–70 Ma). In contrast to silica, the level of Ca^{2+} was much lower than at present. During the Vendian Period, the pCO_2 level in the atmosphere rose, an event that had been correlated with the lowering of the ocean's Mg^{2+} and increase of Ca^{2+} levels (Tucker 1992).

9.2.2 *Emergence of the Animal Organic Hard Skeletons*

Two major processes during the Proterozoic provided the basis for the evolution of a skeleton. First, accumulation/rise of atmospheric oxygen and second the availability of dissolved silica in the ocean. These two parameters gave the basis for the organic (oxygen) and the inorganic (silica) ancient metazoan skeletons. At the beginning of the Proterozoic Eon (2,500 Ma), the atmospheric oxygen level was low or almost not existent, compared to the present atmospheric level (PAL). Following the staging of (Walker 1978/79; Kasting et al. 1992), at Stage I

(>2,000 Ma), the level was in the order of 10^{-14} PAL; during the short overlapping phase of Stage II (2,000–1,700 Ma), the oxygen level rose close to 5×10^{-2} PAL, followed by a sharp increase at 1,500 Ma to 1×PAL (Stage III). Parallel with the turn from the anoxic to the oxic phase, the ocean became depleted of iron, a process which started at an atmospheric oxygen level of 2×10^{-3} PAL (Kasting 1984). In the Meso-Proterozoic (until 1,000 Ma) and especially in the Neo-Proterozoic Oceans (1,000–543 Ma) secular deflections of $\delta^{13}C_{carb}$ could be measured that are indicative of changes in the ratio of organic to inorganic C removed from the oceans by (organic) burial in sediments (Anbar and Knoll 2002). Interestingly, during that period, termed Cryogenian that lasted from 850 to 630 Ma, very cold global climate episodes occurred with more than two major worldwide glaciations. The Sturtian glaciation persisted from 750 to 700 Ma, and the Marinoan/Varanger glaciation terminated at circa 635 Ma. Since the characteristic glacial deposits are also found in places at low latitudes, the hypothesis of deeply frozen planetary oceans and the expression "Snowball Earth" (Hoffman and Schrag 2002; reviewed in Müller et al. 2007b) were coined. Between these two major ice ages, the multicellular animals (Knoll and Carroll 1999) evolved among which only the sponges (Porifera) developed, a genetic repertoire that allowed the survival through those adverse events until today (Xiao et al. 2000); hence, these animals were termed "living fossils" (Müller 1998; see Pilcher 2005).

All multicellular animals have developed both an organic and an inorganic solid skeleton. The organic skeleton is based on collagen and the inorganic solid skeleton comprises silicon or calcium as an inorganic element. Collagen formation requires atmospheric oxygen atoms that are incorporated into Pro-Pro-Gly and Ile-Lys-Gly tripeptides, catalyzed enzymatically by the by prolyl hydroxylase (Towe 1970; Kikuchi et al. 1983). In sponges, collagen molecules exist already in several classes, from the primitive fibrillar collagen to the vertebrate-type basement collagens (see Garrone 1998). Comparative analyses supported the assumption that during the emergence of the sponges in the Meso-Proterozoic, the required atmospheric oxygen pressure had been reached. From collagen formation studies with present-day vertebrate cells, it is known that collagen formation requires a partial oxygen (pO_2) pressure of 16.0 kPa (approx. 120 mmHg) in order to allow a rate of 60% for the synthesis of collagenous proteins; in the human capillaries, a lower pressure of 5 kPa is reached. For the mesopelagic deep-sea fangtooth fish *Anoplogaster cornuta*, the average critical oxygen tension (*Pc*) for the entire group has been determined to be 35 mmHg (Gordon et al. 1976), a tension which had been also measured in the deep sea ocean (Teal and Carey 1967). Hence, this oxygen pressure corresponds to values estimated for those in the air during the Meso-Proterozoic period (Kasting et al. 1992).

Collagen has an important role in the stabilization of the extracellular matrix in the sponge body, the mesohyl (also termed collagenous tissue) (see Garrone 1978; Simpson 1984; Francesco et al. 2001). A few sponge species comprise a cortex consisting mainly of a fibrous internal stroma of collagen bundles like in *Chondrosia reniformis* (Francesco et al. 2001), while most of them have only a poorly developed or completely absent cortex (Garrone 1978). Driven and controlled by the collagen fibrils, the sponges have the capacity of morphogenetic remodeling. This potential is widely documented both in hexactinellids, e.g., during regeneration of grafts in *Rhabdocalyptus dawsoni* (Leys et al. 1999) and also in demosponges, e.g., in *C. reniformis* (Francesco et al. 2001). Those reorganization processes, involving also a temporary plasticisation of the cortex, involve the highly dynamic arrangement of the functional fibrous elements, especially of collagen (Hartman and Reiswig 1973; Gaino and Pronzato 1983). Primarily, the collencytes synthesize the collagen fibers that comprise contractile features (Harrison and De Vos 1991). Even more, those contractile cells have been implicated in neuroid signal conductions, interpreted as "promyoneuroid elements" (Lévi 1970; Garrone 1978). In addition, pinacocytes are capable to undergo contractile responses/amoeboid movements in conjunction with collagen and modulate the water current through the aquiferous canals. The intracellular structural elements, the "histoskeleton" (formed by associated cytoskeletal elements and extracellular components), allow a coordinated cell movement and a controlled contraction/relaxation as well as motility (Pavans de Ceccatty 1986). Within the cell lattices, the collagen fibrils connect the cells also by direct mechanical contact. Finally, and despite of their usual sessile behavior/growth plan (Alexander 1979; Barnes 1987), some sponges have the capacity of active locomotion (Bond and Harris 1988), based on the flexible interaction between the contractile fibrils and collagenous internal stroma (Garrone 1978). Linkages between adjacent collagen fibrils have been identified that are glycoproteins and glycosaminoglycans in nature (Garrone 1978; Simpson 1984).

9.3 Well-Preserved Fossils in Body Preservation at the Ediacaran/Lower Cambrian Border: The Siliceous Sponges from Chengjiang

Equally important as the Burgess Shale fossil sponge fauna is the rank of fossils from the Cambrian/Precambrian/Neoproterozoic period, excavated in the Cambrian "Burgess Shale"-type deposits in South China (Rigby and Collins 2004; Zhang et al. 2008), especially in Chengjiang (Fig. 9.2a). Based on fossil records, the hexactinellids represent the oldest taxon detected and have been described from Australia, China, and Mongolia (older than 540 Ma) (Gehling and Rigby 1996; Brasier et al. 1997; Li et al. 1998). Especially outstanding is the preservation of the fossil hexactinellids from the Niutitang Formation (Sancha) in Hunan (Early Cambrian; China; Steiner et al. 1993; Steiner 1994). There, completely preserved sponge fossils, e.g., *Solactiniella plumata*, were discovered (Steiner et al. 1993), displaying large, 15 mm–100 μm, spicules. These spicules still exhibit the characteristic axial canals as shown by Xiao et al. (2005) and the lamellar organization that is known from modern hexactinellids (Wang et al. 2009); Fig. 9.3f, g. Stratigraphically equivalent are the Chengjiang assemblages found in Yunnan

Fig. 9.2 The Chengjiang area. (**a**) A first geological map from the Chengjiang area compiled by Deprat and Mansuy (1912). North from the Xingyun and the Fuxian Lake and around Chengjiang city (Chengjiang County of Yunnan Province, China), the well-preserved fossils are found in the Lower Cambrian facies. (**b**) View looking north along the north/west shore of the Fuxian Lake (after Deprat and Mansuy (1912)). (**c, d**) Collection of fossils near the Maotianshan mountain. This cornucopia of fossils had been discovered by Hou in 1984 (see Hou et al. 1999)

(China; Rigby and Hou 1995). In our studies of fossils from the Chengjiang formation, we could prove the existence of organic material around the siliceous skeleton by EDX spectroscopy.

The Chengjiang Lagerstätten, also known as Chengjiang Biota or the Maotianshan shales (Yunning Province; China), are well known for their well-preserved soft-bodied metazoan fossils (Fig. 9.2). They are witnesses of the flowering animal radiations from the beginning of the Lower Cambrian animal "explosion". More specific, the biota in the Maotianshan Shales area in Chengjiang County occur in the lower part of the Yu'anshan Member of the Lower Cambrian Heilinpu Formation, about 520–530 Ma (Zhang and Hou 1985; Yang et al. 2007; Chang et al. 2007). Hence these fossils are a little older than the equally well-preserved animals from the Burgess Shale (British Columbia; Canada) with about

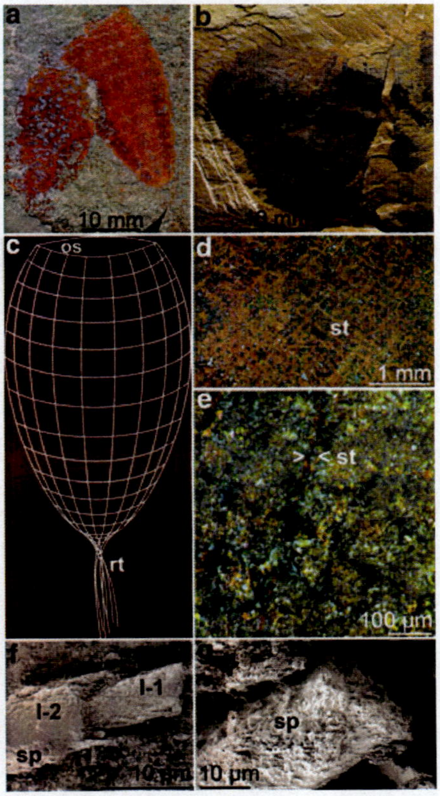

Fig. 9.3 Early hexactinellid sponges: *Diagoniella* sp., *Protospongia* sp. and *Solactiniella plumata*. (**a**) *Diagoniella* sp. from Millard County, Utah. (**b**) *Protospongia* sp. Malong, (Chengjiang biota, South China). (**c**) Reconstruction of the taxon *Diagoniella* displaying the oscule (*os*) and the root tuft spicules (*rt*). (**d**) *Diagoniella* sp. stauractins (*st*). (**e**) *Protospongia* sp. stauractins (*st*). (**f, g**) *Solactiniella plumata* is composed of intact spicules that are surrounded by dark mineralized material. Some spicules (*sp*) are broken and expose the internal axial canals. Analyses of spicules (*sp*) embedded in the surrounding clay. In some spicules, a zonation, into inner layer (*l*–1) and outer layer (*l*–2), of the silica rim of the spicule is seen. Size bars are given

515 Ma. In the Lower Cambrian Chengjiang fauna, the sponges are, after the arthropods, the major taxon with respect to generic and specific diversity. Over 1,000 sponge body fossils, grouped to 15 genera and 30 species, have been collected. Most of the spicules that built their skeletons represent diactins, forming a regular, reticulate skeletal framework (Fig. 9.3a–e). Due to the dermal reticulate organization of the spicules, the fossils are classified as demosponges. These data, gathered from the Chengjiang fauna demonstrate that the main clade of the early sponges was constituted by the monaxonid Demospongiae. The first keratosal demosponges with skeletons composed entirely of spongin fibers have been described likewise from the Lower Cambrian in the Chengjiang fauna (Li et al. 1998). Hexactinellida are less abundant than Demospongiae in Chengjiang, as documented by small-sized sponges, with small triaxons (mainly stauracts),

spicules with their paratangentialia being in a diagonal arrangement (Rigby and Hou 1995; Wu et al. 2005).

The organization of the spicules within the bodies of the Cambrian/Ediacaran sponge classes has been described for the hexactinellid Protospongiidae *Protospongia tetranema* and *Triticispongia diagonata* (see Hou et al. 1999, 2004) as well as for *Diagoniella* sp. (Rigby and Collins 2004). *Protospongia* sp. and *Triticispongia* sp. were considered the basic taxon from which the later groups were derived (Finks 2003a, b); Fig. 9.3b. The thin-walled, vasiform body with one large distal oscule and a basal root tuft is composed of simple diactins. The tissue is stabilized by stauractins or pentactins (Fig. 9.3d, e). The approximately 50 mm large, sessile animals were supported by basal rays (basal spicules) that could reach lengths of 6 mm at a diameters of 0.3 mm. A similar body plan is displayed by other hexactinellids, e.g., *Diagoniella* sp. (Fig. 9.3a). A reconstruction of an ancient sessile animal (see Xiao and Laflamme 2008) is given in Fig. 9.3c. In contrast to the Hexactinellida, the earliest Demospongiae with the family of the Choiidae comprised freely moving species that lived unattached on the seafloor (Rigby and Collins 2004); Fig. 9.4a–c. The Choia species had an elliptical to circular habitus/

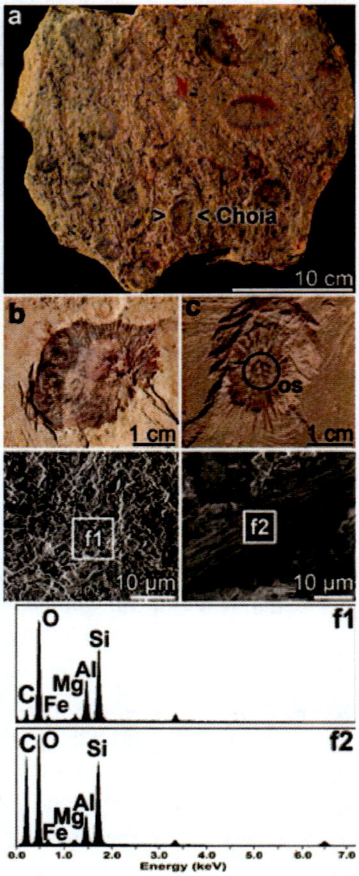

Fig. 9.4 Early demosponge fossils from the Chengjiang biota. (**a**) Assembly of *Choia xiaolantianensis* from Haikou (Chengjiang County). (**b, c**) *C. xiaolantianensis* enlargement of the central region of the body with the dominant oreas and the central oscule (*circle [os]*). (**d, e**) HR-SEM (high-resolution scanning electron microscopic) analysis of the peripheral region of the *Choia* sp. (**f**) Organic preservation of regions within the fossil *Choia* sp. In (**f1**), the corresponding EDX spectrum from this outer part of the organic region from *Choia* sp. is shown. (**f2**) In contrast, the EDX spectrum from the organic areas within that fossil displays a comparatively high peak for carbon (*C*)

body (Fig. 9.4b, c), decorated on their ragged edge with coarse coronal spicules which they used as pillar/support to remain freely exposed on the seafloor. The contact to the substrate on which they lived was held with their basal spicules, without digging into the seafloor. We consider this morphology – free resting on the seafloor – as an ancient living form. Surely, these fossil sponges already produced free-floating larvae (Steiner et al. 2005). Whereas the taxon *Choia* was not firmly attached to the seafloor, other fossil demosponges from that period were sessile, like the species *Paraleptomitella dicytodroma* (Wang et al. 2010a).

Specimens of *Choia* sp. are occasionally highly abundant within the limestone (Fig. 9.4a). They were settling with their basis on the seafloor and left the oscule open at the opposite side (Fig. 9.4b, c). EDX (energy dispersive x-ray) analyses revealed that the regions within the limestone that do not contain fossils (Fig. 9.4d and f1) show low signals for carbon, in contrast to regions with fossils (Fig. 9.4e and f2).

9.4 Morphology and Synthesis of Spicules in Demosponges

In the last few years, motivated by previous ultrastructural analyses (reviewed in Uriz 2006), the development and the morphology of the spicules have been studied thoroughly in the demosponge *Suberites domuncula* (reviewed in Müller et al. 2007c); Fig. 9.5a. The skeleton of *S. domuncula* is composed of only two types of megascleres, monactinal tylostyles and a smaller fraction of diactinal oxeas. The spicules reach lengths of up to 450 μm and diameters of 5–7 μm (Fig. 9.5c); they grow through apposition of lamellar silica layers. While the two ends of the oxeas are pointed, the tylostyles have one pointed end and one swollen knob (Fig. 9.5b). Microscopic analyses showed that all spicules have a 0.3–1.6-μm wide axial canal in their center (Fig. 9.6e–i). Applying the primmorph system (the established 3D cell culture of sponges), it became possible to follow the different steps of spicule formation (Müller et al. 2005). These studies establish unequivocally that the initial steps of spicule formation occur intracellularly in the sclerocytes (Müller et al. 2005); Fig. 9.6a–d. The 15-μm large sclerocytes produce one to three of up to 6-μm long spicules (Fig. 9.6c).

The formation of – at least – the first silica layer around the axial filament starts within the sclerocytes. In the primordial stage, spicule growth begins around the axial filament (Fig. 9.6a–d). Clods with highly electron dense material which represent the first deposits of silica become visible. During growth in the extracellular space, the spicules reach up to 450 μm in length with a diameter of 5 μm. Initially, the 1.6 μm wide axial canal is filled primarily with the axial filament and additional membrane structures, while in the final stage it is almost completely filled with the axial filament, which is homogenous (Fig. 9.6e) and displays the characteristic triangular axial form (Fig. 9.6h, i).

Toward a further understanding of the synthesis of spicules, immunogold labeling/TEM studies with antibodies against silicatein were performed (Müller et al.

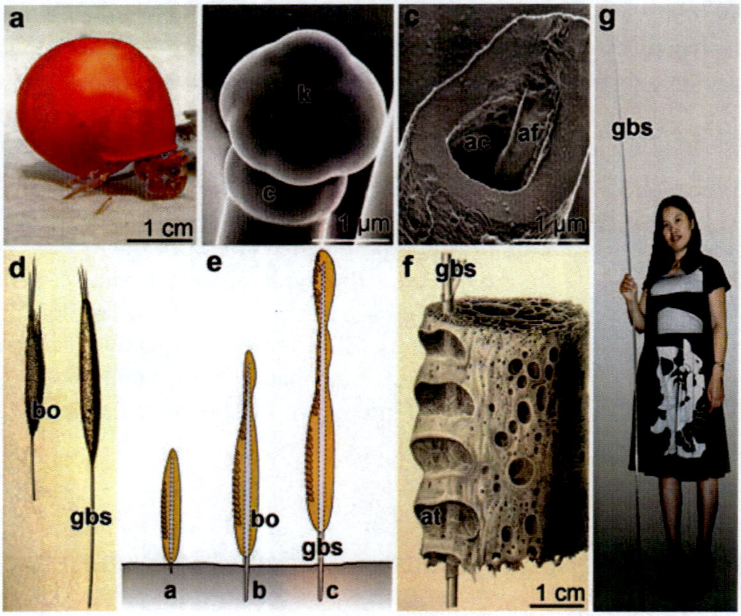

Fig. 9.5 Two silliceous sponges. DEMOSPONGIAE: **(a)** *S. domuncula* specimen. **(b)** spicule from *S. domuncula*; view of the swollen knob (*k*), the tylostyle head, that bases on a collar (*c*); HR-SEM. **(c)** Broken spicule, showing the central axial canal (*ac*) and the proteinaceous axial filament (*af*) located in it; HR-SEM. HEXACTINELLIDA: **(d)** Young specimens of *M. chuni* that are anchored to the muddy substratum by one single giant basal spicule (*gbs*). The body (*bo*) surrounds the spicule as a continuous, round cylinder (Schulze 1904). **(e)** Schematic representation of the growth phases of the sessile animal with its giant basal spicule (*gbs*) which anchors it to the substratum and holds its surrounding soft body (*bo*). The characteristic habitus displays linearly arranged large atrial openings of approximately 2 cm in diameter. With growth, the soft body dies off in the basal region and exposes the bare giant basal spicule (a–c). **(f)** Part of the body (*bo*) with its atrial openings (*at*). The body surface is interspersed with ingestion openings allowing a continuous water flow though canals in the interior which open into oscules that are centralized in atrial openings, the sieve plates. Modified after Schulze (1904). **(g)** Giant basal spicule of *M. chuni*, representing the largest biosilica structure on earth. The sponge body is arranged around one giant basal spicule (*gbs*) per individuum

2005). The immune serum showed a dense accumulation of gold granules in the sclerocytes and the extracellular space. Fine structure analysis revealed that at first concentric rings which are 0.2–0.5 µm apart are seen around the forming spicules (Fig. 9.6f, g). Subsequently, the inner rings fuse and electron-dense clouds become visible. Later, during maturation, the number of the concentric rings increases from two to ten rings with a total diameter 4–6 µm. These data underscore the view that the spicules grow appositionally in both directions, in width and in length (reviewed in Müller et al. 2006b, 2007c, 2009). Limited etching of the spicules with hydrofluoric acid displays that the initially formed and separately existing lamellae can still be distinguished (Fig. 9.6i–k). Finally, the spicules release the axial filament from the bio-silica mantel (Fig. 9.6l).

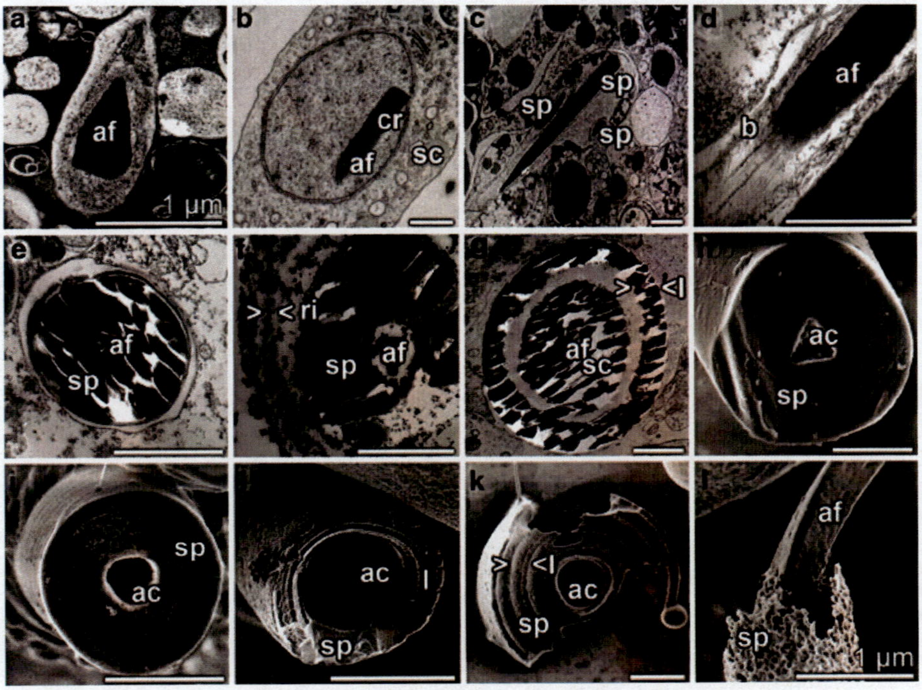

Fig. 9.6 Ultrastructure of tylostyles from *S. domuncula*. (**a–d**) Formation of spicules in sclerocytes; HR-TEM (high-resolution transmission electron microscopic) images. (**a**) Within a vesicle in a sclerocyte, an axial filament (*af*) consisting of silicatein is formed. (**b**) In the initial growth phases, a nano-sized crystal rod (*cr*) is formed which presumably is involved in the growth of the axial filament (*af*). (**c, d**) Three intracellularly located immature spicules (*sp*). (**d**) During the growth, the axial filament (*af*) attaches to filamentous bundles (*b*) of unknown nature. (**e**) Immature spicule (*sp*) harboring a space-filling axial filament (*af*) in the axial canal. (**f**) With antibodies against silicatein, it became possible to demonstrate, by electron immunogold labeling technique, that concentric rings (*ri*) composed of silicatein surround the growing spicules (*sp*), likewise formed of silicatein; the antibodies react also with the axial filament (*af*). This approach also showed that the silicatein molecules are attached to string- and net-like structures. (**g**) Very frequently, the growing spicules break under the creation of ring-formed cylinders, reflecting lamellae (*l*). (**h–k**) Cross sections through *S. domuncula* tylostyles. HR-SEM images. (**h**) Non-etched spicule (*sp*) with its central axial cylinder (*ac*). (**j–l**) Progression of etching results in the exposure of the lamellae (*l*); *ac*: axial canal, *sp*: spicule. (**l**) In the final stage, the axial filament (*af*) is freed of all surrounding spicular material (*sp*). All size bars measure 1 μm

Based on these microscopic data, first clues on the factors involved in morphogenesis of the spicules could be outlined. It is obvious that the size/length of the spicules (megascleres), size >50 μm, exceeds that of a cell. Therefore, several mechanisms might be postulated for the intracellular and subsequent extracellular growth of the spicules, giving rise to a complex morphology. Sollas (1888) assumed that cells migrate along the spicules to allow growth along their axis. This view was strongly supported by Maas (1901). Studying the demosponge *Tethya lyncurium*, he observed (what was forgotten later) that initially the synthesis of the spicules starts intracellularly and is completed in the extracellular space. There, the growth of the spicules is controlled by an ordered arrangement of cells, which – according to his illustrations – surround the spicules. The ordered arrangement of the cells can even be traced after dissolution with hydrofluoric acid (HF) (Müller et al. 2007c).

To unravel the organizing principles underlying spicule formation, specific antibodies were applied. The data revealed in *S. domuncula* that besides silicatein and galectin, collagen fibers surround the spicules in an ordered pattern (Schröder et al. 2006; Eckert et al. 2006). The existence of collagen in extracts from spicules had been recognized already by $NaDodSO_4$/PAGE. Careful time-kinetics recording of the proteins which are released by controlled HF dissolution revealed that prior to the appearance of the axial filament, a proteinaceous coat around the axial filament can be identified. Later, by high-resolution scanning electron microscopic (SEM) analysis, it could be visualized that in the extra-spicular space an ordered network of collagen fibrils surround the spicules. At the tips of the spicules, where the knobs of the tylostyles are formed, a mesh of collagen fibers is seen, suggesting that during spiculogenesis, the enzyme silicatein (which mediates biosilica deposition) is matrix-guided first by galectin (Schröder et al. 2006) and subsequently by collagen (Eckert et al. 2006). Future experiments will address the question for the mechanism and the specificity of interactions of these three molecules and will try to elucidate the controlling genetic machinery switched on during spicule formation. It should be noted here that on the surfaces of the spicules, very frequently clusters of cell fragments, surrounded by collagen, are seen; these structures are very reminiscent of the cell-like structures reported by Maas (1901).

9.5 Morphology and Synthesis of Spicules in Hexactinellids

Among the large hexactinellid sponges, *Monorhaphis chuni* is a giant (Fig. 9.5g). *Monorhaphis chuni* (Schulze 1904) is distributed in the Indo-West Pacific region and found in depths between 516 and 1,920 m. *Monorhaphis* inhabits muddy substrata and is fixed there by a single giant basal spicule. Young specimens are thought to comprise a continuous body, as has been sketched by Schulze (1904) (Fig. 9.5d–f). The cylindrical/oval body of *Monorhaphis* is interspersed with many atrial openings which are located along one side (Fig. 9.5f); the diameter of the body can reach in larger specimens up to 12 cm. During growth, the specimens elongate together with an extension of their giant basal spicules (Figs. 9.5e [a–c] and 9.7a). This growth regime has been deduced from the differently sized fragments of *Monorhaphis* collected during the Valdivia Expedition (Schulze 1904), and during the expeditions organized by the Institute of Oceanography (Qingdao). Older specimens apparently lose the basal portions of their soft body and expose the bare giant basal spicule (Fig. 9.5g).

Like all other hexactinellids, *Monorhaphis* possesses microscleres (<0.1 mm) and megascleres (0.2–30 mm–3 m). Within the oblong, laterally compressed body

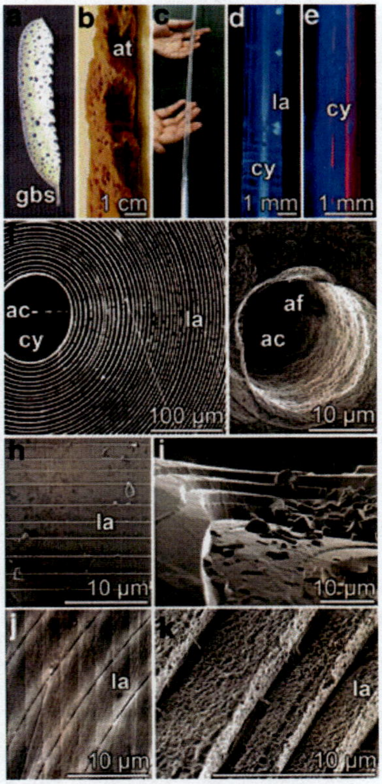

Fig. 9.7 Giant basal spicules from *M. chuni*. (**a**) Drawing of a specimen of *M. chuni*; giant basal spicule (*gbs*). (**b**) View of a dried specimen displaying atrial openings (*at*). (**c**) Lateral view of the largest hitherto collected giant basal spicule having a diameter of 12 mm. (**d, e**) A giant basal spicule displaying the axial cylinder (*cy*) that is surrounded by the concentric lamellae (*la*); inspected by Nomarski interference contrast microscopy. (**f**) Lamellar composition (*la*) of the giant basal spicule, comprising the axial canal (*ac*), the axial cylinder (*cy*) and the lamellar region. (**g**) Higher magnification of the axial cylinder with the central axial canal (*ac*) that harbors the axial filament (*af*). (**h**) Frontal view of the spicule showing the perfectly and concentrically arranged lamellae (*la*). (**i**) A diagonal view of a cross break through a spicule displaying the lamellae. (**j, k**) Higher magnification of the lamellae; polished and untreated (**j**), polished and HF vapor-treated (**k**); *la*: lamella

of *Monorhaphis* (choanosomal body), which is arranged around the single giant basal spicule, 14 further siliceous spicule types with lengths from a few micrometers to 50 mm are found (Schulze 1904; Müller et al. 2007a; Wang et al. 2009); Fig. 9.7a, b. The likewise large comitalia (around 60 mm) stabilize the tissue/body through which particulate food is filtrated via the aqueous canal system. The choanosomal body comprises mainly triactines (tauactines), diactines, and amphidiscs.

GIANT BASAL SPICULES: Very likely due to the relatively small number of giant basal spicules collected so far, their detailed analyses started only recently. The largest hitherto found giant basal spicule had a length of close to 300 cm and a diameter of 12 mm (Figs. 9.5g and 9.7c). Each giant basal spicule is made up of lamellae: Surprisingly, the siliceous lamellae composing the spicules from *Monorhaphis* contain not only a bio-silica matrix but also a proteinaceous scaffold. Initially, Schulze (1904) proposed that the proteinaceous material surrounds the lamellae as fibers. For our initial light microscopic experiments (Müller et al. 2007a, 2008d), all loosely attached organic material was removed from a comitalia by ultrasonication. A distinction between the axial cylinder and the surrounding lamellar zone is possible even by light microscopic inspection in a cross section, using Nomarsky DIC (differential interference contrast) imaging (Fig. 9.7d, e). The axial cylinder occupies about half of the diameter of the comitalia, usually 150 μm. When a sample is further subjected to HF dissolution after about 1 min of exposure, the lamellar zone begins to dissolve. Dissolution starts from cracks in the spicule, primarily following the gaps between the lamellae, thus forming sawtooth edges. The dissolution of the axial cylinder proceeds from the periphery without revealing individual lamellae and the siliceous material is completely dissolved after 90 min. Within the axial canal, the axial filament is located (Fig. 9.7g). Application of different dyes allows visualization of the proteinaceous components in the two zones surrounding the axial canal. Addition of Coomassie Brilliant blue to the HF solution immediately stains the proteins released from the lamellar zone. During the initial phase of dissolution, a proteinaceous coating (lamellar coating) is uncovered which remains only transiently intact until the organized sheets disintegrate to irregular clumps/aggregates. Prior to this disintegration, the lamellar coating blisters. A second tube/sheath is formed from proteins of the axial cylinder. This structure is termed here axial barrel. In contrast to the lamellar coating, the axial barrel is composed of individual ropelike filaments. These can be stained besides by Coomassie Brilliant blue also with Sirius Red; the latter stain does not color proteins from the lamellar zone or from the axial canal. In addition, the axial barrel can be stained with the fluorescent dye Rhodamine 123 (Müller et al. 2008d).

SEM analysis of a broken comitalia (large spicules existing in the body around the atrial openings) shows also the lamellar-wise organization of the silica mantel (Fig. 9.7i) and the perfectly concentrically arranged lamellae around the central axial cylinder (Fig. 9.7f); the central cylinder remains almost intact, while the peripheral lamellar zone undergoes fracturing into concentric piles of chipped lamellae (Wang et al. 2008; Müller et al. 2007a) In center of the spicules lies the axial canal, a structural unit that is often circular in hexactinellids (Pisera 2003; Sandford 2003; Uriz et al. 2003; Uriz 2006); however, toward the tips of the spicules, the axial canal changes its profile and appears square. The axial canal is surrounded by a region of electron-dense homogeneous silica constituting the axial cylinder of a diameter of 100–150 μm. The third and major part of the spicules is composed of 300–800 regularly and concentrically arranged lamellae (each 3–10 μm thick; Fig. 9.7f, h–k). The in average 0.1–0.2 μm wide inter-lamellar space of the spicules does surprisingly not constitute a continuous open slit (Müller et al. 2008a) but is composed of fusion zones and open holes; apparently, the fusion zones allow a continuum between two silica lamellae. The lamellar organization becomes also apparent in longitudinal cuts of the spicules (Fig. 9.7d, e). A closer lateral view of a cross break shows the solid dimension of a lamella. Inside of the axial canal, the axial filament is located (Fig. 9.7g).

Studies, to obtain an insight into the structural organization of the spicules at the nm scale, can be obtained by partial and limited dissolution of the silica using HF with the limitations described (Simpson et al. 1985). A rapid dissolution results in the removal of the inorganic scaffold, while a less-intense exposure of the spicules to HF vapor releases the organic matrix from within the lamellae. Gentle exposure of cross breaks of the spicules to HF vapor results in the dissolution of the silica material under release of the organic component of the lamellae (Müller et al. 2008a, d). The proteinaceous palisade-like scaffold is uncovered; it is composed of fibrous structures, into which holes formed by interconnecting fibers are interspersed. Under avoidance of shear forces during the HF treatment, the complete proteinaceous layer of one lamella can be obtained. At higher magnification, it becomes overt that the rim of each hole is reinforced by densely arranged 10–15 nm large spheres. The protein fibers that are attaching the silica scaffold do not contain any banding pattern reminiscent of collagen.

9.6 Phases of Silica Deposition During Spicule Formation Along the Proteinaceous Filament

Taking into account the data collected (reviewed in Müller et al. 2006b), the process of spicule formation can be divided into phases; the initial intracellular steps and the extracellular final growth and shaping phases.

Intracellular phase in the sclerocytes: Silica is taken up actively by a Na^+/HCO_3^-[$Si(OH)_4$] cotransporter (Schröder et al. 2004). In the first steps, silicatein is synthesized as a proenzyme (signal peptide–propeptide–mature enzyme: 36.3 kDa) and processed via the 34.7 kDa form (propeptide–mature enzyme) to the 23 kDa mature enzyme. Very likely during the transport through the endoplasmic reticulum and the Golgi complex, silicatein undergoes phosphorylation and is transported into vesicles where it forms rods, the axial filaments. After assembly to filaments, the first layer(s) of silica is (are) formed. Silica deposition occurs in two directions; first from the axial canal to the surface (centrifugal orientation) and second from the mesohyl to the surface of the spicule (centripedal). Finally, the spicules are released into the extracellular space where they grow in length and diameter by appositional growth; Fig. 9.8.

Extracellular phase (appositional growth): Silicatein is present also in the extracellular space. It came surprising that also there the silicatein molecules are organized to larger entities. The immunogold electron microscopical analysis showed that the silicatein molecules are arranged along strings, which are organized in parallel to the surfaces of the spicules (Schröder et al. 2006). In the presence of Ca^{2+}, silicatein associates with galectin and allows the appositional growth of the spicules. Since the surface of a new siliceous spicule is also covered with silicatein,

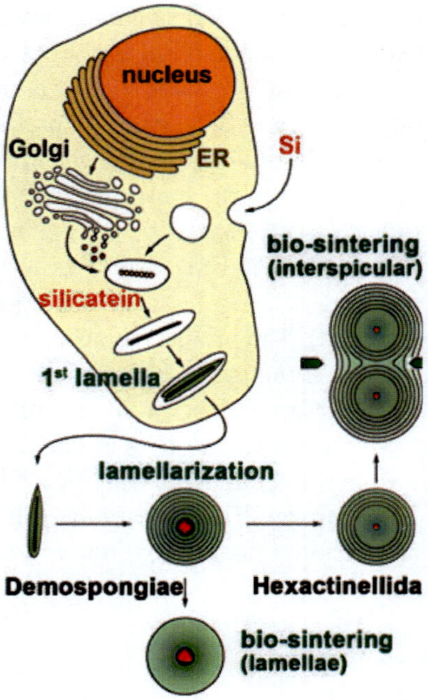

Fig. 9.8 Synthesis of siliceous spicules (scheme). The first lamella is synthesized intracellularly in special organelles, through the catalytic function of silicatein. After extrusion to the extracellular space, the additional appositional layering of the lamellae occurs. In the spicules of demosponges, those lamellae biosinter/fuse together, while the lamellae in hexactinellid spicules remain separated. However, in hexactinellids, a bio-sintering process occurs between individual spicules (interspicular bio-sintering)

we concluded that the appositional growth/thickening of a spicule proceeds in two directions (centrifugal and centripetal).

Extracellular phase (*shaping*): In the next step, the galectin-containing strings are organized by collagen fibers to net-like structures. It is very likely that collagen, which is released by the specialized cells, the collencytes, provides the organized platform for the morphogenesis of the spicules. The longitudinal growth of the spicules can be explained by the assumption that at the tips of the spicules, the galectin/silicatein complexes are incorporated into deposited biosilica under formation and elongation of the axial canal.

All these phases depend on the initial synthesis of bio-silica via the enzyme silicatein.

9.7 Silicatein: Emergence of one Protein Allowed the Establishment of an Organized Body Plan in Sponges

Based on the molecular biological studies was proven that the sponges are true Metazoa and were the first taxon that emerged from the hypothetical ancestor of all Metazoa, the Urmetazoa (Müller 2001; Müller et al. 2004).

The biogenic basis of spicule formation and the turnover of silica in spicules had already been depicted by Duncan (1881). He formulated "The spicule which has lived, has to decay, and may live again in another form". However, it took until 1999 when Cha et al. discovered that the main constituent of the proteinaceous filament within the axial canal of spicules is an enzyme which might be involved in biosilica formation, and that was consequently termed silicatein,. Soon after the discovery of this anabolic enzyme, also the corresponding catabolic enzyme (silicase) was detected (Schröder et al. 2003). The identification of a biosilica degrading enzyme supported the view that the siliceous components in spicules are under metabolic control (Eckert et al. 2006). Studies on the metabolism of spicules on the cellular level became possible after the introduction of a poriferan cell culture system, primmorphs (Imsiecke et al. 1995; Custódio et al. 1998). Already, the first contribution on that topic resolved that spicule formation starts intracellularly in "J" sclerocytes, by formation of an initial organic axial filament, around which the inorganic silica mantel is deposited. This result had later been corroborated by application of more advanced immunochemical and electron microscopy techniques (Müller et al. 2007a).

After the discovery of the cathepsin L (cysteine protease)-related silicatein (Shimizu et al. 1998; Cha et al. 1999) in spicules of the demosponge *Tethya aurantium*, several related genes were elucidated in both marine and freshwater demosponges (reviewed in Müller et al. 2007a). The corresponding deduced polypeptides comprise about 325 amino acids (aa) with a molecular weight of ca. 35 kDa. During maturation, this primary translation product (proenzyme) is processed by cleaving off a signal peptide (aa_1 to aa_{17}; *S. domuncula* [demosponge] silicatein-α) and the adjacent propeptide (aa_{18} to aa_{112};), resulting in the mature enzyme that has a size of 24–25 kDa. Similar to cathepsins, the catalytic center of silicatein contains His and Asn. However, the Cys of the cathepsins' catalytic triad is exchanged by Ser in silicatein. In addition to about ten putative protein kinase phosphorylation sites, silicateins display a cluster of serine residues that is found close to the central aa residue of the catalytic triad, but is otherwise missing in cathepsins. Subsequent phylogenetic analyses revealed that silicateins form a separate branch from cathepsins (Müller et al. 2007a). The alignment and the phylogenetic tree are given in Fig. 9.9a, b.

The difficult accessibility of hexactinellids, which live primarily in depths of more than 300 m, generally results in a very poor sampling. Accordingly, only recently the first hexactinellid silicatein (*Crateromorpha meyeri*) could be identified and characterized (Müller et al. 2008c). This molecule shares high similarity to the demosponge sequences (expect value of $8e^{-58}$) and contains the same catalytic triad amino acids. However, striking in the *C. meyeri* sequence is a second Ser-rich cluster, which is located between the second and the third aa of the catalytic triad; Fig. 9.9b. Strong binding of the protein to the spicule silica surface has been attributed to this cluster (Müller et al. 2008a). The posttranslational modifications of silicatein have been found to be essential for the enzyme activity with respect to (1) association with other structural and functional molecules within the tissue and (2) self-association/self-assembly. For those studies, silicatein had been isolated from spicules in the absence of HF, but in the

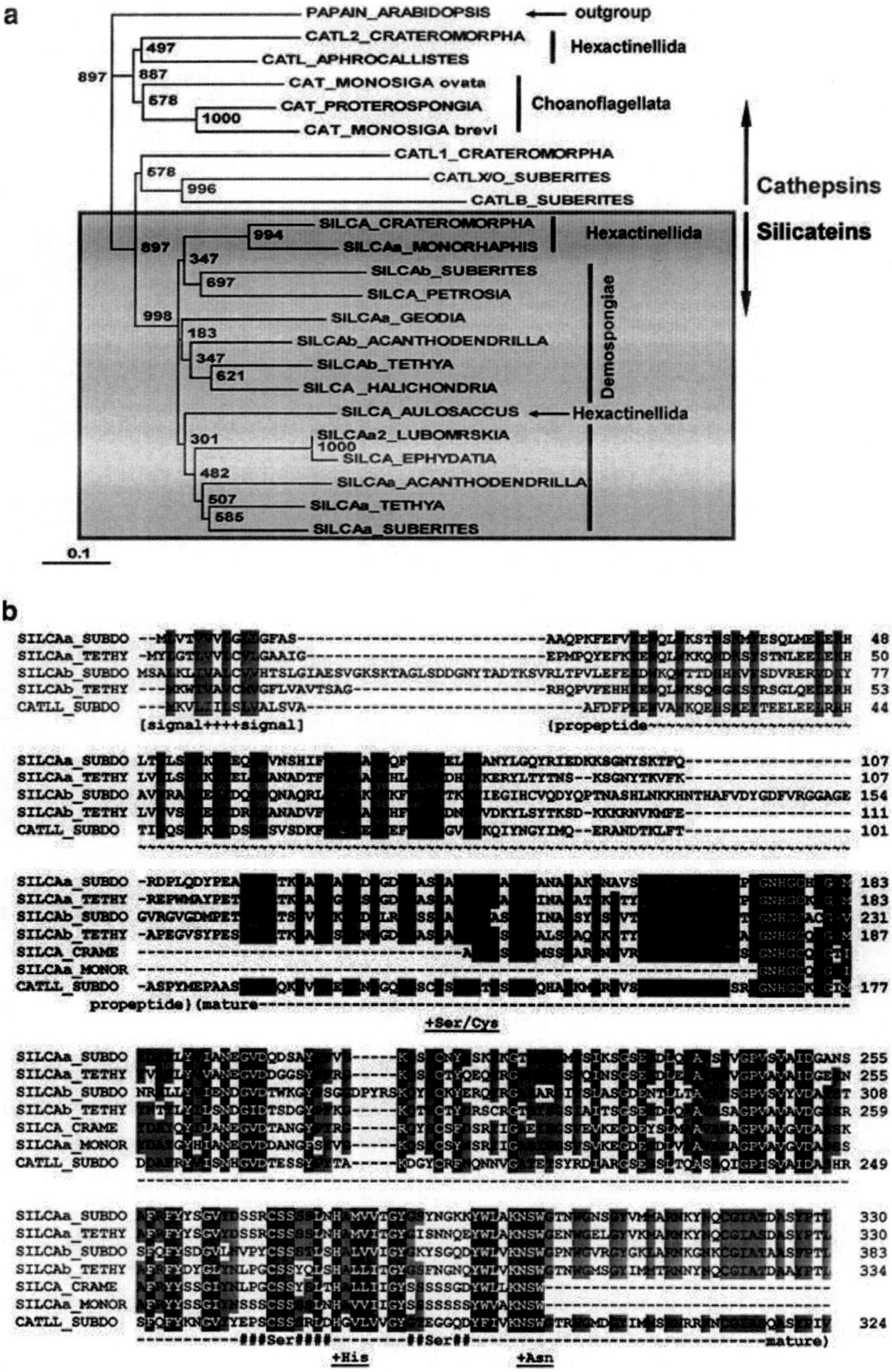

Fig. 9.9 Phylogeny based on the biosilica skeleton. (**a**) Phylogenetic analysis of silicatein within the cathepsin family. The deduced proteins were aligned and the phylogenetic tree was

presence of a glycerol-based buffer. Following this rationale, it could be demonstrated that silicatein exists not only in the axial canal but also in the extraspicular and extracellular space (Müller et al. 2005; Schröder et al. 2006). The enzymatic reaction mechanism of silicatein had been proposed by Cha et al. (1999); the detailed properties of the reaction kinetics have been specified experimentally (Müller et al. 2008b).

9.8 Catabolic Enzyme: Silicase

In the course to further elucidate the metabolism of siliceous spicules, another enzyme, silicase, was identified in the marine sponge *S. domuncula*. Silicase is able to depolymerize amorphous silica. The cDNA was isolated and the deduced

Fig. 9.9 (continued) constructed. The hitherto known three hexactinellid sequences were included; silicatein from *Crateromorpha meyeri* (SILCA_CRATEROMORPHA; AM920776) and from *Monorhaphis chuni* (SILCAa_MONORHAPHIS; FN394978) and the silicatein-like protein *Aulosaccus sp.* (SILCA_AULOSACCUS; ACU86976.1). The bulk of silicatein sequences has been identified in demosponges. First, the silicatein-α sequences from *Suberites domuncula* (SILCAa_SUBERITES; CAC03737.1), *Tethya aurantium* (SILCAa_TETHYA; AAC23951.1), *Geodia cydonium* (SILCAa_GEODIA; CAM57981.1) and *Acanthodendrilla sp. Vietnam* (SILCAa_ACANTHODENDRILLA; ACH92669.1), as well as from *Lubomirskia baicalensis* (SILCAa2_LUBOMIRSKIA; AJ968945) and from *Ephydatia fluviatilis* (SILCA_EPHYDATIA; BAE54434.1). Second, the silicatein-β sequences from *Suberites domuncula* (SILCAb_SUBERITES; CAH04635.1), *Tethya aurantium* (SILCAb_TETHYA; AF098670_1) and *Acanthodendrilla sp. Vietnam* (SILCAb_ACANTHODENDRILLA; FJ013043.1). Third, silicateins that had been identified in marine sponges from which only one isoform had been obtained; silicatein from *Petrosia ficiformis* (SILCA_PETROSIA; AAO23671.1) and from *Halichondria okadai* (SILCA_HALICHONDRIA; BAB86343.1). As reflected in the rooted tree, these silicateins derived from the cathepsins, among which in this tree, the following sequences have been included; cathepsin-like protein 2 *Crateromorpha meyeri* (CATL2_CRATEROMORPHA; CAP17585.1), cathepsin-like protein 1 (*Crateromorpha meyeri*) (CATL1_CRATEROMORPHA; CAP17584.1), mRNA for cathepsin L (catl gene) *Aphrocallistes vastus* (CATL_APHROCALLISTES); AJ968951cathepsin B *Suberites domuncula* (CATLB_SUBERITES; CAH04630.1), cathepsin X/O *Suberites domuncula* (CATLX/O_SUBERITES; lCAH04633.1). The resulting tree was rooted with the sequence from the papain-like cysteine peptidase XBCP3 *Arabidopsis thaliana* (PAPAIN_ARABIDOPSIS; AF388175_1). In addition, the cathepsins from choanoflagellates had been included to show that they derived, according to this tree, from the sponge cathepsins; the cysteine protease from *Proterospongia* sp. (CAT_MONOSIGA ovata), the cathepsin from *Monosiga brevicollis* (CATP_MONOSIGA brevi) and the cathepsin from *Monosiga ovata* MNL00000103. **(b)** A selected set of silicateins, the silicatein-α from *S. domuncula* (SILICAa_SUBDO), and *T. aurantium* (SILICAa_TETHY), as well as the silicatein-β from *S. domuncula* (SILICAb_SUBDO), and *T. aurantium* (SILICAb_TETHY) had been included together with cathepsin L from *S. domuncula* (CATLL_SUBDO). Residues conserved (similar or related with respect to their physicochemical properties) in all sequences are shown in white on black, and those in at least five sequences in black on gray. The characteristic sites in the sequences are marked; the catalytic triad amino acids, Ser in silicateins and Cys in cathepsin, and His and Asn. The borders of the signal peptide (signal), the propeptide (propeptide) and the mature silicatein (mature) are given polypeptide disclosed its relationship to be related to carbonic anhydrases. Recombinant silicase displays besides a carbonic anhydrase activity the ability to dissolve amorphous silica under formation of free silicic acid (Schröder et al. 2003).

9.9 Biosintering

As outlined above, the basic pattern of silica growth around axial filaments is identical in spicules of demosponges and hexactinellids, with one exception: While in demosponges, all silica lamellae fuse to a "solid" structure, a similar process occurs in hexactinellids only in the central part of spicules restricted to Amphidiscosida (e.g., the giant basal spicules from *M. chuni*; Wang et al. 2009); Fig. 9.8. In most other hexactinellid taxa, the lamellae remain separated. In demosponges, the individual lamellae merge by fusion of the 70–300 nm large silica nanospheres that compose the lamellae (Tahir et al. 2004). Whereas fusion of glass of quartz grade by melting processes would require temperatures well above 1,800°C, this process occurs in the living organism at ambient temperature. Accordingly, the product of this biological fusion of silica lamellae resembles the product of a technical process termed sintering, i.e., a thermally activated material transport in a powder or porous compact, decreasing the specific surface by growth of the particle contacts, shrinkage of pore volume, and change of the pore geometry (Thümmler and Oberacker 1993; Wakai and Aldinger 2004). In general, the material is densified below its melting point. Sintering is widely used for the densification of oxide-based ceramic powders including silicon oxide and requires, in general, temperatures above 1,000°C for thermal activation. The free enthalpy (Gibb's energy; ΔG) of sintering is negative, implying that during the reaction energy is released, provided that the activation energy (E_a; reaction minimum energy required to start a chemical reaction) has been overcome. Enzymes work by lowering the activation energy for a reaction and thus dramatically increase the rate of the reaction. Considering the fact that within the silica mantel of spicules, the enzyme silicatein exists [either within (Müller et al. 2008a, d] or between (Woesz et al. 2006) the lamellae), silicatein would be a prime candidate to reduce the activation energy of this exergonic reaction (Fig. 9.10a). Consequently, it acts in principle like the sintering additives used in conventional powder technology processes. Therefore, we propose that the fusion of silica lamellae in demosponge spicules follows a newly defined biocatalytically mediated process, "biosintering". Accordingly, bio-sintering occurs during formation of poriferan siliceous spicules. Similar to demosponges, in hexactinellids fusion between spicules is frequently observed in the orders Hexactinosida, Lyssacinosida, and Lychniscosida (Uriz 2006). There, the initial skeletal elements, composed of hexactine spicules, are subsequently reinforced by additional silica. The large choanosomal spicules from *Euplectella aspergillum* fuse together to a complex silica network (Fig. 9.10b, c).

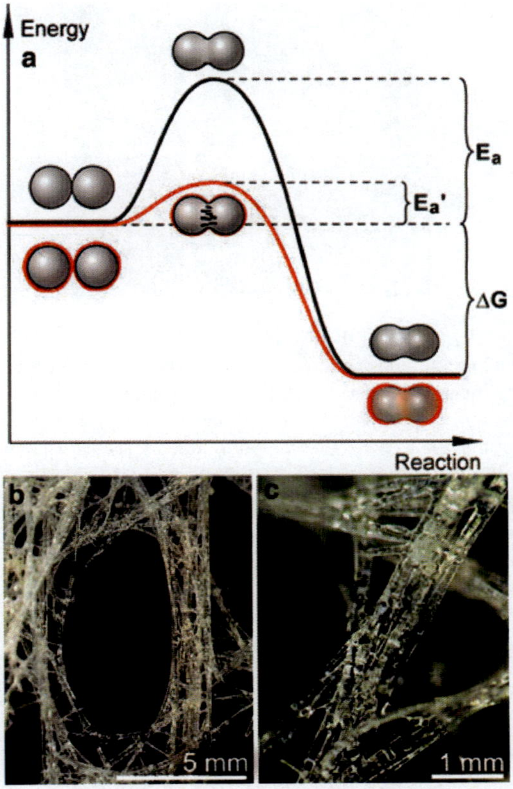

Fig. 9.10 Bio-sintering processes. **(a)** Proposed mechanism of bio-sintering of silica nanospheres within and between siliceous spicules. To initiate conventional sintering processes, activation energy (E_a) is required since fusion of inorganic particles is exergonic (ΔG negative). Due to the presence of the enzyme, silicatein, bio-sintering, however, requires a considerably lower (E_a'). This reduction facilitates and accelerates the fusion process at ambient temperature and allows the free energy (ΔG) for the sintering process to be released. The silica nanoparticles are surrounded by silicatein (*in red*). **(b, c)** This bio-sintering process results in a highly fused network of spicules (synapticulae) as is shown here with the example of *Euplectella aspergillum*

9.10 Implication of the DUF Protein in the Axis Formation

Crucial for a further understanding of this highly complex spicule organization is the answer for the underlying proteinaceous scaffold that organizes the skeletal architecture (Fig. 9.11a, b). At a higher magnification, it becomes apparent that the spicules are held (cemented) together by an organic matrix, which has – in other sponge species – been assumed to be collagen or spongin (Garrone 1978). The molecular nature of this material, spongin, remained unknown. In our approach to solve that question, we purified the skeleton from *L. baicalensis* and subsequently identified the organic matrix, around the spicules (Wang et al. 2010b).

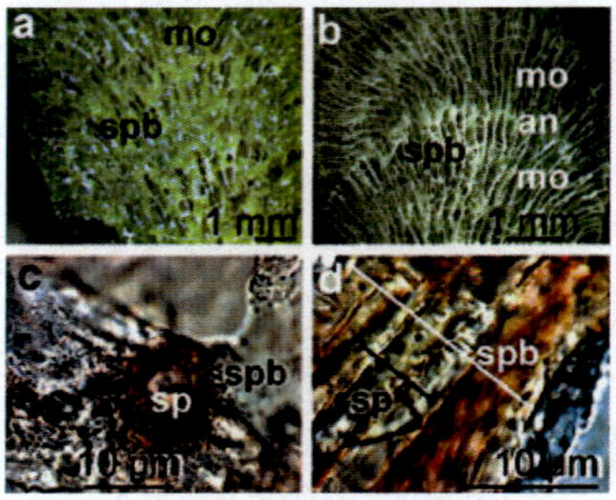

Fig. 9.11 Microscopical analysis of the skeleton of *Lubomirskia baicalensis*. (**a**, **b**) Digital light micrographs through a longitudinal section of a branch. The images show the organized layering of structured tissue along the longitudinal axis stabilized by bundles. Those bundles are composed of an organic matrix that surrounds the centrally located spicules (*spb*). Furthermore, two modules (*mo*) which are separated from each other by an annulus (*an*) are shown. (**c**, **d**) Cross sections through tissue; they had been treated with anti-DUF antibodies that strongly reacted with the organic coat of the sponge bundles (*spb*) that surround the spicules (*sp*)

Subsequently, degenerate primers were designed, directed against those amino acid (aa) sequences. After screening, the cDNA library from *L. baicalensis* highly related sequences were isolated; the deduced polypeptides comprised only 108–110 aa in length and were termed "DUF" (for "domains of unknown function") proteins (Novatchkova et al. 2006). The three sequences obtained from *L. baicalensis* (LBDUF614a, LBDUF614b, and LBDUF614c) comprise, clustered in the center of their sequences, charged aa (Fig. 9.12a). The remaining aa moieties are primarily neutral as reflected also by the respective Kyte–Doolittle hydrophobicity plots (Kyte and Doolittle 1982) (Fig. 9.12b). Computational analysis reveals that each of those three DUF sequences is assembled from three domains of pronounced hydrophobic segments, each spanning around 35 aa. This finding contradicts earlier assumptions that the spongin material in sponges in related to collagen (Gross et al. 1956; Aouacheria et al. 2006). The identification of spongin as hydrophobic DUF sequences opens a new avenue in the understanding of protein–silica interactions and allows a further exploitation for biotechnological applications as well. Antibodies prepared against LBDUF614a reacted to sections through *L. baicalensis* tissue. This immunohistological approach confirmed unambiguously that the organic matrix around the spicules reacts specifically with the antibodies (Fig. 9.11c, d).

9.11 Final Remarks

Silicon biotechnology is to date one of the fastest growing areas to generate/obtain innovative biomaterials. This technology is bioinspired, meaning that nature was used as a model to utilize silicon, the second most abundant element in the Earth's

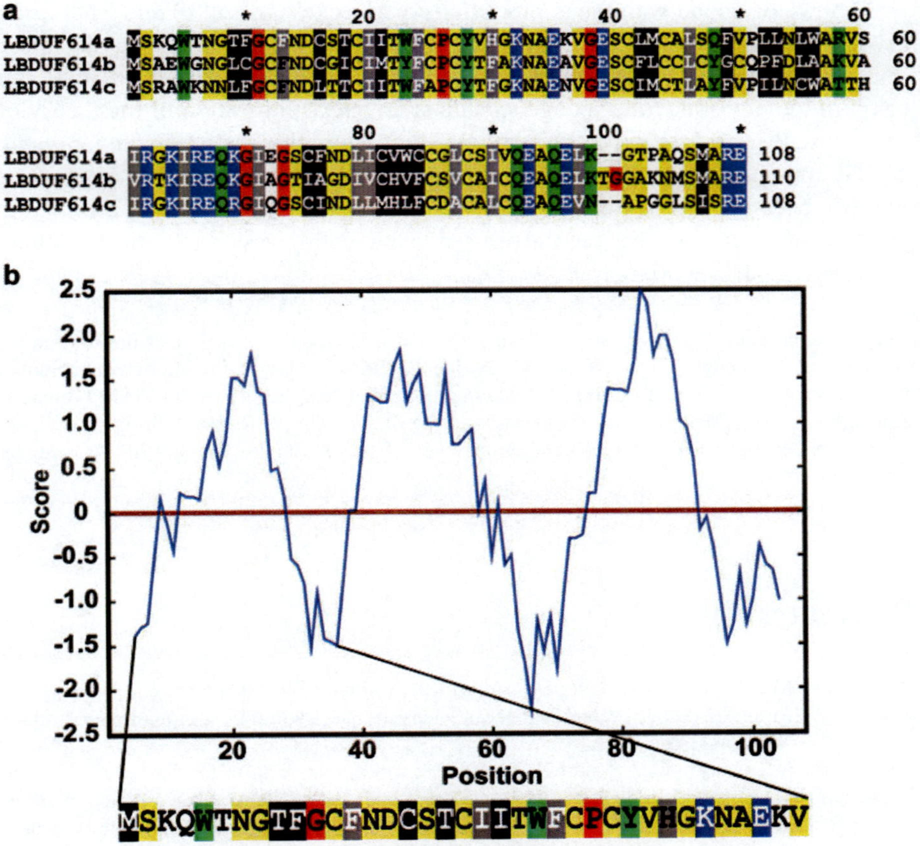

Fig. 9.12 *Lubomirskia baicalensis* DUF protein. (**a**) The organic matrix surrounding the spicules had been partially isolated from the sponge bundles and was then subjected to MALDI-TOF-MS. The sequences obtained were used for the primer design to screen the cDNA library from that sponge. Three highly related sequences were obtained, LBDUF614a, LBDUF614b, and LBDUF614c. In the alignment, the aa have colored according to the groups: proline and glycine (*red*), tiny/small aa (*yellow*), positively charged aa in *blue*, amphoteric/polar aa in *green*, aliphatic/aromatic aa in *grey*, and hydrophobic aa in *black*. (**b**) The Kyte–Doolittle hydrophobicity plot disclosed that the deduced DUF polypeptides comprise three distinct segments of pronounced hydrophobicity. Segment-1 of the sequence LBDUF614a is given in full

crust to form together with oxygen silica and silica polymers. At present, the mechanism by which sponges form the biosilica of their spicules is understood in first outlines and has already enormously stimulated technologies in biomedicine (bone replacement or immuno-masking), in electronics (semiconductor technologies), in optics (light transmission), in lithography, etc. With the application of the silicatein and the silicase enzymes, one could, in principle, envisage the combination of three-dimensional architectures with silica structures to build three-dimensional electronic biosilica chips.

May we conclude with the appreciations of Haeckel (1872a, b) and Thompson (1942), who concluded that "the form of the sponge spicules is all the more important and all the more interesting because it has been discussed time and again in biology" and "the sponge spicule is a typical illustration of the theory of 'bio-crystallisation' to form 'biocrystals' ein Mittelding between an inorganic crystal and an organic secretion." What they did not anticipate is that biosilica is likely to become a prime raw material of this millennium (Müller et al. 2006a); the global market for silica has been estimated to amount to around two billion dollars per year (Kendall 2000).

Acknowledgments This work was supported by grants from the German Bundesministerium für Bildung und Forschung (project "Center of Excellence *BIOTEC*marin"), the International Human Frontier Science Program, the European Commission (project no. 031541 – BIO-LITHO [biomineralization for lithography and microelectronics]), the Basic Scientific Research Program in China (Grant No. 200607CSJ-05), and the International S & T Cooperation Program of China (Grant No. 2008DFA00980).

References

Alexander RM (1979) The invertebrates. Cambridge University Press, Cambridge

Anbar AD, Knoll AH (2002) Proterozoic ocean chemistry and evolution: a bioinorganic bridge? Science 297:137–1142

Aouacheria A, Geourjon C, Aghajari N, Navratil V, Deleage G, Lethias C, Exposito JY (2006) Insights into early extracellular matrix evolution: spongin short chain collagen-related proteins are homologous to basement membrane type IV collagens and form a novel family widely distributed in invertebrates. Mol Biol Evol 23:2288–2302

Barnes RD (1987) Invertebrate zoology. Saunders, Philadelphia

Berner EK, Berner RA (1996) Global environment: water, air, and geochemical cycles. Prentice Hall, New York

Bond C, Harris AK (1988) Locomotion of sponges and its physical mechanism. J Exp Zool 246:271–284

Bramm E, Binderup L, Arrigoni-Martelli E (1980) Inhibition of adjuvant arthritis by intraperitoneal administration of low doses of silica. Agents Actions 10:435–438

Brasier M, Green O, Shields G (1997) Ediacarian sponge spicule clusters from southwest Mongolia and the origins of the Cambrian fauna. Geology 25:303–306

Canfield DE, Farquhar J (2009) Animal evolution, bioturbation, and the sulphate concentration of the oceans. Proc Natl Acad Sci USA 106:8123–8127

Cha JN, Shimizu K, Zhou Y, Christianssen SC, Chmelka BF, Stucky GD, Morse DE (1999) Silicatein filaments and subunits from a marine sponge direct the polymerization of silica and silicones in vitro. Proc Natl Acad Sci USA 96:361–365

Chang XY, Chen LZ, Hu SX, Wang JH, Zhu BQ (2007) Isotopic dating of the Chengjiang fauna-bearing horizon in central Yunnan province, China. Chin J Geochem 26:345–349

Custódio MR, Prokic I, Steffen R, Koziol C, Borojevic R, Brümmer F, Nickel M, Müller WEG (1998) Primmorphs generated from dissociated cells of the sponge *Suberites domuncula*: a model system for studies of cell proliferation and cell death. Mech Ageing Dev 105:45–59

Deprat J, Mansuy H (1912) Etude Géologique du Yun-nan oriental. Géologie générale. Mémoires du Service Géologique de l'Indochine, vol 1, Atlas of 45 geological profiles and maps. Extrème-Orient, Hanoi-Haiphong, 370 pp

Duncan PM (1881) On some remarkable enlargements of the axial canals of sponge spicules and their causes. J R Microsc Soc Ser 2 1:557–572
Eckert C, Schröder HC, Brandt D, Perović-Ottstadt S, Müller WEG (2006) A histochemical and electron microscopic analysis of the spiculogenesis in the demosponge *Suberites domuncula*. J Histochem Cytochem 54:1031–1040
Einsele G (2000) Sedimentary basins: evolution, facies and sediment budget. Springer, Berlin
Finks RM (2003a) Evolution and ecological history of sponges during Paleozoic times. In: Kaesler RL (ed) Treatise on invertebrate paleontology, part E, Porifera, revised, vol 2, Introduction to the Porifera. The Geological Society of America, Boulder, pp 261–274
Finks RM (2003b) Paleozoic Hexactinellida: morphology and phylogeny. In: Kaesler RL (ed) Treatise on invertebrate paleontology, part E, Porifera, revised, vol 2, Introduction to the Porifera. The Geological Society of America, Boulder, pp 135–154
Francesco B, Wilkie IC, Bavestrello G, Cerrano C, Carnevali CMD (2001) Dynamic structure of the mesohyl in the sponge *Chondrosia reniformis* (Porifera, Demospongiae). Zoomorphology 121:109–121
Gaino E, Pronzato R (1983) Étude en microscopie électronique du filament des formes étirées chez *Chondrilla nucula* Schmidt (Porifera, Demospongiae). Ann Sci Nat Zool Paris 5:221–234
Garrone R (1978) Phylogenesis of connective tissue. Morphological aspects and biosynthesis of sponge intercellular matrix. S. Karger, Basel
Garrone R (1998) Evolution of metazoan collagens. Prog Mol Subcell Biol 21:119–139
Gaucher C, Frimmel HE, Ferreira VP, Poire DG (2004) Vendian-Cambrian of western Gondwana: introduction. Gondwana Res 7:659–660
Gehling JG, Rigby JK (1996) Long expected sponges from the neoproterozoic ediacara fauna of South Australia. J Paleontol 2:185–195
Gordon MS, Belman BW, Chow PH (1976) Comparative studies on the metabolism of shallow-water and deep-sea marine fishes. IV. Patterns of aerobic metabolism in the mesopelagic deep-sea fangtooth fish Anoplogaster cornuta. Mar Biol 35:287–293
Gradstein FM, Ogg JG, Smith AG (2005) A geologic time scale. Cambridge University Press, Cambridge, 589 pp
Gross J, Sokal Z, Rougvie M (1956) Structural and chemical studies on the connective tissue of marine sponges. J Histochem Cytochem 4:227–246
Haeckel E (1872a) Atlas der Kalkschwämme. Verlag von Georg Reimer, Berlin
Haeckel E (1872b) Biologie der Kalkschwämme, vol I. Georg Reimer, Berlin
Harrison FW, De Vos L (1991) Porifera. In: Harrison FW, Ruppert EE (eds) Microscopic anatomy of invertebrates, vol 2. Wiley Liss, New York, pp 29–89
Hartman WD, Reiswig H (1973) The individuality of sponges. In: Boardman RS, Cheetham AH, Oliver WA (eds) Animal colonies. Dow, Hutch, Ross, Stroudsburg, pp 567–584
Hoffman PF, Schrag DP (2002) The snowball earth hypothesis: testing the limits of global change. Terra Nova 14:129–155
Hou X, Bergström J, Wang H, Feng X, Chen A (1999) The Chengjiang fauna. Exceptionally well-preserved animals from 530 million years ago. Yunnan Science and Technology Press, Yunnan, 170 pp
Hou XG, Aldridge RJ, Bergström J, Siveter DJ, Siveter DJ, Feng XH (2004) The Cambrian fossils of Chengjiang, China: the flowering of early animal life. Blackwell, Oxford
Hyman LH (1940) Metazoa of the cellular grade of construction phylum Porifera, the sponges; chapter 6. In: Hyman H (ed) Invertebrates: protozoa through Ctenophora. McGraw-Hill, New York, pp 284–364
Iler RK (1979) The chemistry of silica: solubility, polymerization, colloid and surface properties and biochemistry of silica. Wiley, New York
Imsiecke G, Steffen R, Custodio M, Borojevic R, Müller WEG (1995) Formation of spicules by sclerocytes from the freshwater sponge *Ephydatia muelleri* in short-term cultures in vitro. In Vitro Cell Dev Biol 31:528–535
Kasting JF (1984) The evolution of prebiotic atmosphere. Orig Life 14:75–82

Kasting JF, Holland HD, Kump LR (1992) Atmospheric evolution: the rise of oxygen. In: Schopf JW, Klein C (eds) All in the proterozoic biosphere: a multidisciplinary study. Cambridge University Press, New York, pp 159–164

Kazmierczak J, Kempe S, Altermann W (2004) Microbial origin of Precambrian carbonates: lessons from modern analogues. In: Eriksson PG (ed) The Precambrian earth: tempos and events. Elsevier, Amsterdam, pp 545–564

Kempe S, Degens ET (1985) An early soda ocean? Chem Geol 53(95–108):95

Kendall T (2000) Written in sand – the world of specialty silicas. Ind Miner 390:49–59

Kikuchi Y, Suzuki Y, Tamiya N (1983) The source of oxygen in the reaction catalysed by collagen lysyl hydroxylase. Biochem J 213:507–512

Knoll AH, Carroll SB (1999) Early animal evolution: emerging views from comparative biology and geology. Science 284:2129–2137

Kyte J, Doolittle RF (1982) A simple method for displaying the hydrophobic character of a protein. J Mol Biol 157:105–132

Lévi C (1970) Les cellules des éponges. In: Fry WG (ed) The biology of the Porifera. Symp Zool Soc Lond, vol 25. Academic, New York, pp 353–364

Leys SP, Mackie GO, Meech RW (1999) Impulse conduction in a sponge. J Exp Biol 202:1139–1150

Li CW, Chen JY, Hua TE (1998) Precambrian sponges with cellular structures. Science 279:879–882

Maas O (1901) Die Knospenentwicklung der *Tethya* und ihr Vergleich mit der geschlechtlichen Fortpflanzung der Schwämme. Z wiss Zool 70:263–288

Maldonado M, Carmona MC, Uriz MJ, Cruzado A (1999) Decline in Mesozoic reef-building sponges explained by silicon limitation. Nature 401:785–788

Morey RO, Rowe JJ (1964) The solubility of amorphous silica at 25°C. J Geophys Res 69:1995–2002

Müller WEG (1998) Origin of Metazoa: sponges as living fossils. Naturwiss 85:11–25

Müller WEG (2001) How was the metazoan threshold crossed? The hypothetical urmetazoa. Comp Biochem Physiol A 129:433–460

Müller WEG, Wiens M, Adell T, Gamulin V, Schröder HC, Müller IM (2004) Bauplan of urmetazoa: basis for genetic complexity of Metazoa. Intern Rev Cytol 235:53–92

Müller WEG, Rothenberger M, Boreiko A, Tremel W, Reiber A, Schröder HC (2005) Formation of siliceous spicules in the marine demosponge *Suberites domuncula*. Cell Tissue Res 321:285–297

Müller WEG, Belikov SI, Schröder HC (2006a) Biosilica – raw material of the new millennium. Sci First Hand 6:26–35

Müller WEG, Belikov SI, Tremel W, Perry CC, Gieskes WWC, Boreiko A, Schröder HC (2006b) Siliceous spicules in marine demosponges (example *Suberites domuncula*). Micron 37:107–120

Müller WEG, Eckert C, Kropf K, Wang XH, Schloßmacher U, Seckert C, Wolf SE, Tremel W, Schröder HC (2007a) Formation of the giant spicules of the deep sea hexactinellid *Monorhaphis chuni* (Schulze 1904): electron microscopical and biochemical studies. Cell Tissue Res 329:363–378

Müller WEG, Li J, Schröder HC, Qiao L, Wang XH (2007b) The unique skeleton of siliceous sponges (Porifera; Hexactinellida and Demospongiae) that evolved first from the urmetazoa during the proterozoic: a review. Biogeosciences 4:219–232

Müller WEG, Wang XH, Belikov SI, Tremel W, Schloßmacher U, Natoli A, Brandt D, Boreiko A, Tahir MN, Müller IM, Schröder HC (2007c) Formation of siliceous spicules in demosponges: example *Suberites domuncula*. In: Bäuerlein E (ed) Handbook of biomineralization, vol 1, The biology of biominerals structure formation. Wiley-VCH, Weinheim, pp 59–82

Müller WEG, Jochum K, Stoll B, Wang XH (2008a) Formation of giant spicule from quartz glass by the deep sea sponge *Monorhaphis*. Chem Mater 20:4703–4711

Müller WEG, Schloßmacher U, Wang XH, Boreiko A, Brandt D, Wolf SE, Tremel W, Schröder HC (2008b) Poly(silicate)-metabolizing silicatein in siliceous spicules and silicasomes of demosponges comprises dual enzymatic activities (silica-polymerase and silica-esterase). FEBS J 275:362–370

Müller WEG, Wang XH, Kropf K, Boreiko A, Schloßmacher U, Brandt D, Schröder HC, Wiens M (2008c) Silicatein expression in the hexactinellid *Crateromorpha meyeri*: the lead marker gene restricted to siliceous sponges. Cell Tissue Res 333:339–351

Müller WEG, Wang XH, Kropf K, Ushijima H, Geurtsen W, Eckert C, Tahir MN, Tremel W, Boreiko A, Schloßmacher U, Li J, Schröder HC (2008d) Bioorganic/inorganic hybrid composition of sponge spicules: matrix of the giant spicules and of the comitalia of the deep sea hexactinellid *Monorhaphis*. J Struct Biol 161:188–203

Müller WEG, Wang XH, Cui FZ, Jochum KP, Tremel W, Bill J, Schröder HC, Natalio F, Schloßmacher U, Wiens M (2009) Sponge spicules as blueprints for the biofabrication of inorganic–organic composites and biomaterials. Appl Microbiol Biotechnol 83:397–413

Novatchkova M, Schneider G, Fritz R, Eisenhaber F, Schleiffer A (2006) DOUT-finder-identification of distant domain outliers using subsignificant sequence similarity. Nucleic Acids Res 34:W214–W218

Pavans de Ceccatty (1986) Cytoskeletal organisation and tissue patterns of epithelia in the sponge *Ephydatia mülleri*. J Morphol 189:45–65

Pechenik JA (2000) Biology of the invertebrates. McGraw Hill, Boston

Pilcher H (2005) Back to our roots. Nature 435:1022–1023

Pisera A (2003) Some aspects of silica deposition in lithistid demosponge desmas. Microsc Res Tech 62:312–326

Rigby JK, Collins D (2004) Sponges of the Middle Cambrian Burgess Shale and Stephen formations, British Columbia. Royal Ontario Museum, Toronto

Rigby JK, Hou XG (1995) Lower Cambrian demosponges and hexactinellid sponges from Yunnan, China source. J Paleontol 69:1009–1019

Sandford F (2003) Physical and chemical analysis of the siliceous skeleton in six sponges of two groups (Demospongiae and Hexactinellida). Microsc Res Tech 62:336–355

Schröder HC, Krasko A, Le Pennec G, Adell T, Hassanein H, Müller IM, Müller WEG (2003) Silicase, an enzyme which degrades biogenous amorphous silica: contribution to the metabolism of silica deposition in the demosponge *Suberites domuncula*. Progr Molec Subcell Biol 33:249–268

Schröder HC, Perović-Ottstadt S, Rothenberger M, Wiens M, Schwertner H, Batel R, Korzhev M, Müller IM, Müller WEG (2004) Silica transport in the demosponge *Suberites domuncula*: fluorescence emission analysis using the PDMPO probe and cloning of a potential transporter. Biochem J 381:665–673

Schröder HC, Boreiko A, Korzhev M, Tahir MN, Tremel W, Eckert C, Ushijima H, Müller IM, Müller WEG (2006) Co-Expression and functional interaction of silicatein with galectin: matrix-guided formation of siliceous spicules in the marine demosponge *Suberites domuncula*. J Biol Chem 281:12001–12009

Schulze FE (1904) Hexactinellida. Wissenschaftliche Ergebnisse der Deutschen Tiefsee-Expedition auf dem Dampfer "Valdivia" 1898–1899. Fischer, Stuttgart

Seimiya M, Naito M, Watanabe Y, Kurosawa Y (1998) Homeobox genes in the freshwater sponge *Ephydatia fluviatilis*. Prog Mol Subcell Biol 19:133–155

Shanker R, Singh G, Kumar G, Maithy PK (2001) Assembly and break-up of Rodinia and Gondwana – evidence from India. Gondwana Res 4:783–784

Shimizu K, Cha J, Stucky GD, Morse DE (1998) Silicatein alpha: cathepsin L-like protein in sponge biosilica. Proc Natl Acad Sci USA 95:6234–6238

Siever R (1992) The silica cycle in the Precambrian. Geochim Cosmochim Acta 56:3265–3272

Simonson BM (1985) Sedimentology of cherts in the early proterozoic wishart formation, Quebec-newfoundland, Canada. Sedimentology 32:2340

Simpson TL (1984) The cell biology of sponges. Springer, New York

Simpson TL, Langenbruch PF, Scalera-Liaci L (1985) Silica spicules and axial filaments of the marine sponge *Stelletta grubii* (Porifera, Demospongiae). Zoomorphology 105:375–382

Sims PA, Mann DG, Medlin LK (2006) Evolution of the diatoms: insights from fossil, biological and molecular data. Phycologia 45:361–402

Sollas WJ (1888) Report on the Tetractinellida collected by H.M.S. "Challenger", during the years 1873–1876. H.M.S. Challenger Scient Results Zool 25:1–458

Steiner M (1994) Die neoproterozoischen Megaalgen Südchinas. Berl Geowiss Abh E 15:1–146

Steiner M, Mehl D, Reitner J, Erdtmann BD (1993) Oldest entirely preserved sponges and other fossils from the lowermost Cambrian and a new facies reconstruction of the Yangtze Platform (China). Berl Geowiss Abh E 9:293–329

Steiner M, Zhu M, Zhao Y, Erdtmann BD (2005) Lower Cambrian Burgess Shale-type fossil associations of South China. Palaeogeogr Palaeoclimatol Palaeoecol 220:129–152

Street-Perrott FA, Barker PA (2008) Biogenic silica: a neglected component of the coupled global continental biogeochemical cycles of carbon and silicon. Earth Surf Process Land 33:1436–1457

Tahir MN, Théato P, Müller WEG, Schröder HC, Janshoff A, Zhang J, Huth J, Tremel W (2004) Monitoring the formation of biosilica catalysed by histidin-tagged silicatein. ChemComm 24:2848–2849

Teal JM, Carey FG (1967) Respiration of a *Euphausiid* from the oxygen minimum layer. Limnol Oceanogr 12:548–550

Thompson D'Ary W (1942) On growth and form. University Press, Cambridge

Thümmler F, Oberacker R (1993) In: Jenkins IJ, Wood JV (eds) An introduction to powder metallurgy. The Institute of Materials, book 490, Cambridge University Press, Cambridge, pp. 181–188

Towe KM (1970) Oxygen-collagen priority and the early metazoan fossil record. Proc Natl Acad Sci USA 65:781–788

Tucker ME (1992) The Precambrian-Cambrian boundary: seawater chemistry, ocean circulation and nutrient supply in metazoan evolution, extinction and biomineralization. J Geol Soc Lond 149:655–688

Uriz MJ (2006) Mineral spiculogenesis in sponges. Can J Zool 84:322–356

Uriz MJ, Turon X, Becerro MA, Agell G (2003) Siliceous spicules and skeleton frameworks in sponges: origin, diversity, ultrastructural patterns, biological functions. Microsc Res Tech 62:279–299

Wakai F, Aldinger F (2004) Sintering forces in equilibrium and nonequilibrium states during sintering of two particles. Sci Technol Adv Mat 5:521–525

Walker JCG (1978/79) The early history of oxygen and ozone in the atmosphere. Pageoph 117:498–512

Wang XH, Boreiko A, Schloßmacher U, Brandt D, Schröder HC, Li J, Kaandorp JA, Götz H, Duschner H, Müller WEG (2008) Axial growth of hexactinellid spicules: formation of cone-like structural units in the giant basal spicules of the hexactinellid *Monorhaphis*. J Struct Biol 164:270–280

Wang XH, Schröder HC, Müller WEG (2009) Giant siliceous spicules from the deep-sea glass sponge *Monorhaphis chuni*: morphology, biochemistry, and molecular biology. Int Rev Cell Mol Biol 273:69–115

Wang XH, Hu S, Gan L, Wiens M, Müller WEG (2010) Sponges (Porifera) as living metazoan witnesses from the Neoproterozoic: biomineralization and the concept of their evolutionary success. Terra Nova 22:1–11

Wang X, Wiens M, Schröder HC, Hu S, Mugnaioli E, Kolb U, Tremel W, Pisignano D, Müller WEG (2010) Morphology of sponge spicules: silicatein a structural protein for bio-silica formation. Advanced Biomaterials/Advanced Engineering Mat 12:B422–B437

Wiens M, Wrede P, Grebenjuk VA, Kaluzhnaya OV, Belikov SI, Schröder HC, Müller WEG (2009) Towards a molecular systematics of the Lake Baikal/Lake Tuva sponges. In: Müller WEG, Grachev MA (eds) Biosilica in evolution, morphogenesis, and nanobiotechnology.

progress in molecular and subcellular biology [marine molecular biotechnology]. Springer, Berlin, pp 111–144

Woesz A, Weaver JC, Kazanci M, Dauphin Y, Aizenberg J, Morse DE, Fratzl P (2006) Micromechanical properties of biological silica in skeletons of deep-sea sponges. J Mater Res 21:2068–2078

Wu W, Yang AH, Janussen D, Steiner M, Zhu MY (2005) Hexactinellid sponges from the Early Cambrian Black Shale of South Anhui, China. J Paleont 79:1043–1051

Xiao S, Laflamme M (2008) On the eve of animal radiation: phylogeny, ecology and evolution of the Ediacara biota. Trends Ecol Evol 24:31–40

Xiao S, Yuan X, Knoll AH (2000) Eumetazoan fossils in terminal proterozoic phosphorites? Proc Natl Acad Sci USA 97:13684–13689

Xiao S, Hu J, Yuan X, Parsley RL, Cao R (2005) Articulated sponges from the Early Cambrian Hetang formation in southern Anhui, South China: their age and implications for early evolution of sponges. Palaeogeogr Palaeoclimat Palaeoecol 220:89–117

Yang Q, Ma JY, Sun XY, Cong PY (2007) Phylochronology of early metazoans: combined evidence from molecular and fossil data. Geol J 42:281–295

Zhang WT, Hou XG (1985) Preliminary notes on the occurrence of the unusual trilobite Naraoia in Asia. Acta Palaeontol Sin 24:591–595 [In Chinese with English summary]

Zhang X, Liu W, Zhao Y (2008) Cambrian Burgess Shale-type Lagerstätten in South China: distribution and significance. Gondwana Res 14:255–262

Chapter 10
Biosilica-Based Strategies for Treatment of Osteoporosis and Other Bone Diseases

Heinz C. Schröder, Matthias Wiens, Xiaohong Wang, Ute Schloßmacher, and Werner E.G. Müller

Contents

Abstract Osteoporosis is a common disease in later life, which has become a growing public health problem. This degenerative bone disease primarily affects postmenopausal women, but also men may suffer from reduced bone mineral density. The development of prophylactic treatments and medications of osteoporosis has become an urgent issue due to the increasing proportion of the elderly in the population. Apart from medical/hormonal treatments, current strategies for

H.C. Schröder (✉) • W.E.G. Müller (✉)
ERC Advanced Grant Research Group, Institute for Physiological Chemistry, University Medical Center of the Johannes Gutenberg University Mainz, Duesbergweg 6, D-55128 Mainz, Germany

NanotecMARIN GmbH, Duesbergweg 6, D-55128 Mainz, Germany
e-mail: hschroed@uni-mainz.de

W.E.G. Müller (ed.), *Molecular Biomineralization*, Progress in Molecular and Subcellular Biology 52, DOI 10.1007/978-3-642-21230-7_10,

prophylaxis of osteoporosis are primarily based on calcium supplementation as a main constituent of bone hydroxyapatite mineral. Despite previous reports suggesting an essential role in skeletal growth and development, the significance of the trace element silicon in human bone formation has attracted major scientific interest only rather recently. The interest in silicon has been further increased by the latest discoveries in the field of biosilicification, the formation of the inorganic silica skeleton of the oldest still extant animals on Earth, the sponges, which revealed new insights in the biological function of this element. Sponges make use of silicon to build up their inorganic skeleton which consists of biogenously formed polymeric silica (biosilica). The formation of biosilica is mediated by specific enzymes, silicateins, which have been isolated, characterized, and expressed in a recombinant way. Epidemiological studies revealed that dietary silicon reduces the risk of osteoporosis and other bone diseases. Recent results allowed for the first time to understand the molecular mechanism underlying the protective effect of silicic acid/ biosilica against osteoporosis. Biosilica was shown to modulate the ratio of expression of two cytokines involved in bone formation–RANKL and osteoprotegerin. Hence, biosilica has been proposed to have a potential in prophylaxis and therapy of osteoporosis and related bone diseases.

10.1 Introduction

Osteoporosis is the most common metabolic bone disorder (Sambrook and Cooper 2006). Over 200 million people suffer from osteoporosis worldwide (Reginster and Burlet 2006). The prevalence of the disease will further increase as a result of the demographic development in many industrialized countries. Osteoporosis is associated with an enhanced risk of bone fracture (Cummings and Melton 2002). The increased bone fragility is caused by a reduced mineral density of bone tissue and by a deterioration of the bone microarchitecture (Faibish et al. 2006). Osteoporosis is most common in women after menopause but may also develop in men (reviewed in: Patlak 2001; Raisz 2005; Khosla et al. 2008a, b). The disease can be classified as either primary or secondary. Primary osteoporosis is often due to estrogen deficiency in women following menopause (postmenopausal osteoporosis), but may also develop at older age in men (senile osteoporosis, in both females and males). Secondary osteoporosis may occur as a result of hormonal disorders (e.g., hyperparathyroidism) or treatment of patients with glucocorticoids (steroid-induced osteoporosis) (Adachi 1997). In addition, nutritional factors may be involved in the development of osteoporotic disorders (reviewed in: Jugdaohsingh 2007).

The progressive deterioration of the microarchitecture of bone tissue in osteoporosis particularly concerns cancellous (trabecular) bone, resulting in shrinkage of cortical width, rarefaction of the trabecular network and increased occurrence of disconnected trabeculae due to increased osteoclastic resorption. Hence, hip fractures (fractures of the proximal femur) and vertebral fractures (compression fractures) are often observed in osteoporotic patients (Gardner et al. 2006).

The mechanism underlying the pathogenesis of osteoporosis is an imbalance between bone resorption and bone formation (Teitelbaum 2000). The cells responsible for mineralization of bone tissue are osteoblasts, and the cells responsible for bone resorption are osteoclasts, which inhabit the bone surface. The differentiation of pre-osteoclasts to mature osteoclasts and the activation of these cells are regulated by various factors belonging to the tumor necrosis factor (TNF) and TNF receptor superfamily, including RANKL (receptor activator for nuclear factor κB ligand) and osteoprotegerin (OPG) (Wada et al. 2006; Leibbrandt and Penninger 2008). The RANKL/RANK interaction plays a fundamental role in the differentiation and maintenance of osteoclast activity, and hence in the development of osteoporosis. The discovery of the cytokine OPG has significantly contributed to the understanding of the mechanisms controlling bone mineral density (Simonet et al. 1997). OPG is expressed in osteoblasts and inhibits osteoclastogenesis. Hence increased levels of this cytokine are associated with osteosclerosis (Wang et al. 2004), whereas a relative decrease in OPG expression is linked to osteoporosis (Lane and Yao 2009).

The treatment of osteoporosis is mainly based on the use of agents that inhibit bone resorption (Reid 2008; Canalis 2010). Current medications of osteoporosis include the use of bisphosphonates (synthetic analogous of pyrophosphate in which the oxygen of the P–O–P bond has been replaced by carbon; Russell et al. 1999), selective estrogen-receptor modulators (SERMs; raloxifene; Taranta et al. 2002), teriparatide (recombinant parathyroid hormone; Blick et al. 2009), strontium ranelate (Ammann et al. 2004), RANKL inhibitors (Denosumab, a monoclonal antibody mimicking OPG activity; Singer and Grauer 2010), and calcium and vitamin D (used as nutritional supplements; Tang et al. 2007). Among the minerals, used as supplements, silicon/silicate had attracted increasing attention since the pioneering reports of Carlisle (1972) and Schwarz and Milne (1972).

10.2 Bone Formation

Human bone is characterized by a complex hierarchical architecture (Weiner and Traub 1992) based on both inorganic and organic components. The inorganic matrix of bone tissue consists of carbonated hydroxyapatite (HA) [$Ca_{10}(PO_4)_6OH_2$]. This bone mineral is formed by osteoblasts which secrete alkaline phosphatase- containing vesicles. The organic matrix of bone is mainly composed of type I collagen fibrils. These fibrils are intimately involved in the deposition of bone mineral (see Fig. 10.6 in Sect. 10.11). Additional proteins and macromolecules (polysaccharides) constituting the organic component of bone tissue include osteocalcin, osteonectin, osteopontin, bone sialo protein, and glycosaminoglycans. The complex biochemical processes involved in bone formation are driven by a sophisticated network of cytokines/growth factors, which provide, among others, the signals for the differentiation of the progenitor cells, pre-osteoblasts and pre-osteoclasts, to the mature bone-forming cells (osteoblasts) and bone-resorbing cells (osteoclasts). The bone morphogenic proteins (BMPs) are an important group of morphogens controlling these processes (reviewed in: Morgan et al. 2008). Since the studies of Carlisle (1972) and Schwarz and Milne (1972), the close interrelation between silica metabolism and bone formation in mammals has attracted increasing attention (reviewed in: Jugdaohsingh 2007).

10.3 Silicon Chemistry

In the monomeric orthosilicic acid, the silicon atom is tetrahedrally coordinated to four hydroxyl groups [$Si(OH)_4$] (Fig. 10.1a). Polycondensation of this molecule initially results in the formation of dimers which preferentially react with

Fig. 10.1 Biosilica. (**a**) Polycondensation of orthosilicic acid to polysilicate/silica via cyclic siloxane species. (**b**)–(**d**) Enzymatically formed biosilica in siliceous sponges. (**b**) Broken sterraster of the demosponges *Geodia cydonium*, showing the axial canal (ac) and the fusion zones between neighboring spines of this ball-like microsclere. (**c**) Immature tips of the spiny sterrasters. (**d**) Cross section through a giant basal spicule of the hexactinellid *M. chuni*, showing the axial canal (ac) and the surrounding silica lamellae. The most inner lamellae are fused together

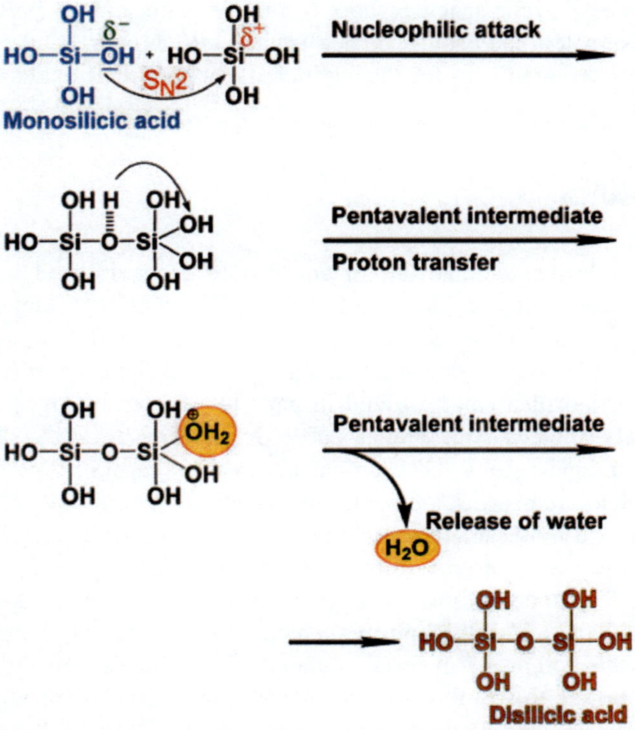

Fig. 10.2 Mechanism of reaction between two silicic acid species at neutral pH. The nucleophilic attack (S_N2 reaction) of the partially negatively charged oxygen of one of the OH ligands of the first silicic acid species at the partially positively charged silicon of the second silicic acid species results in the formation of a pentavalent intermediate. Subsequently, after a proton transfer, a water molecule is released from the intermediate

monomers to trimers and higher oligomers which are linked by siloxane (Si–O–Si) bonds (Iler 1979; Mann 2001). The proximity of the chain ends of the small oligomers which are more reactive than the monomer allows for the transient formation of cyclic siloxane species (Fig. 10.1a). The further polycondensation reaction which may involve an Ostwald ripening process (growth on the expense of silicic acid released from smaller, more soluble particles) finally leads to the formation and deposition of larger, less soluble silica particles (Perry 2003; Perry and Keeling-Tucker 2000).

At neutral pH, the proportion of ionized silicic acid molecules is very small (Perry et al. 2003). Under these conditions, the condensation reaction is based on a nucleophilic substitution (S_N2) reaction which involves formation of a pentacoordinate intermediate, a proton transfer, and release of water (Perry 2003) (Fig. 10.2).

The cyclic oligomers formed at the early phase of silica polycondensation have a higher proportion of ionized silanol groups and a negative charge, as the pKa of the silanol groups decreases with increasing size of the oligomers (Perry 2003); the monomer orthosilicic acid has a pKa of 9.8 and is thus only weakly acidic

(Iler 1979). These cyclic species whose formation is most likely promoted during enzymatic (silicatein-mediated) silica formation (Schröder et al. 2010) hence become the preferential sites for the addition of further silicic acid molecules.

10.4 Biosilica

Biosilica is a biogenic material, in contrast to bioglass used in bone/tissue engineering. The inorganic phase of biosilica glass consists of amorphous silica (SiO_2). This inorganic material had already been used about 800 million years ago (during the Proterozoic) to construct ancient animal skeletal systems (Wang et al. 2010). Today, biosilica can be found in particular in plants, algae, and sponges (Fig. 10.1b–d) (reviewed in: Müller 2003; Schröder et al. 2008). The siliceous sponges are unique among these organisms in their capability of forming their siliceous skeleton (consisting of needle-like spicules and biosintered, higher order structures formed by these elements; Fig. 10.1b, c) through an enzymatic mechanism (see below; reviewed in: Morse 1999; Müller et al. 2007b, 2009b; Schröder et al. 2008). The sponge biosilica is characterized by a quartz-glass-like purity (Müller et al. 2008a). This high purity, in addition to its extreme stability (based on the fact that this composite material contains, besides inorganic silica, an organic component) makes this biomaterial of interest for various applications in nano-biotechnology and nano-biomedicine (Schröder et al. 2007a; Müller et al. 2009b).

In order to form their biosilica skeletons, which may reach a size of up to 3 m (example: basal giant spicules of the glass sponge *Monorhaphis chuni*; a cross section through a spicule with a diameter of 10 mm is shown in Fig. 10.1d; Müller et al. 2008a), the siliceous sponges must accumulate silicon from the environment. In marine waters, which are silicon poor, accumulation of silicon, in the form of orthosilicic acid, requires an active transport mechanism. In the marine demosponge *Suberites domuncula*, a silicic acid transporter has been identified, which acts as a co-transporter of $Si(OH)_4$ and Na^+ ions (Schröder et al. 2004). In diatoms, silicate uptake is also an energy-consuming process, but uses a different transporter (Bhattacharyya and Vulcani 1980; Thamatrakoln et al. 2006; Gröger et al. 2007).

Though increasing evidence demonstrates the importance of silicon in controlling mammalian bone formation (reviewed in: Jugdaohsingh 2007), the mechanism of uptake of silicon in mammalian cells is not yet known. It is likely that only monomeric silica (orthosilicic acid or orthosilicate) is taken up by eukaryotic cells. So far, there are no hints that accumulation of silicon occurs via a passive influx of silicic acid/silicate into cells. Hence, the existence of an energy-dependent silicic acid transporter like in sponges has to be assumed. In this context, it should be mentioned that the sponge silicic acid transporter is highly related to the mammalian Na^+/HCO_3^- co-transporters (Schröder et al. 2004). In addition, a role of aquaporins in silicic acid transport might be conceivable (Sasaki 2008; Bhattacharjee et al. 2008). There are, at present, no hints that silicon accumulation into human cells occurs via uptake of silica nanoparticles. Biosilica nanoparticles formed by sponge silicatein may have a diameter of 50–70 nm (Tahir et al. 2004). Particles of that size are readily taken up by cells through endocytosis (Jin et al. 2009). In sponges, biosilica can be hydrolyzed to orthosilicate by the enzyme silicase which is related to the metazoan carbonic anhydrases (Schröder et al. 2003). Carbonic anhydrases which have a little silica-hydrolyzing activity are also present in the extracellular space (Wetzel et al. 2001; Gao et al. 2007) and hence might allow polymeric silica to be taken up after hydrolysis via an orthosilicate-specific transporter.

Biosilica in sponges has been attributed to exhibit both spiculoinductive and spiculoconductive activity (Wiens et al. 2010b). This assumption is based on the fact that silica triggers differentiation of progenitor cells to spicule-forming sclerocytes (spiculoinductivity) (Müller et al. 2006; Kaandorp et al. 2008; Le Pennec et al. 2003). On the other hand, immature spicules which are released into the extracellular space determine spicular morphogenesis through attracting sclerocytes and triggering collagen synthesis (spiculoconductivity) (Schröder et al. 2006). It should be noted that the development of biomimetic materials that show osteoinductive and osteoconductive potential is a challenging task for bone tissue engineering (Albrektsson and Johansson 2001). Osteoinduction describes processes which involve the differentiation of progenitor cells from hematopoietic stem cells to osteoblasts and osteoclasts, while osteoconduction concerns the growth of bone on a surface directed by the surface structure and mediated by (incorporated) osteogenic agents (Albrektsson and Johansson 2001; Glantz 1987).

10.5 Silicatein

The principle enzyme catalyzing formation of biosilica from soluble precursors is silicatein (Fig. 10.3), an enzyme exclusively found in sponges (Shimizu et al. 1998; Cha et al. 1999; Krasko et al. 2000; Müller et al. 2008b). Silicatein is the major protein component of the axial filament which is present in the axial canal of the sponge spicules (Fig. 10.1b, d). This protein is the first enzyme that has been discovered to be capable of forming an inorganic polymer (silica) from a monomeric precursor; orthosilicic acid or tetraethoxysilane (TEOS; as an orthosilicic acid precursor) are commonly used as a substrate. Harnessing the biocatalytic potential of this unique enzyme is expected to open a variety of new applications of silica in nanotechnology, nanomedicine, and material sciences (Schröder et al. 2007a; Müller et al. 2009b).

Silicatein is related to the cathepsins, a group of proteases, but is characterized by replacement of the Cys residue by a Ser residue in the catalytic center of the molecule (Shimizu et al. 1998; Krasko et al. 2000). In addition, the silicatein sequences comprise a Ser stretch not found in cathepsins. Several genes/cDNAs encoding different isoforms of silicatein have been isolated both from demosponges and hexactinellid sponges, e.g. the marine demosponge *S. domuncula* (two isoforms: silicatein-α and silicatein-β; Shimizu et al. 1998; Cha et al. 1999; Krasko et al. 2000; Schröder et al. 2005b), the freshwater demosponge *Lubomirskia baicalensis*

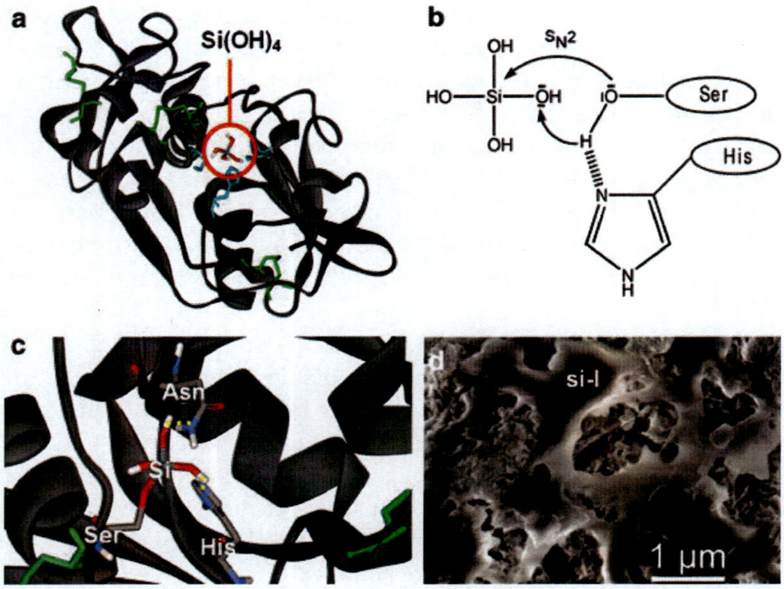

Fig. 10.3 Mode of action of silicatein in biosilica formation. (**a**) Deduced structure of silicatein-α from *S. domuncula* with the orthosilicic acid substrate (red encircled) modeled in the catalytic pocket of the enzyme. The catalytic triad amino acids Ser26, His165, and Asn185 are marked in blue. The cysteine (Cys) residues involved in the formation of the three disulfide bridges of silicatein-α are indicated in green. (**b**) Proposed initial step of the catalytic cycle, consisting of a nucleophilic attack of the negatively charged oxygen atom of the Ser26 hydroxyl group at the positively charged silicon atom of the orthosilicic acid substrate and transfer of a proton (originating from the Ser-His hydrogen bridge) from the imidazole nitrogen of His165 to an OH ligand of the silicic acid molecule. This reaction results in the formation of a covalent bond between the silicic acid molecule and the Ser26 residue of the enzyme. (**c**) Detail of silicatein-α structure showing the interaction of the catalytic triad amino acids with the orthosilicic acid. The close proximity of the free hydroxyl groups of the covalently bound orthosilicic acid molecule to the nitrogen atoms in the side chains of His165 and Asn185 allows for the formation of hydrogen bridges (yellow dots) which position and/or increase the nucleophilicity of the oxygen atoms of the OH ligands of the silicic acid molecule in the subsequent steps (nucleophilic attack to a second orthosilicic acid molecule; not shown). (**d**) Biocatalytically (via silicatein) formed silica layer (si-l) on teeth surface (SEM)

(six silicatein-α isoenzymes; Kaluzhnaya et al. 2005; Wiens et al. 2006), and the hexactinellid sponges, *Crateromorpha meyeri* (Müller et al. 2008c) and *M. chuni* (Müller et al. 2009a). The recombinant proteins can be prepared using both prokaryotic (*Escherichia coli*) and eukaryotic (*Pichia pastori*) systems (bioreactor/laboratory scale). The purified proteins can be applied for the biocatalytic formation of amorphous silica (biosilica) from monomeric precursors at mild (room temperature, near-neutral pH, aqueous buffer systems) conditions (Schröder et al. 2008).

The presumptive 3D structure of silicatein has been obtained by homology modelling (Fig. 10.3a; Müller et al. 2007b; Schröder et al. 2010). Based on this model and docking experiments, a mechanism for silicatein reaction has been proposed, using the physiological substrate, orthosilicic acid (Schröder et al. 2010). This mechanism which starts with a nucleophilic attack at the silicon of the silicic acid substrate under formation of a covalently bound intermediate (Fig. 10.3b, c) explains the increased rate of silica deposition in the presence of silicatein by the enzyme-catalyzed formation of cyclic silicic acid species (Schröder et al. 2010). These reactive intermediates which, at a lower rate, are also generated at nonenzymatic conditions (see Fig. 10.1a) strongly enhance the silica polycondensation reaction. The final product consisting of silica nanospheres may fuse via a sintering-like mechanisms, (biosintering; Müller et al. 2009a, b) forming layered (lamellar; Schröder et al. 2007b) or other biosilica structures (Fig. 10.3d). Silicatein displays not only silica polymerase but also silica esterase activity (Müller et al. 2008b). Besides inorganic substrates, e.g. orthosilicic acid, organic oxysilanes, e.g. bis(p-aminophenoxy)-dimethylsilane (formation of silica; Müller et al. 2008b) and dimethoxy dimethylsilane (formation of silicones; Wolf et al. 2010) can be used as substrates.

Analyses of the composite structure of the biosilica material formed by silicatein revealed that after silica formation, silicatein is present not only on the surface of the silica lamellae which are formed as the result of the appositional growth of the spicules (Müller et al. 2005; Woesz et al. 2006; Schröder et al. 2007b), but also entrapped within biosilica particles (Müller et al. 2010). Recent results demonstrate that the interaction of silicatein with a scaffold protein, silintaphin-1, allows for the generation of spicule-like three-dimensional structures (Müller et al. 2008b; Wiens et al. 2009).

The expression of silicatein is induced by silicate (Krasko et al. 2000; Müller et al. 2006). In addition, silicate induces the expression of myotrophin which triggers collagen synthesis (Schröder et al. 2000a). Moreover, in sponge tissue cultures (primmorphs), silicate stimulates differentiation of stem-like cells into the spicule-forming sclerocytes (Müller et al. 2006).

10.6 Silicon Metabolism

The silicon content of the human body is low and amounts to 1–2 g totally (Jugdaohsingh 2007). The highest concentrations of silicon are found in bone, connective tissue, and blood vessels (Carlisle 1972; Sripanyakorn et al. 2009). In soft tissue, silicon may be complexed to glycosaminoglycans, polyuronides, or silicic acid-binding polysaccharides and proteins (Schwarz 1973). Based on these data, silicon has been proposed to be required for bone formation and the formation of cartilage glycosaminoglycans during development and calcification (Carlisle 1976, 1981). Dietary silicon is taken up through gastrointestinal absorption (Reffitt et al. 1999; Jugdaohsingh et al. 2002). It is excreted from the body via the gastrointestinal tract and via the kidney by glomerular filtration into the urine (Berlyne et al. 1986; Adler and Berlyne 1986; reviewed in: Jugdaohsingh 2007). Tracer experiments with ^{31}Si injected into rats showed an accumulation of silicon in bone, muscle, and skin, but not in the brain (Adler et al. 1986). In blood, silicon is present predominately as $Si(OH)_4$ (Jugdaohsingh 2007), which appears to exist in an unbound form (D'Haese et al. 1995). The plasma silicon levels (7–142 μM; Adler and Berlyne 1986, D'Haese et al. 1995) are lower than the concentration above which polycondensation occurs. To improve its physiological availability which may be limited by poor absorption, organic silicon compounds have been developed that are metabolized in the animal body (Hott et al. 1993; see also Sect. 10.15).

10.7 Silicon and Bone Formation

There is increasing evidence that silicon is beneficial to health of bone and connective tissue (Jugdaohsingh 2007). It has already been established in 1972 that silicon deficiency causes severe defects in connective and skeletal tissue (Carlisle 1972, 1986; Schwarz and Milne 1972). Increased silicon levels are present at active calcification sites during bone formation (Carlisle 1972). It has been proposed that in bone tissue, silicon may have a structural function (Schwarz 1973). Epidemiological studies revealed a positive correlation between the silicon intake and bone mineral density (BMD) at the hip site in men and premenopausal women (Jugdaohsingh et al. 2004). This correlation was not found in postmenopausal women (Jugdaohsingh et al. 2004). Likewise, a positive correlation has also been found during hormone treatment of postmenopausal women (MacDonald et al. 2005). In animal experiments with calcium-deficient ovariectomized rats, supplementation with dietary silicon was found to improve bone mineral density via reduced bone resorption (Kim et al. 2009). These results suggest that enhanced silicon levels are associated with an increase in BMD and bone strength (Jugdaohsingh 2007). Also a potential interaction between the level of silicon and the estrogen status has been discussed (Jugdaohsingh 2007).

10.8 Effect of Biosilica on Cell Proliferation

Using in vitro assays (bone forming SaOS-2 cells), the potential toxicity of orthosilicate has been assessed (Wiens et al. 2010c). SaOS-2 cells are a non-transformed cell line derived from human primary osteosarcoma cells. This cell line is able to differentiate, like osteoblastic cells (Kelly et al. 2010; Hausser and Brenner 2005), and expresses proteins characteristic of osteoblasts, including alkaline phosphatase, type I collagen, and osteocalcin (Hay et al. 2004). Moreover, differentiation of SaOS-2 cells to HA-producing cells can be induced by exposure to cytokines such as granulocyte macrophage colony-stimulating factor (Postiglione et al. 2003). To determine the effect on growth of SaOS-2 cells, the cells were exposed to orthosilicate that had been prepared from prehydrolyzed TEOS (Wiens et al. 2010c). The cell density was assessed by the colorimetric XTT cell proliferation assay. Addition of orthosilicate within the concentration range of 10–1,000 μM resulted in increased absorbance values (incubation period, 72 h), indicating a cell growth-stimulating effect of orthosilicate (Wiens et al. 2010c). A stimulatory (but less significant) effect on cell growth has also been reported for silicon-substituted HA (López-Alvarez et al. 2009; Zou et al. 2009). In parallel to the cell density, cell viability was quantified by the trypan blue exclusion test. This assay did not reveal any silicate-induced cell toxicity during the 72-h incubation period in the presence of a concentration of up to 1,000 μM orthosilicate (Wiens et al. 2010c).

10.9 Effect of Biosilica on HA Formation

In order to determine the effect of biosilica on mineralization activity of SaOS-2 cells, cover slips or bottoms of multi-well plates were coated with a biosilica layer biocatalytically formed by immobilized recombinant silicatein. A His-tagged fusion protein was used (Wiens et al. 2010c). Using the SaOS-2 cell system, we could demonstrate that growth of the cells on the biosilica matrix strongly promotes HA formation (Schröder et al. 2005a; Wiens et al. 2010c). In the experiment shown in Fig. 10.4, SaOS-2 cells were grown on silicatein/biosilica-coated slips placed into 24-well plates for 7 days (Wiens et al. 2010c). The biosilica coating had been prepared by incubation of the immobilized silicatein in the presence of various concentrations of orthosilicate. Subsequently, the slips were stained with Alizarin Red S (staining for HA), or Alcian Blue (staining for cartilaginous proteoglycans and sulfated glycosaminoglycans), or with both dyes. The cells grown on the biosilica-coated cover slips revealed a considerable staining with Alizarin Red S (Fig. 10.4). The intensity of the red color correlated to the orthosilicate concentration that had been applied for biocatalytic silica formation using the immobilized enzyme. In contrast to HA synthesis, formation of cartilaginous proteoglycans and sulfated glycosaminoglycans (staining with Alcian Blue) is downregulated (Wiens et al. 2010c). Cartilage damage has been shown to be inversely correlated with the bone mineral density index (Calvo et al. 2007). Samples that had been double-stained with both dyes showed a dark red/blue color (Wiens et al. 2010c); Fig. 10.4.

Similar results were obtained when cells (SaOS-2) were grown on silicatein/biosilica-modified bone slices or Ca-P-coated cover slips (Wiens et al. 2010b). Immobilization of silicatein to bone HA- or Ca-P-coated cover slips was achieved by using a Glu-tagged silicatein-α (Natalio et al. 2010). The Glu-tag consisting of eight N-terminal glutamic acid residues allows for ionic interaction with Ca^{2+} ions at the HA surface (formation of a coordination complex between the carboxyl groups and the Ca^{2+} ions). Incubation of the immobilized enzyme with 200 μM orthosilicate resulted in the formation of a 50–150 nm thick nano-biosilica layer on the HA surface. Biosilica coating was confirmed by EDX analysis (Wiens et al. 2010b).

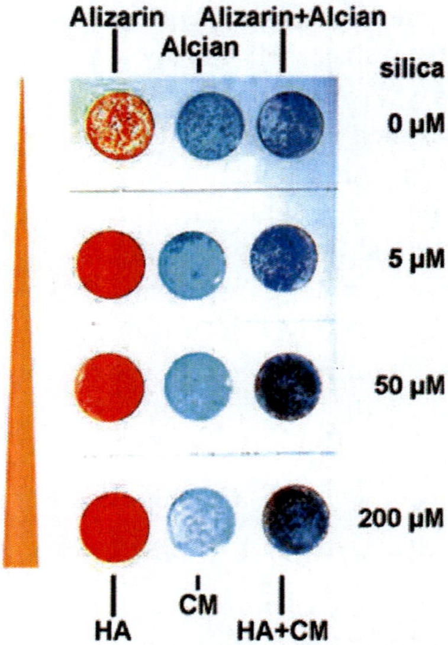

Fig. 10.4 Increased mineralization of osteoblast-like SaOS-2 cells following exposure to biosilica. Cover slips were coated with recombinant silicatein and biosilica was formed by incubation of the slips in the absence (0 µm silica; control) or the presence of various concentrations of orthosilicate (5–200 µM, 4 h). The cells were then incubated on the slips in well plates with medium supplemented with ascorbic acid and ß-glycerophosphate. After an incubation period of 7 days, the cover slips were stained either with Alizarin Red S (for hydroxyapatite [HA]), Alcian Blue (for cartilaginous material [CM]), or a combination of both dyes. From Wiens et al. (2010c)

SaOS-2 cells grown on silicatein/biosilica-modified sections of bone HA in medium containing ascorbic acid and ß-glycerophosphate showed a significant stimulation of mineralization compared to uncoated controls (Wiens et al. 2010b). After 5 days, an enhanced formation of HA nodules, organized in longitudinal arrays or spherical spots, was found on cells grown on the silicatein/biosilica-modified HA surface (Fig. 10.5b), while in control cells grown on uncoated substrates only occasional clusters of small HA rods were seen (Fig. 10.5a) (Wiens et al. 2010b). Nodules of >1 µm in size emerged from cell clusters, consisting of up to eight single cells. HR-SEM analyses of the fine structure of the growing HA nodules revealed that they are initially at least partially covered with cell protrusions (Fig. 10.5c and d) which are later retracted (Fig. 10.5e and f). The nodules consist of irregularly arranged prism-like nanorods (Chen et al. 2005) as resolved at higher magnification (Fig. 10.5g and h) (Wiens et al. 2010b). EDX analyses of the elemental composition of the nodules formed by SaOS-2 cells revealed the presence of Ca and P (HA; spectrum 2 in Fig. 10.5j), while only Si, O, C, Mg, and Na, which originate from both cells and biosilica coating, could be detected in cells in the vicinity of the nodules (spectrum 1 in Fig. 10.5j); the corresponding electron microscopic image showing the two areas analyzed by EDX is shown in Fig. 10.5i (Wiens et al. 2010b).

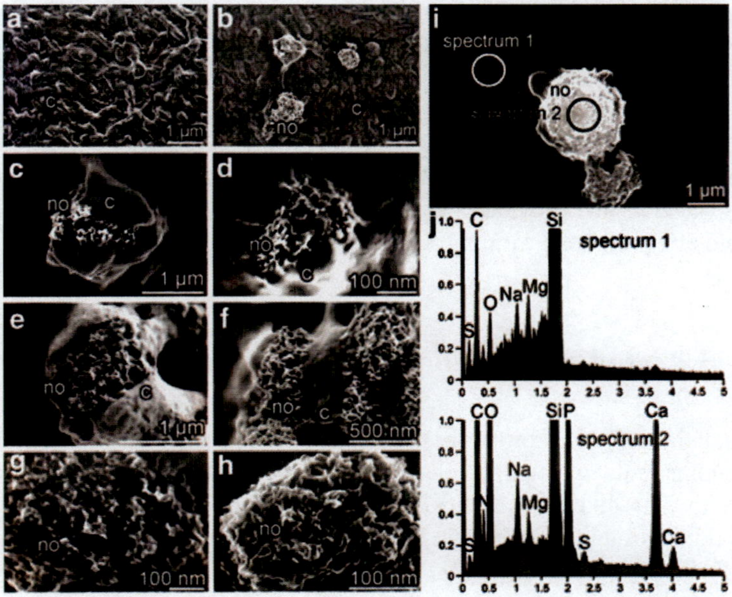

Fig. 10.5 Formation of HA nodules by SaOS-2 cells grown on silicatein/biosilica-coated substrates. SaOS-2 cells were cultivated in medium, supplemented with ascorbic acid and ß-glycerophosphate for 5 days. (**a**) Cells (c) grown on uncoated bone HA (control); only a few HA nodules are visible. (**b**) Cells (c) grown on silicatein/biosilica-coated bone HA; clusters of nodules (no) are formed. (**c**)–(**h**) Different stages of HA nodule formation. (**c**) and (**d**). Initially, the growing nodules are enwrapped by cell protrusions. (**e**) and (**f**) Later, the cell protrusions are retracted. (**g**) and (**h**) Higher magnification of the HA nodules showing an organization of prism-like nanorods. (**i**) and (**j**) SEM and EDX analysis of the nodules (no) formed on cells grown on silicatein/biosilica-coated Ca-P cover slips. (**i**) SEM image showing the areas analyzed by EDX (circles). (**j**) EDX analysis; spectrum 1, cell area; spectrum 2, nodule. Modified from Wiens et al. (2010b)

10.10 Osteoinductive Index

A measure for the osteoinductive activity of a biomaterial is the osteoinductive index which has been defined as the ratio between rate of [^{3}H]dT incorporation into DNA of bone (precursor) cells in vitro and the extent of bone (HA) formation in vivo (Adkisson et al. 2000). To determine the osteoinductive activity in SaOS-2 cells, an in vitro bioassay has been introduced, which is based on the calculation of the ratio of two in vitro parameters, (1) cell proliferative activity, measured as the rate of [^{3}H]dT incorporation, and (2) biomineralization activity, measured by using the Alizarin Red S staining assay of HA (Wiens et al. 2010b). SaOS-2 cells were incubated on silicatein/biosilica-coated or uncoated substrates in medium containing ascorbic acid and ß-glycerophosphate. Then, the cells were exposed to [^{3}H]dT. Finally, the incorporation of [^{3}H]dT into DNA determined in a scintillation counter. Counts were normalized to the total amount of DNA. Applying this assay, cultivation of SaOS-2 cells on silicatein/biosilica-coated substrates was found to elicit a significant increase in [^{3}H]dT incorporation into DNA, indicative of an enhanced cell proliferation compared to uncoated substrates (Wiens et al. 2010b). Calculation of the ratio of cell proliferative activity and biomineralization activity for SaOS-2 cells grown on silicatein/biosilica-coated Ca-P cover slips, uncoated Ca-P cover slips, and glass cover slips revealed the highest ratio for cells grown on biosilica-coated Ca-P cover slips; the lowest ratio was obtained for the glass cover slip cultures (Wiens et al. 2010b). These results provide evidence that enzymatically formed biosilica has a mitogenic effect on bone forming SaOS-2 cells (Wiens et al. 2010b).

10.11 Effect of Biosilica on Gene Expression

The beneficial effects of orthosilicic acid/biosilica on bone metabolism have also been demonstrated in studies of gene expression and in enzymatic studies. In human osteoblast-like cells, orthosilicic acid was found to enhance the expression of several key proteins involved in bone formation, including bone morphogenetic protein-2 (BMP-2; Gao et al. 2001) and collagen type-I (COL1; Reffitt et al. 2003). BMP-2 expression is required for differentiation of osteoblasts (Tanaka et al. 2001; Fromigue et al. 2006; Li et al. 2007). Besides collagen type I, orthosilicic acid has been shown to regulate the expression of alkaline phosphatase and osteocalcin mRNA in human bone-derived osteoblasts (Arumugam et al. 2006). In SaOS-2 cells, biosilica increases the expression of amelogenin and enamelin which are involved in enamel formation (Müller et al. 2007a). Administration of soluble silicate to mice has been shown to affect, in addition to *BMP-2* and *collagen type I*, the expression of *Runx-2* (*runt-related transcription factor 2*) which is involved in the control of skeletal gene expression (Stein et al. 2004), as well as the expression of *OPG* and *RANKL* (Maehira et al. 2008, 2009). Furthermore, silicic acid causes a stimulation of prolyl hydroxylase activity which is involved in collagen synthesis (Carlisle and Alpenfels 1980, 1984; Carlisle and Garvey 1982; Carlisle and Suchil 1983; Carlisle et al. 1981). Hence, soluble silicon causes a significant increase in hydroxyproline content in mice (Maehira et al. 2009). In the tibia of silicon-deficient rats, the amount of hydroxyproline was significantly lower than in silicon-supplemented animals (Seaborn and Nielsen 2002). In addition, silicon deprivation in rats results in a decreased formation of collagen in bone and a lower activity of ornithine transaminase (involved in proline synthesis) in liver (Seaborn and Nielsen 2002). Moreover, zeolite-A, a Si-containing compound, has been reported to stimulate proliferation, differentiation, and protein synthesis in human osteoblast-like cells, and to increase the production of transforming growth-factor-β (TGF-β) in these cells (Brady et al. 1991; Keeting et al. 1992)

The effects of silicon at different stages of collagen formation and mineralization are summarized in Fig. 10.6. In addition to its effect on gene expression and posttranslational modification (formation of hydroxyproline, mediated by prolyl hydroxylase; see above), silicic acid affects the assembly and mineralization of collagen fibrils. Low concentrations of orthosilicic acid promote the collagen

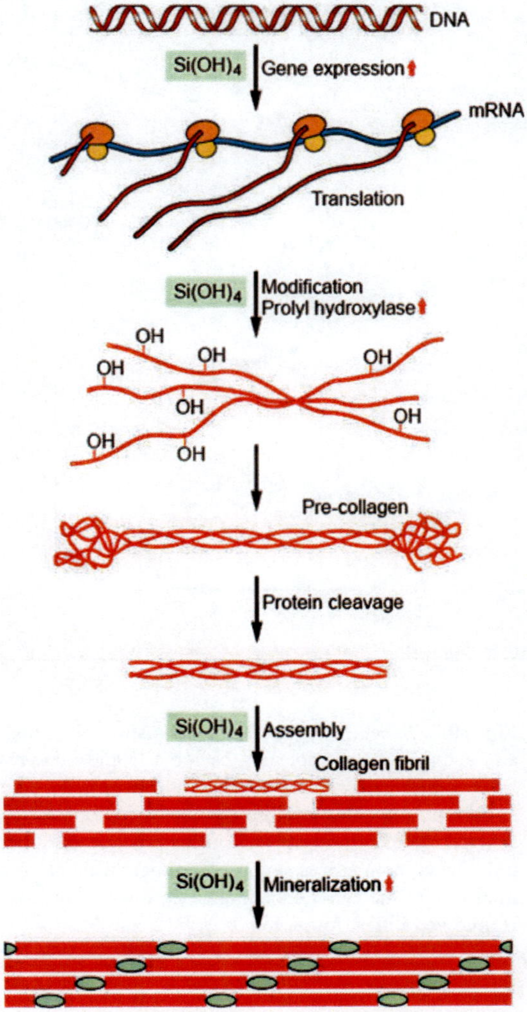

Fig. 10.6 Schematic presentation of the various effects of orthosilicic acid [Si $(OH)_4$] on collagen synthesis, maturation, and mineralization. Orthosilicic acid has been shown to affect both gene expression and posttranslational modification of collagen, as well as the assembly and mineralization of collagen fibrils

self-assembly process, most likely due to a modification of the collagen helix before fibril formation, while more concentrated solutions (presence of polysilicic acid and/or silica nanoparticles) hinder collagen fibrillogenesis, most likely by electrostatic interactions and hydrogen bond formation (Eglin et al. 2006).

10.12 The RANK/RANKL/OPG System

Osteoprotegerin (OPG) and receptor activator for NF-κB ligand (RANKL) are soluble factors released from bone-producing osteoblasts, which are part of one of the most crucial signaling systems that are involved in modulating bone resorption

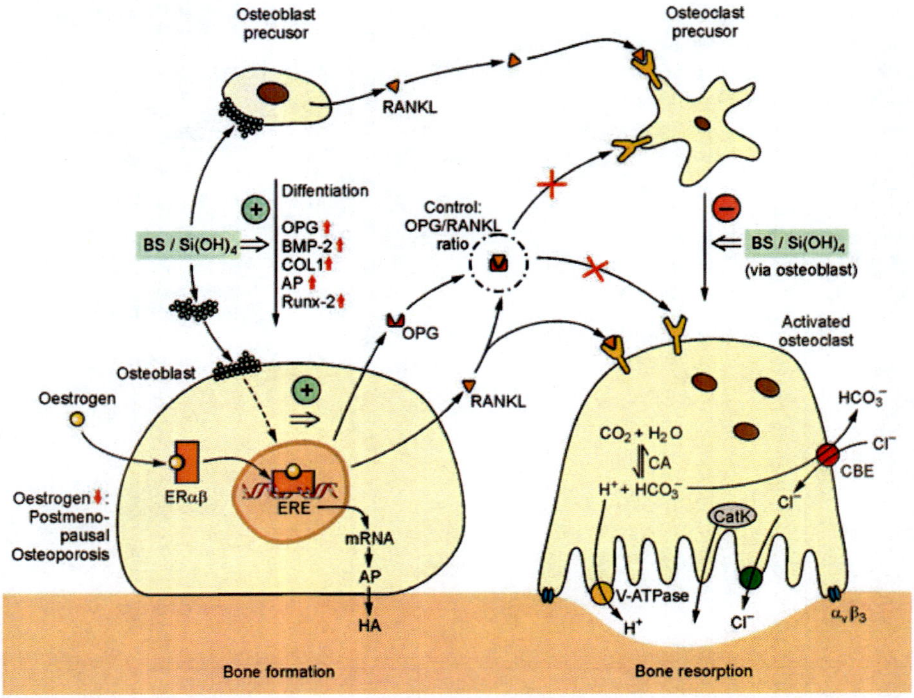

Fig. 10.7 Scheme of the effects of biosilica (BS) and orthosilicic acid [$Si(OH)_4$] on maturation and activity of bone forming osteoblasts and bone-resorbing osteoclasts, and their progenitors. Both monomeric silica [$Si(OH)_4$] and polymeric silica (biosilica, BS) enhance the expression of OPG in osteoblasts, while the expression of RANKL is unaffected. The increased ratio OPG/RANKL results in an impairment of pre-osteoclast maturation and osteoclast activation. On the other hand, both silicas support the progression of precursor osteoblasts to mature osteoblasts by induction of the genes encoding OPG, BMP-2, and alkaline phosphatase (AP). The latter enzyme is involved in HA formation

(reviewed in: Wittrant et al. 2004; Collin-Osdoby 2004; Gallagher 2008; Lane and Yao 2009). These factors play an essential role in the differentiation of pre-osteoclasts to mature osteoclasts (bone-resorbing cells) (Rodan and Martin 1982; Suda et al. 1999; Kanamaru et al. 2004). The following members of the tumor necrosis factor (TNF) and TNF receptor superfamily are involved in the control of bone resorption (Fig. 10.7; reviewed in: Khosla 2001):

1. RANK, the receptor activator of NF-κB, which is expressed on hematopoietic cells and controls maturation of osteoclast (osteoclastogenesis).
2. OPG that is secreted by osteoblasts and blocks osteoclastogenesis and thus inhibits bone resorption.
3. RANKL (ligand of receptor activator for NF-κB, RANK), a protein produced by osteoblasts, which activates osteoclasts through interaction with RANK on osteoclast precursor cells.

The RANKL/RANK interaction plays a key role in the differentiation and maintenance of osteoclast activity. Therefore, this system is fundamentally involved in the pathogenesis of disorders associated with an increased bone resorption (Raisz 2005). The effects of RANKL on osteoclastogensis and activation of osteoclast is regulated by OPG. Hence, the ratio of the levels of OPG and RANKL is crucial for the development of osteoporosis. Transgenic mice overexpressing OPG develop osteopetrosis-like symptoms, while OPG-deficient mice show early onset osteoporosis (Bucay et al. 1998).

10.13 Effect of Biosilica on *OPG* and *RANKL* Expression

The SaOS-2 cell model was used to study the effect of biosilica on the expression of *OPG* and *RANKL* (Wiens et al. 2010c). These cells express both RANK, OPG, and RANKL (Mori et al. 2007; Borsje et al. 2010). RNA was extracted from cells that had been cultivated for 1, 3, or 7 days either on silicatein/biosilica-coated substrates or on substrates that had been coated with silicatein only (control). Quantitative real-time RT-PCR (qRT-PCR) analysis revealed a strong time-depended increase in expression of *OPG* in biosilica-exposed SaOS-2 cells (Fig. 10.8b) while the

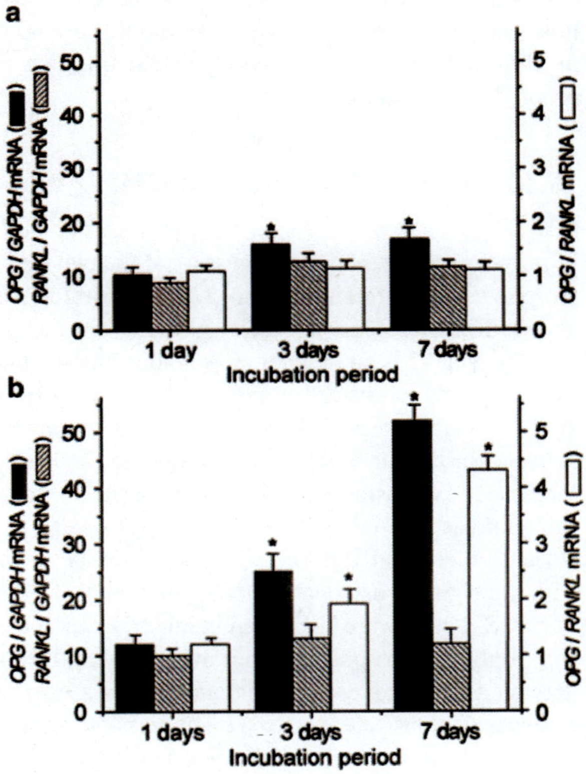

Fig. 10.8 Differential effects of biosilica on expression of *OPG* and *RANKL* in SaOS-2 cells. RNA was extracted from cells that had been cultivated for 1, 3, or 7 days in silicatein-coated (**a**) or silicatein/biosilica-coated (**b**) well plates. The levels of expression of *OPG*, *RANKL*, and *GAPDH* (used as reference for normalization) were determined by qRT-PCR. Closed bars, *OPG*/*GAPDH* expression ratio; hatched bars, *RANKL*/*GAPDH* expression ratio; open bars, *OPG*/*RANKL* (normalized) expression ratio. Error bars show the standard error of the mean (SEM; n = 5); $P < 0.05$. From Wiens et al. (2010c)

steady-state level of expression of *RANKL* remained unchanged (Wiens et al. 2010c). In controls, the expression of *OPG* and *RANKL* changed only unsignificantly during day 1–7 (Fig. 10.8a). Hence, the ratio of *OPG/RANKL* expression markedly increased from day 1 to 7 (Fig. 10.8b). In these experiments, the expression levels of *RANKL* and *OPG* were correlated with the expression of the housekeeping gene *GAPDH* (glyceraldehyde 3-phosphate dehydrogenase) which was used as reference.

The differential expression of *OPG* and *RANKL* induced by biosilica could also be demonstrated on the protein level (Wiens et al. 2010c). ELISA assays revealed a significant increase in the level of OPG released from SaOS-2 cells in the presence of biosilica from 0.8 fg/cell at 12 h to 3.1 fg/cell after 7 days (Wiens et al. 2010c). In contrast, the concentration of RANKL did not markedly change during that time period.

These results indicate that through stimulation of OPG synthesis, biosilica impairs the function of RANKL, i.e. pre-osteoclasts maturation and osteoclasts activation.

A schematic presentation of the RANK/RANKL/OPG system and of the effect of the biosilica on this system is depicted in Fig. 10.7. Biosilica increases the expression of OPG, whereas the expression of RANKL is not affected. By binding to OPG (increased in the presence of biosilica), RANKL is sequestered and unavailable to bind to its receptor, RANK. Hence, biosilica negatively affects, via consequential secondary effects, pre-osteoclast maturation and osteoclast activation. Based on these in vitro data which must be corroborated by future studies in vivo, it is reasonable to assume that biosilica has a potential in therapy and prophylaxis of osteoporosis.

10.14 Effect of Biosilica on *BMP-2* and *TRAP* Expression

In further experiments, the expression of the two further marker genes, *BMP-2* (bone morphogenetic protein 2) and *TRAP* (tartrate-resistant acid phosphatase), was assessed in osteoblast-like cells growing on the biosilica substrate (Wiens et al. 2010b). BMP-2 is an inducer of bone formation. Upregulation of expression of this cytokine is an indicator for osteoinductive activity (Eliseev et al. 2006; Katz et al. 2008). TRAP is a modulator of bone resorption. High levels of *TRAP* are found in osteoclasts, but it is also expressed in SaOS-2 cells (Matsuzaki et al. 1999). Increased expression of *TRAP* has been associated with the development of osteoporosis and other bone diseases (Hollberg et al. 2005; Oddie et al. 2000). On the other hand, TRAP−/− knockout mice have a reduced osteoclast activity with the signs of osteopetrosis (Hayman et al. 1996).

SaOS-2 cells were cultivated in mineralization medium containing ascorbic acid and ß-glycerophosphate, either on silicatein/biosilica-coated Ca-P cover slips, untreated Ca-P cover slips, or glass cover slips (Wiens et al. 2010b). qRT-PCR analyses demonstrated that *BMP2* expression strongly increased at day 3 and day 5 in cultures grown on silicatein/biosilica-coated slips (Fig. 10.9). After 7 days, the

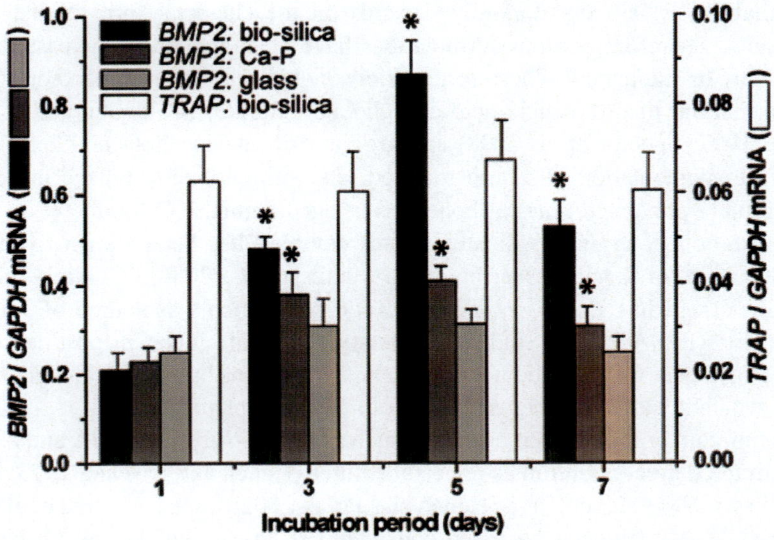

Fig. 10.9 Differential effects of biosilica on expression of *BMP2* and *TRAP* in SaOS-2 cells. RNA was extracted from cells that had been cultivated for 1, 3, 5, or 7 days on different matrices. The levels of expression of *BMP2*, *TRAP*, and *GAPDH* (used as reference for normalization) were determined by qRT-PCR. Filled bars, *BMP2* expression of cells grown on Glu-tagged silicatein/biosilica-coated Ca-P cover slips (*black bars*), on uncoated Ca-P cover slips (*dark grey*), or on glass cover slips (*light grey*). Open (white) bars, *TRAP* expression of cells that had been cultivated on silicatein/biosilica-coated Ca-P cover slips. Error bars show the standard error of the mean (SEM; n = 5); $P < 0.05$. From Wiens et al. (2010b)

expression level decreased. In comparison, *BMP2* expression of SaOS-2 cells grown on uncoated Ca-P cover slips was much lower. In contrast, the expression of *BMP2* did not significantly change in cells that had been grown on plain glass cover slips (Wiens et al. 2010b).

In parallel, *TRAP* transcription was evaluated by qRT-PCR (Wiens et al. 2010b). As shown in Fig. 10.9, no significant change of *TRAP* expression was detected, irrespective of the culture conditions used.

10.15 Silicon Supplementation and Silicon-Containing Implant Materials

Possible strategies for application of silicon (silicic acid and/or biosilica) in treatment and prophylaxis of bone disorders may be based both on silicic acid/biosilica supplementation and silicon/biosilica-containing implant materials.

Bioavailability studies revealed that, based on urinary excretion, orthosilicic acid in water is readily absorbed from the gastrointestinal tract (by about 50%; Reffitt et al. 1999; Jugdaohsingh et al. 2000; Sripanyakorn et al. 2004), while polymeric silicic acid is taken up only marginally (Jugdaohsingh et al. 2000). Hence, the bioavailability of silicon markedly depends on the chemical form of the silicon compound. Therefore, various formulations have been applied to increase the bioavailability of silicic acid. These formulations include choline-stabilized orthosilicic acid (a mixture of orthosilicic acid and choline chloride) (Calomme and Van den Berghe 1997; Spector et al. 2008) and arginine silicate inositol complex (Nielsen 2008). Supplementation of ovariectomized rats with choline-stabilized orthosilicic acid partially prevented femoral bone loss in aged animals (Calomme et al. 2006). Administration of arginine silicate inositol complex has been shown to improve mineralization of bone tissue in quail (Sahin et al. 2006). Choline-stabilized orthosilicic acid has also been added to food supplements as source of silicon in human diet (EFSA 2009). Studies on dietary intake of silicon in postmenopausal women aged over 60 years did not reveal major variations by age (McNaughton et al. 2005), which could affect the usefulness of silicon supplements.

Silicon-containing implant materials such as silicon-substituted HA and bioglass have attracted great attention as bone substitutes (Hench and Paschall 1973; Hench and Wilson 1984; Hench 1998; Hench and Polak 2002; López-Alvarez et al. 2009; Zou et al. 2009). Bioglass has been considered as "bioactive" because it becomes intimately bound to bone tissue (Chen et al. 2006; Bretcanu et al. 2009). This implant material has been reported to show osteoinductive and osteoconductive properties (Hench 2006).

First prototypic bioactive implant materials comprising silicatein and a silica precursor have bee prepared. The results of first preclinical tests are promising (Wiens et al. 2010a). To facilitate the application of silicatein in bone (and dental) replacement materials, a bioengineered recombinant silicatein has been developed, which contains an oligo-glutamate sequence (Glu-tagged silicatein) that allows immobilization of the enzyme onto HA surfaces and the formation of biosilica coatings after addition of substrate (Natalio et al. 2010; Wiens et al. 2010b); Fig. 10.10c.

A schematic presentation of the proposed application of the materials for treatment of vertebral fractures (compression fractures) of osteoporotic patients is shown in Fig. 10.10. The material, comprising either biosilica-formed *ex vivo* or tagged silicatein protein administered together with a suitable biosilica precursor, is injected into the fractured vertebra either alone or together with some other material used in vertebroplasty (Fig. 10.10a). Fig. 10.10b shows the application in kyphoplasty. A balloon (bone tamp) is inserted into the vertebral body by means of two bone biopsy needles. When inflated, the bone tamp re-expands the vertebral body. The restructured vertebral body is then stabilized by the injected polymeric material after hardening. Recruitment of osteoblasts and restoration of the balance between bone formation (osteoblasts) and bone resorption (osteoclasts) by the modulatory effect of the biosilica material on the RANK/RANKL/OPG system and its osteoinductive activity finally result in replacement of the injected material by autologous bone. Figure 10.10c shows a schematic presentation of biosilica formation by the injected Glu-tagged silicatein in the presence of substrate. Silicatein binds via its Glu-tag to the trabeculae. Following its immobilization on the Ca-P substrate (trabecular HA), silicatein facilitates biosilica formation in the

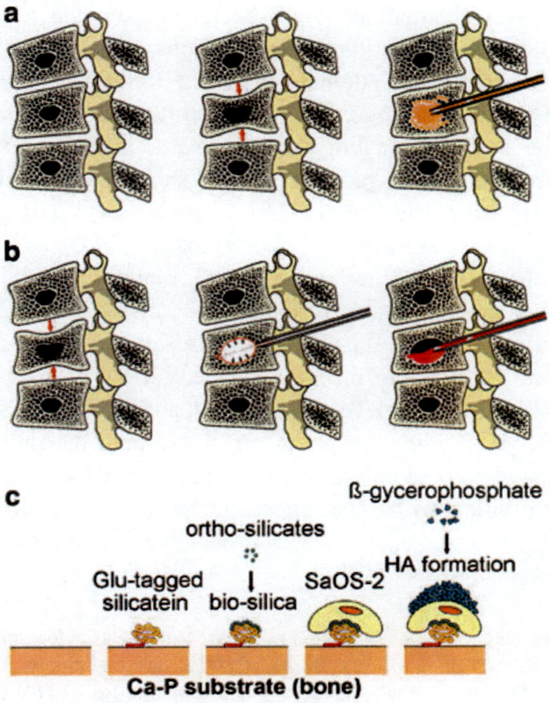

Fig. 10.10 Potential application of biosilica-based material for treatment of vertebral fractures of osteoporotic patients. (**a**) Application in vertebroplasty. (**b**) Application in kyphoplasty. (**c**) Schematic presentation of biosilica formation by Glu-tagged silicatein in the presence of substrate. Silicatein binds via its high-affinity Glu-tag to the trabeculae. Following immobilization, the recombinant protein catalyzes biosilica formation in the presence of a silica source (orthosilicate). The biosilica-modified matrix (trabeculae) then enhances HA formation by mineralizing osteoblasts (or, in vitro, SaOS-2 cells, following addition of ß-glycerophosphate)

presence of a suitable silica precursor (orthosilicate or water glass injected together with the tagged silicatein molecules). The biosilica coating of the trabeculae induces the recruitment of osteoblasts and mineralization (HA formation) of the cells. In addition, it induces OPG expression, resulting in decreased bone-resorbing osteoclast activity due to capturing of RANKL by OPG released from osteoblasts.

10.16 Concluding Remarks

In summary, the data presented in this chapter show that biosilica induces HA formation in vitro. In parallel, biosilica was found to stimulate cell proliferation (Wiens et al. 2010b). Based on these promising in vitro results, first approaches to develop a novel composite which is bioactive and induces new bone formation have been initiated. This material consists of a moldable matrix containing enzymatically active silicatein that had been encapsulated in poly(D,L-lactide)/ polyvinyl pyrrolidone-based microspheres (Wiens et al. 2010a). Application of this material to treat artificial defects in rabbit femurs resulted in a complete restoration of HA and bone regeneration. Therefore, it has been concluded that biosilica might be useful for healing of bone fractures/defects also in vivo (Wiens et al. 2010b). Moreover, the presented results show that biosilica is a selective inducer of *OPG* expression but not of *RANKL* expression (Wiens et al. 2010c). As a consequence, increased amounts of the cytokine OPG are released by the osteoblasts. In the extracellular space, OPG binds to and hence sequesters RANKL which becomes incapable of binding to its receptor RANK. The abolishment of the function of RANKL is expected to result in an inhibition of osteoclast differentiation and of bone resorption. Hence, biosilica has been proposed to possess a considerable potential for treatment and prophylaxis of osteoporosis (Wiens et al. 2010c). Studies are going on to examine the promising biological effects of biosilica in vivo, in animal experiments. In addition, a combined application of polymeric silica with other inorganic polymers, in particular inorganic polyphosphates (Leyhausen et al. 1998; Schröder et al. 2000b; Lorenz and Schröder 2001), will be considered as well.

Acknowledgments This work was supported by grants from the Bundesministerium für Bildung und Forschung (project "Center of Excellence *BIOTEC*marin"), the Deutsche Forschungsgemeinschaft (Schr 277/10-1), the European Commission (no PITN-GA-2008-215507 – BIOMINTEC), the BiomaTiCS consortium of the Mainz University Medical Center, the Johannes Gutenberg University Research Center for Complex Matter (COMATT), and the International S & T Cooperation Program of China (Grant No. 2008DFA00980). W.E.G.M. is a holder of an ERC Advanced Grant (no 268476 BIOSILICA).

References

Adachi JD (1997) Corticosteroid-induced osteoporosis. Am J Med Sci 313:41–49

Adler AJ, Berlyne GM (1986) Silicon metabolism II. Renal handling chronic renal failure patients. Nephron 44:36–39

Adler AJ, Etzion Z, Berlyne GM (1986) Uptake, distribution, and excretion of 31silicon in normal rats. Am J Physiol 251:E670–E673

Adkisson HD, Strauss-Schoenberger J, Gillis M, Wilkins R, Jackson M, Hruska KA (2000) Rapid quantitative bioassay of osteoinduction. J Orthop Res 18:503–511

Albrektsson T, Johansson C (2001) Osteoinduction, osteoconduction and osseointegration. Eur Spine J 10:S96–S101

Ammann P, Shen V, Robin B, Mauras Y, Bonjour JP, Rizzoli R (2004) Strontium ranelate improves bone resistance by increasing bone mass and improving architecture in intact female rats. J Bone Miner Res 19:2012–2020

Arumugam MQ, Ireland DC, Brooks RA, Rushton N, Bonfield W (2006) The effect of orthosilicic acid on collagen type I, alkaline phosphatase and osteocalcin mRNA expression in human bone-derived osteoblasts in vitro. Key Eng Mater 32:309–311

Berlyne GM, Adler AJ, Ferran N, Bennett S, Holt J (1986) Silicon metabolism I: some aspects of renal silicon handling in normal man. Nephron 43:5–9

Bhattacharyya P, Vulcani BE (1980) Sodium-dependent silicate transport in the apochlorotic marine diatom. Proc Natl Acad Sci USA 77:6386–6390

Bhattacharjee H, Mukhopadhyay R, Thiyagarajan S, Rosen BP (2008) Aquaglyceroporins: ancient channels for metalloids. J Biol 7:33

Blick SK, Dhillon S, Keam SJ (2009) Spotlight on teriparatide in osteoporosis. BioDrugs 23:197–199

Borsje MA, Ren Y, de Haan-Visser HW, Kuijer R (2010) Comparison of low-intensity pulsed ultrasound and pulsed electromagnetic field treatments on OPG and RANKL expression in human osteoblast-like cells. Angle Orthod 80:498–503

Brady MC, Dobson PRM, Thavarajah M, Kanis JA (1991) Zeolite A stimulates proliferation and protein synthesis in human osteoblast-like cells and osteosarcoma cell line MG-63. J Bone Miner Res 6:S139

Bretcanu O, Misra S, Roy I, Renghini C, Fiori F, Boccaccini AR, Salih V (2009) In vitro biocompatibility of 45 S5 Bioglass®-derived glass–ceramic scaffolds coated with poly (3-hydroxybutyrate). J Tissue Eng Regen Med 3:139–148

Bucay N, Sarosi I, Dunstan CR, Morony S, Tarpley J, Capparelli C, Scully S, Tan HL, Xu W, Lacey DL, Boyle WJ, Simonet WS (1998) Osteoprotegerin-deficient mice develop early onset osteoporosis and arterial calcification. Genes Dev 12:1260–1268

Calomme MR, Van den Berghe DA (1997) Supplementation of calves with stabilized orthosilicic acid. Effect on the Si, Ca, Mg, and P concentrations in serum and the collagen concentration in skin and cartilage. Biol Trace Elem Res 56:153–165

Calomme M, Geusens P, Demeester N, Behets GJ, D'Haese P, Sindambiwe JB, Van Hoof V, Van den Berghe D (2006) Partial prevention of long-term femoral bone loss in aged ovariectomized rats supplemented with choline-stabilized orthosilicic acid. Calcif Tissue Int 78:227–232

Calvo E, Castañeda S, Largo R, Fernández-Valle ME, Rodríguez-Salvanés F, Herrero-Beaumont G (2007) Osteoporosis increases the severity of cartilage damage in an experimental model of osteoarthritis in rabbits. Osteoarthr Cartil 15:69–77

Canalis E (2010) New treatment modalities in osteoporosis. Endocr Pract 29:1–23

Carlisle EM (1972) Silicon: an essential element for the chick. Science 178:619–621

Carlisle EM (1976) *In vivo* requirement for silicon in articular cartilage and connective tissue formation in the chick. J Nutr 106:478–484

Carlisle EM (1981) Silicon in bone formation, vol 4. In: Simpson TL, Volcani BE (eds) Springer Verlag, New York, pp 69–94

Carlisle EM (1986) Silicon as an essential trace element in animal nutrition. In: Ciba Foundation symposium 121. Wiley, Chichester, UK, pp 123–139

Carlisle EM, Alpenfels WF (1980) A silicon requirement for normal growth for cartilage in culture. Fed Proc 39:787

Carlisle EM, Alpenfels WF (1984) The role of silicon in proline synthesis. Fed Proc 43:680

Carlisle EM, Garvey DL (1982) The effect of silicon on formation of extracellular matrix components by chondrocytes in culture. Fed Proc 41:461

Carlisle EM, Berger JW, Alpenfels WF (1981) A silicon requirement for prolyl hydroxylase activity. Fed Proc 40:886

Carlisle EM, Suchil C (1983) Silicon and ascorbate interaction in cartilage formation in culture. Fed Proc 42:398

Cha JN, Shimizu K, Zhou Y, Christianssen SC, Chmelka BF, Stucky GD, Morse DE (1999) Silicatein filaments and subunits from a marine sponge direct the polymerization of silica and silicones in vitro. Proc Natl Acad Sci USA 96:361–365

Chen H, Clarkson BH, Sun K, Mansfield JF (2005) Self-assembly of synthetic hydroxyapatite nanorods into an enamel prism-like structure. J Colloid Interface Sci 288:97–103

Chen QZ, Thompson ID, Boccaccini AR (2006) 45 S5 Bioglass®-derived glass-ceramic scaffolds for bone tissue engineering. Biomaterials 27:2414–2425

Collin-Osdoby P (2004) Regulation of vascular calcification by osteoclast regulatory factors RANKL and osteoprotegerin. Circ Res 95:1046–1057

Cummings SR, Melton LJ (2002) Epidemiology and outcomes of osteoporotic fractures. Lancet 359:1761–1767

D'Haese PC, Shaheen FA, Huraid SO, Djukanovic L, Polenakovic MH, Spasovski G, Shikole A, Schurgers ML, Daneels RF, Lamberts LV, Van Landeghem GF, De Broe ME (1995) Increased silicon levels in dialysis patients due to high silicon content in the drinking water, inadequate water treatment procedures, and concentrate contamination: a multicentre study. Nephrol Dial Transplant 10:1838–1844

EFSA (2009) Choline-stabilised orthosilicic acid added for nutritional purposes to food supplements scientific opinion of the panel on food additives and nutrient sources added to food. The EFSA J 948:1–23

Eglin D, Shafran KL, Livage J, Coradin T, Perry CC (2006) Comparative study of the influence of several silica precursors on collagen self-assembly and of collagen on 'Si' speciation and condensation. J Mater Chem 16:4220–4230

Eliseev RA, Schwarz EM, Zuscik MJ, O'Keefe Regis J, Drissi H, Rosier RN (2006) Smad7 mediates inhibition of Saos2 osteosarcoma cell differentiation by NFκnB. Exp Cell Res 312:40–50

Faibish D, Ott SM, Boskey AL (2006) Mineral changes in osteoporosis: a review. Clin Orthop Relat Res 443:28–38

Fromigue O, Hay E, Modrowski D, Bouvet S, Jacquel A, Auberge P, Marie PJ (2006) RhoA GTPase inactivation by statins induces osteosarcoma cell apoptosis by inhibiting p42/p44-MAPKs-Bcl-2 signaling independently of BMP-2 and cell differentiation. Cell Death Differ 13:1845–1856

Gallagher JC (2008) Advances in bone biology and new treatments for bone loss. Maturitas 60:65–69

Gao T, Aro HT, Ylänen H, Vuorio E (2001) Silica-based bioactive glasses modulate expression of bone morphogenetic protein-2 mRNA in Saos-2 osteoblasts in vitro. Biomaterials 22:1475–1483

Gao BB, Clermont A, Rook S, Fonda SJ, Srinivasan VJ, Wojtkowski M, Fujimoto JG, Avery RL, Arrigg PG, Bursell SE, Aiello LP, Feener E (2007) Extracellular carbonic anhydrase mediates hemorrhagic retinal and cerebral vascular permeability through prekallikrein activation. Nat Med 13:181–188

Gardner MJ, Demetrakopoulos D, Shindle MK, Griffith MH, Lane JM (2006) Osteoporosis and skeletal fractures. HSS J 2:62–69

Glantz PO (1987) Comment. In: Williams DF (ed) Progress in biomedical engineering, vol 4. definitions in biomaterials. Elsevier, Amsterdam, p 24

Gröger C, Sumper M, Brunner E (2007) Silicon uptake and metabolism of the marine diatom *Thalassiosira pseudonana*: solid-state ^{29}Si NMR and fluorescence microscopic studies. J Struct Biol 161:55–63

Hausser HJ, Brenner RE (2005) Phenotypic instability of SaOS-2 cells in long-term culture. Biochem Biophys Res Commun 333:216–222

Hay E, Lemonnier J, Fromigue O, Guenou H, Pierre JM (2004) Bone morphogenetic protein receptor IB signaling mediates apoptosis independently of differentiation in osteoblastic cells. J Biol Chem 279:1650–1658

Hayman AR, Jones SJ, Boyde A, Foster D, Colledge WH, Carlton MB, Evans MJ, Cox TM (1996) Mice lacking tartrate-resistant acid phosphatase (Acp 5) have disrupted endochondral ossification and mild osteopetrosis. Development 122:3151–3162

Hench LL (1998) Bioceramics. J Am Ceram Soc 81:1705–1728

Hench LL (2006) The story of bioglass. J Mater Sci Mater Med 17:967–978

Hench LL, Paschall HA (1973) Direct chemical bond of bioactive glass-ceramic materials to bone and muscle. J Biomed Mater Res 4:25–42

Hench LL, Wilson J (1984) Surface-active biomaterials. Science 226:630–636

Hench LL, Polak JM (2002) Third-generation biomedical materials. Science 295:1014–1017

Hollberg K, Nordahl J, Hultenby K, Mengarelli-Widholm S, Andersson G, Reinholt FP (2005) Polarization and secretion of cathepsin K precede tartarate-resistant acid phosphatase secretion to the ruffled border area during the activation of matrix-resorbing clasts. J Bone Miner Metab 23:441–449

Hott M, de Pollak C, Modrowski DMPJ (1993) Short-term effects of organic silicon on trabecular bone in mature ovariectomized rats. Calcif Tissue Int 53:174–179

Iler RK (1979) Solubility, polymerisation, colloid and surface properties, and biochemistry. Wiley, New York

Jin H, Heller DA, Sharma R, Strano MS (2009) Size-dependent cellular uptake and expulsion of single-walled carbon nanotubes: single particle tracking and a generic uptake model for nanoparticles. Nano 3:149–158

Jugdaohsingh R (2007) Silicon and bone health. J Nutr Health Aging 11:99–110

Jugdaohsingh R, Reffitt DM, Oldham C, Day JP, Fifield LK, Thompson RPH, Powell JJ (2000) Oligomeric but not monomeric silica prevents aluminum absorption in humans. Am J Clin Nutr 71:944–949

Jugdaohsingh R, Anderson SH, Tucker KL, Elliott H, Kiel DP, Thompson RPH, Powell JJ (2002) Dietary silicon intake and absorption. Am J Clin Nutr 75:887–893

Jugdaohsingh R, Tucker KL, Qiao N, Cupples LA, Kiel DP, Powell JJ (2004) Silicon intake is a major dietary determinant of bone mineral density in men and pre-menopausal women of the Framingham offspring cohort. J Bone Miner Res 19:297–307

Kaandorp JA, Blom JG, Verhoef J, Filatov M, Postma M, Müller WEG (2008) Modelling genetic regulation of growth and form in a branching sponge. Proc Biol Sci 275:2569–2575

Kaluzhnaya OV, Belikov SI, Schröder HC, Wiens M, Giovine M, Krasko A, Müller IM, Müller WEG (2005) Dynamics of skeleton formation in the Lake Baikal sponge *Lubomirskia baicalensis* Part II. Molecular biological studies. Naturwissenschaften 92:134–138

Kanamaru F, Iwai H, Ikeda T, Nakajima A, Ishikawa I, Azuma M (2004) Expression of membrane-bound and soluble receptor activator of NF-kappa B ligand (RANKL) in human T cells. Immunol Lett 94:239–246

Katz JM, Nataraj C, Jaw R, Deigl E, Bursac P (2008) Demineralized bone matrix as an osteoinductive biomaterial and in vitro predictors of its biological potential. J Biomed Mater Res 89B:127–134

Kelly SE, Di Benedetto A, Greco A, Howard CM, Sollars VE, Primerano DA, Valluri JV, Claudio PP (2010) Rapid selection and proliferation of CD133(+) cells from cancer cell lines: chemotherapeutic implications. PLoS ONE 5:e10035. doi:10.1371/journal.pone.0010035

Keeting PE, Oursler MJ, Wiegand KE, Bonde SK, Spelsberg TC, Riggs BL (1992) Zeolite-A increases proliferation, differentiation, and transforming growth-factor-b production in normal adult human osteoblast-like cells-in vitro. J Bone Miner Res 7:1281–1289

Khosla S (2001) Minireview: the OPG/RANKL/RANK system. Endocrinology 142:5050–5055

Khosla S, Amin S, Orwoll E (2008a) Osteoporosis in men. Endocr Rev 29:441–464

Khosla S, Westendorf JJ, Oursler MJ (2008b) Building bone to reverse osteoporosis and repair fractures. J Clin Invest 118:421–428

Kim M-H, Bae Y-J, Choi M-K, Chung Y-S (2009) Silicon supplementation improves the bone mineral density of calcium-deficient ovariectomized rats by reducing bone resorption. Biol Trace Elem Res 128:239–247

Krasko A, Batel R, Schröder HC, Müller IM, Müller WEG (2000) Expression of silicatein and collagen genes in the marine sponge *Suberites domuncula* is controlled by silicate and myotrophin. Eur J Biochem 267:4878–4887

Lane NE, Yao W (2009) Developments in the scientific understanding of osteoporosis. Arthritis Res Ther 11:228

Leibbrandt A, Penninger JM (2008) RANK/RANKL: regulators of immune responses and bone physiology. Ann NY Acad Sci 1143:123–150

Le Pennec G, Perovic S, Ammar SMA, Grebenjuk VA, Steffen R, Brümmer F, Müller WEG (2003) Cultivation of primmorphs from the marine sponge *Suberites domuncula*: morphogenetic potential of silicon and iron. A review J Biotechnol 100:93–108

Leyhausen G, Lorenz B, Zhu H, Geurtsen W, Bohnensack R, Müller WEG, Schröder HC (1998) Inorganic polyphosphate in human osteoblast-like cells. J Bone Miner Res 13:803–812

Li Q, Kannan A, Wang W, Demayo FJ, Taylor RN, Bagchi MK, Bagchi IC (2007) Bone morphogenetic protein 2 functions via a conserved signaling pathway involving Wnt4 to regulate uterine decidualization in the mouse and the human. J Biol Chem 282:31725–31732

López-Alvarez M, Solla EL, González P, Serra J, León B, Marques AP, Reis RL (2009) Silicon-hydroxyapatite bioactive coatings (Si-HA) from diatomaceous earth and silica. Study of adhesion and proliferation of osteoblast-like cells. J Mater Sci Mater Med 20:1131–1136

Lorenz B, Schröder HC (2001) Mammalian intestinal alkaline phosphatase acts as highly active exopolyphosphatase. Biochim Biophys Acta 1547:254–261

MacDonald HM, Hardcastle AE, Jugdaohsingh R, Reid DM, Powell JJ (2005) Dietary silicon intake is associated with bone mineral density in premenopasual women and postmenopausal women taking HRT. J Bone Miner Res 20:S393

Maehira F, Iinuma Y, Eguchi Y, Miyagi I, Teruya S (2008) Effects of soluble silicon compound and deep-sea water on biochemical and mechanical properties of bone and the related gene expression in mice. J Bone Miner Metab 26:446–455

Maehira F, Miyagi I, Eguchi Y (2009) Effects of calcium sources and soluble silicate on bone metabolism and the related gene expression in mice. Nutrition 25:581–589

Mann S (2001) Biomineralization: principles and concepts in bioinorganic materials chemistry. Oxford University Press, Oxford

McNaughton SA, Bolton-Smith C, Mishra GD, Jugdaohsingh R, Powell JJ (2005) Dietary silicon intake in post-menopausal women. Br J Nutr 94:813–817

Matsuzaki K, Katayama K, Takahashi Y, Nakamura I, Udagawa N, Tsurukai T, Nishinakamura R, Toyama Y, Yabe Y, Hori M, Takahashi N, Suda T (1999) Human osteoclast-like cells are formed from peripheral blood mononuclear cells in a coculture with SaOS-2 cells transfected with the parathyroid hormone (PTH)/PTH-related protein receptor gene. Endocrinology 140:925–932

Morgan EF, Barnes GL, Einhorn TA (2008) The bone organ system: form and function. In: Marcus R, Feldman D, Nelson DA, Rosen CJ (eds) Osteoporosis., 3rd edn. Elsevier, San Diego, pp 3–25

Mori K, Berreur M, Blanchard F, Chevalier C, Guisle-Marsollier I, Masson M, Rédini F, Heymann D (2007) Receptor activator of nuclear factor-κB ligand (RANKL) directly modulates the gene expression profile of RANK-positive Saos-2 human osteosarcoma cells. Oncol Rep 18:1365–1371

Morse DE (1999) Silicon biotechnology: harnessing biological silica production to construct new materials. Trends Biotechnol 17:230–232

Müller WEG (2003) Silicon biomineralization: biology-biochemistry-molecular biology-biotechnology. Springer, Berlin

Müller WEG, Rothenberger M, Boreiko A, Tremel W, Reiber A, Schröder HC (2005) Formation of siliceous spicules in the marine demosponge *Suberites domuncula*. Cell Tissue Res 321:285–297

Müller WE, Belikov SI, Tremel W, Perry CC, Gieskes WW, Boreiko A, Schröder HC (2006) Siliceous spicules in marine demosponges (example *Suberites domuncula*). Micron 37:107–120

Müller WEG, Boreiko A, Wang XH, Krasko A, Geurtsen W, Custódio MR, Winkler T, Lukić-Bilela L, Link T, Schröder HC (2007a) Morphogenetic activity of silica and bio-silica on the expression of genes, controlling biomineralization using SaOS-2 cells. Calcif Tissue Int 81:382–393

Müller WEG, Wang XM, Belikov SI, Tremel W, Schloßmacher U, Natoli A, Brandt D, Boreiko A, Tahir MN, Müller IM, Schröder HC (2007b) Formation of siliceous spicules in demosponges: example *Suberites domuncula*. In: Bäuerlein E (ed) Handbook of biomineralization; Vol. 1: biological aspects and structure formation. Wiley-VCH, Weinheim, pp 59–82

Müller WEG, Jochum K, Stoll B, Wang XH (2008a) Formation of giant spicule from quartz glass by the deep sea sponge *Monorhaphis*. Chem Mater 20:4703–4711

Müller WEG, Schloßmacher U, Wang XH, Boreiko A, Brandt D, Wolf SE, Tremel W, Schröder HC (2008b) Poly(silicate)-metabolizing silicatein in siliceous spicules and silicasomes of demosponges comprises dual enzymatic activities (silica-polymerase and silica-esterase). FEBS J 275:362–370

Müller WEG, Wang X, Kropf K, Boreiko A, Schloßmacher U, Brandt D, Schröder HC, Wiens M (2008c) Silicatein expression in the hexactinellid *Crateromorpha meyeri*: the lead marker gene restricted to siliceous sponges. Cell Tissue Res 333:339–351

Müller WEG, Wang X, Burghard Z, Bill J, Krasko A, Boreiko A, Schloßmacher U, Schröder HC, Wiens M (2009a) Bio-sintering processes in hexactinellid sponges: fusion of biosilica in giant basal spicules from *Monorhaphis chuni*. J Struct Biol 168:548–561

Müller WE, Wang X, Cui FZ, Jochum KP, Tremel W, Bill J, Schröder HC, Natalio F, Schlossmacher U, Wiens M (2009b) Sponge spicules as blueprints for the Biofabrication of inorganic-organic composites and biomaterials. Appl Microbiol Biotechnol 83:397–413

Müller WEG, Wang X, Sinha B, Wiens M, Schröder HC, Jochum KP (2010) NanoSIMS: Insights into the organization of the proteinaceous scaffold within hexactinellid sponge spicules. Chembiochem 11:077–1082

Natalio F, Link T, Müller WEG, Schröder HC, Cui FZ, Wang XH, Wiens M (2010) Bioengineering of the silica-polymerizing enzyme silicatein-α for a targeted application to hydroxyapatite. Acta Biomater 6:3720–3728

Nielsen FH (2008) A novel silicon complex is as effective as sodium metasilicate in enhancing the collagen-induced inflammatory response of silicon-deprived rats. J Trace Elem Med Biol 22:39–49

Oddie GW, Schenk G, Angel NZ, Walsh N, Guddat LW, de Jersey J, Cassady AI, Hamilton SE, Hume DA (2000) Structure, function, and regulation of tartrate-resistant acid phosphatase. Bone 27:575–584

Patlak M (2001) Bone builders: the discoveries behind preventing and treating osteoporosis. FASEB J 15:1677E–E

Perry CC (2003) Silicification: the processes by which organisms capture and mineralize silica. Rev Mineral Geochem 54:291–327

Perry CC, Keeling-Tucker T (2000) Biosilification: the role of the organic matrix in structure control. J Biol Inorg Chem 5:537–550

Perry CC, Belton D, Shafran K (2003) Studies of biosilicas: structural aspects, chemical principles, model studies and the future. In: Müller WEG (ed) Silicon biomineralization: Biology – Biochemistry – Molecular biology – Biotechnology. Prog Mol Subcell Biol 33:269–299

Postiglione L, DiDomenico G, Montagnani S, Di Spigna G, Salzano S, Castaldo C, Ramaglia L, Sbordone L, Rossi G (2003) Granulocyte macrophage colony-stimulating factor (GM-CSF) induces the osteoblastic differentiation of the human osteosarcoma cell line SaOS-2. Calcif Tissue Int 72:85–97

Raisz LG (2005) Pathogenesis of osteoporosis: concepts, conflicts, and prospects. J Clin Invest 115:3318–3325

Reffitt DM, Jugdoahsingh R, Thompson RPH, Powell JJ (1999) Silicic acid: its gastrointestinal uptake and urinary excretion in man and effects on aluminium excretion. J Inorg Biochem 76:141–147

Reffitt DM, Ogston N, Jugdaohsingh R, Cheung HF, Evans BA, Thompson RP, Powell JJ, Hampson GN (2003) Orthosilicic acid stimulates collagen type I synthesis and osteoblastic differentiation in human osteoblast-like cells in vitro. Bone 32:127–135

Reginster JY, Burlet N (2006) Osteoporosis: a still increasing prevalence. Bone 38(2 Suppl 1): S4–S9

Reid IR (2008) Anti-resorptive therapies for osteoporosis. Stem Cell Dev Biol 19:473–478

Rodan GA, Martin TJ (1982) Role of osteoblasts in hormonal control of bone resorption - hypothesis [letter]. Calcif Tissue Int 34:311

Russell RG, Croucher PI, Rogers MJ (1999) Bisphosphonates: pharmacology, mechanisms of action and clinical uses. Osteoporos Int 9(Suppl 2):S66–S80

Sahin K, Onderci M, Sahin N, Balci TA, Gursu MF, Juturu V, Kucuk O (2006) Dietary arginine silicate inositol complex improves bone mineralization in quail. Poult Sci 85:486–492

Sambrook P, Cooper C (2006) Osteoporosis. Lancet 367:2010–2018

Sasaki S (2008) Introduction for special issue for aquaporin expanding the world of aquaporins: new members and new functions. Pflügers Arch Eur J Physiol 456:647–649

Schröder HC, Krasko A, Batel R, Skorokhod A, Pahler S, Kruse M, Müller IM, Müller WEG (2000a) Stimulation of protein (collagen) synthesis in sponge cells by a cardiac myotrophin-related molecule from *Suberites domuncula*. FASEB J 14:2022–2031

Schröder HC, Kurz L, Müller WEG, Lorenz B (2000b) Polyphosphate in bone. Biochemistry (Moscow) 65:296–303

Schröder HC, Krasko A, Le Pennec G, Adell T, Hassanein H, Müller IM, Müller WEG (2003) Silicase, an enzyme which degrades biogenous amorphous silica: Contribution to the metabolism of silica deposition in the demosponge *Suberites domuncula*. In: Müller WEG (ed) Silicon biomineralization: biology-biochemistry-molecular biology-biotechnology. Springer, Berlin, Prog Mol Subcell Biol 33:249–268

Schröder HC, Perović-Ottstadt S, Rothenberger M, Wiens M, Schwertner H, Batel R, Korzhev M, Müller IM, Müller WEG (2004) Silica transport in the demosponge *Suberites domuncula*: Fluorescence emission analysis using the PDMPO probe and cloning of a potential transporter. Biochem J 381:665–673

Schröder HC, Borejko A, Krasko A, Reiber A, Schwertner H, Müller WEG (2005a) Mineralization of SaOS-2 cells on enzymatically (Silicatein) modified bioactive osteoblast-stimulating surfaces. J Biomed Mat Res B Appl Biomater 75B:387–392

Schröder HC, Perović-Ottstadt S, Grebenjuk VA, Engel S, Müller IM, Müller WEG (2005b) Biosilica formation in spicules of the sponge *Suberites domuncula*: Synchronous expression of a gene cluster. Genomics 85:666–678

Schröder HC, Boreiko A, Korzhev M, Tahir MN, Tremel W, Eckert C, Ushijima H, Müller IM, Müller WEG (2006) Co-Expression and functional interaction of silicatein with galectin: matrix-guided formation of siliceous spicules in the marine demosponge *Suberites domuncula*. J Biol Chem 281:12001–12009

Schröder HC, Brandt D, Schlossmacher U, Wang X, Tahir MN, Tremel W, Belikov SI, Müller WEG (2007a) Enzymatic production of biosilica glass using enzymes from sponges: basic aspects and application in nanobiotechnology (material sciences and medicine). Naturwissenschaften 94:339–359

Schröder HC, Natalio F, Shukoor I, Tremel W, Schloßmacher U, Wang XH, Müller WEG (2007b) Apposition of silica lamellae during growth of spicules in the demosponge *Suberites domuncula*: biological/biochemical studies and chemical/biomimetical confirmation. J Struct Biol 159:325–334

Schröder HC, Wang XH, Tremel W, Ushijima H, Müller WEG (2008) Biofabrication of biosilica-glass by living organisms. Nat Prod Rep 25:455–474

Schröder HC, Wiens M, Schloßmacher U, Brandt D, Müller WEG (2010) Silicatein-mediated polycondensation of orthosilicic acid: Modeling of catalytic mechanism involving ring formation. Silicon, in press (DOI: 10.1007/s12633-010-9057-4)

Schwarz K (1973) A bound form of silicon in glycosaminoglycans and polyuronides. Proc Natl Acad Sci USA 70:1608–1612

Schwarz K, Milne DB (1972) Growth promoting effects of silicon in rats. Nature 239:333–334

Seaborn CD, Nielsen FH (2002) Silicon deprivation decreases collagen formation in wounds and bone, and ornithine transaminase enzyme activity in liver. Biol Trace Elem Res 89:251–261

Shimizu K, Cha J, Stucky GD, Morse DE (1998) Silicatein alpha: cathepsin L-like protein in sponge biosilica. Proc Natl Acad Sci USA 95:6234–6238

Simonet WS, Lacey DL, Dunstan CR, Kelley M, Chang MS, Lüthy R, Nguyen HQ, Wooden S, Bennett L, Boone T, Shimamoto G, DeRose M, Elliott R, Colombero A, Tan HL, Trail G, Sullivan J, Davy E, Bucay N, Renshaw-Gegg L, Hughes TM, Hill D, Pattison W, Campbell P, Sander S, Van G, Tarpley J, Derby P, Lee R, Boyle WJ (1997) Osteoprotegerin: a novel secreted protein involved in the regulation of bone density. Cell 89:309–319

Singer A, Grauer A (2010) Denosumab for the management of postmenopausal osteoporosis. Postgrad Med 122:176–187

Spector TD, Calomme MR, Anderson SH, Clement G, Bevan L, Demeester N, Swaminathan R, Jugdaohsingh R, Berghe DA, Powell JJ (2008) Choline-stabilized orthosilicic acid supplementation as an adjunct to calcium/vitamin D3 stimulates markers of bone formation in osteopenic females: a randomized, placebo-controlled trial. BMC Musculoskelet Disord 9:85

Sripanyakorn S, Jugdaohsingh R, Elliott H, Walker C, Mehta P, Shoukru S, Thompson RPH, Powell JJ (2004) The silicon content of beer and its bioavailability in healthy volunteers. Brit J Nutr 91:403–409

Sripanyakorn S, Jugdaohsingh R, Dissayabutr W, Anderson SH, Thompson RP, Powell JJ (2009) The comparative absorption of silicon from different foods and food supplements. Br J Nutr 102:825–834

Stein GS, Lian JB, van Wijnen AJ, Stein JL, Montecino M, Javed A, Zaidi SK, Young DW, Choi J-Y, Pockwinse SM (2004) Runx2 control of organization, assembly and activity of the regulatory machinery for skeletal gene expression. Oncogene 23:4315–4329

Suda T, Takahashi N, Udagawa N, Jimi E, Gillespie MT, Martin TJ (1999) Modulation of osteoclast differentiation and function by the new members of the tumor necrosis factor receptor and ligand families. Endocr Rev 20:345–357

Tahir MN, Théato P, Müller WEG, Schröder HC, Janshoff A, Zhang J, Huth J, Tremel W (2004) Monitoring the formation of biosilica catalysed by histidin-tagged silicatein. Chem Commun 24:2848–2849

Tanaka H, Nagai E, Murata H, Tsubone T, Shirakura Y, Sugiyama T, Taguchi T, Kawai S (2001) Involvement of bone morphogenic protein-2 (BMP-2) in the pathological ossification process of the spinal ligament. Rheumatology 40:1163–1168

Tang BM, Eslick GD, Nowson C, Smith C, Bensoussan A (2007) Use of calcium or calcium in combination with vitamin D supplementation to prevent fractures and bone loss in people aged 50 years and older: a meta-analysis. Lancet 370:657–666

Taranta A, Brama M, Teti A, De luca V, Scandurra R, Spera G, Agnusdei D, Termine JD, Migliaccio S (2002) The selective estrogen receptor modulator raloxifene regulates osteoclast and osteoblast activity in vitro. Bone 30:368–376

Teitelbaum SL (2000) Bone resorption by osteoclasts. Science 289:1504–1508

Thamatrakoln K, Alverson AJ, Hildebrand M (2006) Comparative sequence analysis of diatom silicon transporters: towards a mechanistic model of silicon transport. J Phycol 42:822–834

Wada T, Nakashima T, Hiroshi N, Penninger JM (2006) RANKL-RANK signaling in osteoclastogenesis and bone disease. Trends Mol Med 12:17–25

Wang JC, Hemavathy K, Charles W, Zhang H, Dua PK, Novetsky AD, Chang T, Wong C, Jabara M (2004) Osteosclerosis in idiopathic myelofibrosis is related to the overproduction of osteoprotegerin (OPG). Exp Hematol 32:905–910

Wang XH, Hu S, Gan L, Wiens M, Müller WEG (2010) Sponges (Porifera) as living metazoan witnesses from the neoproterozoic: biomineralization and the concept of their evolutionary success. Terra Nova 22:1–11

Weiner S, Traub W (1992) Bone structure: from angstroms to microns. FASEB J 6:879–885

Wetzel P, Hasse A, Papadopoulos S, Voipio J, Kaila K, Gros G (2001) Extracellular carbonic anhydrase activity facilitates lactic acid transport in rat skeletal muscle fibres. J Physiol 531:743–756

Wiens M, Belikov SI, Kaluzhnaya OV, Krasko A, Schröder HC, Perovic-Ottstadt S, Müller WEG (2006) Molecular control of serial module formation along the apical-basal axis in the sponge

Lubomirskia baicalensis: silicateins, mannose-binding lectin and Mago Nashi. Dev Genes Evol 216:229–242
Wiens M, Bausen M, Natalio F, Link T, Schlossmacher U, Müller WEG (2009) The role of the silicatein-α interactor silintaphin-1 in biomimetic biomineralization. Biomaterials 30:1648–1656
Wiens M, Wang X, Natalio F, Schröder HC, Schloßmacher U, Wang S, Korzhev M, Geurtsen W, Müller WEG (2010a) Bioinspired fabrication of bio-silica-based bone substitution materials. Adv Eng Mater 12:B438–B450
Wiens M, Wang X, Schloßmacher U, Lieberwirth I, Glasser G, Ushijima H, Schröder HC, Müller WEG (2010b) Osteogenic potential of biosilica on human osteoblast-like (SaOS-2) cells. Calcif Tissue Int 87:513–524
Wiens M, Wang X, Schröder HC, Kolb U, Schloßmacher U, Ushijima H, Müller WEG (2010c) The role of biosilica in the osteoprotegerin/RANKL ratio in human osteoblast-like cells. Biomaterials 31:7716–7725
Woesz A, Weaver JC, Kazanci M, Dauphin Y, Aizenberg J, Morse DE, Fratzl P (2006) Hierarchical assembly of the siliceous skeletal lattice of the hexactinellid sponge *Euplectella aspergillum*. J Mater Res 21:2068–2078
Wittrant Y, Theoleyre S, Chipoy C, Padrines M, Blanchard F, Heymann D, Redini F (2004) RANKL/RANK/OPG: new therapeutic targets in bone tumours and associated osteolysis. Biochim Biophys Acta 1704:49–57
Wolf SE, Schlossmacher U, Pietuch A, Mathiasch B, Schröder HC, Müller WEG, Tremel W (2010) Formation of silicones mediated by the sponge enzyme silicatein-α. Dalton Trans 39:9245–9249
Zou S, Ireland D, Brooks RA, Rushton N, Best S (2009) The effects of silicate ions on human osteoblast adhesion, proliferation, and differentiation. J Biomed Mater Res B Appl Biomater 90:123–130

Part IV
Nacre

Chapter 11
Structure and Function of Matrix Proteins and Peptides in the Biomineral Formation in Crustaceans

Hiromichi Nagasawa

Contents

Abstract Crustaceans have hard cuticle with layered structure, which is composed mainly of chitin, proteins, and calcium carbonate. Crustaceans grow by shedding the old cuticle and replacing it with a new one. Decalcification in the cuticle during the pre-molt stage and concomitant calcification in the stomach to form gastroliths observed in some crustacean species are triggered by the molting hormone. Various proteins and peptides have been identified from calcified cuticle and gastroliths, and their functions have been examined in terms of calcification and interaction with chitin. Acidic nature of matrix proteins is important for recruitment of calcium ions and interaction with calcium carbonate. Examination of the relationship between amino acid sequence containing acidic amino acid residues and calcification inhibitory activity revealed that the potency did not depend on the sequence but on the number of acidic amino acid residues. Calcium carbonate in the calcified

H. Nagasawa (✉)
Department of Applied Biological Chemistry, Graduate School of Agricultural and Life Sciences, The University of Tokyo, 1-1-1 Yayoi, Bunkyo, Tokyo 113-8657, Japan
e-mail: anagahi@mail.ecc.u-tokyo.ac.jp

W.E.G. Müller (ed.), *Molecular Biomineralization*, Progress in Molecular and Subcellular Biology 52, DOI 10.1007/978-3-642-21230-7_11,

tissues of crustaceans is amorphous in many cases. Crustaceans take a strategy to induce and maintain amorphous calcium carbonate by using low-molecular-weight phosphorus compounds.

11.1 Introduction

Molting is essential for crustaceans to grow. Crustaceans become bigger by shedding the old cuticle, which entirely surrounds the soft body. The cuticle is a hard tissue and is therefore mechanically less flexible, which interrupts crustaceans to grow. The hardness of the cuticle is maintained by a high content of mineral, calcium carbonate, and is useful for protection from enemies and maintenance of body structure. This property is one of the characteristics of crustaceans among arthropods, while insects, spiders, and other arthropods have almost no minerals in the cuticle. The cuticle consists mainly of chitin, proteins and calcium carbonate, whose proportions depend on the species, the site of cuticle, and molting stage. The cuticle has a layered structure, consisting of epicuticle, exocuticle, endocuticle, and membrane layer from outside to inside (Roer 1984, Simkiss and Wilbur 1989) (Fig. 11.1). Among the layers, exocuticle and endocuticle are calcified, and these layers are generally thicker than the other layers. Before molting, new epicuticle and exocuticle are formed under the old cuticle, but the exocuticle has not yet been calcified until molting. After molting, endocuticle starts to be formed and calcified from the outer part of exocuticle. Some crustaceans feed the old shedded cuticle of their own, probably to save nutrients and minerals.

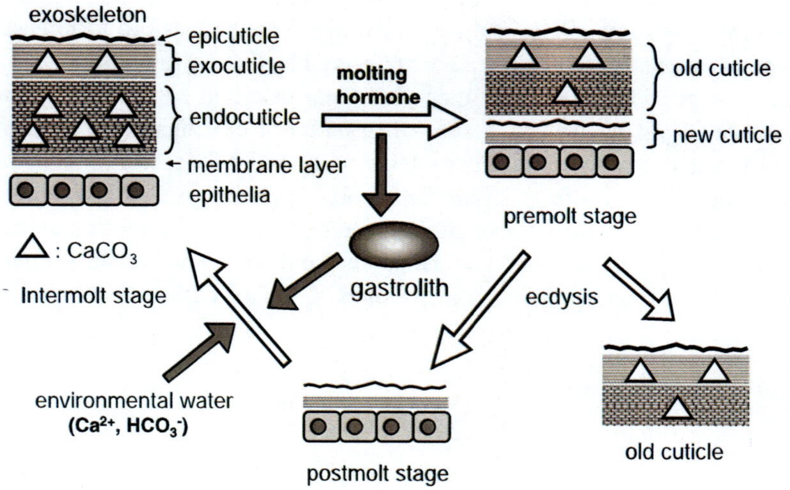

Fig. 11.1 Calcification and decalcification in crustaceans.Schematic representation of cuticle structure during molting cycle and movement of calcium carbonate under endocrine control

11.2 Molting and Calcification

Before molting, crustaceans have to make new larger cuticle for the next growing stage. Molting process is under endocrine control (Fig. 11.2) and is directly triggered by the molting hormone, ecdysteroid, as in other arthropods (Hampshire and Horn 1966). The molting hormone is produced by a pair of Y-organs located at the ventral part of the thorax. The Y-organs produce 3-dehydroecdysone and/or ecdysone, which are finally converted to 20-hydroxyecdysone, the most active form, in other tissues or in hemolymph. The Y-organs are negatively controlled by molt-inhibiting hormone (MIH) produced by the X-organ/sinus gland complex located in the eyestalk. The presence of MIH was first estimated over 100 years ago by a simple experiment, where bilateral eyestalk ablation caused precocious molting (Zeleny 1905). Therefore, it is thought that MIH may be secreted by the X-organ/sinus gland complex and inhibit the Y-organs to produce molting hormone during the inter-molt stage, but when the secretion of MIH is lowered or stopped by an unknown mechanism, the Y-organs are activated to begin production of molting hormone and crustaceans enter the pre-molt stage. Later, MIH was characterized as a peptide consisting of some 70 amino acid residues (Keller 1992; Nagasawa et al. 1996; Ohira et al. 1997; Yang et al. 1996) (Fig. 11.3).

As mentioned above, molting is an essential event for crustaceans to grow. At molting, they shed calcium carbonate-containing cuticle. However, in order to make the cuticle soft for easy shedding, they perform partial resorption of calcium carbonate, and the dissolved calcium carbonate is transferred to other tissues such as the hepatopancreas and stomach, where calcium carbonate is deposited. The proportion of resorbed calcium carbonate varies from 25% to 75% depending on the species (Simkiss and Wilbur 1989). This process is triggered by the molting hormone. After molting, the calcium carbonate once deposited temporarily before molting in hepatopancreas, stomach, or other tissues is resorbed again and transferred to a newly formed cuticle to harden it (Luquet and Marin 2004). Also, an individual having finished molting takes up calcium and bicarbonate ions from the environmental water to compensate the lack of calcium carbonate by shedding.

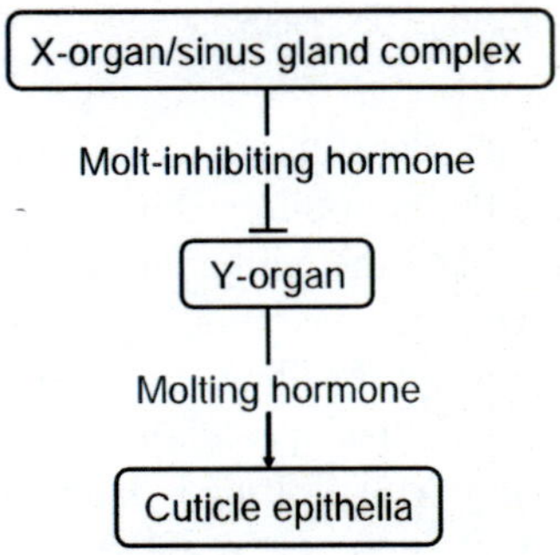

Fig. 11.2 Endocrine control of molting in crustaceans.Molting hormone is produced by Y-organs when inhibition by molt-inhibiting hormone is released

```
              1        10        20        30       40
Pej-SGP-IV    SFIDNTCRGVMGNRDIYKKVVRVCEDCTNIFRLPGLDGMC
Cam-MIH       RV-NDE-PNLI----L----EWI----S----KT-MASL-
Prc-MIH       -YVFEE-P------AVHG--T------Y-V--DTDVLAG-

              41       50        60        70
              RNRCFYNEWFLICLKAANREDEIEKFRVWISILNAGQ(OH)
              -RN--F--D-VW-VH-TE-SE-LRDLEE-VG--G--RD(OH)
              -KG--SS-M-KL--L-ME-VE-FPD-KR--G----(NH2)
```

Fig. 11.3 Amino acid sequences of molt-inhibiting hormones.Pej-SGP-IV is MIH of the kuruma prawn, *Penaeus* (*Marsupenaeus*) *japonicus*. Cam-MIH and Prc-MIH are MIHs from the crab, *Carcinus maenus*, and the crayfish, *Procambarus clarkii*, respectively. Hyphen means the same residue as that of Pej-SGP-IV. The C-terminus of only Prc-MIH is amidated. Six cysteine residues form three intramolecular disulfide bridges between Cys7 and Cys41, Cys24 and Cys40, and Cys27 and Cys53

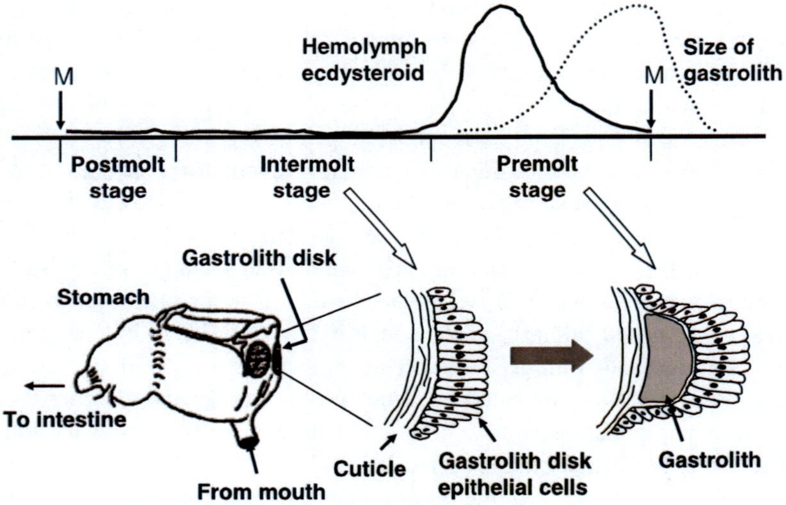

Fig. 11.4 The site and molting stage of gastrolith formation.Gastrolith is formed only at the gastrolith disk located at the frontal part of stomach. Inner wall of stomach is covered with cuticle, and gastrolith is formed between cuticle and epithelia

In the crayfish, e.g., a pair of gastroliths is formed in the stomach for temporary storage of calcium carbonate before molting (Fig. 11.4). The gastroliths are formed at the site between the cuticle and the epithelial cells, called the gastrolith disk, at the frontal part of the stomach (Travis 1960), grow quickly toward the ecdysis, and reach the maximum size weighing about 1 g in total just before ecdysis in case of the body size of about 10 cm. Thus, calcium carbonate moves as calcium and bicarbonate ions synchronously with molting in the body, which is regulated by the molting hormone. This indirect endocrine regulation of calcification and decalcification probably through the function of epithelia, direct target tissues of the molting hormone, is one of the most peculiar phenomena in crustaceans. It is also of great interest that calcification and decalcification occur concurrently at the different tissues in an individual; at the pre-molt stage, decalcification and calcification occur in the cuticle and stomach, respectively, while at the post-molt stage they do in a completely converse manner (Shechter et al. 2008a). The former process is triggered by the molting hormone, but it is not clear whether the latter process is under endocrine control or not.

11.3 Identification of Matrix Proteins in the Tissues for Temporary Storage of Calcium Carbonate

Gastroliths are formed only prior to molting as a temporary storage of calcium carbonate and do not contain any cells. A gastrolith of the crayfish, *Procambarus clarkii*, was decalcified with 1 M acetic acid, and insoluble materials remained with almost the same shape as the original gastrolith like a sponge rubber. The insoluble material was then treated with a solution containing 1% SDS and 10 mM dithiothreitol in a boiling water bath for 10 min. The extract was applied to reverse-phase HPLC, which gave almost one peak on the chromatogram. The peak material was a protein, and the yield was about 460 μg/g gastrolith. This protein was a novel protein and was named gastrolith matrix protein (GAMP). The time-of-flight mass spectrum of GAMP showed that the molecular mass was around 55 kDa. The N-terminal sequencing failed to identify any sequence, indicating that the N-terminus was blocked. So, GAMP was digested with some proteolytic enzymes. The resulting fragment peptides were separated by reverse-phase HPLC, and their amino acid sequences were determined (Ishii et al. 1998). By combining the sequences overlapped, five fragments with a total of 225 amino acid residues were obtained.

The residue after SDS extraction was dried, dissolved in [^{2}H]-formic acid, and subjected to ^{1}H NMR measurement. The spectrum was almost identical with authentic chitin commercially available, indicating that the residual material was chitin (Ishii et al. 1996). Thus, GAMP was tightly bound to chitin in the gastrolith and was not released with dilute acetic acid but extracted with a hot SDS solution, although we could not find any known chitin-binding domains in the amino acid sequence of GAMP. Therefore, GAMP may have a novel chitin-binding domain.

Based on the partial amino acid sequences thus determined, a cDNA encoding GAMP was cloned by combination of RT-PCR with 5′ and 3′ RACEs (Tsutsui et al. 1999). An open reading frame of 1,515 bp encoded a precursor protein consisting of a signal peptide (18 amino acid residues) and GAMP (487 amino acid residues) (Fig. 11.5). MS/MS analyses of the N-terminal 10-residue peptide obtained by digestion with endoproteinase Asp-N demonstrated that the N-terminus was blocked by a pyroglutamic acid residue. The deduced amino acid sequence of GAMP indicated that it contained two kinds of tandem repeated sequences; 17 repeats of 10 amino acid residues near the N-terminus and 15 repeats of 5 amino acid residues near the C-terminus. Although repeated sequences were found in many other matrix

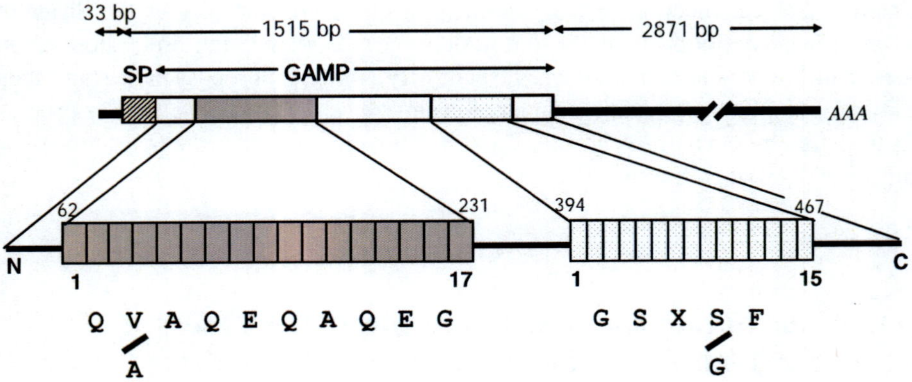

Fig. 11.5 Schematic representation of the structure of GAMP cDNA and protein.Two kinds of characteristic repeated sequences are present in GAMP

proteins identified, the sequences were different from one another and the significance of the repeated sequences was unclear. Gene expression analyses showed that GAMP mRNA was present only during pre-molt stage at the gastrolith disk. The expression was significantly enhanced by 20-hydroxyecdysone (molting hormone) in in vitro incubation of the gastrolith disk (Tsutsui et al. 1999). Immunohistochemistry using an anti-GAMP antiserum showed that GAMP was present in gastrolith disk epithelial cells and uniformly in the gastrolith, indicating that GAMP is produced by and secreted from the gastrolith disk cells constantly during the formation of gastrolith (Takagi et al. 2000). GAMP inhibited calcium carbonate precipitation from its supersaturated solution at more than 5×10^{-8} M. All these results strongly indicate that GAMP plays an important role in gastrolith formation.

An EDTA-soluble matrix protein in the gastroliths was also studied using a different species of the crayfish, *Cherax quadricarinatus* (Shechter et al. 2008b). A major component in the EDTA-soluble fraction was detected on an SDS-PAGE gel as a band stained with Coomassie brilliant blue at an apparent molecular mass of 65 kDa and named GAP 65 because of a novel protein. The deduced amino acid sequence of GAP 65 from the nucleotide sequence of a cDNA encoding GAP 65 revealed that it consists of 528 amino acid residues and contains three known domains, chitin-binding domain, a low-density lipoprotein receptor class A domain, and a polysaccharide deacetylase domain. It is not clear whether the latter two domains are related to calcification or not. The *GAP 65* gene was expressed at the gastrolith disk and subepidermal tissue, but not at stomach wall. The expression was greatly enhanced only at the gastrolith disk during the pre-molt stage, suggesting the association of GAP 65 with gastrolith formation and the induction of gene expression of GAP 65 by molting hormone. The RNAi knockdown experiment on GAP 65 indicated that it induced an abnormal ultrastructure of gastrolith. Although many matrix proteins have been identified chemically, their function is poorly understood. Application of molecular biological techniques such as RNAi and anti-sense RNA (Söllner et al. 2003; Murayama et al. 2005) will help better understanding of the function of matrix proteins.

Orchestia cavimana, the terrestrial crustacean, stores calcium carbonate in the posterior ceca of the midgut during the pre-molt stage to keep calcium carbonate for recalcification of the new cuticle at the next molting stage. A phosphoprotein with an apparent molecular mass of 23 kDa on a gel of SDS-PAGE was purified from the EDTA-soluble fraction of the posterior ceca, and named Orchestin (Luquet et al. 1996; Testenière et al. 2002; Hecker et al. 2003, 2004), because it has no sequence homology with known proteins. Orchestin can bind calcium, and this ability depends on phosphorylation of Ser residues. The molecular mass calculated from the amino acid sequence deduced from the nucleotide sequence of a cDNA was 12.5 kD. The discrepancy may be due to the strong acidity of this protein with a high proportion of acidic amino acid residues (30%) and phosphorylated serine residues. In situ hybridization, Northern blot analysis, and immunohistochemistry showed that Orchestin was synthesized specifically not only during the pre-molt stage but also during the post-molt stage as a component of the organic matrix of calcium carbonate precipitates. Thus, Orchestin is probably a key molecule in calcium carbonate precipitation process in this species.

11.4 Identification of Matrix Peptides and Proteins in Exoskeleton

The main calcification site in crustaceans is the cuticle of exoskeleton. By using almost the same method as that used for identification of GAMP in gastroliths, organic matrices seemingly responsible for calcification were searched in the extract of crayfish exoskeleton. The difference in the searching strategy from that for identification of GAMP was that we aimed at searching compounds with calcification inhibitory activity. For this purpose, we modified the method for assessing calcification inhibitory activity developed previously (Wheeler et al. 1981) to miniaturize the scale of experiment to about a tenth for purification of inhibitory compounds (Fig. 11.6). In this case, we followed the increase in the turbidity of the solution caused by precipitation instead of following the decrease of the pH value (Inoue et al. 2001).

Extraction with an SDS/dithiothreitol solution from insoluble carapace exoskeleton after decalcification with dilute acetic acid, followed by three steps of HPLC purification afforded two calcification inhibitory peptides, named calcification-associated peptide (CAP)-1 and -2 (Inoue et al. 2001, 2004). The yields of CAP-1 and -2 were 7 and 12 μg, respectively, from 1 g of dried exoskeleton, each being far less than that of GAMP. Sequence analyses of CAP-1 and -2 clarified that they consisted of 78 and 65 amino acid residues (Fig. 11.7). These peptides are rich in acidic amino acid residues and share about 60% sequence similarity. They both

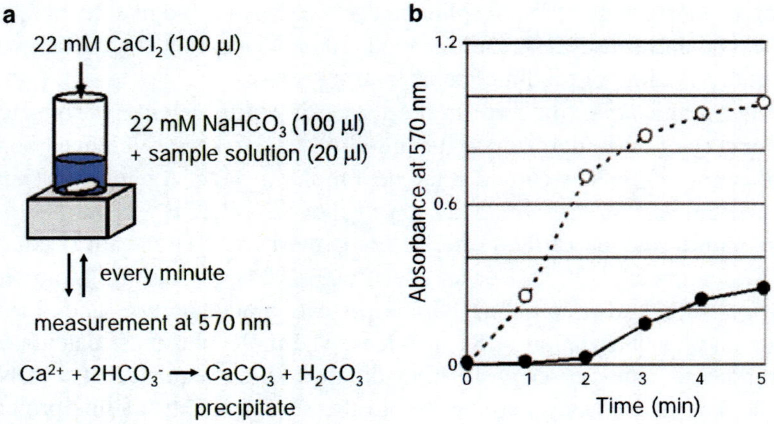

Fig. 11.6 Assay method of calcification inhibitory activity.(**a**) Calcification inhibitory activity of a sample was assessed by measuring the turbidity (absorbance at 570 nm) of the supersaturated solution of calcium carbonate. (**b**) An example of the result. *Closed circle*: a crude extract of exoskeleton. *Open circle*: control

```
            1        10        20        30        40
CAP-1       DVDLDEIHQEQNIDDDNTITGSYRWTSPEGVEYFVKYIAD
            :.. : .  ::       . : : :  :.: :: :::.::
CAP-2   SDIIDIEEDHLEHEQEGVPGTAVEGEYSWVAPDGNEYKVKYVAD
 RR                         G--------G------Y-A-

            41       50        60        70      78
CAP-1       EDGYRVLESNAVPATADGVRADGAQGSFVpSSEDDDDDD
              :::::: : ::   .
CAP-2       HLGYRVLEDNVVPEVPELEDY
 RR         E-GY--------P--P
```

Fig. 11.7 Amino acid sequences of CAP-1 and -2.Acidic amino acid residues are underlined. Symbols, (:) and (.), between the two sequences represent identical and similar residues, respectively. RR means Rebers-Riddiford consensus sequence for chitin binding

contain chitin-binding consensus sequence, called the Rebers-Riddiford (R-R) consensus sequence (Rebers and Riddiford 1988; Rebers and Willis 2001) spanning about 30 amino acid residues, at the central part of each peptide. The R-R sequence is found in many cuticle proteins and peptides in arthropods including insects and crustaceans (Andersen et al. 1995; Andersen 1999; Endo et al. 2000; Faircloth and Shafer 2007; Ikeya et al. 2001; Shafer et al. 2006; Wynn and Shafer 2005). However, the similarity is confined only to the R-R sequence, and there is almost no similarity in the other parts of peptides and proteins of insect cuticle proteins. Thus, the other part than the R-R sequence may be related to calcification. In CAP-1, only Ser at the 70th position was phosphorylated among 6 Ser residues.

As described later, phosphorylation enhances calcification inhibitory activity and therefore seems to be important in calcification of the cuticle. The phosphorylation of CAP-1 influenced the shape of crystals, when in vitro calcium carbonate precipitation was performed (Sugawara et al. 2006; Yamamoto et al. 2008).

Cloning of cDNAs encoding CAP-1 and -2 were attained by the conventional method (Inoue et al. 2003, 2004). The precursor peptide of CAP-1 consists of a signal peptide, CAP-1 and two basic amino acid residues at the C-terminus. Therefore, the mature CAP-1 is produced through posttranslational processing of cleavage of a signal peptide and removal of the C-terminal dibasic residues and phosphorylation of Ser^{70}. On the other hand, the precursor peptide of CAP-2 is simple and consists of only a signal peptide and CAP-2. RT-PCR analyses showed that the *CAP-1* and *-2* genes were expressed only in the epidermis during the time from the end of the pre-molt stage to the early post-molt stage, indicating their production precedent to and concomitant with calcification of new cuticle.

Using these cDNAs, recombinant peptides of CAP-1 and -2 were prepared using an *Escherichia coli* expression system and used for their functional analyses (Inoue et al. 2007). The chitin-binding ability of these peptides was demonstrated by the experiment of incubation with commercially available chitin. This ability seems to be consistent with the fact that CAP-1 and -2 were not easily extracted with dilute acetic acid but extracted with an SDS-containing solution at high temperature. This also means that these peptides exist as a complex with chitin in the cuticle. The calcium-binding ability of recombinant CAP-1 (dephosphorylated CAP-1 with an additional Ala at the N-terminus) was verified by incubation with $^{45}CaCl_2$. Further analyses indicated that the calcium binding comprised high-affinity and low-affinity bindings. Considering that CAP-1 shows calcification inhibitory activity and has abilities of chitin binding and calcium binding, CAP-1 may serve as a bifunctional molecule to bind to chitin for formation of scaffold for calcification and to regulate calcification including the initiation of calcium carbonate deposition.

11.5 Structure-Activity Relationship of a Cuticle Matrix Peptide

Structure-function relationship study has rarely been performed in the research of biomineralization, perhaps partly because the function of organic matrices still remains obscure and partly because peptide chemists are not aware of the importance of peptides and proteins in biomineralization. Recently, detailed studies on the relationship between peptide sequence and calcification-inhibitory activity of a matrix peptide in the cuticle were conducted. The following is an example of structure-function relationship study on CAP-1.

As mentioned previously, CAP-1 is a 78-amino acid residue peptide and has a chitin-binding sequence (the R-R consensus sequence) well conserved among

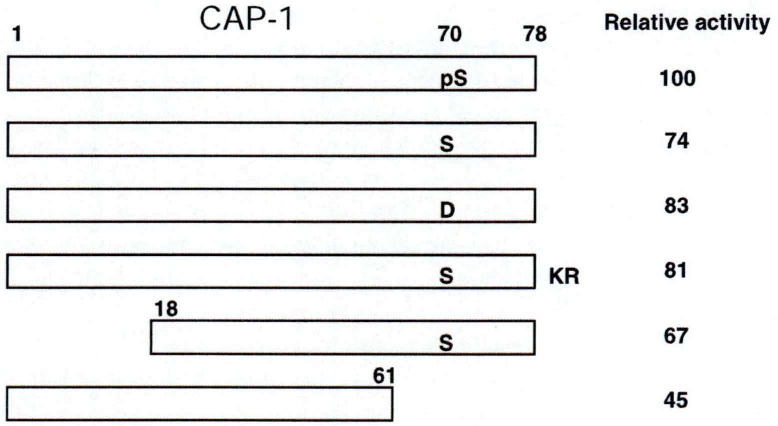

Fig. 11.8 Structure-activity relationship of CAP-1.Various recombinant peptides related to CAP-1 were prepared and assessed for their calcification inhibitory activity at 300 nM. Relative activity of each peptide to that (defined as 100%) of natural CAP-1 is presented

arthropod cuticle proteins at the central part of the molecule. The N- and C-terminal parts are rich in acidic amino acid residues, especially aspartate, glutamate, and phosphoserine, making this peptide highly acidic. The calcification-inhibitory activity was adopted for evaluation of function. This inhibition assay simply means that the peptide has an ability to bind to solid calcium carbonate, and it is presumed that it may regulate calcification by inhibition of crystal growth. Some recombinant peptides were prepared by a bacterial expression system (Fig. 11.8). They included (1) a peptide with the same sequence as natural CAP-1 but without a phosphate group at the 70th position, (2) a peptide with the same sequence but bearing additional residues, Lys-Arg, at the C-terminus, which is a precursor peptide deduced from the nucleotide sequence of a cDNA encoding CAP-1, (3) a peptide having the same sequence as natural CAP-1 except for the replacement of a phosphoserine residue with an Asp residue, (4) a peptide lacking the 16 N-terminal residues, and (5) a peptide lacking 16 C-terminal residues. Lacking the phosphate group decreased the activity by 22%, indicating the importance of the phosphate group, perhaps in terms of its highly acidic nature. Replacing phosphoserine with Asp decreased the activity, but did less than the peptide with Ser at the same position, suggesting again that acidic nature strengthens the inhibitory activity. Unexpectedly, the addition of two basic amino acid residues at the C-terminus did not reduce the activity, but the reason is unclear, because the role of basic residues has never been presented thus far. Lacking the C-terminal part decreased the activity more than lacking the N-terminal part, indicating that the C-terminal acidic part is more important than the N-terminal acidic part for the inhibitory activity.

Since the C-terminal part was found to be most potent in the inhibitory activity, we next focused on the C-terminal acidic part. Many peptides with similar amino

Table 11.1 Structure-activity relationship of various synthetic peptides

Peptide	Calcification inhibitory activity
1 YVSSEDDDDDD	100
2 YVSEDDDSDDD	97 ± 3
3 YVEDDDSSDDD	94 ± 1
4 YEDDDVSSDDD	84 ± 2
5 EDDDYVSSDDD	90 ± 7
6 EDDYVDDSSDD	78 ± 1
7 DSDYDVDSDSD	44 ± 2
8 YVCEDDDDDCD[a]	101 ± 2
9 YVEDCDDDDCD[a]	99 ± 2
10 YVEDDCDDDCD[a]	100 ± 1
11 YVEDDDCDDCD[a]	101 ± 2
12 YVSSEEEEEEE	45 ± 5

All peptides were assayed at 2 μM
Activity is expressed as relative activity compared with peptide 1 (100)
Acidic amino acid residues are underlined
[a]An intramolecular disulfide bond is formed to be a cyclic peptide

acid sequences were chemically synthesized (Sugisaka et al. 2009) (Table 11.1). In order to examine the importance of the Asp repeat, we compared the activities of some peptides, **1**, **2**, **3**, **4**, **5**, and **6**, which had the same number of Asp residues but different sequences. All the peptides had a Tyr residue instead of a Phe residue in the natural peptide in order that peptide amount could be calculated by measuring the absorbance at 274 nm. The results showed that they were almost equally potent, indicating that the sequence was not important for the activity. Then, the cyclic peptides, **8**, **9**, **10**, and **11**, with a different ring size prepared by an intramolecular disulfide bridge were compared for assessing their activity. Interestingly, the activities were almost comparable irrespective of ring size and also comparable to corresponding linear peptides with the same number of Asp. Since the cyclization makes the peptide more rigid, these results suggest that peptide conformation is not important for the activity. However, it is not clear whether this conclusion is applicable to larger molecules which may have a definite three-dimensional structure. If we compare the activities of **1** and **12**, it is clear that Asp is more effective than Glu. This may be due that Asp is more acidic than Glu.

The above data propose an important problem about molecular evolution of matrix peptides and proteins in biominerals, because we cannot find sequence homology among matrix proteins from phylogenetically far species. Shell matrix proteins, e.g., have almost no sequence similarity to one another, although a few proteins from phylogenetically close species are similar. The rate of molecular evolution, mutation of each amino acid residue, may be much faster than that of ordinary functional proteins. The reason why matrix proteins permit mutation more easily may be weak, less specific interaction with calcium carbonate (Sugisaka et al. 2009).

11.6 Regulation of Amorphous Calcium Carbonate

Calcium carbonate has three crystal polymorphs; calcite, aragonite, and vaterite in the order of thermodynamic stability. Calcifying organisms do not necessarily adopt the most stable calcite but often form unstable aragonite or vaterite. Crustaceans usually deposit calcium carbonate as an amorphous form (Lowenstam and Weiner 1989; Addadi et al. 2003). Amorphous calcium carbonate (ACC) is the least stable form and is apt to be converted to more stable crystal. There are two types of ACC; one is stable ACC which is maintained for a long time and the other is a transition state to the crystalline form (Aizenberg et al. 2003). Crustacean ACC belongs to the former type. Crustaceans may have an unknown mechanism to induce and maintain ACC. ACC is more soluble than any other crystalline forms of calcium carbonate. It seems reasonable that crustaceans adopt ACC as a precipitating form, because calcium carbonate in the cuticle and gastroliths should be dissolved easily before and after molting, respectively, as mentioned previously.

Previous works reported that crustacean calcified exoskeleton contains phosphorus, which may be related to the stability of ACC (Simkiss and Wilbur 1989). Indeed, an in vitro experiment indicated that phosphate has an ability to inhibit crystallization and stabilize ACC at high pH (Hikida et al. 2003). We have recently tried to identify phosphorus-containing compounds from exoskeleton and gastroliths. They were decalcified with a dilute acetic acid solution, and the resulting solution was separated by ultrafiltration into two fractions, a high-molecular-weight fraction (>10 kDa) and a low-molecular-weight fraction (<10 kDa). Both fractions were assessed for ACC inducing ability in vitro. The result showed that the low-molecular-weight fraction induced ACC at a lower concentration. The low-molecular-weight fraction was then passed through a cation-exchange column, from which a flow-through fraction containing phosphorus compounds was obtained. ^{31}P NMR spectra of these fractions showed that they both contained phosphate, phosphoenolpyruvate (PEP) and 3-phosphoglycerate (3PG) with different proportions. The latter two compounds were chemically identified by ^{1}H and ^{13}C NMR, and two-dimensional NMR spectral analyses combined with mass spectral analyses (unpublished results). These two compounds inhibited crystallization of calcium carbonate and stabilized ACC at the concentration of 1 mM in vitro.

Interestingly, both PEP and 3-PG are intermediates of glycolysis and therefore occur in every cell. Considering the large amount of PEP and 3-PG in the gastroliths and cuticle, metabolism is changed in the gastrolith disk probably by the function of the molting hormone and in the epithelial cells after molting, respectively. A part of these compounds in the cell seem to be actively secreted by the gastrolith disk cells and epithelial cells of the exoskeleton by an unknown mechanism. These compounds were more potent than phosphate in induction and maintenance of ACC, perhaps because they have an additional carboxyl group, possibly serving as a chelator in conjunction with a phosphate group.

11.7 Conclusion

Calcification and decalcification in crustaceans are characteristic in that (1) they occur synchronously with molting, (2) at least decalcification in the cuticle and calcification in the stomach are under endocrine control, (3) they occur concurrently in different tissues in an individual, and (4) calcified tissues contain ACC. A set of calcification and decalcification can be thought as one of various events that happen during the molting process. Until now, many organic matrices including high-molecular-weight and low-molecular-weight compounds have been identified. Concerning matrix proteins and peptides in crustacean calcified tissues, most of them have the R-R consensus sequence responsible for chitin binding. As in other biominerals, calcium carbonate precipitation in crustaceans occurs on organic scaffold, which is mainly composed of chitin and proteins. So, most matrix proteins and peptides with chitin-binding ability are produced before calcification and are waiting for supply of calcium and bicarbonate ions. Most matrix proteins are acidic, and may concentrate calcium ions leading to the formation of calcium carbonate precipitation. On the other hand, low-molecular-weight phosphorus compounds recently identified may be responsible for ACC formation. The role of such low-molecular-weight compounds has not yet been studied, and will be an important subject in the future. Another important subject to be solved is how a large amount of calcium and bicarbonate ions are passed through epithelial cells of exoskeleton and gastrolith disk, although almost no studies have ever been done.

Acknowledgment A part of this article was supported by a Grant-in-Aid for Creative Basic Research (17GS0311) from the Ministry of Education, Culture, Sports, Science, and Technology of Japan.

References

Addadi L, Raz S, Weiner S (2003) Taking advantage of disorder: amorphous calcium carbonate and its roles in biomineralization. Adv Mater 15:959–970

Aizenberg J, Weiner S, Addadi L (2003) Coexistence of amorphous and crystalline calcium carbonate in skeletal tissues. Connect Tissue Res 44(Suppl 1):20–25

Andersen SO (1999) Exoskeletal proteins from the crab, *Cancer pugrus*. Comp Biochem Physiol A 123:203–211

Andersen SO, Hojrup P, Roepstorff P (1995) Insect cuticular proteins. Insect Biochem Mol Biol 25:411–425

Endo H, Persson P, Watanabe T (2000) Molecular cloning of the crustacean DD4 cDNA encoding a Ca^{2+}-binding protein. Biochem Biophys Res Commun 276:286–291

Faircloth LN, Shafer TH (2007) Differential expression of eight transcripts and their roles in the cuticle of the blue crab, *Callinectes sapidus*. Comp Biochem Physiol B 146:370–383

Hampshire F, Horn DHS (1966) Structure of crustecdysone, a crustacean moulting hormone. Chem Commun (2), 37–38

Hecker A, Testenère O, Marin F, Luquet G (2003) Phosphorylation of serine residues is fundamental for the calcium binding ability of Orchestin, a soluble matrix protein from crustacean calcium storage structures. FEBS Letts 535:49–54

Hecker A, Quennedey B, Testenière O, Quennedey A, Graf F, Luquet G (2004) Orchestin, a calcium-binding phosphoprotein, is a matrix component of two successive transitory calcified biomineralizations cyclically elaborated by a terrestrial crustacean. J Struct Biol 146:310–324

Hikida T, Nagasawa H, Kogure T (2003) Characterization of amorphous calcium carbonate in the gastrolith of crayfish, *Procambarus clarkii*. In: Kobayashi, I. and Ozawa, H (ed) Biomineralization (BIOM 2001): formation, diversity, evolution and application, Proceedings of the 8th International Symposium on Biomineralization, Tokai University Press, Kanagawa, Japan, pp. 81–84

Ikeya T, Persson P, Kono M, Watanabe T (2001) The DD5 gene of the decapods crustacean Penaeus japonicas encodes a putative exoskeletal protein with a novel tandem repeat structure. Comp Biochem Physiol B 128:379–388

Inoue H, Ozaki N, Nagasawa H (2001) Purification and structural determination of a phosphorylated peptide with anti-calcification and chitin-binding activities in the exoskeleton of the crayfish, *Procambarus clarkii*. Biosci Biotechnol Biochem 65:1840–1848

Inoue H, Ohira T, Ozaki N, Nagasawa H (2003) Cloning and expression of a cDNA encoding a matrix peptide associated with calcification in the exoskeleton of the crayfish. Comp Biochem Physiol B 136:755–765

Inoue H, Ohira T, Ozaki N, Nagasawa H (2004) A novel calcium-binding peptide from the cuticle of the crayfish, *Procambarus clarkii*. Biochem Biophys Res Commun 318:649–654

Inoue H, Ohira T, Nagasawa H (2007) Significance of the C-terminal acidic region of CAP-1, a cuticle calcification-associated peptide from the crayfish, for calcification. Peptides 28:566–573

Ishii K, Yanagisawa T, Nagasawa H (1996) Characterization of a matrix protein in the gastroliths of the crayfish *Procambarus clarkii*. Biosci Biotechnol Biochem 60:1479–1482

Ishii K, Tsutsui N, Watanabe T, Yanagisawa T, Nagasawa H (1998) Solubilization and chemical characterization of an insoluble matrix protein in the gastroliths of a crayfish, *Procambarus clarkii*. Biosci Biotechnol Biochem 62:291–296

Keller R (1992) Crustacean neuropeptides: structures, functions and comparative aspects. Experientia 48:439–448

Lowenstam HA, Weiner S (1989) On biomineralization. Oxford University Press, New York

Luquet G, Marin F (2004) Biomineralisations in crustaceans: storage strategies. C R Palevol 3:515–534

Luquet G, Testenière O, Graf F (1996) Characterization and N-terminal sequencing of a calcium-binding protein from the calcareous concretion organic matrix of the terrestrial crustacean *Orchestia cavimana*. Biochim Biophys Acta 1293:272–276

Murayama E, Herbomel P, Kawakami A, Takeda H, Nagasawa H (2005) Otolith matrix proteins OMP-1 and Otolin-1 are necessary for normal otolith growth and their correct anchoring onto the sensory maculae. Mech Develop 122:791–803

Nagasawa H, Yang WJ, Shimizu H, Aida K, Tsutsumi H, Terauchi A, Sonobe H (1996) Isolation and amino acid sequence of a molt-inhibiting hormone from the American crayfish, *Procambarus clarkii*. Biosci Biotechnol Biochem 60:554–556

Ohira T, Watanabe T, Nagasawa H, Aida K (1997) Molecular cloning of a molt-inhibiting hormone cDNA from the kuruma prawn *Penaeus japonicus*. Zool Sci 14:785–789

Rebers JE, Riddiford L (1988) Structure and expression of a *Manduca sexta* larval cuticle gene homologous to *Drosophila* cuticle genes. J Mol Biol 203:411–423

Rebers JE, Willis JH (2001) A conserved domain in arthropod cuticular proteins binds chitin. Insect Biochem Mol Biol 31:1083–1093

Roer RD, Dillaman RM (1984) The structure and calcification of the crustacean cuticle. Amer Zool 24:893–909

Shafer TH, McCartney M, Faircloth LM (2006) Identifying exoskeleton proteinsin the blue crab from an expressed sequence tag (EST) library. Integr Comp Biol 46:978–990

Shechter A, Berman A, Singer A, Freiman A, Gristein M, Erez J, Aflalo ED, Sagi A (2008a) Reciprocal changes in calcification of the gastrolith and cuticle during the molt cycle of the red claw crayfish *Cherax quadricarinatus*. Biol Bull Woods Hole 214:122–134

Shechter A, Glazer L, Cheled S, Mor E, Weil S, Berman A, Bentov S, Aflado ED, Khalaila I, Sagi A (2008b) A gastrolith protein serving a dual role in the formation of an amorphous mineral containing extracellular matrix. Proc Natl Acad Sci USA 105:7129–7134

Simkiss K, Wilbur KM (1989) Biomineralization: cell biology and mineral deposition. Academic, San Diego

Söllner C, Burghammer M, Busch-Nentwich E, Berger J, Schwarz H, Riekel C, Nicolson T (2003) Control crystal size and lattice formation by starmaker in otolith biomineralization. Science 302:282–286

Sugawara A, Nishimura T, Yamamoto Y, Inoue H, Nagasawa H, Kato T (2006) Self-organization of oriented calcium carbonate/polymer composites: Effects of a matrix peptide isolated from the exoskeleton of a crayfish. Angew Chem Int Ed 45:2876–2879

Sugisaka A, Inoue H, Nagasawa H (2009) Structure-activity relationship of CAP-1, a cuticle peptide of the crayfish *Procambarus clarkii*, in terms of calcification inhibitory activity. Front Mater Sci China 3:183–186

Takagi Y, Ishii K, Ozaki N, Nagasawa H (2000) Immunolocalization of gastrolith matrix protein (GAMP) in the gastroliths and exoskeleton of crayfish, *Procambarus clarkii*. Zool Sci 17:179–184

Testenière O, Hecker A, Le Gurun S, Quennedey B, Graf F, Luquet G (2002) Characterization and spatiotemporal expression of orchestin, a gene encoding an ecdysone-inducible protein from a crustacean organic matrix. Biochem J 361:327–335

Travis DF (1960) The deposition of skeletal structures in the Crustacea. I. The histology of the gastrolith skeletal tissue complex and the gastrolith in the crayfish, *Orconectes (cambarus) vileris* Hagen-Decapoda. Biol Bull 118:137–149

Tsutsui N, Ishii K, Takagi Y, Watanabe T, Nagasawa H (1999) Cloning and expression of a cDNA encoding an insoluble matrix protein in the gastroliths of a crayfish, *Procambarus clarkii*. Zool Sci 16:619–628

Wheeler AP, George JW, Evans CA (1981) Control of carbonate nucleation and crystal growth by soluble matrix of oyster shell. Science 212:1397–1398

Wynn A, Shafer TH (2005) Four differentially expressed cDNAs in *Callinectes sapidus* containing Rebers-Riddiford consensus sequence. Comp Biochem Physiol B 141:294–306

Yamamoto Y, Nishimura T, Sugawara A, Inoue H, Nagasawa H, Kato T (2008) Effects of peptides on $CaCO_3$ crystallization: mineralization properties of an acidic peptide isolated from exoskeleton of a crayfish and its derivatives. Cryst Growth Des 8:4062–4065

Yang WJ, Aida K, Terauchi A, Sonobe H, Nagasawa H (1996) Amino acid sequence of a peptide with molt-inhibiting activity from the kuruma prawn *Penaeus japonicus*. Peptides 17:197–202

Zeleny C (1905) Compensatory regulation. J Exp Zool 2:1–102

Chapter 12
Molecular Approaches to Understand Biomineralization of Shell Nacreous Layer

Li-ping Xie, Fang-jie Zhu, Yu-juan Zhou, Chao Yang, and Rong-qing Zhang

Contents

Abstract The nacreous layer of molluskan shells, which consists of highly oriented aragonitic crystals and an organic matrix (including chitin and proteins), is a product of biomineralization. This paper briefly introduces the recent research advances on nacre biomineralization of shells from bivalves and gastropods, which mainly focus on analysis of the micro- and nano-structure and components of shell nacreous layers, and investigations of the characteristics and functions of matrix proteins from nacre. Matrix proteins not only participate in construction of the organic nacre framework, but also control the nucleation and growth of aragonitic crystals, as well as determine the polymorph specificity of calcium carbonate

L.-p. Xie (✉)
Protein Science Laboratory of the Ministry of Education, Institute of Marine Biotechnology, School of Life Sciences, Tsinghua University, Beijing 100084, P. R. China
e-mail: lpxie@mail.tsinghua.edu.cn

W.E.G. Müller (ed.), *Molecular Biomineralization*, Progress in Molecular and Subcellular Biology 52, DOI 10.1007/978-3-642-21230-7_12,

in nacre. Moreover, the inorganic aragonite phase also plays an active role in organizing nacre microstructure. Based on these studies, several models to illustrate the formation mechanism related to lamellar nacre in bivalves, and columnar nacre in gastropods are introduced.

12.1 Introduction

The nacreous layer (mother-of-pearl) of the molluskan shell is a microlamellar composite of highly oriented aragonitic crystals and proteins, exhibiting exceptional regularity, high mechanical strength (Meyers et al. 2008), and great thermal stability (Balmain et al. 1999). Its structure and properties gives it potential for the treatment of bone deficiencies (Westbroek and Marin 1998; Balmain et al. 1999; Lamghari et al. 1999) and in materials synthesis. Biomineralization, by definition a multidisciplinary field, especially the biomineralization of the nacreous layer, has attracted researchers from various scientific fields such as, biology, biotechnology, physics, chemistry, geology, and material science etc. The study of nacre biomineralization not only offers valuable insights into the scope and nature of materials, but also may provide new ideas for improved design of synthetic materials (Mann 1993; Aksay et al. 1996; Weiner and Addadi 1997).

Recently, many new techniques have been used in the study of the nacreous layer, such as FESEM (field-emission scanning electron microscopy), FETEM (field-emission transmission electron microscopy) (Oaki and Imai 2005), Confocal and IR (Dauphin et al. 2008), environmental and cryo-scanning electron microscopy (Nudelman et al. 2008), and AFM (atomic force microscopy) (Bezares et al. 2008). With the development and use of new techniques and biotechnology, more and more advances have been made in the illustration of the structure of the nacreous layer at a nanoscale, and in the characterization of increasing numbers of matrix proteins (Marin et al. 2008). Information on the following four aspects facilitates speculation on the primary and secondary structure of matrix proteins and their functions: (1) the gene sequences encoding the matrix proteins, (2) the deduced amino acid sequences, (3) gene expression character in the mantle, and (4) their distribution in the interlamellar sheets of the nacreous layer. Several researchers have reviewed investigations on matrix proteins (Marin and Luquet 2004; Samata 2004; Cusack and Freer 2008; Marin et al. 2008) and biomineralization of shell (Cusack and Freer 2008).

This paper briefly introduces the recent research advances at the molecular level on the nacre layers of shells (mainly from bivalves and gastropods), which range from analyzing the micro- and nano-structure and components of shell nacreous layers, to illustrating the characteristics and functions of matrix proteins from nacre. Matrix proteins not only participate in construction of the organic nacre framework, but also control the nucleation and growth of aragonitic crystals, as well as the polymorph specificity of calcium carbonate in nacre. Moreover, the inorganic aragonite phase also plays an active role in organizing nacre microstructure.

Based on these studies, several new models to illustrate the formation mechanism related to lamellar nacre in bivalves, and columnar nacre in gastropods are introduced. All these explorations on the mechanism of nacre biomineralization may greatly influence the strategies not only to improve the quality of culture pearls and promote their production, but also to design some specific biomaterial mimics.

12.2 The Structure of the Nacreous Layer

The nacreous layer of mollusks, which is composed of 95% calcium carbonate and less than 5% organic matrix in weight, is the best-known aragonitic structure and is the usual model for biomineralization. SEM images show that the nacreous layer of shells is made of polygonal aragonitic tablets (Kobayashi and Samata 2006), which are interspaced by thin interlamellar organic matrix sheets and finely accumulated lamellae parallel to the inner shell surface. The thickness of platelets is about 0.25 μm, and organic interlamellar matrix sheets between the tablets are 10~50 nm in thickness (Sarikaya and Aksay 1992), depending on the site of the shell from which the sample is extracted. Recent investigation by intermittent-contact AFM revealed that each crystal within which an intracrystalline organic matrix forms the foam-like structure, is composed of coherent flat nanograins (45 nm mean size) which share the same crystallographic orientation (Rousseau et al. 2005b). Similar observations were obtained using FESEM and FETEM (Oaki and Imai 2005). They proposed a three-level hierarchical architecture model (Fig. 12.1). The nacreous layer (tier 1) is composed of oriented aragonite plates (tier 2), each of which is an assembly of nanobuilding blocks (tier 3). This model will facilitate further understanding of the overall architecture in the nacreous layer from a nanoscopic to a macroscopic scale, and to further direct the generation of specific mimetic biominerals.

Two types of the nacre structure (Fig. 12.2) are discriminated according to their different formation methods. The first is called the sheet nacre structure. In most bivalves, the shell is built of tablets which, during development, form one or a few layers at a time, usually arranged like a “brick-wall” when seen in vertical cross section, or a “stair-step” pattern in horizontal view (Fig. 12.2a, b). The second is called the columnar nacre structure. The crystal tablets of nacre in gastropods are “stacked up” vertically on each other, and each stack resembles the shape of a pyramid. However, these pyramidal structures develop only at the surface layers, and the major portion of the layer is similar to that of bivalves (Fig. 12.2c), giving a brick-wall appearance (Watabe 1981).

The interlamellar organic matrix, mainly composed of biomacromolecules between calcium carbonate tablets, has a double function. On the one hand, some Asp-rich proteins adsorbed to the matrix surface provide nucleation sites for the next crystal layer. On the other hand, some proteins on the matrix inhibit and terminate crystal growth to ensure uniform thickness of the nacreous layers (Addadi and Weiner 1985; Mann 2001). Because the interlamellar organic matrix plays such

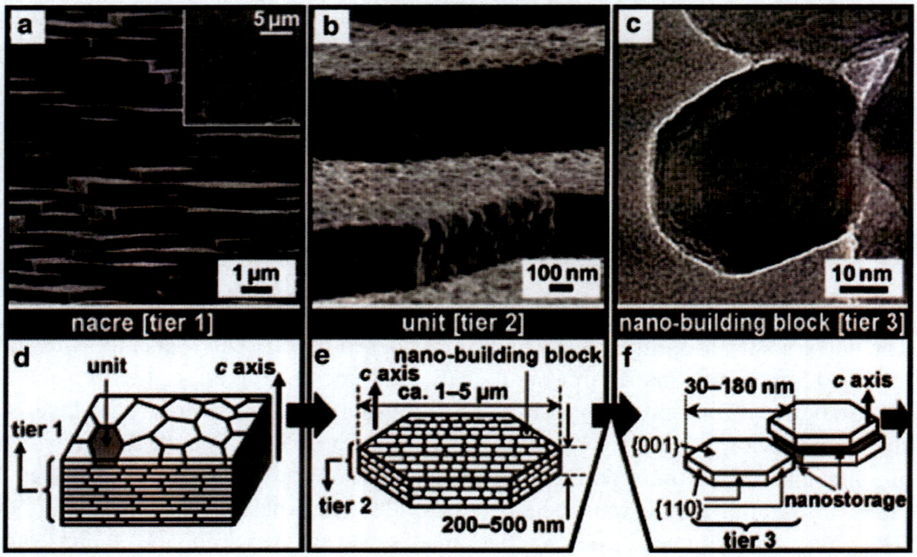

Fig. 12.1 Three-level hierarchical structure of the nacreous layer from the pearl oyster, *P. fucata*. **(a, b)** FESEM (field-emission scanning electron microscopy) and **(c)** FETEM (field-emission transmission electron microscopy) images of tiers 1–3; **(d–f)** Schematic representations of tiers 1–3. The nacre lamella (tier 1, Fig. 12.1**a, d**) consists of aragonite plates (tier 2, Fig. 12.1**b,e**). A single plate (tier 2) is also an oriented assembly of pseudohexagonal aragonite nanograins (tier 3, Fig. 12.1**c, f**) (The figure is reproduced from Oaki and Imai. 2005 with permission from the author and Wiley-VCH Verlag GmbH & Co)

an important role in nacre biomineralization, researchers have paid more attention to the interlamellar matrix sheets.

Examination of stained sections of individual matrix sheets from gastropod nacreous layers showed that these may be composed of five layers (Weiner and Traub 1984). The core of the matrix sheet consists of a layer of β-chitin. The latter is sandwiched by layers of silk fibroin-like proteins which are coated by Asp-rich proteins (Fig. 12.3a). The different organic layers show various levels of mechanical resilience, which may be important factors contributing to the ductility and toughness of nacre (Yao et al. 2009). However, the matrix from bivalves, such as *Atrina,* and the pearl oyster *Pinctada*, has a different structure. Under cryo-transmission electron microscopy (Cryo-TEM), the interlamellar sheets of *Atrina* nacreous layer appear to consist of three layers only (Fig. 12.3b) (Levi-Kalisman et al. 2001), with highly ordered and aligned beta-chitin fibrils sandwiched by Asp-rich proteins. The silk fibroin-like proteins are thought to be located between sheets in the form of a hydrated gel at least prior to mineralization, and are finally incorporated into the interlamellar or intertabular matrix within mature nacreous layers. A similar matrix structure also exists in the pearl oysters, *Pinctada maxima* (Pereira-Mouries et al. 2002) and *P. margaritifera* (Nudelman et al. 2008).

Fig. 12.2 Two types of nacreous layer structure, (**a**) vertical fracture (scale, 2 μm) and (**b**) growth surface (scale, 20 μm) of the nacreous layer in the pearl oyster *Pinctada fucata* showing the stepwise sheet nacreous structure in horizontal view. Each nacre lamella is composed of polygonal aragonitic tablets, which are sealed by an intertabular matrix (white arrow); (**c**) Vertical fracture through the nacre showing the columnar stack of crystals in *Calliostoma unicum* (Gastropoda). Each stack resembles the shape of a pyramid at the surface layer (scale, 400 μm). ((**a**) and (**b**) are reproduced from Gong et al. 2008c with permission from Elsevier Inc. (**c**) is reproduced from Nakahara 1981 with permission from Fukuyama: Malacological Society of Japan)

12.3 Nacreous Organic Matrix

12.3.1 Chitin

Chitin is an insoluble component in the shells of mollusks, which is synthesized via a complex transmembrane chitin synthase with an intracellular myosin motor domain (Schonitzer and Weiss 2007). Chitin exists in the organic framework of nacreous and prismatic layers, e.g., from *Pinctada* (Suzuki et al. 2007), *Atrina* (Levi-Kalisman et al. 2001), to *Haliotis* (Bezares et al. 2008), and thought to play important roles in shell biomineralization (Furuhashi et al. 2009). As the main components of organic interlamellar sheets in nacre, highly ordered and aligned beta-chitin constructs the scaffold as the template for nucleation of the calcium carbonate aragonite crystals (Levi-Kalisman et al. 2001). *In vitro* experiments

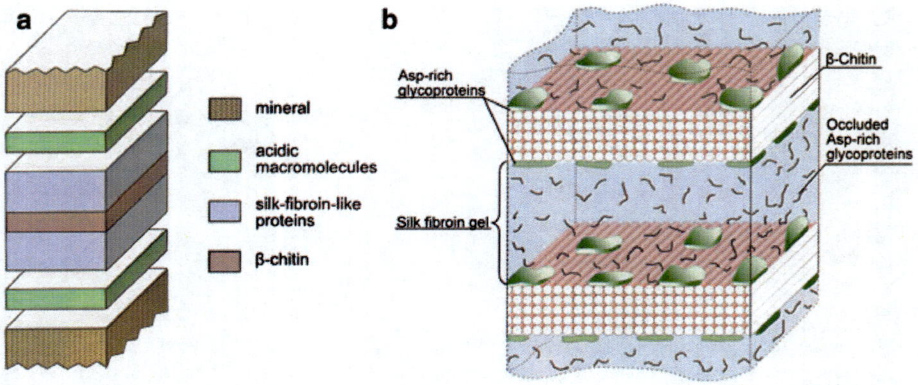

Fig. 12.3 A schematic composite section of one individual nacreous matrix sheet of a gastropod (**a**) and a schematic structure of the demineralized *Atrina* nacreous layer organic matrix (**b**). (**a**) β-chitin forms a thin layer sandwiched by silk-fibroin-like proteins and acidic macromolecules; (**b**) The interlamellar organic sheets are composed mainly of β-chitin on which aspartic acid-rich proteins adsorb discontinuously. The silk-fibroin-like proteins are putatively located between sheets in a gel phase prior to mineralization. Some acidic glycoproteins are occluded within the silk gel, and will finally become components of crystal as intracrystalline matrix in mature nacreous layers. ((**a**) is reproduced from Weiner and Traub, 1984 with permission from the Royal Society.(**b**) is reproduced from Levi-Kalisman et al.2001 with permission from Elsevier Inc.)

revealed that silk fibroin (framework protein) and chitin interact mainly through the chitin acetyl groups, and it is thought that there is an interfacial plane between them, in which the interactions are through the amide groups (Falini et al. 2003).

Another important function of chitin is in the development of the larval shell of bivalve mollusks. The cDNA encoding chitin synthase, the enzyme responsible for synthesis of chitin, has been cloned from *Pinctada fucata* (Suzuki et al. 2007), *Atrina rigida*, and *Mytilus galloprovincialis* (Weiss et al. 2006). Chitinous material, which exists in the larval shell of mussels, changes with the development of the larvae (Weiss and Schonitzer 2006). During larval development, when the activity of chitin synthase is partially inhibited by Nikkomycin Z in vivo, the structure of the larval shell at various growth fronts is dramatically changed (Schonitzer and Weiss 2007).We also screened and obtained a partial fragment of the gene encoding chitin synthase at the early stage of larval development of *P. fucata* (data not shown). All these observations indicate that chitin fulfills an important function in the formation and functionality of larval bivalve shells.

Moreover, the expression levels of genes involved in synthesis of chitin show significant differences between the pearl oyster and abalone. A chitin synthase and several chitin deacetylation genes are expressed at high levels in *P. maxima*, whereas only one protein involved in chitin interactions is present in the *H. asinina* dataset, suggesting that the organic matrix upon which nacre biomineralization proceeds differs fundamentally between these species (Jackson et al. 2010).

12.3.2 *Matrix Proteins*

12.3.2.1 General Features

Predominant Amino Acids

Analysis of the amino acid composition of matrix proteins shows that some amino acids are predominant, e.g., Asp, Gly, and Ser. These amino acids organize some repeat regions or modular structure of the proteins to perform specific functions involved in biomineralization. Different proteins may have different predominant amino acid residues which determine their different functions. The acidic protein N16 (Pearlin) has a high content of Gly, Tyr, Asn, and Cys, whereas basic Lustrin A contains a high proportion of Ser, Pro, Gly, and Cys. Alanine-rich is one of the features of silk-like proteins, whereas Gly-rich represents one of the characteristics of framework proteins. Asp-rich acidic proteins are thought to be the sites of crystal nucleation (Weiner and Traub 1984) on which Asp presumably binds and interacts with calcium by providing carboxyl groups, and may therefore initiate crystal nucleation (Addadi and Weiner 1985). This is why the soluble matrix of shells contains a high content of aspartic residues (Simkiss and Wilbur 1989). Extremely acidic shell proteins ($pI < 4.5$) with a high proportion of Asp are preferentially associated with calcitic prismatic layers rather than with aragonitic nacreous layers (Marin et al. 2008). Only one Asp-rich protein, Pif, which has been identified from nacreous layers, specifically binds to aragonite crystal and induces aragonite formation (Suzuki et al. 2009).

Sequence Repeats and Modular Structure

One of the most remarkable structural features of matrix proteins is the existence of modular units, which means that their primary structure contains one or more functional domains. Several proteins have a modular structure. For example, Nacrein (Miyamoto et al. 1996) and N66 (Kono et al. 2000) contain a carbonic anhydrase domain, Perlucin (Mann et al. 2000) has a C-type functional lectin domain, Perlustrin (Weiss et al. 2001) is similar to insulin-like growth factor binding protein, and Lustrin A (Shen et al. 1997) contains a similar domain of protease inhibitors and another domain that is similar to the sequence of extracellular matrix proteins. These modular multidomain proteins are probably produced by exon shuffling (Patthy 1999), which is assumed to be a "fast tool" of evolution. This is why similar domains were found to be widespread among seemingly unrelated extracellular proteins (Bork 1991).

Having sequence repeats is another common structural feature of matrix proteins, which also has some specific roles. For example, MSI60 contains 11 poly(alanine) blocks and 39 poly(glycine) blocks. Because polyalanine is one of the features of fibroin (Guerette et al. 1996), MSI60 is therefore referred to as a typical silk-like framework protein. However, poly(glycine) is suggested not only to participate in the formation of crystalline β-sheets, but also to bind calcium ions (Sudo et al. 1997). Gly-X-Asn repeat domain of Nacrein is linked to the inhibition of calcium carbonate precipitation (Miyamoto et al. 1996).

Acting as a Protein Complex

The presence of a protein complex is a prerequisite for aragonite crystallization (Matsushiro et al. 2003). Only the complex of pearlin and pearl keratin, and the reconstituted complex, induces aragonite crystallization in vitro within a calcium-carbonate-saturated solution containing Mg^{2+}, whereas the mixture of the individual components has no function. Other known protein complexes present in the nacreous layer of the pearl oyster *Pinctada*, include P60 (complex of Nacrein and its two derivatives, N28 and N35, by disulfide bridges) (Lao et al. 2007) and Pif. The latter, a complex of Pif 97and Pif 80, is assumed to assemble with N16 and other proteins to control nacre formation (Suzuki et al. 2009). Because the proteins always form complexes, it is difficult to purify them from nacreous layers. Characterization of the protein complexes will facilitate an understanding of their functions.

Posttranslational Modifications

Matrix proteins usually exhibit several posttranslational modifications, such as phosphorylations, glycosylations, and sulfations, *etc*. (Marin et al. 2008). Both Nacrein and Pearlin are sulfated glycoproteins, the former containing N-linked sialic acid at its terminus (Takakura et al. 2008), whereas the latter has mucopolysaccharides (Miyashita et al. 2000). ACCBP (amorphous calcium carbonate binding protein), purified from extrapallial fluid of *Pindata fucata*, has two potential glycosylation sites at Asn29 and Asn184 (Ma et al. 2007), and is heavily glycosylated with peculiar glycans (data not shown). These posttranslational modifications may have important roles in nacre biomineralization. Structured polysaccharide moieties of glycoproteins are putatively important in controlling the crystal growth *in vivo* (Albeck et al. 1996), while the negative charge on sulfate groups linked to the mucopolysaccharide of Pearlin is assumed to bind to calcium (Miyashita et al. 2000). The covalently bound sulfated polysaccharides supposedly act as "antennae" that help to accumulate sufficient numbers of Ca^{2+} ions around the asp-rich binding/nucleation sites on the antiparallel β–pleated sheet of the matrix (Mann 2001). Nevertheless, glycosylations of ACCBP are also a prerequisite for performing functions in the control of nacre morphology (data not shown). Matrix protein Pif regulates nacre formation by its posttranslational modification products, Pif 97 and Pif 80 (Suzuki et al. 2009). These kinds of posttranslational modifications probably exist in other nacreous matrix proteins, and it is necessary to investigate their functions in nacre biomineralization.

Distribution Differences

Different genes that encode nacreous and prismatic matrix proteins are expressed in different regions of the mantle (Takeuchi and Endo 2006), and even in different developmental stage of the larva shell (Miyazaki et al. 2010). Matrix proteins are distributed in different layers of the shell or in different regions in the same layer, which is related to their functions. For instance, proteins involved in aragonitic formation occur in the nacreous layer of the shell, whereas proteins involved in calcite formation are found in prismatic layers (Falini et al. 1996). In each individual interlamellar sheet of the nacreous layer, proteins occur in different zones underlying a single aragonite crystal, which means they have a different function. Immunofluorescence staining has revealed that aragonite-nucleating fractions are primarily localized in the center of the crystal imprint and the intertabular matrix of interlamellar sheets (Addadi et al. 2006; Nudelman et al. 2006).

12.3.2.2 Framework Proteins

In the nacreous layer, the framework proteins are rich in Ala and Gly. They are thought to fill the space between two interlamellar sheets in a hydrated gel-like state before biomineralization (Nudelman et al. 2008). So far, only one silk-like framework matrix protein, MSI60, has been identified from the nacreous layers of the pearl oyster *Pinctada fucata*. This protein participates in the formation of $CaCO_3$ crystals, presumably by binding soluble Asp-rich matrix glycoproteins and carbonic anhydrase (Sudo et al. 1997). Its messenger RNA is preferentially expressed only in the more dorsal outer epithelia of the mantle (Takeuchi and Endo 2006) and is highly expressed in the pearl sac (Wang et al. 2009) in which its relative expression level is related to pearl quality (Inoue et al. 2010).

12.3.2.3 Regulative Proteins

Proteins from Pearl Oyster Nacre

Nacrein is the first identified molluskan matrix protein from which a complete amino acid sequence has been obtained. It is thought to be specifically involved in forming the nacreous layer of the pearl oysters *Pinctada fucata* (Miyamoto et al. 1996) and *P. maxima* (Kono et al. 2000). The protein, secreted by epithelial cells of the mantle (Gong et al. 2008a, b), is a soluble matrix glycoprotein containing sulfite and N-linked sialic acid (Takakura et al. 2008), whose transcript is expressed in dorsal and ventral regions of the mantle (Takeuchi and Endo 2006), especially in the epithelial cells (Miyamoto et al. 2005). Having a carbonic anhydrase domain, Nacrein acts as an enzyme to catalyze HCO_3^- formation, thus participating in calcium carbonate crystal formation of the nacreous layer (Miyamoto et al. 1996). It has another acidic Gly-Xaa-Asn (Xaa = Asp, Asn, or Glu) repeat domain, which possibly binds calcium and plays a role in inhibiting shell formation (Miyamoto et al. 2005). Nacrein acts as a negative regulator in calcification by inhibiting the precipitation of $CaCO_3$ in vitro (Miyamoto et al. 2005) and growth of an aragonitic tablet in vivo (Gong et al. 2008c). Nacrein has been shown to be distributed within aragonitic tablets and the intertabular matrix by immunolabeling (Gong et al. 2008c). Nacrein probably exists and functions as a complex P60 in nacreous layer (Lao et al. 2007).

N16 is an acidic EDTA-insoluble nacreous matrix protein found in *P. fucata* (Samata et al. 1999). Its homologues, Pearlin and N14, were identified from *P. fucata* (Miyashita et al. 2000) and *P. maxima* (Kono et al. 2000), respectively. The protein contains a sulfated mucopolysaccharide, with high proportions of Gly, Tyr, and Asn together with NG repeat sequences. Its messenger RNA is expressed in the dorsal region of the mantle as demonstrated by Northern blot analysis (Miyashita et al. 2000). In vitro crystallization experiments revealed that N16-induced aragonite crystals once adsorbed onto the water-insoluble matrix membrane (Samata et al. 1999), and the mixture of N66 and N14 could induce flat aragonite layers very similar to the nacreous layer under similar conditions (Kono et al. 2000). N-and C-terminal sequence regions of N16 most likely play a key role in regulating the crystal growth of calcium carbonate in the nacre layers (Kim et al. 2004). N16 is also thought to act as a linker to connect with fibrion-like proteins and more highly acidic proteins, and to participate in the formation of aragonite (Mann 2001).

Recently, an acidic nacreous matrix protein complex, Pif, has been identified from *P. fucata*. It consists of Pif 97 and Pif 80, and specifically binds to aragonite crystals. The results from immunolocalization, RNA interference, and in vitro calcium carbonate crystallization experiments strongly indicate that Pif may aggregate with N16 and other proteins to regulate nacre formation (Suzuki et al. 2009).

Other known nacreous proteins include the positive regulator P10 (Zhang and Zhang 2006), N40, and negative regulator alkaline N19 (Yano et al. 2007). Especially N40 is an exclusive protein that can nucleate aragonite by itself, without the need for adsorption to a substrate. Thus, this study has proposed the possibility that the nonacidic shell protein can also directly participate in aragonite nucleation and even act as a nucleation site, which is different from earlier theories (Yan et al. 2007). Another acidic glycoprotein, ACCBP, purified from extrapallial fluid rather than from nacre, could also modify the morphology of nacre lamellae by inhibiting the growth of undesired aragonite crystal phases and meanwhile maintain the stability of $CaCO_3$-supersaturated body fluid by ceasing the nucleation and growth of calcite (Ma et al. 2007).

Proteins from Abalone

Nacreous matrix proteins from the abalone *Haliotis laevigata* significantly differ from those of pearl oysters. Most of them are alkaline, and possess unique characteristics, such as the insulin-like growth factor binding protein Perlustrin

(Weiss et al. 2001), Perlwapin with an acidic protein domain (Treccani et al. 2006), and Perlucin containing a C-type functional lectin domain (Weiss et al. 2000). Perlucin is able to nucleate calcium carbonate on calcite, and to be occluded into calcium carbonate crystals as an intracrystalline protein (Blank et al. 2003). Perlinhibin induces the formation of aragonite on calcite substrate by inhibiting the growth of calcite crystals (Mann et al. 2007), whereas Perlwapin inhibits the growth of certain crystallographic planes in the mineral phase of nacre (Treccani et al.2006).

Lustrin A is the largest protein found in the aragonite layer of abalone shell, identified from *Haliotis rufescens* by screening the mantle cDNA library (Shen et al. 1997), with a high proportion of Ser, Pro, Gly, and Cys. Having a protease inhibitor-like domain and a similar extracellular matrix domain, Lustrin A is thought to be a multifunctional protein. Its 24AA Asp-rich (D4) sequence domain influences crystal growth in a concentration-dependent manner during in vitro mineralization (Wustman et al. 2003).

12.4 Function of Matrix Proteins

Generally, at least five strategies are used to study the functions of matrix proteins: (1) Analyzing in situ hybridization to display the gene expression characteristics in different mantle regions that are known to secrete components to form shell (Lowenstam and Weiner 1989). Genes encoding nacreous matrix proteins are preferentially expressed in more dorsal regions of the mantle (Sudo et al. 1997; Takeuchi and Endo 2006) where outer epithelia have a high proliferation rate (Fang et al. 2008), whereas genes encoding prismatic ones usually are expressed in the edge of the mantle (Takeuchi and Endo 2006). (2) Cloning the genes encoding matrix proteins, from which the complete amino acid sequence and the secondary structure are deduced. Based on this information, the proteins' functions involved in nacre biomineralization are speculated. (3) Using in vitro calcium carbonate crystallization experiments to study the effect of native nacreous matrix proteins (Yan et al. 2007), recombinant ones, or synthesized functional polypeptides (Michenfelder et al. 2003; Kim et al. 2006; Evans 2008) from these proteins, on aragonitic crystals formation and growth; (4) Mapping the distribution of organic matrix proteins or their components on the surface of interlamellar matrices from nacreous layer of shell using immunohistochemical or immunofluorescence methods (Addadi et al. 2006; Nudelman et al. 2006; Bezares et al. 2008), their presence at the location is usually consistent with their supposed role in nacre formation. (5) Knocking out the genes encoding matrix proteins by RNA interference (Suzuki et al. 2009), or restraining the physiological functions of these proteins by injecting the antibodies against them into the extrapallial space (Ma et al. 2007; Gong et al. 2008c; Kong et al. 2009), then checking the formation of the nacreous layer of shell in vivo by SEM. Therefore, the functions of proteins in nacre biomineralization are mainly for the construction of an organic framework, and the control of the nucleation and growth of crystals.

12.4.1 Constructing the Organic Framework

Matrix proteins are one of the most important components of the organic framework in the nacreous layers of shell. These proteins have been conventionally separated into "water-soluble" and "water-insoluble" fractions, according to their solubility in aqueous solutions after decalcification with acid or EDTA (Pereira-Mouries et al. 2002). Generally, insoluble proteins, such as fibrion-like proteins from *Atrina* nacre and MSI60 from *Pinctada*, participate in the construction of the framework (together with chitin) on which the nucleation and growth of aragonite crystals occur under the regulation of soluble matrix proteins (Addadi et al. 2006, Nudelman et al. 2006). Moreover, intracrystalline organic matrix even forms the network within aragonite crystal (Rousseau et al. 2005b), which is presumably related with mediating nacre's mechanical response. As well as acting as a framework, these proteins may have other functions which should be investigated in future.

12.4.2 Controlling the Nucleation and Growth of Crystals

Acidic Asp-rich proteins on the organic sheet of the nacreous layer induce the nucleation of aragonitic crystals (Weiner and Traub 1984; Addadi et al. 2006; Nudelman et al. 2006). The fundamental principle governing nucleation of crystals is interfacial molecular recognition at the surface of an organic matrix. Because of molecular complementation between Ca atoms in the aragonite ab face and aspartic acid residues organized in the Asp-X-Asp repeat domains along the β-sheet matrix interface, aspartic acid residues play important roles in Ca^{2+} binding and oriented nucleation in shell nacre (Mann 2001). Acting as a component of the nucleation site of the mineral crystals, acidic proteins are more effective crystal modulators than other proteins from the same biomineralized material (Fu et al. 2005). However, there is no direct correlation between the acidity of soluble shell proteins and shell structure (Furuhashi et al. 2010).

Some proteins induce the precipitation of aragonite crystals, such as N16 (Samata et al. 1999), P10 (Zhang et al. 2006), N40 (Yan et al. 2007), and Pif 80 protein (Suzuki et al. 2009) from *P. fucata*. N16 may also react with N66 (the homologue of Nacrein from *P. maxima*) or aggregate with Pif 97 to promote the nucleation of aragonite crystals (Kono et al. 2000; Suzuki et al. 2009). By contrast, other nacreous matrix proteins inhibit the crystallization of $CaCO_3$ in vitro as a negative regulator, such as Nacrein and N19 (Yano et al. 2007) from *P. fucata*, and Perlwapin from abalone. The formation of exquisite nacreous layer structures may be the result of balance between positive factors and negative ones.

12.4.3 Calcium Carbonate Polymorph Specificity

Calcite and aragonite are the two most stable polymorphs of calcium carbonate. Although they have very similar structures, gastropods and bivalves are able to selectively deposit calcite in the outer layer and aragonite in inner layers. Which is the key factor determining their polymorph specificity?

It has been demonstrated that it is the soluble proteins from different shell layers that specifically control calcium carbonate crystal morphology in an appropriate microenvironment. The macromolecules extracted from the aragonite and calcitic shell layers of mollusks induced formation of aragonite and calcite in vitro, respectively (Falini et al. 1996). The amino acid composition of the organic matrix, their location, and the manner in which they affect crystal nucleation, differed markedly in relation to the shell microstructures, especially between the nacreous and prismatic matrix (Samata 1990; Nudelman et al. 2007). Moreover, the switch between aragonite and calcite can also be controlled in vitro by soluble polyanionic proteins extracted from abalone shell (Belcher et al. 1996). However, proteins extracted from the abalone aragonitic nacreous layer first induce (in vitro) the formation of amorphous calcium carbonate, and then transform into the aragonitic crystalline form (Gotliv et al. 2003). Further studies show that some proteins control the morphology of nacre either through selective adsorption and spatial constraint on the growing crystal (Giles et al. 1995), or by inhibiting the growth of certain aragonite crystal planes (Ma et al. 2007). Others induce formation of aragonite on calcite substrate by inhibiting the growth of calcite (Treccani et al.2006; Mann et al. 2007).

Like the soluble matrix, the insoluble matrix also has an influence on the growth of $CaCO_3$. The EDTA insoluble matrix of abalone nacre induces the growth of flat and roughly polygonal $CaCO_3$ crystals, and the growth of three-dimensional parallel sheets of densely packed platelets, shown to consist of oriented aragonite by XRD (Heinemann et al. 2006).

Further investigations indicated that the polymorph specificity of calcium carbonate is dependent upon the amino acid sequence, the conformation of specific protein(s) in the mollusk shell, and the microenvironment in which crystal nucleation and growth takes place. Synthetic polypeptides (Asp-Leu)n have been demonstrated to be capable of specifically inducing aragonite formation in vitro (Levi et al. 1998). Calcium carbonate polymorphs change in vivo from aragonite to calcite accompanied by significant alterations in protein conformation in the aragonitic layer (Choi and Kim 2000).

12.4.4 *Pearl Quality*

Although several genes encoding matrix proteins from pearl oyster nacreous layers and the gene expression profile in different region of mantle (Takeuchi and Endo 2006) and in the pearl sac (Wang et al. 2009) have been characterized by real-time PCR, research on the function of matrix proteins has mainly focused on their effects on calcium carbonate crystallization in vitro. Research on the functions of matrix proteins rarely concern pearl quality. Recently, Inoue et al. (2010) used real-time PCR analysis to investigate the relationship between pearl quality and gene expression patterns of six shell matrix proteins involved in nacreous and prismatic layer formation in the pearl sac. They discovered that the relative expression level of *MSI31*, which encodes a framework protein in the prismatic layer, is higher in the pearl sac producing lower-quality pearls than in that producing high-quality ones (Inoue et al. 2010). Their findings provide a new possibility to improve the quality of pearls by inhibiting the expression level of some specific genes involved in prismatic layers in the early stage of pearl formation.

12.5 The Molecular Mechanism Involved in Nacreous Biomineralization

Investigations of the mechanism of nacre biomineralization mainly focus on two aspects: (1) the nucleation and growth of aragonite crystal; (2) the orientation of crystal growth.

12.5.1 *The Nucleation and Growth of Aragonite Crystal*

Several models related to nacre biomineralization have been proposed. According to the early compartments model, compartments are first formed on lamellae parallel to the surface of the epithelium in which crystal nucleation is initiated in contact with a crystal in an adjacent layer (Bevelander and Nakahara 1969). The uniform thickness, orientation, and other features exhibited in the mature nacre are determined by the compartments. This model is partly supported and modified by the Voronoi model, which suggests that the mature nacre layer first existed as a film (opened compartment), on which the crystal nucleation occurs in time and space probably stimulated by a signal coming from an underlying layer. Then there is progressive lateral crystallization, and consequently formation of polygonal tablets of bio-aragonite. The growth of the tablets is controlled by an "aggregation-like" process of small "crystallites" (Rousseau et al. 2005a).

The "generally accepted" template theory indicates that the organic matrix constructs a framework which acts as a template for calcium carbonate crystal formation by heteroepitaxial nucleation growth (Weiner and Traub 1984). Nacreous layer formation in the shell may process in four stages, such as assembly of the matrix, the first formation of the mineral phase, nucleation of individual aragonite tablets, and formation of the mature nacreous layer (Addadi et al. 2006). Formation of the nacre starts with the assembly of the organic matrix. Two sheets of β-chitin construct the framework onto which the acidic proteins are adsorbed and form aragonite nucleation sites. The silk-like proteins, which inhibit nonspecific crystallization, fill the space between the two sheets of β-chitin in a hydrated gel-like state prior to mineral formation, and are finally incorporated into the intertablets and interlamellar matrices in mature nacreous layers (Nudelman et al. 2008). Therefore, the high uniform thickness of the crystals is controlled by the distance between two sheets of chitin (Nudelman et al. 2007, 2008). Fig. 12.4 shows a schematic model of the nacreous layer structure prior to, and after mineralization, which is compatible with the nacre of *Pinctada* and *Atrina*.

However, gastropod (abalone) nacre is also assumed to be formed on the organic interlamellar sheets, built upon a fibrous chitin core. Aragonite crystals grow on the nucleation sites penetrating the interlamellar layer, and form multiple mineral bridges through which successive aragonite tablets grow (Bezares et al. 2008),

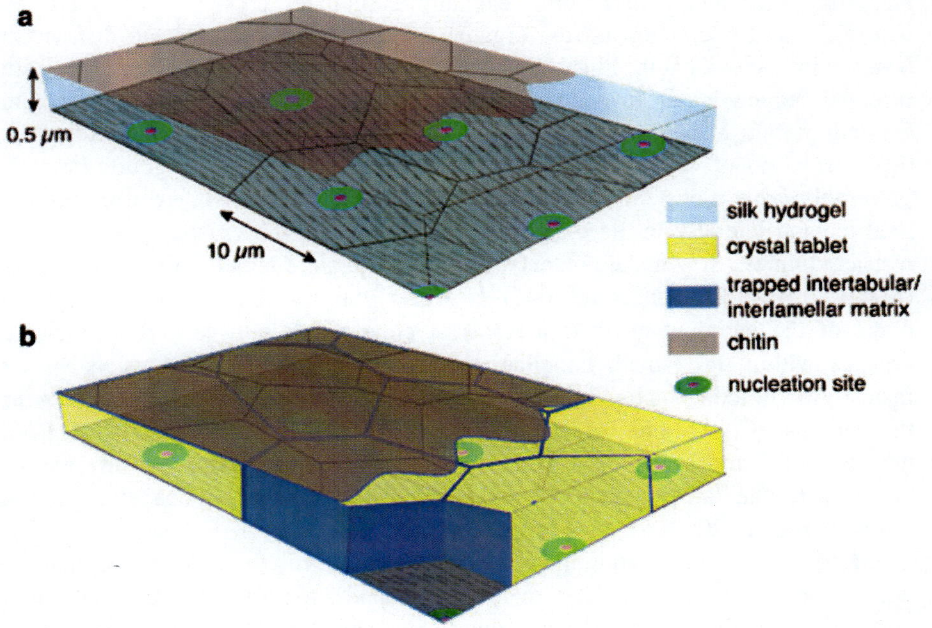

Fig. 12.4 Scheme of a nacreous layer before mineralization (**a**) and after mineralization (**b**). (**a**) Two sheets of β-chitin are interspaced by the silk-like proteins in a hydrated gel-like state. The acidic proteins adsorb on a β-Chitin scaffold and form the crystal nucleation site, while the silk-like proteins inhibit nonspecific crystallization. (**b**) Nucleation of aragonite is induced by the acidic proteins on the crystal nucleation site. As the mineral tablets grow, the silk-like proteins are pushed aside and are integrated into an intertabular and interlamellar matrix (The figure is reproduced from Addadi et al. 2006 with permission from Wiley-VCH Verlag GmbH & Co)

rather than on heteroepitaxial nucleation (Schaffer et al. 1997). The growth through mineral bridges may ensure perfect alignment of the crystal axes along a stack without disrupting the basic brick and mortar alternation which provides the material with elasticity and increased resistance to fracture (Addadi and Weiner 1997). The multiple nucleation sites on any given layer are assumed discrete with random distribution which may lead to the nano-polycrystalline structure of individual layers (Bezares et al. 2008). Although the mineral bridges are mainly located in the nacre layers of gastropods (Song et al. 2002; Lin et al. 2008), and absent in mature nacreous layers of bivalves (Rousseau et al. 2005), the mineral bridge model is still suitable to explain the microstructure of incipient nacre in pearl oyster shell (Saruwatari et al. 2009).

The different growth mechanisms of gastropods and bivalves may be related to their different nacreous microstructures.

12.5.2 The Orientation of Crystal Growth

According to template theory, calcium carbonate crystal is formed by heteroepitaxial nucleation on the organic matrix. X-ray and electron diffraction observations show that the chitin fibers and the protein polypeptide chains (they are oriented perpendicular to each other) are aligned with the a and b aragonite crystallographic axes, respectively, which means that the organic matrix determines the orientation of crystal growth (Weiner and Traub 1984). Aragonite crystals preferably grow along their c axis until they reach the upper layer of chitin and then expand laterally in the plane of the sheets where the crystal growth is less restricted unless they make contact with other aragonite tablets in the same layer (Mann 2001; Nudelman et al. 2007). However, the crystal orientation in the a-b plane of the incipient nacre is a result of competition between adjacent nacre crystals within the growth lamellae, which favors selection of crystals whose fastest growth axis (b-axis) is oriented parallel to the direction of propagation of the lamella (Checa and Rodriguez-Navarro 2005; Checa et al. 2006). Both heteroepitaxy and selection by competition possibly superimpose in bivalve nacre, with the latter probably having a prevalent effect (Checa et al. 2006; Saruwatari et al. 2009).

Gastropod nacre formation may involve the spiral growth model rather than the competition model mentioned above. In gastropods, the aragonite platelets grow vertically via helices that surround numerous screw dislocation cores, and extend horizontally simultaneously (Yao et al. 2009). These new findings may aid in creating novel organic–inorganic micro/nano composites through synthetic or biomineralization pathways (Yao et al. 2009)

Further nanoscopic and cross-sectional investigation shows that the aragonitic crystal growth of incipient nacre is not only a biotic process, but also an inorganic one, such as geometrical selection and mineral bridges (Saruwatari et al. 2009).

In summary, both the organic matrix component and the mineral phase play an active role in organizing the final microstructure (Checa and Rodriguez-Navarro 2005).

Although the matrix proteins from nacre have the potential to regulate calcium carbonate deposition and crystallization, even to determine the crystal polymorph specificity, this point of detail requires further clarification. How matrix proteins recognize and interact with calcium carbonate will be crucial for illustrating the molecular mechanism. Control of crystal growth seems to depend on the control of matrix protein secretion or activation processes in the mantle cells (Jolly et al. 2004); however, there is very little research on this. How the genes encoding matrix proteins are regulated to express in an appropriate time and space during the development of nacreous layers is still far from understood. A better understanding of the secretory mechanism that results in the progressive formation of the true nacreous layer may contribute to improving the quality of culture pearls.

12.6 Conclusion

Nacre, consisting of calcium carbonate aragonite crystals and matrix, is produced by biomineralization, having a specific structure and property. Although the organic matrix (including chitin and matrix proteins) constitutes less than 5% of the shell weight, it plays a crucial role in nacre biomineralization. Beside participating in construction of the organic framework, matrix proteins not only govern the nucleation and growth of aragonitic crystals in nacre, but also control their crystallization morphology in an appropriate microenvironment. The polymorph specificity of calcium carbonate is presumably linked to the amino acid sequence, the conformation of specific protein(s) in the mollusk shell, and the microenvironment in which crystal nucleation and growth takes place. Moreover, expression level of genes encoding shell matrix proteins in the pearl sac, especially the genes encoding shell prismatic matrix proteins, is related to the quality of culture pearls.

There are two principal types of nacre structure, lamellar and columnar. The former exists in the shell of bivalves, whereas the latter is present in that of gastropods. The different nacre microstructures arise from their different formation mechanisms. Formation of nacre in bivalve shell starts by constructing the organic sheets, consisting of beta-chitin, as a template for nucleation. The orientation of crystal growth is a result of competitive selection between adjacent crystals and control of the organic matrix. Gastropod nacre crystal growth may be compatible with the spiral growth model and consequently successive nacreous lamellae are formed through a mineral bridge. Not only the organic matrix, but also the calcium carbonate mineral phase play important roles in the process of nacre biomineralization. The more detailed molecular mechanisms involved need to be clarified in future.

Acknowledgments This work was financially supported by the National Basic Research Program of China (973 Program) (2010CB126405), the National High Technology Research and Development Program of China (2006AA09Z441), and the National Natural Science Foundation of China (30871911).

References

Addadi L, Weiner S (1985) Interactions between acidic proteins and crystals: stereochemical requirements in biomineralization. Proc Natl Acad Sci USA 82:4110–4114

Addadi L, Weiner S (1997) Biomineralization: a pavement of pearl. Nature 389:912–915

Addadi L, Joester D, Nudelman F, Weiner S (2006) Mollusk shell formation: a source of new concepts for understanding biomineralization processes. Chem Eur J 12:981–987

Aksay IA, Trau M, Manne S, Honma I, Yao N, Zhou L, Fenter P, Eisenberger PM, Gruner SM (1996) Biomimetic pathways for assembling inorganic thin films. Science 273:892–898

Albeck S, Addadi L, Weiner S (1996) Regulation of calcite crystal morphology by intracrystalline acidic proteins and glycoproteins. Connect Tissue Res 35:365–370

Balmain J, Hannoyer B, Lopez E (1999) Fourier transform infrared spectroscopy (FTIR) and X-ray diffraction analyses of mineral and organic matrix during heating of mother of pearl (nacre) from the shell of the mollusc *Pinctada maxima*. J Biomed Mater Res 48:749–754

Belcher AM, Wu XH, Christensen RJ, Hansma PK, Stucky GD, Morse DE (1996) Control of crystal phase switching and orientation by soluble mollusc-shell proteins. Nature 381:56–58

Bevelander G, Nakahara H (1969) An electron microscope study of the formation of the nacreous layer in the shell of certain bivalve molluscs. Calcif Tissue Int 3:84–92

Bezares J, Asaro RJ, Hawley M (2008) Macromolecular structure of the organic framework of nacre in Haliotis rufescens: implications for growth and mechanical behavior. J Struct Biol 163:61–75

Blank S, Arnoldi M, Khoshnavaz S, Treccani L, Kuntz M, Mann K, Grathwohl G, Fritz M (2003) The nacre protein perlucin nucleates growth of calcium carbonate crystals. J Microsc Oxf 212:280–291

Bork P (1991) Shuffled domains in extracellular proteins. FEBS Lett 286:47–54

Checa AG, Rodriguez-Navarro AB (2005) Self-organisation of nacre in the shells of Pterioida (Bivalvia: Mollusca). Biomaterials 26:1071–1079

Checa AG, Okamoto T, Ramirez J (2006) Organization pattern of nacre in Pteriidae (Bivalvia: Mollusca) explained by crystal competition. Proc R Soc Lond B Biol Sci 273:1329–1337

Choi CS, Kim YW (2000) A study of the correlation between organic matrices and nanocomposite materials in oyster shell formation. Biomaterials 21:213–222

Cusack M, Freer A (2008) Biomineralization: elemental and organic influence in carbonate systems. Chem Rev Wash DC 108:4433–4454

Dauphin Y, Ball AD, Cotte M, Cuif JP, Meibom A, Salomé M, Susini J, Williams CT (2008) Structure and composition of the nacre-prisms transition in the shell of Pinctada margaritifera (Mollusca, Bivalvia). Anal Bioanal Chem 390(6):1659–1669

Erben HK (1972) Über die bildung und das wachstum von perlmutt. Biomineralization 4:16–46

Evans JS (2008) "Tuning in" to mollusk shell nacre- and prismatic-associated protein terminal sequences: implications for biomineralization and the construction of high performance inorganic-organic composites. Chem Rev (Washington DC) 108:4455–4462

Falini G, Albeck S, Weiner S, Addadi L (1996) Control of aragonite or calcite polymorphism by mollusk shell macromolecules. Science 271:67–69

Falini G, Weiner S, Addadi L (2003) Chitin-silk fibroin interactions: relevance to calcium carbonate formation in invertebrates. Calcif Tissue Int 72:548–554

Fang Z, Feng Q, Chi Y, Xie L, Zhang R (2008) Investigation of cell proliferation and differentiation in the mantle of *Pinctada fucata* (Bivalve, Mollusca). Mar Biol 153:745–754

Fu G, Valiyaveettil S, Wopenka B, Morse DE (2005) $CaCO_3$ Biomineralization: acidic 8-kDa proteins isolated from aragonitic abalone shell nacre can specifically modify calcite crystal morphology. Biomacromolecules 6:1289–1298

Furuhashi T, Beran A, Blazso M, Czegeny Z, Schwarzinger C, Steiner G (2009) Pyrolysis GC/MS and IR spectroscopy in chitin analysis of molluscan shells. Biosci Biotechnol Biochem 73:93–103

Furuhashi T, Miksik I, Smrz M, Germann B, Nebija D, Lachmann B, Noe C (2010) Comparison of aragonitic molluscan shell proteins. Comp Biochem Phys B 155:195–200

Giles R, Manne S, Mann S, Morse DE, Stucky GD, Hansma PK (1995) Inorganic overgrowth of aragonite on molluscan nacre examined by atomic-force microscopy. Biol Bull US 188:8–15

Gong N, Li Q, Huang J, Fang Z, Zhang G, Xie L, Zhang R (2008a) Culture of outer epithelial cells from mantle tissue to study shell matrix protein secretion for biomineralization. Cell Tissue Res 333:493–501

Gong N, Ma Z, Li Q, Yan Z, Xie L, Zhang R (2008b) Characterization of calcium deposition and shell matrix protein secretion in primary mantle tissue culture from the marine pearl oyster *Pinctada fucata*. Mar Biotechnol 10:457–465

Gong N, Shangguan J, Liu X, Yan Z, Ma Z, Xie L, Zhang R (2008c) Immunolocalization of matrix proteins in nacre lamellae and their in vivo effects on aragonitic tablet growth. J Struct Biol 164:33–40

Gotliv BA, Addadi L, Weiner S (2003) Mollusk shell acidic proteins: in search of individual functions. Chembiochem 4:522–529

Guerette PA, Ginzinger DG, Weber BHF, Gosline JM (1996) Silk properties determined by gland-specific expression of a spider fibroin gene family. Science 272:112–115

Heinemann F, Treccani L, Fritz M (2006) Abalone nacre insoluble matrix induces growth of flat and oriented aragonite crystals. Biochem Bioph Res Co 344:45–49

Inoue N, Ishibashi R, Ishikawa T, Atsumi T, Aoki H, Komaru A (2011) Can the quality of pearls from the Japanese pearl oyster (*Pinctada fucata*) be explained by the gene expression patterns of the major shell matrix proteins in the pearl sac? Mar Biotechnol (NY) 13(1):48–55

Jackson DJ, McDougall C, Woodcroft B, Moase P, Rose RA, Kube M, Reinhardt R, Rokhsar DS, Montagnani C, Joubert C, Piquemal D, Degnan BM (2010) Parallel evolution of nacre building gene sets in molluscs. Mol Biol Evol 27:591–608

Jolly C, Berland S, Milet C, Borzeix S, Lopez E, Doumenc D (2004) Zonal localization of shell matrix proteins in mantle of Haliotis tuberculata (Mollusca, Gastropoda). Mar Biotechnol 6:541–551

Kim IW, DiMasi E, Evans JS (2004) Identification of mineral modulation sequences within the nacre-associated oyster shell protein, n16. Cryst Growth Des 4:1113–1118

Kim IW, Darragh MR, Orme C, Evans JS (2006) Molecular "tuning" of crystal growth by nacre-associated polypeptides. Cryst Growth Des 6:5–10

Kobayashi I, Samata T (2006) Bivalve shell structure and organic matrix. Mat Sci Eng C Bio S 26:692–698

Kong Y, Jing G, Yan Z, Li C, Gong N, Zhu F, Li D, Zhang Y, Zheng G, Wang H, Xie L, Zhang R (2009) Cloning and characterization of Prisilkin-39, a novel matrix protein serving a dual role in the prismatic layer formation from the oyster *Pinctada fucata*. J Biol Chem 284:10841–10854

Kono M, Hayashi N, Samata T (2000) Molecular mechanism of the nacreous layer formation in *Pinctada maxima*. Biochem Bioph Res Co 269:213–218

Lamghari M, Almeida M, Berland S, Huet H, Laurent A, Milet C, Lopez E (1999) Stimulation of bone marrow cells and bone formation by nacre: in vivo and in vitro studies. Bone 25:91S–94S

Lao YX, Zhang XQ, Zhou J, Su WW, Chen RJ, Wang YG, Zhou WH, Xu ZF (2007) Characterization and in vitro mineralization function of a soluble protein complex P60 from the nacre of *Pinctada fucata*. Comp Biochem Phys B 148:201–208

Levi Y, Albeck S, Brack A, Weiner S, Addadi L (1998) Control over aragonite crystal nucleation and growth: an in vitro study of biomineralization. Chem Eur J 4:389–396

Levi-Kalisman Y, Falini G, Addadi L, Weiner S (2001) Structure of the nacreous organic matrix of a bivalve mollusk shell examined in the hydrated state using Cryo-TEM. J Struct Biol 135:8–17

Lin AYM, Chen PY, Meyers MA (2008) The growth of nacre in the abalone shell. Acta Biomater 4:131–138

Lowenstam HA, Weiner S (1989) On biomineralization. Oxford University Press, New York/Oxford, pp 99–110

Ma ZJ, Huang J, Sun J, Wang GN, Li CZ, Xie LP, Zhang RQ (2007) A novel extrapallial fluid protein controls the morphology of nacre lamellae in the pearl oyster, *Pinctada fucata*. J Biol Chem 282:23253–23263

Mann S (1993) Molecular tectonics in biomineralization and biomimetic materials chemistry. Nature 365:499–505

Mann S (2001) Biomineralization: principles and concepts in bioinorganic materials chemistry. Oxford University Press, Oxford, pp 8, 78, 103–106

Mann K, Weiss IM, Andre S, Gabius HJ, Fritz M (2000) The amino-acid sequence of the abalone (Haliotis laevigata) nacre protein perlucin – Detection of a functional C-type lectin domain with galactose/mannose specificity. Eur J Biochem 267:5257–5264

Mann K, Siedler F, Treccani L, Heinemann F, Fritz M (2007) Perlinhibin, a cysteine-, histidine-, and arginine-rich miniprotein from abalone (Haliotis laevigata) nacre, inhibits in vitro calcium carbonate crystallization. Biophys J 93:1246–1254

Marin F, Luquet G (2004) Molluscan shell proteins. CR Palevol 3:469–492

Marin F, Luquet G, Marie B, Medakovic D (2008) Molluscan shell proteins: primary structure, origin, and evolution. Curr Top Dev Biol 80:209–276

Matsushiro A, Miyashita T, Miyamoto H, Morimoto K, Tonomura B, Tanaka A, Sato K (2003) Presence of protein complex is prerequisite for aragonite crystallization in the nacreous layer. Mar Biotechnol 5:37–44

Meyers MA, Lin AYM, Chen PY, Muyco J (2008) Mechanical strength of abalone nacre: role of the soft organic layer. J Mech Behav Biomed 1:76–85

Michenfelder M, Fu G, Lawrence C, Weaver JC, Wustman BA, Taranto L, Evans JS, Morse DE (2003) Characterization of two molluscan crystal-modulating biomineralization proteins and identification of putative mineral binding domains. Biopolymers 70:522–533

Miyamoto H, Miyashita T, Okushima M, Nakano S, Morita T, Matsushiro A (1996) A carbonic anhydrase from the nacreous layer in oyster pearls. Proc Natl Acad Sci USA 93:9657–9660

Miyamoto H, Miyoshi F, Kohno J (2005) The carbonic anhydrase domain protein nacrein is expressed in the epithelial cells of the mantle and acts as a negative regulator in calcification in the mollusc *Pinctada fucata*. Zool Sci 22:311–315

Miyashita T, Takagi R, Okushima M, Nakano S, Miyamoto H, Nishikawa E, Matsushiro A (2000) Complementary DNA cloning and characterization of pearlin, a new class of matrix protein in the nacreous layer of oyster pearls. Mar Biotechnol 2:409–418

Miyazaki Y, Nishida T, Aoki H, Samata T (2010) Expression of genes responsible for biomineralization of *Pinctada fucata* during development. Comp Biochem Phys B 155:241–248

Nakahara H (1981) The Formation and Fine Structure of the Organic Phase of the Nacreous Layer in Mollusc Shell. In: Mollusk Research (Collection of Assays Contributed in Celebration of Professor Masae Omori's 60th Birthday), pp 21–27

Nudelman F, Gotliv BA, Addadi L, Weiner S (2006) Mollusk shell formation: mapping the distribution of organic matrix components underlying a single aragonitic tablet in nacre. J Struct Biol 153:176–187

Nudelman F, Chen HH, Goldberg HA, Weiner S, Addadi L (2007) Spiers memorial lecture: lessons from biomineralization: comparing the growth strategies of mollusc shell prismatic and nacreous layers in Atrina rigida. Faraday Discuss 136:9–25

Nudelman F, Shimoni E, Klein E, Rousseau M, Bourrat X, Lopez E, Addadi L, Weiner S (2008) Forming nacreous layer of the shells of the bivalves Atrina rigida and *Pinctada margaritifera*: an environmental- and cryo-scanning electron microscopy study. J Struct Biol 162:290–300

Oaki Y, Imai H (2005) The hierarchical architecture of nacre and its mimetic material. Angew Chem Int Ed Engl 44:6571–6575

Patthy L (1999) Genome evolution and the evolution of exon-shuffling – a review. Gene 238:103–114

Pereira-Mouries L, Almeida MJ, Ribeiro C, Peduzzi J, Barthelemy M, Milet C, Lopez E (2002) Soluble silk-like organic matrix in the nacreous layer of the bivalve *Pinctada maxima*. Eur J Biochem 269:4994–5003

Rousseau M, Lopez E, Coute A, Mascarel G, Smith DC, Naslain R, Bourrat X (2005a) Sheet nacre growth mechanism: a Voronoi model. J Struct Biol 149:149–157

Rousseau M, Lopez E, Stempfle P, Brendle M, Franke L, Guette A, Naslain R, Bourrat X (2005b) Multiscale structure of sheet nacre. Biomaterials 26:6254–6262

Samata T (1990) Ca-binding glycoproteins in molluscan shells with different types of ultrastructure. Veliger 33:190–201

Samata T (2004) Recent advances in studies on nacreous layer biomineralization: molecular and cellular aspects. Thalassas 20:25–44

Samata T, Hayashi N, Kono M, Hasegawa K, Horita C, Akera S (1999) A new matrix protein family related to the nacreous layer formation of *Pinctada fucata*. FEBS Lett 462:225–229

Sarikaya M, Aksay I (1992) Nacre of abalone shell: a natural multifunctional nanolaminated ceramic-polymer composite material. In: Case ST (ed) Results and problems in cell differentiation, vol 19. Springer, Berlin/London, pp 1–25

Saruwatari K, Matsui T, Mukai H, Nagasawa H, Kogure T (2009) Nucleation and growth of aragonite crystals at the growth front of nacres in pearl oyster, *Pinctada fucata*. Biomaterials 30:3028–3034

Schaffer TE, IonescuZanetti C, Proksch R, Fritz M, Walters DA, Almqvist N, Zaremba CM, Belcher AM, Smith BL, Stucky GD, Morse DE, Hansma PK (1997) Does abalone nacre form by heteroepitaxial nucleation or by growth through mineral bridges? Chem Mater 9:1731–1740

Schonitzer V, Weiss IM (2007) The structure of mollusc larval shells formed in the presence of the chitin synthase inhibitor Nikkomycin Z. BMC Struct Biol 7:71

Shen XY, Belcher AM, Hansma PK, Stucky GD, Morse DE (1997) Molecular cloning and characterization of lustrin A, a matrix protein from shell and pearl nacre of Haliotis rufescens. J Biol Chem 272:32472–32481

Simkiss K, Wilbur KM (1989) Biomineralization: cell biology and mineral deposition. Academic, San Diego, pp 230–250

Song F, Zhang XH, Bai YL (2002) Microstructure and characteristics in the organic matrix layers of nacre. J Mater Res 17:1567–1570

Sudo S, Fujikawa T, Nagakura T, Ohkubo T, Sakaguchi K, Tanaka M, Nakashima K, Takahashi T (1997) Structures of mollusc shell framework proteins. Nature 387:563–564

Suzuki M, Sakuda S, Nagasawa H (2007) Identification of chitin in the prismatic layer of the shell and a chitin synthase gene from the Japanese pearl oyster, *Pinctada fucata*. Biosci Biotech Biochem 71:1735–1744

Suzuki M, Saruwatari K, Kogure T, Yamamoto Y, Nishimura T, Kato T, Nagasawa H (2009) An acidic matrix protein, Pif, is a key macromolecule for nacre formation. Science 325:1388–1390

Takakura D, Norizuki M, Ishikawa F, Samata T (2008) Isolation and characterization of the N-linked oligosaccharides in nacrein from *Pinctada fucata*. Mar Biotechnol 10:290–296

Takeuchi T, Endo K (2006) Biphasic and dually coordinated expression of the genes encoding major shell matrix proteins in the pearl oyster *Pinctada fucata*. Mar Biotechnol 8:52–61

Treccani L, Mann K, Heinemann F, Fritz M (2006) Perlwapin, an abalone nacre protein with three four-disulfide core (whey acidic protein) domains, inhibits the growth of calcium carbonate crystals. Biophys J 91(7):2601–2608

Wang N, Kinoshita S, Riho C, Maeyama K, Nagai K, Watabe S (2009) Quantitative expression analysis of nacreous shell matrix protein genes in the process of pearl biogenesis. Comp Biochem Phys B 154:346–350

Watabe N (1981) Crystal-growth of calcium-carbonate in the invertebrates. Prog Cryst Growth Charact Mater 4:99–147

Weiner S, Addadi L (1997) Design strategies in mineralized biological materials. J Mater Chem 7:689–702

Weiner S, Traub W (1984) Macromolecules in mollusk shells and their functions in biomineralization. Philos Tr Soc B 304:425–433

Weiss IM, Schonitzer V (2006) The distribution of chitin in larval shells of the bivalve mollusk Mytilus galloprovincialis. J Struct Biol 153:264–277

Weiss IM, Kaufmann S, Mann K, Fritz M (2000) Purification and characterization of perlucin and perlustrin, two new proteins from the shell of the mollusc Haliotis laevigata. Biochem Biophys Res Commun 267:17–21

Weiss IM, Gohring W, Fritz M, Mann K (2001) Perlustrin, a Haliotis laevigata (abalone) nacre protein, is homologous to the insulin-like growth factor binding protein N-terminal module of vertebrates. Biochem Biophys Res Commun 285:244–249

Weiss IM, Schonitzer V, Eichner N, Sumper M (2006) The chitin synthase involved in marine bivalve mollusk shell formation contains a myosin domain. FEBS Lett 580:1846–1852

Westbroek P, Marin F (1998) A marriage of bone and nacre. Nature 392:861–862

Wustman BA, Weaver JC, Morse DE, Evans JS (2003) Structure-function studies of the lustrin A polyelectrolyte domains, RKSY and D4. Connect Tissue Res 44:10–15

Yan ZG, Jing G, Gong NP, Li CZ, Zhou YJ, Xie LP, Zhang RQ (2007) N40, a novel nonacidic matrix protein from pearl oyster nacre, facilitates nucleation of aragonite in vitro. Biomacromolecules 8:3597–3601

Yano M, Nagai K, Morimoto K, Miyamoto H (2007) A novel nacre protein N19 in the pearl oyster *Pinctada fucata*. Biochem Biophys Res Commun 362:158–163

Yao N, Epstein AK, Liu WW, Sauer F, Yang N (2009) Organic-inorganic interfaces and spiral growth in nacre. J R Soc Interface 6:367–376

Zhang C, Zhang RQ (2006) Matrix proteins in the outer shells of molluscs. Mar Biotechnol 8:572–586

Zhang C, Li S, Ma ZJ, Xie LP, Zhang RQ (2006) A novel matrix protein p10 from the nacre of pearl oyster (*Pinctada fucata*) and its effects on both $CaCO_3$ crystal formation and mineralogenic cells. Mar Biotechnol 8:624–633

Chapter 13
Acidic Shell Proteins of the Mediterranean Fan Mussel *Pinna nobilis*

Frédéric Marin, Prabakaran Narayanappa, and Sébastien Motreuil

Contents

F. Marin (✉)
UMR CNRS 5561 "Biogéosciences", Université de Bourgogne, 6, Boulevard Gabriel, Dijon 21000, France
e-mail: Frederic.marin@u-bourgogne.fr

W.E.G. Müller (ed.), *Molecular Biomineralization*, Progress in Molecular and Subcellular Biology 52, DOI 10.1007/978-3-642-21230-7_13,

Abstract In molluscs, the shell secretion process is controlled by a set of extracellular macromolecules collectively called the shell matrix. The shell matrix, which is produced by the mantle epithelial cells during mineralization, is predominantly composed of proteins, glycoproteins, acidic polysaccharides, and chitin that precisely regulate the deposition of calcium carbonate outside the mantle cells. In the present paper, we focus on the shell of *Pinna nobilis*, the giant Mediterranean fan mussel, usually considered as a model for studying molluscan biomineralization processes. *P. nobilis* exhibits indeed a nacro-prismatic shell, the outer layer of which is constituted of the so-called "regular simple calcitic prisms," according to Carter and Clark (1985). We review here the microstructural characteristics of the prisms and nacre and the biochemical properties of their associated matrices. In particular, the calcitic prisms of *P. nobilis* are characterized by a cortege of unusually acidic intraprismatic proteins, while the ones of the nacreous layer seem less acidic. A brief description of the molecular characterization of three acidic proteins, caspartin, calprismin and mucoperlin, is given. In particular, we show that extremely acidic intracrystalline proteins such as caspartin interact with calcium carbonate at different scales, from micrometric to crystal lattice levels.

13.1 Biomineralization of the Molluscan Shell, a Brief Overview of the Mechanism

In the metazoan world, molluscs are often considered as the master controllers of biomineralization processes, and their shell represents the most remarkable example of a biologically controlled mineralization produced outside living tissues by a calcifying epithelium (Simkiss and Wilbur 1989).

The organ responsible for the mineral deposition and shell formation is the mantle, the thin epithelium that envelopes the soft body of molluscs, and which is in contact with the growing shell. More precisely, according to classical views, the calcification takes place in a closed space, the extrapallial space, located at the interface between the mantle tissues, the leathery periostracum and the growing shell itself. The shell emerges from a subtle chemistry between the precursor mineral ions, i.e., calcium, bicarbonate, and minor elements such as magnesium or strontium that are extruded in the extrapallial space, and the numerous macromolecular extracellular components produced by the mantle epithelial cells, i.e., proteins and polysaccharides. All these components react in a well-coordinated manner. This self-assembling process produces solid and compact microstructures such as nacre, prisms, crossed-lamellar, foliated, or homogeneous textures (Carter 1990).

In the recent years, numerous new data have completed this classical view. The most recent advances deal more particularly with the structure of shell biominerals at the nanoscale (Oaki and Imai 2005), and with the identification of an increasing number of shell matrix proteins (Marin et al. 2008). Although the organic matrix represents usually less than 5% of the shell weight, it is however the major component that controls different aspects of the shell formation processes (Addadi et al. 2006). Until recently, the classical paradigm was to consider that the control of shell synthesis at the biocrystal scale was performed primarily by two main mechanisms: crystal nucleation (and subsequent elongation) and growth inhibition (Wheeler et al. 1988). New concepts and emerging models try now to translate a more complex and dynamic reality, which is remarkably illustrated by the wide variety of shell proteins, characterized in the last few years (Marin et al. 2008). These proteins cover a broad spectrum of pI, from very acidic to very basic. The primary structure of a number of them is composed of different modules, suggesting that these proteins are multifunctional. Some of them exhibit enzymatic activities. Others may be involved in cell signaling. Some others have remarkable crystal-binding properties. Nowadays, the extracellular calcifying shell matrix appears as a whole integrated system, which regulates protein–mineral and protein–protein interactions as well as feedback interactions between the biominerals and the calcifying epithelium that synthesized them.

In the present paper, we focus on a peculiar model organism, *Pinna nobilis*, the Mediterranean fan mussel. By many aspects, the choice of this species may appear odd. This species cumulates indeed few handicaps: first, it is a strongly protected species, a fact which does not facilitate field sampling; second, studies on the physiology and reproduction of *Pinna nobilis* are sparse, and several field and laboratory data have still to be generated on these topics; third, there are no extensive genetic data available, and, as far as we know, there are no coming initiative aiming at sequencing the genome, or, at least, aiming at obtaining transcriptomic data (ESTs). At last, and this last fact explains the situation, the world scientific community, which works – temporarily or full-time – on *Pinna nobilis* is extremely limited: a rough estimation oscillates between two and three dozens of scientists on a world scale. These obstacles would appear unsurmountable, if they were not compensated by the fact that *Pinna nobilis*, from a historical viewpoint, is considered as a model organism in biomineralization, in particular, by the production of its outer shell layer, made of "regular simple calcitic prisms," the structural and molecular characteristics of which are detailed in the present review. Because the calcitic prisms of *Pinna nobilis* are large and exhibit remarkable optical properties, they represent fascinating objects, for trying to understand how molluscs are able to control their biocrystal synthesis over molecular to millimeter scales. We believe that they may be a source of inspiration for the analysis of structural hierarchy in biomineralization and for biomimicry purpose.

13.2 *Pinna nobilis*, a Model for Understanding Molluscan Shell Formation

13.2.1 Presentation of Pinna nobilis

Also called noble pen mussel, rough pen shell, fan mussel, fan shell, razor fish, sea-wings or wing-shell, *Pinna nobilis* is the biggest Mediterranean bivalve, and together with the Giant clam *Tridacna gigas*, one of the biggest molluscs in the world: its size currently exceeds 80 cm and may reach 1 m. Its longevity is estimated to be around 30 years.

This species is endemic to the Mediterranean sea and can be observed along different coasts, in particular along the "côte d'Azur" and around Corsica in France (Moreteau and Vicente 1982; Medioni and Vicente 2003), the Spanish coast and around the Balearic Islands (Garcia-March 2003; Garcia-March et al. 2007), on the Tunisian (Rabaoui et al. 2008), Italian (Centoducati et al. 2007), Greek (Katsanevakis 2007), and Crete (Katsanevakis and Thessalou-Legaki 2009) coasts. *Pinna nobilis* is also well represented in the Adriatic, along the Croatian coast (Zavodnik 1967; Siletic and Peharda 2003). This species, which lives in the infralittoral domain between 0.5 m and 30 m depth, grows almost exclusively in the meadows of the seagrass *Posidonia oceanica*. In its natural environment, the shell of *Pinna nobilis* has a very characteristic living position: the pointed anterior end of the shell – about one-third – is buried in the soft sediment and the elongated fan-shaped shell is erected more or less vertically. Byssal threads anchor the animal to underlying stones or other objects. *Pinna nobilis* lives in small grouping: the populations are usually patchy and sparse (Katsanevakis 2007).

Although its size rendered it attractive, *Pinna nobilis* had limited uses throughout history as a marine resource: during the Phenician period, and later, during the Roman times and until the first world war in Italy and in Malta, the long and solid byssus (up to 20 cm) – the"sea-silk" – was cut and collected for the making of tissues, of gloves, or of hats of priceless value (Brisou 1985; Maeder and Halbeisen 2001). Although of moderate taste value, the soft tissues of *Pinna nobilis* were consumed in Malta, in Corsica and in former Yougoslavia. The thick nacre of adult specimens was used for manufacturing clothes buttons, in particular in Sicilia, in Malta, and in the South of Italy. Natural brown-red pearls of respectable size, known as the pen shell pearl can be found in adult specimens, but they have no commercial value (Gauthier et al. 1994). These pearls, known as the prism pearls (Schmidt 1932), are naturally stained by carotenoid pigments (Karampelas et al. 2009). Until recently, the full empty shell of juvenile or adult specimens was collected for souvenir purposes. Nowadays, *Pinna nobilis* is considered as an endangered species: since 1992, it is registered in the official list established by the European Union for the conservation of natural habitats and of wild fauna and flora (*Off. J. E. C.* L206, 22.7.1992). In some spots (Port-Cros field, mission "Poseidon") where populations could be followed during three decades (Vicente 2003), the situation was particularly dramatic few years ago. In protected areas, nowadays, populations tend to reconstitute again (Medioni and Vicente 2003; Foulquié and Dupuy de la Grandrive 2003). However, this growth is fragile, and populations are vulnerable to catastrophic climatic events such as the recent storm (May 2010) that broke on Côte d'Azur (S. Motreuil, personal communication, 2010). Attempts to capture and recruit larvae on artificial devices, to let them grow in controlled conditions and to reimplant juveniles in natural environment may be the best option for reconstituting natural populations (De Gaulejac and Vicente 1990; Vicente 2003; Cabanellas-Reboredo et al. 2009).

13.2.2 *Physiology, Development, and Reproduction of* Pinna nobilis

There has been a limited number of recent studies on the physiology and reproduction of *Pinna nobilis* (De Gaulejac 1993; Riva 2003). Most of them are due to the CERAM (Centre d'Etude des Ressources Animales Marines) and to the Institut océanographique Paul Ricard (Vicente, De Gaulejac, Riva, and coworkers). The general metabolism of *Pinna nobilis*, in particular respiration and filtration, seems to be rather low compared to other bivalves like the mussel, the edible oyster or the clam (Vicente et al. 1992). To our knowledge, no physiological studies were performed in relation with shell calcification: in particular, although hemocytes of *Pinna nobilis* were precisely described (Henry et al. 1992), no link was established between the presence of these cells in mantle tissues and the transport of calcium or of amorphous calcium carbonate granules for shell mineralization.

Pinna nobilis is a successive hermaphrodite, with an asynchronous gamete maturation (De Gaulejac et al. 1995a, b; Vicente 2003), which avoids self-fertilization. Its sexual cycle is divided into two phases: a sexual repose from October to March, an active sexual period for the rest of the year, with in particular a succession of alternate spawning and fast gametogenesis from June to August. Ovocytes are fertilized in full seawater. After fecondation, the development of *P. nobilis* follows the classical developmental pathway for bivalves, namely, a trochophore larval stage (ciliated larva), and a veliger stage (swimming larva). When the veliger larva gets fixed of a substratum, the metamorphosis operates (loss of the velum, reorganization of the internal organs), and the larva transforms into a juvenile specimen. The larval shell starts to be secreted early during development, during the trochophore stage. However, unlike works on other bivalves like *Mytilus* or *Ostrea*, there were no precise studies on the shell ontogeny of *P. nobilis*.

13.2.3 *Systematic Position of* Pinna nobilis *and Ancestry of the Pinnid Family*

For the taxonomical position of *Pinna nobilis* within the Mollusca phylum, we refer to the classification of Giribet (2008), based on morphological and molecular characters, and to the molecular phylogeny established by Steiner and Hammer (2000) for Pteriomorphian bivalves. *P. nobilis* is a bivalve which belongs to the Pteriomorphia subclass. This heterogeneous clade is nowadays considered as the sister group of the heteroconchs, which itself comprises paleoheterodont (freshwater mussels) and heterodont (clams) bivalves. The Pteriomorphia subclass comprises nine superfamilies (Mytiloidea, Pterioidea, Pinnoidea, Arcoidea, Limoidea, Ostreoidea, Plicatuloidea, Pectinoidea, and Anomioidea, see Steiner and Hammer 2000), the three first of them being characterized by typical nacro-prismatic shell texture. *P. nobilis* belongs to the Pinnoidea superfamily, and to the Pinnidae family, which comprises three living genera, *Atrina*, *Pinna*, and *Streptopinna*, and seven fossil ones. The Pinnidae family comprises 123 living and fossil species and subspecies. The *Pinna* genus itself comprises 75 extant and extinct species and subspecies. The living species all together have a broad geographic repartition all over the world, with a predilection for warm waters. For a detailed list of the species that belong to the *Pinna* genus, we refer to the following Web site: http://zipcodezoo.com/Key/Animalia/Pinna_Genus.asp.

According to Taylor et al. (1969) and Carter (1990), the Pinnidae family is of Paleozoic origin, and appeared at the basal Carboniferous. Genera like *Aviculopinna*, *Meekopinna*, or *Pteronites*, and species like *Pinna peracuta or Pinna flexistria* are attested to be of Carboniferous age. The Pinnid family may derive from Devonian pterineids, such as Leptodesma (Carter 1990). The family has a discontinuous fossil record. The well-known *Trichites* genus, characterized by an extremely thick calcitic prismatic layer, is frequent in the calcareous sediments of the Upper Jurassic (Tithonian).

13.3 Shell Formation Process

13.3.1 Shell Growth

As shown in Fig. 13.1, the shell of *Pinna nobilis* has a typical elongated triangular shape with a sharp anterior. The shell is equivalve. The color of the shell is light brown for juvenile specimens and becomes brown-red when the shell thickens. This coloration, localized exclusively in the outer shell layer, is due to a mixture of carotenoid pigments (Gauthier et al. 1994), the exact molecular characterization of which has still to be done. In juvenile specimens, the shell is translucent (Fig. 13.1b), but becomes opaque with age.

Like several bivalves of Mediterranean or temperate zones, the shell growth of *P. nobilis* follows a seasonal pattern: the shell grows slowly during the cold months and has a maximal growth during late spring and early summer (Katsanevakis 2007). This optimum may result from a combination of temperature and food availability. When the water temperatures exceed a threshold (around 29°C) during August, the shell growth slows down or stops. The growth of the shell, which has been monitored during the entire life of several specimens, follows a classical Von Bertalanffy model, with a rapid almost linear growth during the first 3 years, and a progressive decrease of the growth rate until reaching a plateau (Moreteau and Vicente 1982). An equation was formulated to describe the mean shell growth:

$H_t = 86.3\left[1 - e^{-0.053(t+0.22)}\right]$ where H_t is the full length, and t, the age in months.

According to this equation, a 1-year-old specimen reaches 12 cm, a 2-year-old, 22 cm, a 3-year-old, 30 cm. At these early stages, the shell remains very thin; it is extremely fragile and translucent. 20-year-old specimens are about 80 cm long, and the shell thickness is about 1 cm at the tip. More recent studies on different

Fig. 13.1 The shell of *Pinna nobilis*. (**a**) shell of a juvenile specimen, approx. age 3-year-old, size 300 mm; note that the internal nacreous layer, in shiny *white*, is restricted to the first half of the shell. The outer calcitic prismatic layer is covered with spines. (**b**) shell of a very young specimen, age about 5–6 months old, size 60 mm; the shell is almost transparent. (**c**) detailed view of the delicate calcitic hollow spines of the outer shell layer, among juvenile specimens; with aging, the spines progressively erode

populations showed that the growth kinetics is submitted to slight variations (Garcia-March et al. 2002; Rabaoui et al. 2007), according to environmental parameters, including temperatures, water turbidity, depth, nutrient fluxes.

Because the growth of the shell is rapid during the first years (a daily growth above 0.3 mm for a 1-year-old specimen), it is likely that juvenile specimens allocate an important part of their energetic budget to the calcification of their shell, which, in other words, may mean that other metabolic activities are kept low. However, to our knowledge, there are no precise data in the literature on this aspect of the physiology of *Pinna nobilis*.

The shells of young specimens are characterized by the formation of thin calcitic hollow spines, produced by the pleated border of the mantle (Fig. 13.1a–c). These spines, which are extremely sharp, fragile and brittle, are regularly distributed on all the outer surface of the two valves (Fig. 13.1c), according to the shell growth lines. When aging, they are progressively eroded (Cosentino and Giacobbe 2006). In 3-year-old shells, one observes a gradient of spine erosion from the shell border where spines are newly formed to the shell tip, where they are fully abraded (Fig. 13.1a). In specimens older than 4–5 years, the spines are fully eroded. The spines do not have a biomechanical function for the shell. They constitute however a defensive tool against predation and biting of young specimens by fishes. They also constitute anchoring points for the settlement of epibiontic algal communities, which provide a camouflage to the shell in *Posidinia* meadows. At last, the spines shelter a large community of epizoobiontic molluscs (Giacobbe 2002; Cosentino and Giacobbe 2007a, b; Rabaoui et al. 2007, 2009). Thus, the spines constitute an efficient adaptation to passively and actively prevent predation on juvenile specimens, the shell of which is particularly fragile.

13.3.2 Shell Microstructures

The shell of *P. nobilis* exhibits the typical bilayered calcified structure (in addition to the outer organic periostracum), found in several pteriomorphid bivalves. The periostracal layer is visible particularly in juvenile specimens, but tends to quickly erode in older specimens. The outer mineralized layer is calcitic and composed of "simple" prisms, developed perpendicularly to the surface of the shell. The "simple prisms" terminology is doubly misleading. First, "simple" would refer to an elementary structure, but we show below (Sect. 3.4) that the prisms of *P. nobilis* exhibit a structural hierarchy, which is everything but simple. Second, "prism" designates very different objects in molluscan shell microstructures, like the tiny oblique prisms of the edible mussel, the prisms of the freshwater mussel, the composite prisms of the clam, and those of the pearl oyster (Carter 1990). These objects are brought together under a single term, but it seems obvious that they are not synthesized through a unique pathway.

The internal layer is light grey and aragonitic. It exhibits the iridescence characteristic of nacreous textures. While the outer prismatic layer covers the entire shell, the nacreous layer is restricted to the area between the adductor muscle scars, which represents about one-half of the shell height. By analogy with more studied nacro-prismatic bivalve models such as *Pinctada* (Sudo et al. 1997), the fact that the shell is bilayered means that the mantle epithelium that produces the shell is not homogeneous: the epithelial cells located at the border of the mantle are dedicated to the synthesis of the prismatic layer while epithelial cells situated more proximally from the shell tip secrete the nacre layer. However, histological studies should be done to confirm this finding.

From a historical perspective, it is remarkable to notice that the shell microstructure of *Pinna* sp., which is a rather uncommon genus, was extensively described as early as the beginning of the ninetieth century, with the successive works of De Bournon (1808), Gray (1835), Bowerbank (1844), Carpenter (1844), Leydolt (1856), Rose (1858). During the early twentieth century, other microstructural studies followed, in particular those of Biedermann (1901), Römer (1903), Karny (1913), Cayeux (1916), Schmidt (1923, 1924) and of course, the classical study of Boggild (1930).

With the development of the electron microscopy in the thirties, and the conception of the scanning electron microscope after WWII, *Pinna* sp. was also the subject of different studies and monographies, including papers from Grégoire (1967), Wise (1970), Mutvei (1970), Wada (1972, 1980). Let us cite the extensive work of Taylor and coworkers (1969) on bivalves. In the eighties, the shell microstructures of *Pinna* sp. were described through series of papers of Cuif and coworkers (1980, 1983a, b, 1985, 1987a, b), Cuif et Raguideau (1982), Nakahara et al. (1980), Carter and Clark (1985) and finally, the voluminous treatise on microstructures of Carter (1990). Recent microstructural, mineralogical, and physical studies include the works of Dauphin et al. (2003), of Checa et al. (2005), and of Esteban-Delgado et al. (2008).

13.3.3 The Calcitic Prisms of Pinna nobilis

The data on the shell ontogeny of *Pinna nobilis* are not known, and all our knowledge derives from developmental studies performed on bivalves with similar shell texture, like *Mytilus* (Medakovic 2000). Following this general scheme, the early shell appears during the trochophore stage, and is secreted by the transitory shell gland, the group of cells resulting from the invagination of the ectoderm (Kniprath 1981). After the production of an organic lamella – the future periostracum – by the shell gland, the first mineralization is deposited between the lamella and the shell gland itself. By analogy with other better-studied bivalves, it is likely that the first minerals produced are amorphous (Mao Che et al. 2001; Weiss et al. 2002). Then, the prismatic layer appears, by growing inward from the organic lamella. At a later stage (veliger or post-metamorphosis?), the deposition of the nacreous layer starts in the hinge area, while the prismatic layer continues its lateral extension on the two valves.

Poorly investigated from an ontogenetic viewpoint, the prismatic layer has been well studied in juvenile or adult specimens. As described by Boggild (1930), the prisms, when seen in transverse section, "are very regular and, at the same time, distinct, with an interprismatic substance. The orientation of the optic axis is unusually regular, parallel to the direction of the prisms." Taylor and coworkers (1969) distinguished two types of prisms, the simple ones and the composite ones. Those of *P. nobilis* belong to the first category. These authors gave a more precise definition of the prisms: "simple calcite prisms, arranged with their crystallographic c axes normal to the layer surface. The prisms are larger than in any other family... and the prismatic layer as a whole can be several millimeters thick." In their scrupulous attempt to classify molluscan shell microstructures, Carter and Clark (1985) recognized several types of prismatic structures, defined as "mutually parallel, adjacent structural units (1st order prisms) that do not strongly interdigitate along their mutual boundaries. The first order prisms are generally longer than they are wide." Those of *Pinna nobilis* are defined as *simple prismatic* structures, "in which each 1st order prism shows a non-spherulitic prismatic and non-composite prismatic substructure. The boundaries of adjacent 1st order prisms are well defined and generally non-interdigitating"...and further "they are commonly oriented with their length axes perpendicular to the plane of the shell layer." Furthermore, Carter and Clark (1985) introduced a subcategory, of *regular simple prismatic* structures, in which "each 1st order prism is columnar and has a more or less equidimensional (non-elongated) polygonal cross sectional shape. The prism shapes and their diameter appear rather uniform in section perpendicular to the prism length axes, although the prism

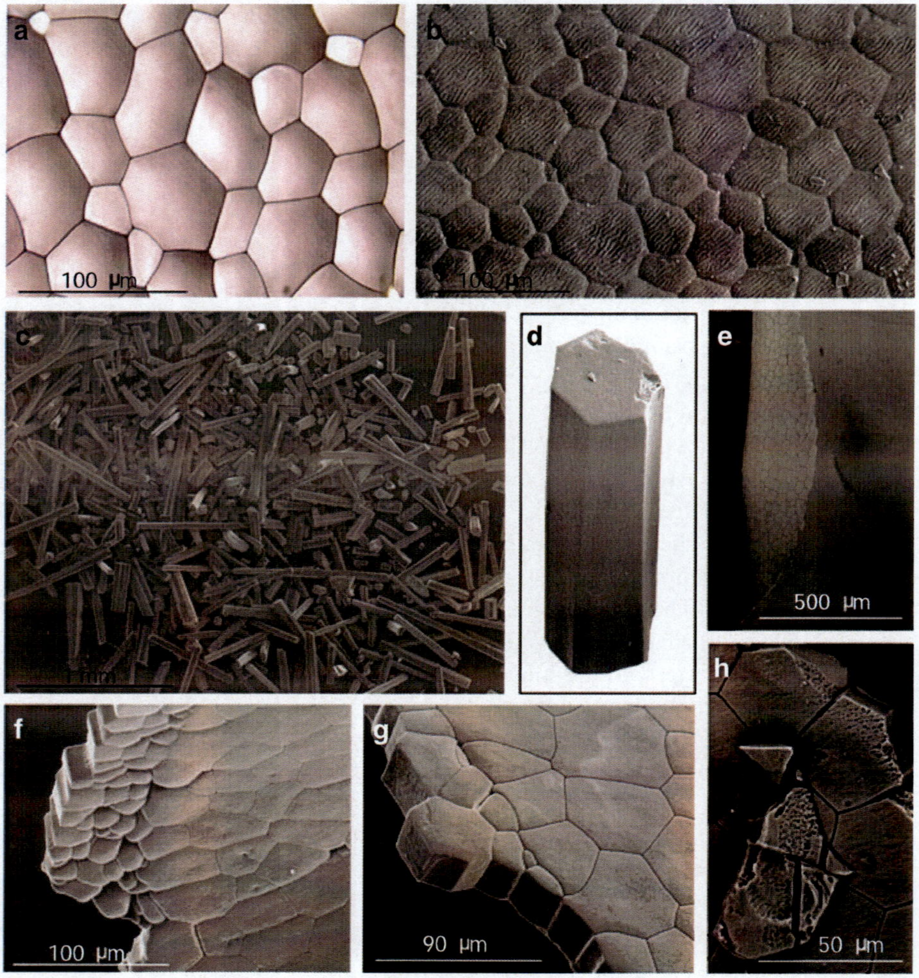

Fig. 13.2 The prismatic calcitic layer of *Pinna nobilis*. (**a**) thin section (about 30 μm) of the prismatic layer, observed with an optical microscope. Ridges are visible on the prisms. (**b**) similar view, taken with a SEM. The ridges are clearly visible. Note that they all have the same orientation. (**c**) isolation of single calcitic prisms with sodium hypochlorite. (**d**) one isolated prism. (**e**) macroscopic view of a hollow spine of the shell of a few-months-old specimen. The whole spine is constituted of flat prisms. (**f–h**) newest synthesized prisms at the edge of a 6-month-old shell. Prisms look like tiles. (**h**) slight etching with EDTA 1% (w/vol), 5 min

diameters generally increase from the outer to the inner parts of the shell through geometric selection." The calcitic prisms of *P. nobilis* belong to this subcategory.

When observed in cross section, the prisms of *P. nobilis* exhibit a polygonal section, generally penta- to heptagonal (Fig. 13.2). Consequently, in surface view, the prismatic layer exhibits a typical "honeycomb" structure pattern (Taylor et al. 1969), with the prisms maintained together by an interprimatic insoluble organic sheath, that old authors assimilated to conchiolin (Grégoire 1967).

The thickness of this sheath is less than 1 μm (we estimate that it varies between 0.8 and 1 μm, except at the triple junctions where it is thicker). This sheath is structurally coherent and flexible: when the prisms are partly or totally dissolved with EDTA or with weak acid, the sheath remains in place. Because the sheath represents a high proportion of the shell (about 4% wt/wt) and because of its honeycomb structure, it confers to the shell a surprising bending capability, in particular, in juvenile specimens. On the other hand, the prismatic layer of *P. nobilis* can be entirely dissociated in single prism units, by selectively degrading the organic sheath. The most classical chemical reagent used for that purpose is dilute sodium hypochlorite. In our lab, we currently treat fragments of prismatic layers for few days with a 0.26% of active chlorine solution (dilution ten times from a commercial solution). The single prisms can be subsequently collected and extensively washed with water on a 5-μm filter. Removal of the organic sheath has also been successfully performed by enzymatic proteolysis with an enzyme – like pronase B – that exhibits a broad spectrum of proteolytic activity (Keller 1981; Keller and Dauphin 1983).

13.3.4 *Ultrastructure of the Prisms of* P. nobilis *and Complexity of the Organo-mineral Interactions*

By many aspects, the calcitic prisms of *Pinna nobilis* appear as paradoxical objects, because they conciliate an apparent crystallographic simplicity and an ultrastructural complexity. The nature of the calcitic prisms of *P. nobilis* is revealed by investigating the crystallographic properties on single prisms, or on prisms preparations observed in cross sections, in combination with analysis of their substructures. As mentioned by Carter (1990, p 213), "the calcitic, regular simple prisms in pinnid shells are commonly optically homogeneous with perfect, monocrystalline, non-grainy extinction under crossed polarized light." This means that each prism behaves like a single crystal, with a single extinction (Wada 1961; Cuif and Raguideau 1982). This optical property is confirmed by X-ray diffraction on single prism (F. Marin, unpublished data, 2003). This technique demonstrates that one prism diffracts like a single crystal.

Although the prisms of *P. nobilis* are optically homogeneous, they are neither "single" nor "simple" crystals, and many studies have underlined their complex substructures that superimpose at different scales. What makes their peculiar ultrastructural complexity is the presence of an intraprismatic organic matrix, the characteristics of which are detailed in Sect. 4.

Decalcification experiments aiming at dissolving the prisms and keeping the organic sheaths show that these sheaths exhibit a transverse striation (Grégoire 1967). To this striation, perpendicular to the long axis of the prisms, corresponds a striation on the prisms themselves (Fig. 13.2), visible in longitudinal view. This striation marks the successive growth plans. In spite of the presence of the sheath that separates each prism from its neighbors, the growth lines can be followed for millimiters throughout the length of the shell section (Fig. 13.2).

Pyrolysis experiments performed on single prisms showed that they could be entirely dissociated in minute flat crystals (Frérotte 1987). In an oversimplified view, one could consider that each single prism is constituted of a pile of numerous flat crystallites – intercalated with layers of organic material – that are deposited layer per layer. However, as noted by Cuif and coworkers (1983a, b), this view is completely deceptive, for at least two reasons: (1) The relationships between the organic and the mineral phases are much more complex than a simple alternation between organic and mineral layers. (2) The so-called "elementary" crystals produced by pyrolysis are not homogeneous, but they have a micro-granular texture; this suggests that they are themselves constituted of nanocrystals, which exhibit the same crystallographic orientation. Such an organization of "single crystals" in elementary nanoblocks has been observed for nacre tablets (Oaki and Imai 2005), and may represent a universal rule in biomineralization.

Whatever the organization of the intraprismatic matrix, whatever the size and shape of the elementary units that constitute each prism, all these units exhibit exactly the same optical orientation of their three axes, which is, by itself, absolutely remarkable. This explains the "monocrystal" appearance of each single prism.

The set of experiments performed by Cuif and coworkers, by using different surface treatments (proteolytic digestion, bacterial treatment, slight decalcification, and fixation) showed different ultrastructural aspects of the distribution of the intracrystalline matrix (Cuif et al. 1981; Cuif and Raguideau 1982; Cuif et al. 1983a, b, 1985, 1987a). First, in accordance with the striation observed on the surface of each prism, the intracrystalline organic network exhibits a periodicity, which is perpendicular to the prism axis. In cross section, although periprismatic sheaths separate contiguous prisms, this periodicity, which corresponds to cycles of mineral deposition, is found from prisms to prisms. More surprising was the visualization by Cuif and coworkers of a longitudinal organization of the prisms. By using appropriate treatments with glutaraldehyde/acetic acid/alcian blue, they observed a longitudinal periodicity of the organic network. This periodicity is not strictly parallel to the axis of the prisms, but slightly oblique. This longitudinal organization of the matrix is correlated to the presence of ridges (that look like "ripple-marks") in the internal side of the prisms observed in section perpendicular to the prisms axis (Fig. 13.2b). It is striking to observe that these ridges are parallel in a single prism, but also parallel from prism to prism. Cuif suggested that these ridges result from the intersection of the periodic plans of the intracrystalline network and of the plan perpendicular to the prism axis. So far, the significance of the longitudinal organization of the intracrystalline matrix remains obscure.

13.3.5 The Prism/Nacre Transition and the Nacreous Layer

The transition zone between the prismatic and the nacreous layer is rather abrupt (Cuif et al. 1985). It is marked by the formation of an intermediate layer, about 50 μm thick (Fig. 13.3a, b). We sometimes observed that the thickness of this layer

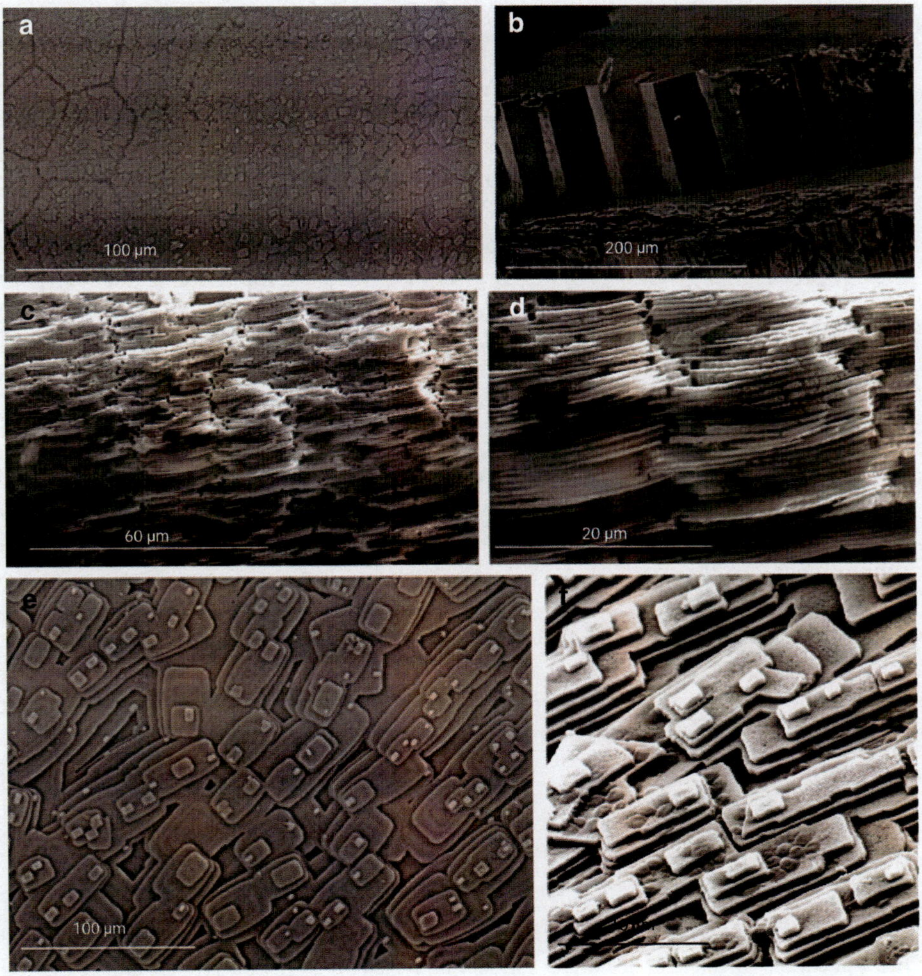

Fig. 13.3 Prisms–nacre transition and the nacreous layer. (**a**) prisms–nacre transition, observed from the internal shell surface. The prisms are on the *left*, the nacre, on the *right*. The prisms are progressively covered by a nacreous layer. (**b**) cross section of a 6-month-old shell. The prisms are on the top, the nacre below. (**c–d**) cross section through the nacreous layer. The nacre exhibits the typical "row stack nacre" microstructure, observed in the Pinnid family. (**e–f**) nacre tablets, seen from the internal shell surface (photo (**f**) Caseiro J, Gauthier JP). The nacre tablets of *Pinna nobilis* are *rectangular*. They do not show the brickwall pattern observed in the pearl oyster nacre, for example

is reduced to 10–20 μm. This intermediate brownish layer is essentially organic and insoluble. Although of homogeneous aspect once sealed by the subjacent nacreous layer, it has been shown that this layer results from the coalescence of polygonal organic elements during its synthesis (Cuif et al. 1985). By many aspects (insolubility, color), the intermediate layer exhibits similarities with the periostracal layer, and may correspond to chitino-proteinaceous complexes resulting from a quinone-tanning process. This layer seals the prismatic layer and represents the template for the deposition of nacreous crystallites. As we will see in the next paragraph, the transition layer marks the profound difference of the secretory regime between the mantle cells involved in prisms formation and those involved in nacre deposition.

In *Pinna nobilis*, as mentioned before, the nacreous layer is restricted to the area comprised between the two muscle scars, about one-half of the shell. In young specimens (2–3 years old), the nacre layer is extremely thin (not more than 2–3 mm) and translucent. In old specimens (>20 years), the nacreous layer is opaque and reaches a thickness of about 10 mm in the anterior tip. Taylor et al. (1969) proposed a general evolutionary trend toward the reduction of the nacreous layer in the Pinnidae family: the ancestral type, which would be nowadays represented by *Atrina*, exhibits indeed a more expanded nacre layer, while this layer is residual in the most derived *Streptopinna* genus, *Pinna* constituting an intermediary stage. This trend toward a reduction of the nacreous layer, if it is verified, would be in agreement with the finding of Palmer (1992), i.e., microstructures like nacre, which have a higher energetic cost to produce, are consequently disadvantaged from an evolutionary viewpoint. Alternatively, it may also correspond to different modes of life, implying different mechanical requirements for the shell.

The fact that the nacreous layer covers only one-half of the shell has an important implication on the geometry of the calcification process: it means that the mineralization fronts of nacre and of prisms are distant from each other. In other words, the mantle epithelial cells that secrete the ionic and macromolecular components for nacre deposition are physically remote (several centimeters in a 2-year-old specimen) from those that initiate the prisms at the shell edge. It also means that the cells involved in the prisms extension, i.e., in the increase of the thickness of the prismatic layer, constitute an extremely large zone of the mantle tissue.With these geometrical constraints, *Pinna nobilis* differs considerably from *Mytilus, Pinctada*, or *Unio* models.

The nacreous layer of *P. nobilis* is rather atypical, and does not correspond to the classical "brickwall texture" (sheet nacre) with consecutive tablets arranged in staggered rows, found in most pterioid, mytiloid, and unionoid bivalves (Wise 1970). The nacre of *Pinna nobilis* is indeed defined as "row-stack nacre" (Wise 1970; Carter and Clark 1985) and is described as a "nacreous structure in which mutually parallel elongate tablets show vertical stacking in vertical sections perpendicular to their length axes, and brick wall and/or stair step stacking in vertical sections parallel to their length axes. Row stack nacreous structure is well developed in the bivalve *Pinna*, where it enhances directional flexibility of the shell." In cross section, as shown in Fig. 13.3c–d, nacre elements appear as flexuous flat tablets, 0.5–0.8 μm thick and 15–20 μm long, which exhibit a pseudo-columnar disposition. This vertical disposition, which was remarkably described a long time ago by Rose (1858), then by Schmidt (1924, Fig. 48, Tafel 4, "Vertikalschichtung"), appears different, larger, and slightly less structurally regular than that found in haliotid gastropods e.g., but has some similarities with that of *Nucula*. Vertical sections show the more or less columnar alignments of the successive nacre laminae (Fig. 13.3c–d). When observed from above on the mineralization front (Fig. 13.3d), newly formed minute nacre crystals have a typical quadrangular shape and are oriented in the same direction. Proteolytic treatments (Cuif et al. 1983a, b, 1985) revealed the ultrastructural organization of the nacre crystallite and highlighted their composite nature. In particular, it showed that each nacre crystallite is composed of minute granular elements (about 100 nm), which may represent the basic mineral unit. For each tablet, it also showed a typical pattern, characterized by the formation of opposite triangular domains of mineral dissolution.

13.3.6 Minor Elements in Prisms and Nacre

Beside the shapes of their crystallites and their mineralogy, another aspect that marks the difference between the prismatic and the nacreous layer is their bulk elemental composition obtained by microprobe analysis (Masuda and Hirano 1980; Cuif et al. 1985). Prisms exhibit high levels of magnesium and sulfur, respectively 6,000 and 4,000/4,500 ppm, while the concentrations of these two elements are about 500 ppm in nacre. On the contrary, the concentrations of sodium and of strontium are higher in nacre, respectively >6,000 and 1,500 ppm, in comparison to prisms (4,500 and less than 1,000 ppm). The case of sulfur is interesting (Cuif et al. 1985, 1986, 1988a): its concentration, measured along the prisms, is constant throughout the prisms length, but drastically collapses at the prisms/nacre transition. Few years ago, FTIR, WDX, and XANES data (Dauphin 2002; Dauphin 2003; Dauphin et al. 2003) showed first that the sulfur in the prisms is present as sulfate, and second, demonstrated that the intracrystalline prism matrix is particularly enriched in sulfated polysaccharides. This corroborated the previous finding of Wada (1980), who underlined the importance of sulfur (under the form of polysaccharides) deposited on a glass coverslip preparation placed in *Pinna attenuata*.

13.4 The Shell Matrices of *Pinna* sp. and of *Pinna nobilis*

13.4.1 Early Biochemical Studies

The discovery of "conchiolin"– the strongly insoluble organic fraction extracted from nacre – and its subsequent naming by Frémy (1855) was the starting point of the analysis of several insoluble organic fractions associated to diverse molluscan shell microstructures. For almost a century, due to technical limitations, the biochemical information retrieved from shell proteins was circumscribed to that on the composition in their elementary bricks, amino acids. Consequently, most of the studies performed on shell matrices, between the beginning of the twentieth century and the late sixties, focused on the amino acid compositions, usually obtained after hydrochloric acid hydrolysis of the matrix (for an old – but rather complete – review, see Grégoire 1972).

The first biochemical characterization of the shell matrix of *Pinna* sp. was performed by Wetzel (1900), who clearly established the proteinaceous nature of "conchiolin," by detecting glycine, leucine, tyrosine, and arginine residues in "conchiolin" hydrolyzates from prisms and nacre. Interestingly, he also showed that the prisms conchiolin, i.e., the substance that composes the interprismatic walls, was different in its amino acid composition from the nacre conchiolin.

Half a century later, Roche et al. (1951) and Ranson (1952, 1966) observed that the "conchiolin" of the calcitic prisms of *Pinna* sp. had a higher content in Tyr and Gly and lower content in arginine, serine, and glutamic acid residues than the nacre conchiolin. Akiyama (1966) also observed a high amount of glycine in the prismatic matrix of *Pinna attenuata*.

Amino acid analyses on prisms and nacre "conchiolin" of the pinnid bivalves *Pinna nobilis* and *Atrina nigra* were performed by Bricteux-Grégoire et al. (1968). For the nacre of *P. nobilis*, they obtained a composition dominated by alanine (29%), glycine (21%), aspartic acid (13%), then serine (9%), while, for prisms, glycine (37%) and aspartic acid (23%) constituted the dominant amino acids. The proportions of amino acid residues varied in the same proportions for *A. nigra*. Today, we know that the "conchiolin" terminology defines a mixture of different insoluble constituents that include hydrophobic proteins and chitin.

The discovery that molluscan shell matrices comprise also soluble *intracrystalline* components (Crenshaw 1972) and the emphasis of their role in calcification process (Weiner et Hood 1975) led many investigators in biomineralization to focus on this previously undetected organic fraction: not only was the soluble fraction accessible to amino acid analysis but also to aqueous phase fractionation techniques, such as chromatography or electrophoresis. At that time, the general perspective was to purify single proteins and obtain sequence information.

Nakahara et al. (1980) performed series of amino acid analyses on prism-associated matrices of *Pinna carnea, Atrina pectinata*, and pearl oysters. They distinguished the interprismatic insoluble walls from an organic fraction that they considered to be intraprismatic. For the first one (*P. carnea*), they partly confirmed previous findings: enrichment in glycine and aspartic acid. They however found that this matrix was also enriched in tyrosine and cysteine residues, the high amount of tyrosine suggesting that the interprismatic walls were insolubilized by a quinone-tanning process. For the intraprismatic matrix, they measured a striking 72% molar for aspartic acid (62% in the case of *A. pectinata)*, a fact that fully confirmed the previous finding of Weiner and Hood (1975).

Few years later, Cuif et al. (1987b, 1988b, 1991), Frérotte (1987), Kervadec (1990), and Marin (1992) analyzed the soluble matrices from prisms and nacre of *Pinna nobilis* by gel permeation or ion-exchange chromatography, both fractionation techniques coupled with subsequent amino acid analysis (after 6 N–HCL hydrolysis and PITC pre-column derivatization). They clearly evidenced that the two soluble matrices were heterogeneous in their chromatography profiles, a fact which was confirmed by the heterogeneity of amino acid compositions of isolated peaks (Marin 1992). Furthermore, Marin (1992) showed that the intraprismatic matrix was much more acidic than the soluble matrix associated to nacre. However, none of the fractionation technique used was able to obtain single macromolecular components that could be sequenced further. The reasons of this failure were that the macromolecules associated to prisms and nacre – and more generally to all calcified tissues – are not globular and that many of them are negatively charged. In addition, they tend to be polydisperse, and are frequently submitted to posttranslational modifications, such as glycosylation or phosphorylation. Consequently, they behave anomalously during chromatographical fractionation (for technical informations on acidic proteins, see the review of Marin and Luquet 2007). These combined technical obstacles precluded for several years the possibility to purify a single protein and to obtain partial or full amino acid sequences.

13.4.2 Electrophoresis and Serology on the Shell Matrix of Pinna nobilis

Decisive analytical improvements were brought in biomineralization investigations, by using polyacrylamide gel electrophoresis in denaturing conditions. By extracting acido-soluble matrices from prism and nacre of *Pinna nobilis*, Marin et al. (1994) showed that the two matrices exhibited similarities and differences: none of them stained accurately with the classical Coomassie Brilliant Blue for the reasons indicated here above; when stained with silver nitrate, the two electrophoretic profiles were different but both were characterized by few discrete macromolecular components (from 10 to above 50 kDa), embedded in smearing polydisperse macromolecules; in the prisms matrix, one prominent band was observed around 15 kDa, and another one, at higher molecular weight; the biggest difference was observed when employing alcian blue for revealing polyanionic substances, such as sulfated sugars: the soluble matrix associated to nacre did not stain at all, whereas the one extracted from prisms gave a strong staining. This clearly suggested that the prism matrix was extremely acidic ("polyanionic"). We correlated this finding with the high content of sulfur of the prisms, this finding being also correlated to the high quantity of soluble sulfated polysaccharides of the intraprismatic matrix.

Another approach, which was also very effective and brought a significant contribution to the analysis of the shell matrices from *Pinna nobilis*, was serology, also called serotaxonomy. As described elsewhere (Muyzer et al. 1984; Collins et al. 1991; Marin et al. 1999), the method exploits the ability of the mammalian immune system to produce antibodies against foreign macromolecules, e.g., macromolecules extracted from shells. Usually, antibodies recognize short domains, called epitopes (or antigenic determinants), which, in the case of proteins, are not longer than 5–8 amino acid residues. Collected antibodies can be tested against a variety of shell matrices, via different techniques (ELISA, dot blot, Western blots, in situ localization), the aim of which is to provide a comparison between matrices. Thus, although the method does not provide any structural information on the studied shell matrices, it gives indirectly an informative picture on the degree of structural similarity between macromolecules.

In the present case, polyclonal antibodies were produced in rabbits against non-fractionated acetic acid-soluble matrix macromolecules extracted from prisms and from nacre of *Pinna nobilis*. These antibodies were tested against prisms and nacre matrices (Fig. 13.4a for the western blot results obtained with the antibodies raised against prisms matrix). They cross-reacted in a more or less symmetrical way, i.e., antibodies elicited against the nacre matrix recognized, in a lesser extent, epitopes in the prisms matrix, and antibodies elicited against the prisms matrix crossed-reacted with nacre matrix. This suggested that both matrices shared several epitopes. This idea was – and still is – very appealing: it may indicate that shell microstructures as different as prisms and nacre can be produced from different secretory protein repertoires, which contain however some identical epitopes (short domains). It is interesting to notice that before our study, by using a different approach – a combination of ion-exchange and HPLC chromatography on the soluble nacre and prisms matrices of the mussel *Mytilus californianus* – Weiner (1983) observed that some of the numerous peaks were shared by the two matrices. He consequently came to the conclusion of a partial similarity between the two matrices.

As the developed anti-prism/nacre matrices antibodies exhibited good titers and gave satisfactory results, we tested them by ELISA technique against a broad range of shell matrices extracted from numerous molluscs, including bivalves, gastropods, and cephalopods. In addition, the antibodies were tested against skeletal matrices of extra-group samples, brachiopods, and corals (Marin 1992; Marin et al. 1999, 2007a). The results were surprising (Fig. 13.4b). First, none of the two antibodies gave immunological patterns, which could be simply correlated to mollusc phylogeny. Consequently, these antibodies could not be used for sero-taxonomy purposes. To give an example, both antibodies gave strong cross-reactivities with the shell matrices of the two gastropods *Haliotis* and *Littorina*, while they did not cross-react with the pteriomorphid bivalve *Arca* (Fig. 13.4). One of the two antibodies (anti-nacre) also reacted strongly with the shell matrix of the cephalopod *Nautilus*. Second, the obtained signals were not strictly correlated to shell microstructures, neither to the mineralogy (calcite vs. aragonite). For the anti-prisms antiserum, we observed that it strongly cross-reacted with the matrices of nacro-prismatic pteriomorphid bivalves (*Mytilus, Pteria, Pteria, Isognomon, Brachidontes*), with those of foliated pteriomorphids (*Pecten*, *Crassostrea*) but also with the fully aragonitic nacro-prismatic paleoheterodont bivalves (*Unio*, *Anodonta*, *Neotrigonia*). There were no cross-reactivities with the matrices of "crossed-lamellar" aragonitic veneroids (*Venus*, *Mercenaria*, *Tridacna, Ensis*), but curiously, cross-reactivities with aragonitic crossed-lamellar myoids (*Mya*, *Pholas*). This pattern was not expressed with the second antibody (anti-nacre). Of course, we were fully aware that the serological approach has some drawbacks: (a) it did not give any indication on the chemical nature of the target antigens; (b)

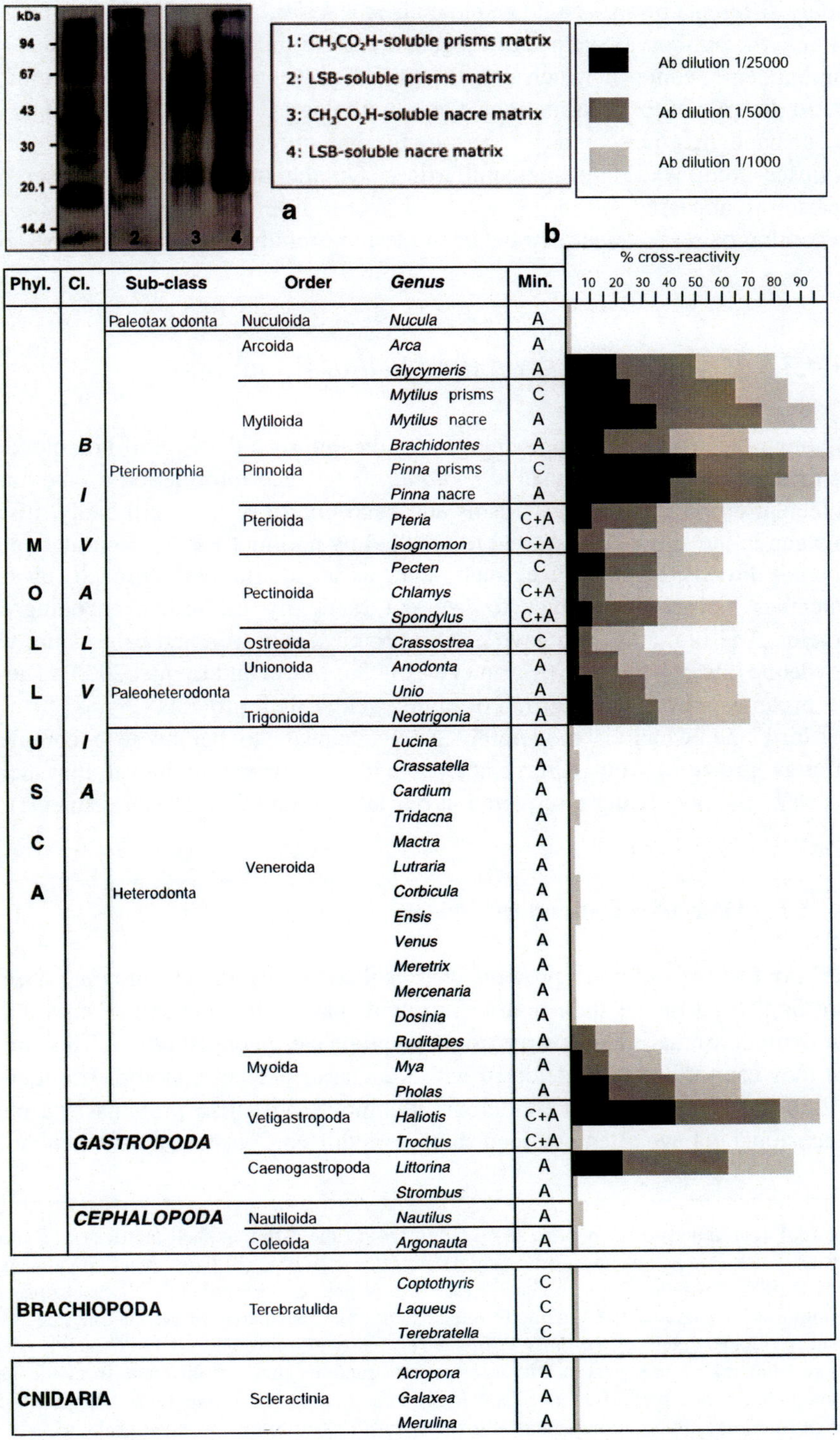

Fig. 13.4 Cross-reactivities of the soluble skeletal matrices of several invertebrates with a polyclonal antibody elicited against the acetic acid-soluble matrix of the prismatic layer

widely different epitopes could be recognized at a similar level by a given antibody; (c) because antibodies were raised against protein mixtures, some of which were immunogenic, some of which not, the final signal may have been contorted and biased. In spite of these limitations, the immunological pattern obtained suggested, on one hand, that the secretory repertoire used by different phylogenetically distant molluscs could exhibit some similarities; On the other hand, some secretory repertoires appeared to be extremely different from group to group, the best illustration being the clear gap between pteriomorphids and heterodonts bivalves.

13.4.3 Molecular Data on the Shell of P. nobilis

Although the serological approach gave indications on the overall homologies of matrices in comparison to that of *P. nobilis*, sequence information on the macromolecular constituents of the prisms and nacre matrices were still badly missing. This gap in the knowledge started to be filled by finding tricks to accurately purify proteins directly from shell extracts, and, as an alternative approach, by using molecular biology techniques to fish out randomly transcripts encoding shell proteins. The first approach gave us two proteins, caspartin and calprismin, while the second one, mucoperlin (Marin et al. 2003b; Marin and Luquet 2005). The first two proteins belong to the group of highly acidic shell proteins (Fig. 13.5), while the third one, which is moderately acidic, appears to belong to a completely different protein family (Fig. 13.6). In addition, new proteins of the shell of *P. nobilis* are now being discovered in our lab by one of us (P. Narayanappa).

13.4.3.1 Molecular Data on the Prisms

The way two prism matrix proteins were isolated relies on the following observation: first, most of the shell matrices proteins have a low content of amino acids that contain aromatic groups (tyrosine, tryptophane, phenylalanine). This implies that they cannot be easily detected with the classical spectrophotometric methods at 280 nm. Second, in our numerous attempts to visualize proteins in a monodimensional gel, we often obtained sharper pictures on Western blots when the blot

Fig. 13.4 (continued) of *P. nobilis*. (**a**) western blot of the different shell matrices of *P. nobilis* (prisms, lanes 1 and 2; nacre, lanes 3 and 4) with this antibody. (**b**) Cross-reactivities measured with ELISA. Although the highest cross-reactivities are recorded with nacro-prismatic pteriomorphid bivalves, the obtained signal cannot be considered to be "taxonomic": some distantly related genera (*Haliotis, Littorina*) cross-react strongly. The signal is neither "microstructural": some genera with shell microstructures that are different from the nacro-prismatic microstructure of *P. nobilis* (*Glycymeris, Pecten*) also give high cross-reactivities. Except *Ruditapes*, none of the veneroid bivalves cross-react. Similarly, none of the tested extra-groups (cnidarian, brachiopod) give cross-reactivity. *Min* mineralogy, *A* aragonite, *C* calcite

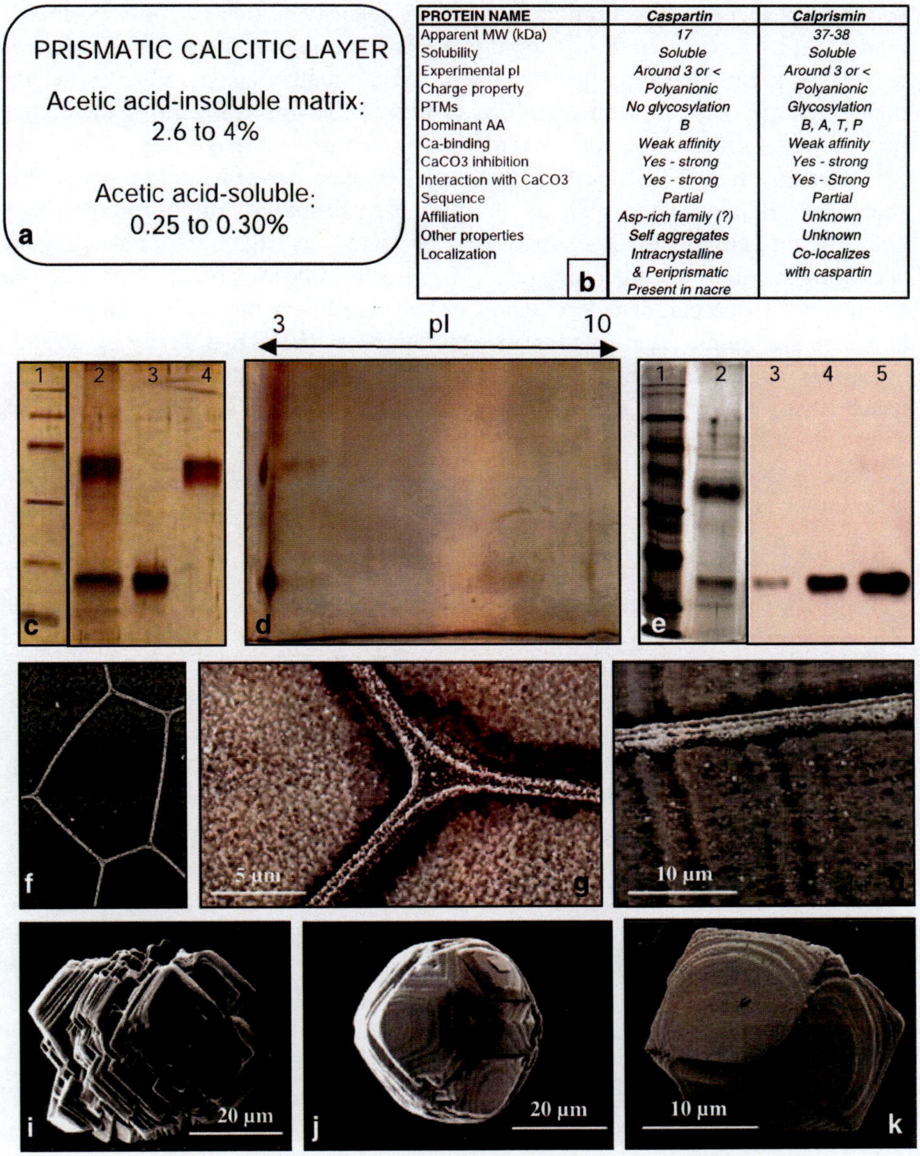

Fig. 13.5 The calcifying matrix of the shell prismatic layer of *Pinna nobilis*. (**a**) Proportions of soluble and insoluble matrices. (**b**) Summary of the biochemical properties of caspartin and calprismin; B is the one-letter symbol for Asx, i.e., Asp or Asn. (**c**) SDS-PAGE of the intracrystalline prisms matrix and isolated fractions; lane 1, molecular weight markers, from *top* to *bottom* 94, 67, 43, 30, 20, and 14.4 kDa; lane 2, acetic acid-soluble matrix of isolated prisms; lane 3, purified caspartin; lane 4, purified calprismin. (**d**) 2D-gel electrophoresis of the acetic acid-soluble matrix of isolated prisms. (**e**) Western blot of the acetic acid-soluble matrix of isolated prisms. The blotted extract was incubated with the polyclonal antibody (dilution 1:3,000) raised against the purified caspartin; lane 1, molecular weight markers, from *top* to *bottom* 116, 97, 66, 45, 31, 21.5, 14.4 and 6.5 kDa; lane 2, acetic acid-soluble matrix of isolated prisms, silver stained; lanes 3–5, western blot of the acetic acid-soluble matrix of isolated prisms, 5 µg (lane 3), 10 µg

was priorly incubated with antibodies raised against the non-fractionated matrix than by using silver nitrate or Coomassie Blue staining on the gel. For unknown reasons, it seems that discrete components of the matrix are more immunogenic than smearing macromolecules, which results in the improvement of the signal/background ratio (if we consider that the smearing materials are part of the "background effect"). We used this property to isolate two prominent prism matrix proteins as follows: after the isolation of single prisms with dilute sodium hypochlorite, the intraprismatic matrix was extracted by dilute cold acetic acid, according to our standard procedure (Marin et al. 2001). Then, the matrix was fractionated by preparative electrophoresis, which allowed the "blind" collection of numerous fractions. Each fraction was blotted on a PVDF membrane by dot-blot apparatus, and the membrane was subsequently treated with anti-prisms or anti-nacre matrix antibodies, according to a classical procedure used in Western blot. This allowed detecting in which fractions were eluted the proteins of interest. Proteins were collected, and extensively dialyzed before being freeze-dried. The technique was described in Marin et al. (2001) and in Marin (2003).

In this manner, we identified two new proteins, that we subsequently named caspartin and calprismin, respectively (Marin and Luquet 2005; Marin et al. 2005). Calprismin is an acidic intracrystalline soluble protein of 38 kDa of apparent molecular weight. It is enriched in Ala, Asx, Thr, and Pro residues, these four residues constituting 55% of the sequence. The N-terminus of calprismin (61 a residues, about 18% of the sequence) was sequenced. It is enriched in Glu residues and presents a pattern containing four cysteine residues that is also observed in different extracellular matrix (ECM) proteins. A blast search did not allow finding affiliation of calprismin with other shell proteins, neither with other ECM proteins. Calprismin is a very weak calcium-binding protein, and this property is extremely labile. On 2D-gel, calprismin migrates as a thick spot that is localized below *p*I 3. Two other diffuse spots, which are slightly less acidic, suggest that calprismin may have different isoforms/glycoforms. The enzymatic deglycosylation of calprismin with a cocktail of endoglycosidases led to a loss of few kDa of apparent molecular weight on a gel. This weight loss is correlated with the disappearance of Alcian blue staining of the protein after deglycosylation. Calprismin may be rather unstable since it degrades when performing a chemical deglycosylation with TFMS acid. The analysis of its N-terminal sequence, and the comparison with other N/C-terminal sequences of prisms-associated proteins allowed obtaining general features (Evans 2008): abundance of anionic residues, low amount of cationic residues, variable amounts of hydrogen-bonding donor/acceptor (HBDA). Like other proteins associated to prisms, calprismin may be intrinsically disordered,

Fig. 13.5 (continued) (lane 4), 20 μg (lane 5). **(f–h)** Immunogold localization of caspartin. **(f)** cross section, low magnification. **(g)** cross section, high magnification of a triple junction. **(h)** longitudinal section of a prism. **(i–k)** in vitro $CaCO_3$ crystallization experiment in the presence of purified caspartin. **(i,j)** polycrystalline aggregates formed at 2 μg/ml of caspartin. **(k)** "single" crystal formed at 3 μg/ml of caspartin

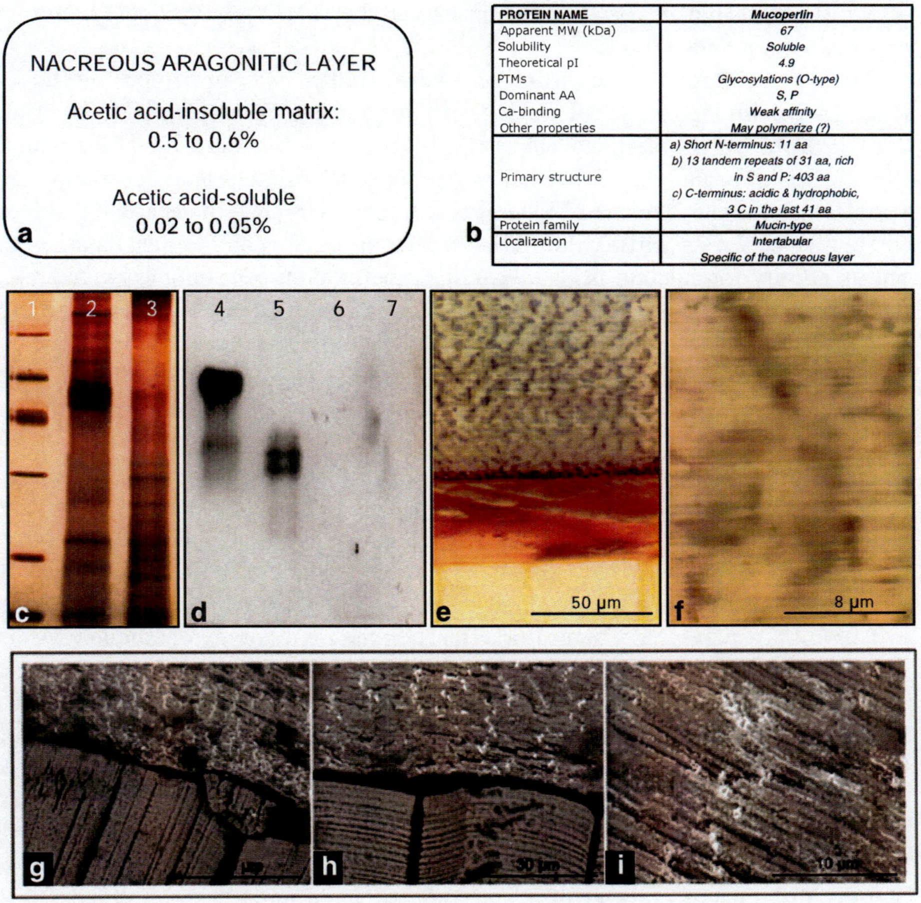

PROTEIN NAME	*Mucoperlin*
Apparent MW (kDa)	*67*
Solubility	*Soluble*
Theoretical pI	*4.9*
PTMs	*Glycosylations (O-type)*
Dominant AA	*S, P*
Ca-binding	*Weak affinity*
Other properties	*May polymerize (?)*
Primary structure	*a) Short N-terminus: 11 aa b) 13 tandem repeats of 31 aa, rich in S and P: 403 aa c) C-terminus: acidic & hydrophobic, 3 C in the last 41 aa*
Protein family	*Mucin-type*
Localization	*Intertabular Specific of the nacreous layer*

Fig. 13.6 The calcifying matrix of the shell nacreous layer of *Pinna nobilis*. (**a**) Proportions of soluble and insoluble matrices. (**b**) Summary of the biochemical properties of mucoperlin, an acidic nacre-specific protein. (**c**) SDS-PAGE of the acetic acid-soluble (lane 2) and of the acetic acid-insoluble/Laemmli-soluble (lane 3) nacre extracts. Mucoperlin is the thick band in lane 2. (**d**) western blot of the different shell matrices of *P. nobilis*. The blot was incubated with a polyclonal antibody elicited against doubly purified recombinant mucoperlin; lane 4, acetic acid-soluble nacre extract; lane 5, acetic acid-insoluble/Laemmli-soluble nacre extract; lane 6, acetic acid-soluble prisms extract; lane 7, acetic acid-insoluble/Laemmli-soluble prisms extract. There is no signal in the prisms extracts, indicating that mucoperlin is nacre specific. (**e, f**) immunolocalization of mucoperlin in cross sections of nacro-prismatic preparations, observation with an optical microscope; (**e**) lower part, prisms, upper part, nacre. (**f**) nacre, at higher magnification. (**g–i**) immunogold staining of a cross section (fresh fracture), observed with a SEM. (**i**) the white spots, which localize mucoperlin, are situated between nacre tablets (intertabular location). Patterns in (**e, f**) and in (**g–i**) are identical

and may require external interactions to stabilize its secondary structure (Evans 2008). One of us (P. Narayanappa) works on obtaining the full sequence of calprismin to confirm this prediction. Other ongoing characterizations include the effect of calprismin *in vitro*, its localization in the shell by immunogold, and its expression in mantle tissues.

The second protein, caspartin, has been studied *in vitro* more in detail (Fig. 13.5), although its sequence is still far from being complete (Marin et al. 2005, 2007a, b). Similarly to calprismin, caspartin is an intracrystalline acidic protein, which is abundant in the acid-soluble extract of the prisms. Caspartin has a molecular weight of about 17 kDa. It is extremely enriched in aspartate residues (more than 60%). Caspartin exhibits better calcium-binding ability than calprismin, but its affinity for calcium is also weak, i.e., the fixation of calcium is reversible, and calcium can be easily removed by gentle rinsing of the membrane where caspartin has been blotted. Caspartin does not seem to be glycosylated, but we do not exclude the possibility that it may be phosphorylated. An interesting finding was that caspartin self-agregates in solution: in non-denaturing gels, we were able to show that caspartin forms di-, tri-, tetra-, penta-, hexa-, and heptamers. On 2D gels, caspartin migrates as a big spot at *p*I below 3. However, three tenuous minor spots of the same molecular weight, one in the acidic domain, and two, basic, are also observed, which may indicate that caspartin may not be absolutely pure, or that it exhibits isoforms. *In vitro*, caspartin is a powerful inhibitor of the precipitation of calcium carbonate, since it induces an inhibitory effect for concentrations inferior to 1 μg/ml. Caspartin has also drastic effects on the morphology of calcite crystals grown *in vitro*: effects are already observed around 0.2 μg/ml, and they increase with the concentration of the protein in solution. Between 1 and 3 μg/ml, caspartin induces the formation of polycrystalline aggregates, together with the formation of "single crystals," which exhibit a foliation. When the concentration of caspartin is increased, the size of the crystals decreases, and above a threshold, caspartin acts as an inhibitor of calcification and no crystals are produced. A polyclonal antibody, specific of caspartin, indicated that caspartin, or a similar 17-kDa protein, is also present in the nacreous layer of *Pinna nobilis*, but in much lesser amounts (6–8 times less). This antibody was used to localize caspartin directly in the shell of *Pinna nobilis* (Marin et al. 2007b). In cross sections, perpendicular to the prisms axis, we observe that caspartin has two locations: first, it is distributed at the surface of the prisms (= within the prisms), a predictable finding. We do not observe a particular pattern, but rather, a uniform distribution of the signal at the prisms surface. Second, more surprisingly, caspartin occupies also an intercristalline location: it forms a continuous film at the interface between the prisms themselves and the interprismatic insoluble walls. In other words, the insoluble hydrophobic interprismatic wall is sandwiched between two caspartin-containing layers. In longitudinal sections, caspartin is located discontinuously along the prisms. Its distribution in that dimension is not strictly correlated with the growth lines observed for each prism, although we observed that caspartin may be more concentrated on some growth plans (Marin et al. 2007b).

Following the first biochemical characterization of caspartin (Marin et al. 2005), *de novo* sequencing of caspartin extracts was performed and different short peptides were obtained, by using trypsin, pepsin, or AspN digests (Marin et al. 2007b). However, so far, by using different couples of degenerated oligo primers, we did not retrieve yet the full sequence of caspartin, and more work has still to be done before we elucidate its complete primary structure.

13.4.3.2 Molecular Data on the Nacre

The second strategy we chose was to construct a cDNA library from mantle tissues of non-stressed juvenile specimens of *Pinna nobilis*, which were actively calcifying. Briefly, the technique implies the extraction of total RNAs from a tissue, the purification of the messenger RNAs by Oligo-dT resin, the conversion of the transcripts into double-stranded cDNAs, the cloning of "adaptors" in the two termini of the cDNAs, the selection of the longest cDNA by size-exclusion chromatography, the cloning of the cDNAs into a phage vector, and the packaging of the construct into phage capsids. The resulting primary library, which yielded about 3.5×10^5 clones, was amplified once.

At that point, because we did not have sequence information on shell proteins of *Pinna nobilis* – let us recall that the first full sequence of a shell protein from the pearl oyster *Pinctada* was published at the end of 1996 – we could not develop degenerate probes (biotin or digoxigenin-labeled) for screening the constructed library. As our constructed cDNA library was an expression library, we chose to screen it with "our" homologous antibodies that we used before, those elicited against the acid-soluble shell matrices of prisms and of nacre of *Pinna nobilis*.

This operation described elsewhere (Marin et al. 2000, 2003a) generated series of clones, which were re-screened twice (secondary and tertiary screening) to purity. The pure clones exhibited long cDNA inserts (above 2 kb), which encoded a putative protein of 636 amino acids. The analysis of this ORF showed that the corresponding acidic protein (theoretical pI = 4.9) possesses three domains: a short N-terminus, then a long central domain constituted of a succession of 13 almost identical repeats put in tandem (403 residues in total), which represents two-third of the protein, then a C-terminus (222 residues). Each tandem repeat, 31 amino acids long, is enriched in proline and serine residues, and many of these serine residues are presumably glycosylated. Although similarity search indicated only limited homologies with known proteins, the most convincing hit was found with PGM, a pig gastric mucin. Similarities with a mucin may not be fortuitous. Indeed, our protein possesses at least two properties found in mucins, the main protein components found in mucus secretions: the presence of tandem repeats, and the richness of these repeats in proline and serine residues. In addition, like most of the mucins, our protein exhibits, in the C-terminus, few cysteine residues. In mucins, these residues potentially allow the formation of dimers first, and then, of multimeric insoluble gels. Because of these features, we consequently chose to name this protein mucoperlin (Fig. 13.6).

Mucoperlin was overexpressed in a bacterial strain, and after a double purification, the single obtained band was used for generating a very specific polyclonal antibody (Marin et al. 2000, 2003a) and for different in vitro experiments. Because the overexpressed mucoperlin did not exhibit posttranslational modifications, it was not active in "inhibition tests" e.g., neither in calcium-binding assay. The antibody elicited against the recombinant mucoperlin reacted only with nacre extracts, but not at all with prisms extracts, on Western blots, on dot blots or on ELISA (Fig. 13.6d). This clearly suggested that mucoperlin is specific of the nacreous layer of *Pinna nobilis*. To our knowledge, this was the first convincing demonstration of the existence of a nacre-specific protein among molluscs. Our finding was confirmed by screening mucoperlin clones with anti-prisms matrix and anti-nacre matrix antibodies, separately (Marin et al. 2003a), and by histological staining: using conventional staining of shell cross sections, we showed that mucoperlin was localized in the nacre layer exclusively, and that it was particularly concentrated in the intertabular zones, i.e., the short lateral sides of the tablets, also defined as the zone that is comprised between adjacent tablets (Fig. 13.6e, f). In addition, a staining was also observed in the interlamellar matrix, but this staining was much weaker. Later, we repeated the immunohistological experiments by using the more precise immunogold technique (Marin et al. 2007b). We then observed a completely superimposed staining pattern, a concentration of white spots localizing mucoperlin in the intertabular zones (F. Marin, unpublished data, 2006; Fig. 13.6g–i; Marin 2009).

The identification and molecular characterization of mucoperlin, an acidic mucin-like protein, in a shell was a "première" and established a bridge between molecular physiology, especially that of vertebrates, and biomineralization in general. Mucins are indeed involved in several functions, such as lubrication, protection of epithelial tissues against chemical or microbial aggressors. They are also involved in cell signaling (Bafna et al. 2008). Interestingly, in vertebrates, mucins are also associated with fluids that calcify or that prevent calcification. In the buccal cavity, the saliva is a supersaturated fluid, which – fortunately for us – never calcifies. The main reason is that saliva contains a cortege of proteins that prevent the deposition of calcium salts. Salivary mucins, when free in solution, do not allow calcium salts crystals to grow further (Tabak et al. 1985; Nieuw-Amerongen et al. 1989; Tabak 1995). Furthermore, these mucins can adhere on tooth hydroxyapatite with a strong affinity, and prevent their demineralization (Meyer-Lueckel et al. 2006). Similarly, in the bladder, urine comprises a set of proteins, among which the Tamm-Horsfall Protein (THP) and urinary mucins, the function of which is to prevent the precipitation of calcium oxalates or of uric acid (Grases and Llobera 1998). Finally, the gallbladder is also an organ in contact with supersaturated fluids (Afdhal et al. 1995). In pathological situations (lithiasis), gallbladder mucins are frequently involved in the deposition of mineralized concretions (Lechene de la Porte et al. 1996), an observation which suggests that they may be also involved in nucleation processes. Clearly, mucins in general seem to play important – although probably underestimated – functions associated with calcification.

13.4.3.3 Other Shell Proteins of *P. nobilis*

Although caspartin and calprismin represent the two dominant intracrystalline proteins of the isolated prisms of *P. nobilis*, we detected about eight minor protein bands in the acetic acid-soluble prisms extract, below 10 kDa, and above 60 kDa, on 1D electrophoretic profiles (Marin et al. 2007b). However, sequence information on these bands is missing, so far. When the matrix of the complete prismatic layer – and not only the sodium hypochlorite-isolated prisms – is extracted (P. Narayanappa, unpublished, 2010), a huge set of additional protein bands are visualized on gel, probably comprising proteins which are weakly linked to the mineral phase. Similarly, in the nacre extracts, several proteins bands have been detected in the acetic acid-soluble and in the Laemmli-soluble fractions (Marin et al. 2000). For none of these proteins, we possess sequence information.

Other putative shell proteins have been retrieved by one of us (P. Narayanappa un published, 2010) by screening our cDNA expression library of the mantle of *P. nobilis* with anti-caspartin and anti-calprismin polyclonal antibodies. This operation yielded additional clones encoding putative proteins that have a "shell" signature: short hydrophobic or acidic motifs, predominance of few amino acids, such as glutamic acid, serine or leucine in the overall composition of the protein. Among them, the translated product of clone CLP2T7 is enriched in glutamic acid (18%), in leucine (10%), and in serine (10.3%), while the one of clone CSP3 is characterized by an extremely acidic C-terminus (with a poly-D domain). Interestingly, in parallel to the immunoscreening, a proteomic analysis of the acetic acid-soluble prisms matrix of *P. nobilis* yielded different peptides. One of them belongs to CSP3 translated sequence (Narayanappa et al., in preparation), proving that this putative protein is truly a shell matrix protein of the prismatic layer. Our results are being compiled for their subsequent publication.

13.4.4 Effect of the Acidic Proteins of **P. nobilis** *on "Calcification," Sensu Lato*

13.4.4.1 "Classical" Biochemical Effects

As described in Sect. 4.3, the identified shell proteins of *P. nobilis* exert different biochemical effects in relation with calcification. These effects, such as the calcium-binding effect, the inhibition effect, or the interference with $CaCO_3$ crystallization, can be quantified in a "test tube." Although they give us precious information on the putative functions of the tested protein, we are fully aware that, by no means, they substitute to the "function." In the present paragraph, we clearly dissociate the effects – which are known – from the true functions – which are still elusive.

The calcium-binding property of the shell proteins of *P. nobilis* is an interesting property, prone to discussion. In our hands, we always observed that the calcium-binding ability of caspartin, of calprismin, and of the nacre and prisms matrices in general, was labile. We observed that the level of calcium-binding was better for the whole matrices than for its separate constituents, like caspartin or calprismin, when equivalent amounts of materials were tested. We suggested that these two proteins, and probably many more proteins of the prisms and nacre matrices, belong to the class of low affinity–high capacity calcium-binding proteins (Maurer et al. 1996; Marin and Luquet 2007). The binding of calcium ions would be performed via the negatively charged side chains of aspartate and glutamate residues, and maybe via some of their posttranslational modifications, many of which may exhibit anionic properties: glycosylations, in particular those involving sulfated sugars, and phosphorylation of serine and threonine residues. Although the full sequences of caspartin and of calprismin are not known, it is very likely that these proteins do not exhibit canonical calcium-binding domains, such as EF-hands (Kretsinger 1976), which are characteristic of the class of low capacity–high affinity calcium-binding protein. For a shell protein, we believe that binding reversibly numerous calcium ions is probably more compatible with a crystal synthesis process than sequestering strongly few calcium ions. We do not exclude the possibility that bound calcium ions also stabilize shell proteins *in vivo* and help them to maintain a precise conformation.

The *in vitro* inhibition property is another point in case. In our hands, we always observed that the acetic acid-soluble matrices of prisms and nacre of *P. nobilis* had good inhibitory effect. However, we observed that the inhibitory effect of the prisms matrix (and of caspartin and calprismin) was much higher than that of nacre matrix, for similar tested concentrations (Marin 1992). This effect seems to be correlated with the "acidity" of both matrices, the prisms matrix being much more "acidic" than the nacre matrix. The higher acidity of proteins associated to calcitic prisms is supported by two observations: first, we have shown in a previous review paper (Marin et al. 2008) that molluscan shell proteins associated to calcite (foliated or prismatic) and the ones associated to aragonite (mainly nacre) exhibit different distributions, when plotted on a diagram showing their theoretical calculated *p*I vs. their molecular weight. The first ones have a bimodal distribution, either very acidic, either very basic, while the second ones are characterized by weakly acidic to weakly basic *p*Is. As the very basic proteins of the prisms may be associated to the insoluble framework, the soluble prism matrix proteins are necessarily extremely acidic. Second, anionic posttranslational modifications, such as sulfated sugars, are present in high quantity in the prisms of *P. nobilis* but are almost absent from the nacre matrix proteins. In the case of *P. nobilis*, these two combined factors explain the difference of inhibitory effect of the prisms and the nacre matrices.

At last, proteins of the shell matrix of *P. nobilis* interfere with the precipitation of calcium carbonate *in vitro*. This effect is dose dependent, up to a certain threshold above which no crystals are formed due to inhibitory effects (Marin et al. 2005). Among the most common crystal figures, one can see the formation of polycrystalline aggregates of calcite and the formation of apparent "single" crystals, which exhibit a layering and truncated corners. Although these effects can be observed with several shell proteins in general, we notice that they are particularly drastic when acidic proteins such as caspartin and calprismin are employed.

13.4.4.2 "Nonclassical Effects" of Acidic Proteins of *P. nobilis*: Effects at the Crystal Lattice Level

Besides these effects, which can be measured at the micrometric scale (interference tests), or at the macroscopic levels on gels (calcium-binding) or with an electrode (inhibition), far less classical effects have been recorded few years ago. These remarkable findings, which have been generated by our colleagues B. Pokroy and E. Zolotoyabko, both from the Technion in Haïfa, Israel, require high-precision diffraction equipment, such as the one of the ESRF, in Grenoble. The goal of their study was to evidence or not whether intracrystalline proteins had an effect on the crystallographic properties of biogenic calcite.

In a first set of experiments, Pokroy and Zolotoyabko measured the lattice parameters of five calcitic molluscan shells microstructures, namely, *Pinna nobilis* (prisms), *Atrina rigida* (prisms), *Ostrea edulis* (foliated), *Crassostrea gigas* (foliated), *and Haliotis rufescens* (prisms). By applying corrections dues to the incorporation of magnesium and of sulfur in the crystalline lattice, they calculated the theoretical a, b, c, parameters of these biogenic calcitic crystals. Without any exception, they observed higher lattice parameter values than that of chemical calcite. This clearly suggests a slight distortion of the crystal lattice, the maximal distorsion being according to the c axis (0.2%).

In a second set of experiments, they heated the biogenic calcitic structures, in order to burn the intracrystalline matrix. They observed a relaxation of the crystal lattice, i.e., a reduction of the crystal lattice parameters. This indirectly proved that the intracrystalline molecules induce the lattice distortion.

To demonstrate without any ambiguity the effect of intracrystalline proteins on calcite, Pokroy crystallized in vitro pure calcite and also calcite in the presence of caspartin, via the ammonium bicarbonate diffusion technique. A comparison of the lattice parameters of both calcites shows again a slight distorsion for the calcite grown with caspartin. These remarkable results were published in 2006 (Pokroy et al. 2006). Recently, Zolotoyabko et al. (2010), by using high-resolution neutron powder diffraction, demonstrated that some of the atomic bonds of biogenic calcite have significantly different lengths as compared to those of geological calcite. Of course, in their measurements, they took in account the incorporation of magnesium ions in the calcite crystal lattice.

Not less remarkable was the following finding of Pokroy and Zolotoyabko. This time they focused on the phenomenon of twinning. In crystallography, twinning is a well-known phenomenon. It is the oriented association of two or more crystals of the same chemical nature. These crystals are linked together via a symmetry operation, according to a plan (twinning plan, twinning by reflection), to an axis (twinning axis, twinning by rotation), or to a center (twinning center, twinning by inversion). In more than 150 years of research in crystallography, the different twinning patterns of calcite had been identified. They are four, corresponding to the following twinning plans: (001), (012), (104), (018).

By crystallizing *in vitro* calcite in the presence of caspartin, Pokroy performed X-ray diffraction measurements on the produced crystals. He evidenced a fifth twinning plan, according to (108), which had never been observed before. The results, which were published one year after the first one (Pokroy et al. 2007), have passed largely unnoticed.

So far, it is difficult to measure the impact of such discoveries and their future potential applications for nanotechnology and nanomaterials science. However, from what has been described here above, it becomes clear that biocrystals, such as the calcitic prisms of *Pinna nobilis* cannot be simply envisaged as "standard chemical crystals with a little bit of organics around."

13.5 Putative Functions of *P. nobilis* Shell Proteins: Toward a Dynamic View of the Shell Fabrication

It would appear extremely ambitious – not to say risky – to explain how the prismatic and nacreous layers of *Pinna nobilis* are formed, only with the *a posteriori* knowledge of the ultrastructure of these layers, once formed, and only with an extremely limited set of identified proteins. Indeed, for evident technical reasons, there is no continuous record of the calcification process neither in *P. nobilis*, neither in any other type of mollusc and we are limited to sample shells at different growth stages and to check how they look like. Furthermore, we should keep in mind that for a well-studied model, the pearl oyster *Pinctada,* more than two dozens of shell proteins have been described so far (Marin et al. 2008), but this is simply not sufficient to understand the emergence of calcitic prisms and brickwall nacre of *Pinctada* shell. We do not have yet the key for understanding what happens at the supramolecular level, and how shell matrix proteins interact with each other. Worse, the transcriptomic analysis performed by Jackson and coworkers on the abalone *Haliotis asinina* clearly suggests that the elaboration of the molluscan shell microstructures requires some additional protein constituents that are not incorporated in the shell during its growth (Jackson et al. 2006), and consequently, do not belong to the shell matrix. These "silent" proteins may be extremely important contributors to the shell fabrication but they are completely ignored until now. In the case of *Pinna nobilis,* we do not have the slightest idea of what these proteins could be, and only a transcriptomic approach (ESTs), performed on actively calcifying mantle tissues, will tell us more about the complete cortege of proteins required for shell formation.

In the absence of a complete theory on the mineralization of the molluscan shell, we can only propose hypotheses that take in account geometrical, crystallographic, and physiological constraints, as well as the biochemical properties of the secreted matrices.

13.5.1 The Prismatic Layer

As we have seen in Sect. 3.4, the prisms of *Pinna nobilis* have a paradoxical status: on one side, they behave like monocrystals, having a single extinction under polarized, analyzed light and a single crystal diffraction pattern; on the other side, they are complex biominerals that exhibit a structural hierarchy. The basic elements are mineral nanograins, which are spatially extremely well oriented, and which form, together with intracrystalline protein an intimate organo-mineral association. A great part of these two contradicting views disappears when considering the formation of calcitic prisms of *P. nobilis* via a "mesocrystal" pathway.

The concept of "mesocrystal" was introduced by Cölfen and Antonietti (2005) and by Cölfen (2007). The formation of mesocrystals typically follows a "nonclassical"crystallization pathway. In short, the starting point is identical to that of classical chemical crystallization pathway, ions, which, by concentration, form nucleation clusters. In natural environments, these clusters can grow or disintegrate again. When they grow, they can reach the size of the critical crystal nucleus. In the case of nonclassical crystallization pathways, the formed primary nanoparticles are temporarily stabilized by organic polymers, which adsorb on their faces. The later stage implies that the nanoparticles, on which organic polymers are adsorbed, assemble and orient in a superstructure, a mesocrystal, which, by fusion of its oriented nanoparticles, becomes a "single" crystal. The polymers associated to this "crystal" remain entrapped after the fusion. When applied to the calcitic prisms of *P. nobilis*, the concept of mesocrystal conciliates the apparent simplicity of each prism with their truly ultrastructural complexity.

How are the calcitic prisms of *P. nobilis* formed? In natural environments, prismatic crystalline structures can be produced through a completely abiotic pathway, which has been described for a long time. Grigor'ev (1965) explained the growth of prismatic structures via a process of competition for space. On an uneven surface, spherulites can be more or less regularly seeded; when they grow in a centrifugal manner, they enter into contact with neighbor spherulites, and consequently compete for space. They tend to grow in one direction, the fast-growing spherulites "absorbing" the slow-growing ones. After a while, the growth of the elongated structures is unidirectional and perpendicular to the initial plane where spherulites were seeded.

Up to a certain point, such a description might apply to the prisms of *P. nobilis*, and, more generally, to prismatic microstructures in bivalve shells (Ubukata 1994). Several analogies may be found between the two processes. First, similarly to the abiotic competition process, prisms of *P. nobilis* grow in one direction (inward) from an uneven surface, the periostracal layer. Their growth is perpendicular to that surface. Second, Cuif and coworkers (1983a, b) observed that the early stage of prisms formation at the edge of the shell was characterized by the formation of nodules (spherulites) that concentrically develop from a center and completely fill the polygonal spaces delimited by the periprismatic sheaths that form prior mineral deposition. In juvenile specimens of *P. nobilis*, we fully confirm, for each newly formed "prism" (at that stage, it looks like a tile), the existence of concentric patterns when the edge of the shell is observed from the outer surface, "through" the thin periostracal layer (Marin, unpublished data). This clearly suggests that nodules (or spherulites) are the first event of mineralization, similarly to what happens during the abiotic competition process. Third, Checa and coworkers observed, in cross sections of *P. nobilis* shells, competition for space, in a very narrow zone, comprised between 50 and 100 μm below the periostracal surface (Checa et al. 2005). However, we cannot subscribe to the view of an analogy between the abiotic competition process and the growth of *P. nobilis* prisms for the following reason: as underlined before, the initial "prisms" – or tiles, to be more exact – develop in a preformed honeycomb network, which consequently, antedates the early mineralization. At that stage, no competition for space occurs, since the polygons are already constrained by the periprismatic framework. As underlined by Checa and coworkers (2005) referring to Cuif et al. (1983a, b) the "unmineralized stage extends for at least some 100 μm backward from the margin and that empty organic cells (polygons) may attain some 5 μm in height at the onset of mineralization." Clearly, the competition for space observed previously seems to intervene at a later stage, during the early elongation phase of the prisms between 50 and 100 μm. How this competition occurs and why it occurs at that stage is still unclear, but would indicate that the growth of the mineral phase (prisms) catches up that of the organic periprismatic membranes.

In parallel to these observations, a remarkable idea has been developed by Checa and coworkers (2005) for explaining the reason of the polygonal shape of the prisms. These authors observed that the honeycomb-like organic framework has the structure of foam. Foams are gas-in-liquid dispersion mainly stabilized by interfacial tensions, in which the liquid phase is the thin film that separates gas bubbles. In ideal foams, the thin liquid films join at angles of about 120° to form triple junctions. In the present case, Checa et al. suggested that, rather than foam, the initial stage of the precursors of the mineralization, including the prisms matrix and the extrapallial fluid, may be a liquid–liquid emulsion, i.e., a mixture of two non-miscible liquids. Similarly to foam, an emulsion is patterned by interfacial tension. In this case, the matrix represents the viscoelastic hydrophobic fluid that makes a continuous mesh at the contact of the inner surface of the periostracum. The aqueous discontinuous solution – the extrapallial fluid – would be entrapped in the polygonal cells.

We can try to relate what is described here to the putative functions of calprismin and caspartin, according to their biochemical properties and to their location in the prismatic layer. In our previous study, we have shown that caspartin is intracrystalline and intercrystalline as well (Marin et al. 2007). A similar location seems to occur with calprismin (Narayanappa, in preparation). Because they are extremely acidic, both proteins may act as nucleators of nanocrystals. We can imagine that clusters – rather than films – of caspartin–calprismin complexes may locally catalyze the formation of nanocrystals that self-orient in the forming mesocrystal. The intercrystalline caspartin (and calprismin as well) may play an active role in stabilizing the emulsion, being at the surface of extrapallial fluid droplets, while the more hydrophobic organic network polymerizes in a solid and flexible honeycomb-like structure. To summarize, we tentatively dissect the whole sequence of prism formation in *P. nobilis* in short elementary chronological steps.

- In step one, the periostracal layer is secreted by the periostracal groove and hardens when in contact with seawater. The hardening is performed via a quinone-tanning process. Whether the internal surface of the periostracum is "pre-patterned" or not for mineralization is not known. Maybe some spots are "pre-designed" that correspond later to the nucleation centers from which nodules/spherulites grow.
- In step two, a mixture of heterogeneous organic components – the prisms matrix – is secreted together with the precursor fluid that contains mineral ions, between the newly formed periostracum and the mantle. Because some of these components are hydrophobic while the others are hydrophilic, a liquid–liquid emulsion is formed, in close contact with the internal surface of the periostracum. This labile structure, governed by interfacial tension, self-organizes in a viscoelastic mesh composed of polygonal cells. It may be transiently stabilized by acidic proteins at the interface between the two phases.
- In step three, the organic mesh polymerizes similarly to the periostracum, becoming insoluble and flexible.
- In step four, crystal nucleation occurs within each individual polygon, and the newly formed crystalline aggregates, which can be assimilated to nodules/spherulites, expand concentrically ("hemispherically") until reaching the polygonal boundaries. They form then flat "tiles." Whether the step of crystallization of calcite is preceded by a step of formation of amorphous mineral is not known but can be suspected. The nucleating macromolecules may be acidic proteins such as caspartin or calprismin, or sulfated polysaccharides.
- In step five, while the polygonal boundaries continue to grow in height, the newly formed tiles elongate perpendicularly to the periostracal surface. The extension of periprismatic boundaries may occur by successive accretion of organic elements to their edge. The process involved might be contact recognition. The prism elongation is accomplished "layer-per-layer." For each prism, one "layer" results from the fusion of nano-grains, which are surrounded and crystallographically oriented by acidic proteins (caspartin and calprismin), according to the scheme of Cölfen. Each "layer" can be consequently considered as a flat mesocrystal, which itself fuses with the subjacent mineral receptacle. Whether nano-grains together with their surrounding acidic macromolecules are imported by exocytosis from vesicles of some mantle cells or whether they form directly in the growing structure, at the interface between the mantle cells and the periostracum, is not known.
- In step six, while the prisms continue to grow, competition for space – or something, which looks like it – occurs in the range of 50–100 μm: some slow-growing prisms are absorbed by their fast-growing neighbors. Because the periprismatic boundaries are elastic and deformable, they accompany the lateral expansion of the fast-growing prisms.

– In step seven, the prisms continue to grow layer per layer, according to the same process described here above. Competition for growth is extremely marginal and most of the prisms grow in parallel. The growth is definitively stopped when the organic interface between the prismatic and the nacreous layers covers the inward surface of the prisms.

In this general scheme, very acidic proteins like caspartin and calprismin may display different functions, and their role is coordinated with that of the hydrophobic constituents of the periprismatic walls. We are fully aware that the steps described above are speculative, and that the true process may be completely different.

13.5.2 The Nacreous Layer

Similarly to the formation of prisms, the synthesis of nacre in *P. nobilis* gives rise to a broad range of speculations, which cannot be easily verified. Indeed, our knowledge of the nacre constituents of *P. nobilis* is limited to one single nacre-specific protein, mucoperlin, and most of the knowledge accumulated on nacre derives from other models, such as *Nautilus*, *Pinctada*, or *Haliotis*. Thus, we cannot guarantee that the different aspects of the nacre formation referred below can strictly apply the nacre of *P. nobilis*: as described in Sect. 3.5, the nacre of *P. nobilis* is indeed rather atypical in the bivalvian world, since it represents the row-stack type, and not the brickwall type.

A paper published few years ago by Addadi and coworkers (2006) gives an excellent review of critically important elements for nacre formation. These elements can be summarized as follows: chitin is the structural polymer that pre-organizes the interface between the mineralization front and the mantle cells, before mineralization. In nacre, chitin is supposed to define the interlamellar template on which nacre tablets grow. Between successive chitin sheets, a silklike gel composed of hydrophobic proteins is present. This gel also contains acid-soluble proteins that may work as mineral nucleators.

Concerning the process of nacre growth itself, the paper from Addadi et al. (2006) underlines the following aspects: nacre tablets grow from their center, which is enriched in polyanionic polymers (carboxylate-rich, or sulfate-rich). Carboxylate groups may be involved in the nucleation itself, while sulfate groups may concentrate calcium ions at the vicinity of the carboxylate groups (Addadi et al. 1987; Nudelman et al. 2006). The initial minerals formed are amorphous, but they convert to aragonite. Nacre tablets grow vertically, then laterally. By doing so, they entrap acidic proteins and push laterally the hydrophobic gel. When neighboring nacre tablets come in confluence, the squeezed gel polymerizes and forms the intertabular insoluble matrix. Whether nacre tablets grow by heteroepitaxy or by mineral bridges is still unclear and may depend on the studied model.

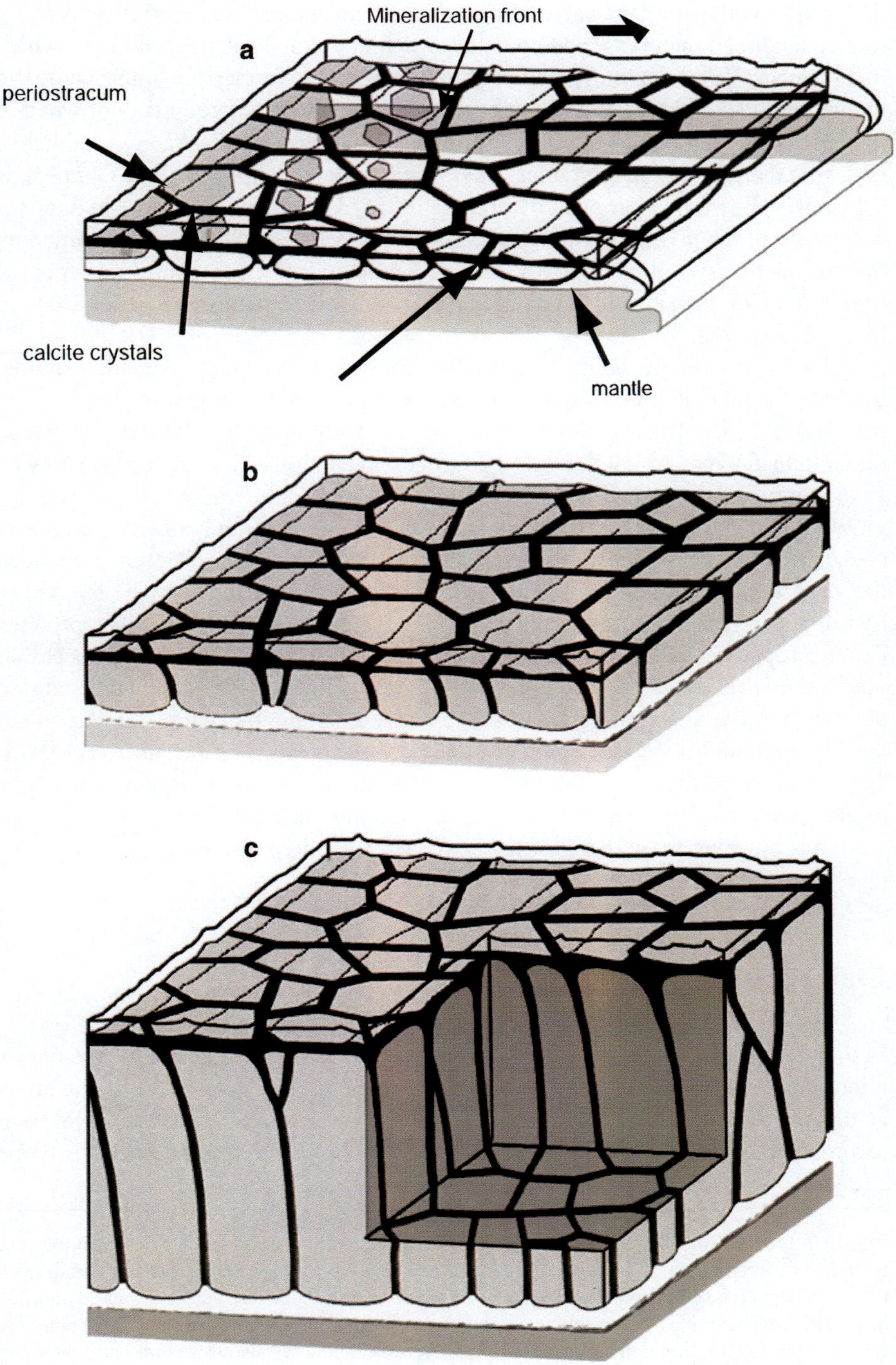

Fig. 13.7 Mode of formation of the calcitic prisms of *Pinna nobilis*, at different stages (Drawing adapted from a diagram published by Checa et al. (2005, fig. 9 p 6,413). (**a**) early synthesis stage.

For the synthesis of the nacre of *P. nobilis*, some elements are seriously missing to give the broad picture: we do not know whether chitin is present, and if so, where it is located. We do not know whether the first minerals formed are amorphous. We do not know whether the row-stack nacre tablets grow by heteroepitaxy or whether they grow on each other via mineral bridges. What remains accurately established concerns the fact that nacre tablets of *P. nobilis* are constituted of nano-grains, as previously shown by Cuif and coworkers (1985). Again, similarly to what happens in the case of the prism, nacre tablets can be considered as mesocrystals, formed by the fusion of crystallographically oriented nanograins. Once more, the concept developed Cölfen seems to be valid and probably corresponds to a general law in biomineralization.

Another important point that partly converges with the Addadi's model described above is the location of mucoperlin. As explained in Sect. 4.3.2, mucoperlin is localized at the periphery of the tablets, in the intertabular matrix, rather than in the interlamellar one. Thus, our explanation is that mucoperlin may be a macromolecular constituent of the gel-like polymer mixture in which nacre tablets grow. When the tablets extend laterally, they push aside mucoperlin, which is consequently squeezed in the intertabular matrix. The single difference between the model and our finding is that mucoperlin is primarily soluble and rather hydrophilic. Thus, mucoperlin is markedly different from the silklike proteins that are supposed to form the hydrophobic gel. Being acidic, one may expect that mucoperlin is involved in nucleating nacre tablets. However, in none of our staining did we observe that mucoperlin was localized within the nacre tablets. What appears extremely consistent is that mucoperlin is specific of the nacreous layer. Thus, we believe that, together with other components, mucoperlin may contribute to the selection of the aragonite polymorph, among other putative functions. More work has however to be done for testing *in vitro* this hypothesis.

13.6 Conclusion

In the present paper, we have tried, as far as we could, to synthesize all the data – micro- and nanostructural and biochemical – on the shell of *Pinna nobilis* and on its acidic macromolecular constituents. Although remarkable advances have been recorded, our story is partial and "holey." Fundamental researches have to be

Fig. 13.7 (continued) The polygonal shape of the periprismatic membranes suggests that they are patterned by interfacial tension of an emulsion. If so, the periprismatic membranes may result from the polymerization of the continuous phase of a liquid–liquid emulsion, consisting of an aqueous discontinuous phase and a hydrophobic continuous one. Note that the periprismatic membranes are formed prior to their mineralization. For each polygon, calcite crystals grow from one center and expand laterally until reaching the periprismatic membranes. (**b**) later stage. The prisms grow in thickness inward (toward the mantle epithelium), perpendicularly to the periostracal surface. (**c**) advanced stage. The prisms continue to grow vertically. Locally, crystal competition may occur dedicated to the full protein assemblages that shape the prismatic and nacreous layers of *Pinna nobilis*. This may be performed by combining transcriptomics on mantle tissues that secrete the prismatic and the nacreous layers and proteomics on the shell matrices of the separated layers. In addition, important efforts should also focus on the glycomics aspect of *P. nobilis* shell synthesis and on the supramolecular chemistry that allows the emergence of these beautiful microstructures.

Acknowledgments This chapter is a contribution to the ANR project ACCRO-Earth, ref. BLAN06-2_159971, coordinator Gilles Ramstein, LSCE, Gif/Yvette, for the period 2007–2010. A complementary support was provided by the INTERRVIE program from INSU for the year 2010. Frédéric Marin thanks Professor Jean-Pierre Gauthier for providing the SEM picture of newly formed nacre tablets (Fig. 13.3f) and for scrupulous rereading. Alain Godon (UMR CNRS 5561) is also acknowleged for redesigning Fig. 13.7 from the paper of Checa et al. (2005).

References

Addadi L, Moradian J, Shay E, Maroudas NG, Weiner S (1987) A chemical model for the cooperation of sulfates and carboxylates in calcite crystal nucleation: relevance to biomineralization. Proc Natl Acad Sci USA 84:2732–2736

Addadi L, Joester D, Nudelman F, Weiner S (2006) Mollusk shell formation: a source of new concepts for understanding biomineralization processes. Chem Eur J 12:980–987

Afdhal NH, Ostrow JD, Koehler R, Niu N, Groen AK, Veis A, Nunes DP, Offner GD (1995) Interaction of bovine gallbladder mucin and calcium-binding protein: effect on calcium phosphate precipitation. Gastroenterology 109:1661–1672

Akiyama M (1966) Conchiolin-constituent amino acids and shell structures of bivalved shells. Proc Jpn Acad 2:800–805

Bafna S, Singh AP, Moniaux N, Eudy JD, Meza JL, Batra SK (2008) MUC4, a multifunctional transmembrane glycoprotein, induces oncogenic transformation of NIH3T3 mouse fibroblast cells. Cancer Res 68:9231–9238

Biedermann W (1901) Untersuchungen über Bau und Entstehung der Molluskenschalen. Jena Z Naturwiss 36:1–164

Boggild OB (1930) The shell structure of the mollusks. D Kgl Danske Vidensk Selsk Skrifter, Naturvidensk Og Mathem Afd, 9. Raekke, II.2:231–326

Bowerbank JS (1844) On the structure of the shell of molluscous and conchiferous animals. Trans Microsc Soc Lond 1:123–154

Bricteux-Grégoire S, Florkin M, Grégoire C (1968) Prism conchiolin of modern or fossil molluscan shells. An example of protein paleization. Comp Biochem Physiol 24:567–572

Brisou J (1985) Les coquillages dans l'histoire des hommes. Ed. Ouest France, p 140

Cabanellas-Reboredo M, Deudero S, Alos J, Valencia JM, March D, Hendriks IE, Alvarez E (2009) Recruitment of *Pinna nobilis* (Mollusca: Bivalvia) on artificial structures. Mar Biodiv Rec 2:1–5

Carpenter WB (1844) Report on the microscopic structure of shells. Rep Brit Assoc Adv Sci 14th meeting, pp 1–24

Carter JG (1990) Skeletal Biomineralization: Patterns, Processes and Evolutionary Trends, vol I and II "Atlas and index". Van Nostrand Reinhold, New York

Carter JG, Clark GR II (1985) Classification and phylogenetic significance of molluscan shell microstructure. In: Broadhead TW (ed) Mollusks – Notes for a short course, University of Tennessee, Dept of Geological Sciences, Studies in Geology 13, pp 50–71

Cayeux L (1916) Introduction à l'étude pétrographique des roches sédimentaires: mémoire pour servir à l'explication de la carte géologique détaillée de la France, vol 1&2. Ministère des Travaux Publics/Imprimerie Nationale, Paris

Centoducati G, Tarsitano E, Bottalico A, Marvulli M, Lai OR, Crescenzo G (2007) Monitoring of the endangered *Pinna nobilis* Linné 1758 in the Mar Grande of Taranto (Ionian sea, Italy). Environ Monit Assess 131:339–347

Checa AG, Rodriguez-Navarro AB, Esteban-Delgado FJ (2005) The nature and formation of calcitic columnar prismatic shell layers in pteriomorphian bivalves. Biomater 26:6404–6414

Cölfen H (2007) Non classical crystallization. In: Arias JL, Fernandez MS (eds) Biomineralization, from paleontology to materials science. Editorial Universitaria, Santiago de Chile, pp 515–526

Cölfen H, Antonietti M (2005) Mesocrystals: inorganic superstructures made by highly parallel crystallization and controlled alignment. Angew Chem Int Ed 44:5576–5591

Collins MJ, Muyzer G, Curry GB, Sandberg P, Westbroek P (1991) Macromolecules in brachiopod shells: characterization and diagenesis. Lethaia 24:387–397

Cosentino A, Giacobbe S (2006) Shell ornament in *Pinna nobilis* and *Pinna rudis* (Bivalvia: Pteriomorpha). J Conchiol 39:135–140

Cosentino A, Giacobbe S (2007a) Aspects of epizoobiontic mollusk assemblages on *Pinna* shells. Composition and structure. Cah Biol Mar 48:187–197

Cosentino A, Giacobbe S (2007b) Aspects of epizoobiontic mollusk assemblages on *Pinna* shells. II. Does the Mediterranean *P. nobilis* represent an isle of biodiversity? Cah Biol Mar 49:161–173

Crenshaw MA (1972) The soluble matrix from *Mercenaria mercenaria* shell. Biomineralization 6:6–11

Cuif JP, Raguideau A (1982) Observation sur l'individualité cristallographique des prismes de *Pinna nobilis* L. C R Acad Sci Paris, sér II, 295:415–418

Cuif JP, Dauphin Y, Denis A, Gaspard D, Keller JP (1980) Continuité et périodicité du réseau organique intraprismatique dans le test de *Pinna muricata* Linné (Lamellibranche). C R Séanc Acad Sci Paris, sér D, 290:759–762

Cuif JP, Denis A, Gaspard D (1981) Recherche d'une méthode d'analyse ultrastructurale des tests carbonatés d'invertébrés. Bull Soc Geol Fr 9, XXIII, 5:525–534

Cuif JP, Dauphin Y, Denis A, Gaspard D, Keller JP (1983a) Etude des caractéristiques de la phase minérale dans les structures prismatiques du test de quelques mollusques. Bull Mus Natn Hist Nat Paris 4e sér. 5, section A, 3:679–717

Cuif JP, Denis A, Raguideau A (1983b) Observations sur les modalités de mise en place de la couche prismatique du test de *Pinna nobilis* L. par l'étude des caractéristiques de la phase minérale. Haliotis 13:131–141

Cuif JP, Denis A, Flamand D, Frérotte B (1985) Etude ultrastructurale de la transition prismes/nacre dans le test de *Pinna nobilis* L (mollusque, lamellibranche). Sci Rep Port Cros natl Park Fr 11:95–107

Cuif JP, Dauphin Y, Flamand D, Frérotte B, Gautret P (1986) La mesure localisée du taux de soufre comme indicateur de l'origine et de l'état diagénétique des biocristaux carbonatés. C R Acad Sci Paris, sér II, 303(3):251–256

Cuif JP, Dauphin Y, Denis A, Gautret P, Lawniczak A, Raguideau A (1987a) Résultats récents concernant l'analyse des biocristaux carbonatés; implications biologiques et sédimentologiques. Bull Soc Geol Fr 8, t III, 2:269–288

Cuif JP, Flamand D, Frérotte B, Chabin A, Raguideau A (1987b) Fractionnement de la matrice protéique intraprismatique chez *Pinna nobilis* L et composition en acides aminés des différentes phases. C R Acad Sci Paris, sér II, 304(9):475–478

Cuif JP, Denis A, Frérotte B, Rekkab D (1988a) Gradient de concentration d'élements mineurs et séquence microstructurale dans le test de mollusques. C R Acad Sci Paris, sér II, 307:837–842

Cuif JP, Dauphin Y, Gautret P (1988b) Corrélation entre l'organisation cristallographique des unités microstructurales formant le test des Mollusques et la masse moléculaire moyenne de leur phase organique soluble. C R Acad Sci Paris, sér II, 307:1943–1948

Cuif JP, Gautret P, Marin F (1991) Correlation between the size of crystals and the molecular weight of organic fractions in the soluble matrices of mollusc, coral and sponge carbonate skeletons. In: Suga S, Nakahara H (eds) Mechanisms and Phylogeny of Mineralization in Biological Systems. Springer, Tokyo, pp 391–395

Dauphin Y (2002) Comparison of the soluble matrices of the calcitic prismatic layer of *Pinna nobilis* (Mollusca, Bivalvia, Pteriomorpha). Comp Biochem Physiol A 132:577–590

Dauphin Y (2003) Soluble organic matrices of the calcitic prismatic shell layers of two pteriomorphid bivalves: *Pinna nobilis* and *Pinctada margaritifera*. J Biol Chem 278:15168–15177

Dauphin Y, Cuif JP, Doucet J, Salomé M, Susini J, Williams CT (2003) In situ chemical speciation of sulfur in calcitic biominerals and the simple prism concept. J Struct Biol 142:272–280

De Bournon E (1808) Traité complet de la chaux carbonatée et de l'aragonite, William Phillips (ed.) vol I. London

De Gaulejac B (1993) Etude écophysiologique du mollusque bivalve méditerranéen *Pinna nobilis* L. Reproduction, croissance, respiration. Thèse 3ème cycle, Université d'Aix-Marseille III, p 220

De Gaulejac B, Vicente N (1990) Ecologie de *Pinna nobilis* (L.) mollusque bivalve sur les côtes de Corse. Essais de transplantation et expériences en milieu contrôlé. Haliotis 10:83–100

De Gaulejac B, Henry M, Vicente N (1995a) An ultrastructural study of gametogenesis of the marine bivalve *Pinna nobilis* (Linnaeus 1758) I. Oogenesis. J Mollus Stud 61:375–392

De Gaulejac B, Henry M, Vicente N (1995b) An ultrastructural study of gametogenesis of the marine bivalve *Pinna nobilis* (Linnaeus 1758) II. Spermatogenesis. J Mollus Stud 61:393–403

Esteban-Delgado FJ, Harper EM, Checa AG, Rodriguez-Navarro AB (2008) Origin and expansion of foliated microstructures in pteriomorph bivalves. Biol Bull 214:153–165

Evans JS (2008) "Tuning in" to mollusk shell nacre- and prismatic-associated protein terminal sequences. Implications for biomineralization and the construction of high performance inorganic-organic composites. Chem Rev 108:4455–4462

Foulquié M, Dupuy de la Grandrive R (2003) Mise en place d'un suivi des grandes nacres (*Pinna nobilis*) dans la zone Natura 2000 des "Posidonies du Cap d'Agde", Hérault, France. In: Vicente N (ed) Mémoires de l'Institut Océanographique Paul Ricard, 1er Séminaire International sur la grande Nacre de Méditerranée : Pinna nobilis, 10–12 Octobre 2002, Institut Océanographique Paul Ricard, pp 49–55

Frémy ME (1855) Recherches chimiques sur les os. Annales Chim Phys, 3ème sér. 43:47–107

Frérotte B (1987) Etude de l'organisation et de la composition des biocristaux du test des lamellibranches. Thèse de 3ème Cycle, Laboratoire de Paléontologie, Université Paris XI, Orsay

Garcia-March JR (2003) Contribution to the knowledge of the status of *Pinna nobilis* (L.) 1758 in Spanish Coasts. In: Vicente N (ed) Mémoires de l'Institut Océanographique Paul Ricard, 1er Séminaire International sur la grande Nacre de Méditerranée : Pinna nobilis, 10–12 Octobre 2002, Institut Océanographique Paul Ricard, pp 29–41

Garcia-March JR, Carrascosa AMG, Pena AL (2002) *In situ* measurement of *Pinna nobilis* shells for age and growth studies: a new device. Mar Ecol 23:207–217

Garcia-March JR, Garcia-Carrascosa AM, Pena Cantero AL, Wang YG (2007) Population structure, mortality and growth of *Pinna nobilis* Linnaeus, 1758 (Mollusca, Bivalvia) at different depths in Moraira bay (Alicante, Western Mediterranean). Mar Biol 150:861–871

Gauthier JP, Caseiro J, Lasnier B (1994) Les perles rouges de *Pinna nobilis*. Revue de Gemmologie, A.F.G., 118:2–4; 119:2–4

Giacobbe S (2002) Epibiontic mollusc communities on *Pinna nobilis* L. (Bivalvia, Mollusca). J Nat Hist 36:1385–1396

Giribet G (2008) Bivalvia. In: Ponder WF, Lindberg DR (eds) Phylogeny and Evolution of the Mollusca. University of California Press, Berkeley, pp 105–141

Grases F, Llobera A (1998) Experimental model to study sedimentary kidney stones. Micron 29:105–111

Gray JE (1835) Remarks on the difficulty of distinguishing certain genera of testaceous mollusca by their shell alone, and on the anomalies in regard to habitation observed in certain species. Phil Trans R Soc Lond 125:301–310

Grégoire C (1967) Sur la structure des matrices organiques des coquilles de mollusques. Biol Rev 42:653–688

Grégoire C (1972) Structure of the molluscan shell. In: Florkin M, Scheer BT (eds) Chemical Zoology, vol VII, mollusca. Academic, New York, pp 45–102

Grigor'ev DP (1965) Ontogeny of Minerals. Israel Program for Scientific Translation, Jerusalem, 250 pp

Henry M, Vicente N, Houache N (1992) Caractérisation des hémocytes d'un mollusque bivalve marin, la nacre, *Pinna nobilis* L. 1758. In: Aspects Récents de la Biologie des Mollusques, Ifremer, Actes de Colloques 13, pp 97–106

Jackson DJ, McDougall C, Green K, Simpson F, Wörheide G, Degnan BM (2006) A rapidly evolving secretome builds and patterns a sea shell. BMC Biol 4:40–49

Karampelas S, Gauthier JP, Fritsch E, Notari F (2009) Characterization of some pearls of the Pinnidae family. Gems Gemol 45:221–223

Karny H (1913) Optische Untersuchungen zur Aufklärung der Struktur der Muschenschalen. I. Aviculidae, II. Unionidae. Sitzungsberichte der Akademie der Wissenschaften, Mathematisch-Naturwissenschaftliche Klasse. Wien 122:207–259

Katsanevakis S (2007) Growth and mortality rates of the fan mussel *Pinna nobilis* in Lake Vouliagmeni (Korinthiakos Gulf, Greece): a generalized additive modelling approach. Mar Biol 152:1319–1331

Katsanevakis S, Thessalou-Legaki M (2009) Spatial distribution, abundance and habitat use of the protected fan mussel *Pinna nobilis* in Souda Bay, Crete. Aquat Biol 8:45–54

Keller JP (1981) Le dégagement du matériel minéral des tests d'invertébrés (Bivalves) par protéolyse enzymatique de la trame organique. Geobios 14:269–273

Keller JP, Dauphin Y (1983) Methodological aspects of the ultrastructural analysis of the organic and mineral components in mollusc shells. In: Westbroek P, De Jong EW (eds) Biomineralization and biological metal accumulation. D Reidel Publishing, Dordrecht, pp 255–260

Kervadec G (1990) Estimation de la validité taxonomique du critère minéralogique par l'analyse des phases organiques solubles des biocristaux carbonatés des mollusques. Thèse de 3ème Cycle, Laboratoire de Paléontologie, Université Paris XI, Orsay

Kniprath E (1981) Ontogeny of the molluscan shell field. Zool Scr 10:61–79

Kretsinger RH (1976) Calcium-binding proteins. Annu Rev Biochem 45:239–266

Lechene de la Porte P, Domingo N, van Wijland M, Groen AK, Ostrow JD, Lafont H (1996) Distinct immuno-localization of mucin and other biliary proteins in human cholesterol gallstones. J Hepatol 25:339–348

Leydolt F (1856) Über die Struktur und Zusammensetzung der Krystalle des prismatischen Kalkhaloides nebst einem Anhang über die Struktur der kalkigen Teile einiger wirbellosen Tiere. Sitzungsberichte Mathematisch Naturwiss Klasse Kaiserlichen Akad Wiss Wien 19:10–32

Maeder F, Halbeisen M (2001) Muschelseide: Auf der Suche nach einel vergessenen Material. Waffen Kostumkunde 43:33–41

Mao Che L, Golubic S, Le Campion-Alsumard T, Payri CE (2001) Developmental aspects of biomineralization in the polynesian pearl oyster *Pinctada margaritifera* var. *cumingii*. Oceanol Acta 24:S37–S49

Marin F (1992). Essai de caractérisation chromatographique et immunologique des constituants organiques associés aux biocristaux carbonatés des squelettes de mollusques, cnidaires et spongiaires. Thèse de 3ème Cycle, Laboratoire de Paléontologie, Université Paris XI, Orsay

Marin F (2003) Molluscan shell matrix characterization by preparative SDS-PAGE. Sci World J 3:342–347

Marin F (2009) Biominéralisation de la coquille des mollusques : origine, évolution, formation. Mémoire d'Habilitation à Diriger des Recherches. Université de Bourgogne, Dijon, p 243

Marin F, Luquet G (2005) Molluscan biomineralization: the proteinaceous shell constituents of *Pinna nobilis* L. Mater Sci Eng C 25:105–111

Marin F, Luquet G (2007) Unusually acidic proteins in biomineralization. In: Baeuerlein E (ed) Handbook of Biomineralization, vol 1, The Biology of Biominerals Structure Formation. Wiley-VCH, Weinheim, pp 273–290, Chapter 16

Marin F, Muyzer G, Dauphin Y (1994) Caractérisation électrophorétique et immunologique des matrices organiques solubles de deux Bivalves Ptériomorphes actuels, *Pinna nobilis* L. et *Pinctada margaritifera* (L.). C R Acad Sci Paris II 318:1653–1659

Marin F, Gillibert M, Wesbroek P, Muyzer G, Dauphin Y (1999) Evolution: disjunct degeneration of immunological determinants. Geol Mijnbouw 78:135–139

Marin F, Corstjens P, de Gaulejac B, de Vrind-De JE, Westbroek P (2000) Mucins and molluscan calcification: molecular characterization of mucoperlin, a novel mucin-like protein of the nacreous shell-layer of the fan mussel *Pinna nobilis* (Bivalvia, Pteriomorphia). J Biol Chem 275:20667–20675

Marin F, Pereira L, Westbroek P (2001) Large-scale purification of molluscan shell matrix. Prot Expres Purif 23:175–179

Marin F, de Groot K, Westbroek P (2003a) Screening molluscan cDNA expression libraries with anti-shell matrix antibodies. Prot Expres Purif 30:246–252

Marin F, Westbroek P, de Groot K (2003b) The proteinaceous constituents of the shell of *Pinna nobilis* L. In: Vicente N (ed) Mémoires de l'Institut Océanographique Paul Ricard, 1er Séminaire International sur la grande Nacre de Méditerranée: *Pinna nobilis*, 10–12 Octobre 2002, Institut Océanographique Paul Ricard, pp 77–90

Marin F, Amons R, Guichard N, Stigter M, Hecker A, Luquet G, Layrolle P, Alcaraz G, Riondet C, Westbroek P (2005) Caspartin and calprismin, two proteins of the shell calcitic prisms of the Mediterranean fan mussel *Pinna nobilis*. J Biol Chem 280:33895–33908

Marin F, Morin V, Knap F, Guichard N, Marie B, Luquet G, Westbroek P, Medakovic D (2007a) Caspartin: thermal stability and occurrence in mollusk calcified tissues. In: Arias JL, Fernandez MS (eds) Biomineralization, from paleontology to materials science. Editorial Universitaria, Santiago de Chile, pp 281–288

Marin F, Pokroy B, Luquet G, Layrolle P, de Groot K (2007b) Protein mapping of calcium carbonate biominerals by immunogold. Biomater 28:2368–2377

Marin F, Luquet G, Marie B, Medakovic D (2008) Molluscan shell proteins: primary structure, origin and evolution. Curr Top Dev Biol 80:209–276

Masuda F, Hirano M (1980) Chemical composition of some modern marine pelecypod shells. Sci Rep Inst Geosci Univ Tsukuba section B1:163–177

Maurer P, Hohenester E, Engel J (1996) Extracellular calcium-binding proteins. Curr Opin Cell Biol 8:609–617

Medakovic D (2000) Carbonic anhydrase activity and biomineralization in embryos, larvae and adult blue mussels *Mytilus edulis* L. Helgol Mar Res 54:1–6

Medioni E, Vicente N (2003) Etude de la cinétique des populations de Pinna nobilis L. 1758 sur le littoral méditerranéen français. In: Vicente N (ed) Mémoires de l'Institut Océanographique Paul Ricard, 1er Séminaire International sur la grande Nacre de Méditerranée: *Pinna nobilis*, 10–12 Octobre 2002, Institut Océanographique Paul Ricard, pp 43–48

Meyer-Lüeckel H, Tschoppe P, Hopfenmüller W, Stenzel WR, Kielbassa AM (2006) Effect of polymers used in saliva substitutes on demineralized bovine enamel and dentin. Am J Dent 19:308–312

Moreteau JC, Vicente N (1982) Evolution d'une population de *Pinna nobilis* L. (Mollusca, Bivalvia). Malacologia 22:341–345

Mutvei H (1970) Ultrastructure of the mineral and organic components of molluscan nacreous layers. Biomineral Res Rep 2:48–72

Muyzer G, Westbroek P, De Vrind JPM, Tanke J, Vrijheid T, De Jong EW, Bruning JW, Wehmiller JF (1984) Immunology and organic geochemistry. Org Geochem 6:847–855
Nakahara H, Kakei M, Bevelander G (1980) Fine structure and amino acid composition of the organic "envelope" in the prismatic layer of some bivalve shells. Venus 39:167–177
Nieuw-Amerongen AV, Oderkerk CH, Veerman ECI (1989) Interaction of human salivary mucins with hydroxyapatite. J Biol Buccale 17:85–92
Nudelman F, Gotliv BA, Addadi L, Weiner S (2006) Mollusk shell formation: mapping the distribution of organic matrix components underlying a single aragonite tablet in nacre. J Struct Biol 153:176–187
Oaki Y, Imai H (2005) The hierarchical architecture of nacre and its mimetic material. Angew Chem Int Ed Engl 44:6571–6575
Palmer AR (1992) Calcification in marine molluscs: how costly is it? Proc Natl Acad Sci USA 89:1379–1382
Pokroy B, Fitch AN, Marin F, Kapon M, Adir N, Zolotoyabko E (2006) Anisotropic lattice distorsions in biogenic calcite induced by intra-crystalline organic molecules. J Struct Biol 155:96–103
Pokroy B, Kapon M, Marin F, Adir N, Zolotoyabko E (2007) Protein-induced, previously unidentified twin form of calcite. Proc Natl Acad Sci USA 104:7337–7341
Rabaoui L, Tlig-Zouari S, Ben Hassine OK (2007) Description de la faune épibionte de *Pinna nobilis* sur les côtes nord et est de la Tunisie. Rapp Comm Int Mer Medit 38:578
Rabaoui L, Tlig-Zouari S, Ben Hassine OK (2008) Distribution and habitat of the fan mussel *Pinna nobilis* Linnaeus, 1758 (Mollusca: Bivalvia) along the northern and eastern Tunisian coasts. Cah Biol Mar 49:67–78
Rabaoui L, Tlig-Zouari S, Cosentino A, Ben Hassine OK (2009) Associated fauna of the fan shell *Pinna nobilis* (Mollusca: Bivalvia) in the northern and eastern Tunisian coasts. Sci Mar 73:129–141
Ranson G (1952) Les huîtres et le calcaire. Calcaire et substratum organique chez les mollusques et quelques autres invertébrés marins. C R Acad Sci Paris 234:1485–1487
Ranson G (1966) Substratum organique et matrice organique des prismes de la couche prismatique de la coquille de certains mollusques lamellibranches. C R Acad Sci Paris 262:1280–1282
Riva A (2003) Approche méthodologique de quelques paramètres bioénergétiques chez *Pinna nobilis*. In: Vicente N (ed) Mémoires de l'Institut Océanographique Paul Ricard, 1er Séminaire International sur la grande Nacre de Méditerranée : *Pinna nobilis*, 10–12 Octobre 2002, Institut Océanographique Paul Ricard, pp 91–101
Roche J, Ranson G, Eysseric-Lafon M (1951) Sur la composition des scléroprotéines des coquilles des mollusques (conchiolines). C R Séanc Soc Biol 145:1474–1477
Römer O (1903) Untersuchungen über den feineren Bau einiger Muschelschalen. Z Wiss Zool 75:437–472
Rose G (1858) Über die heteromorphen Zustände der Kohlensauren Kalkerde. II. Vorkommen des Aragonits und Kalkspaths in der organischen Natur: Physikalische Abhandlungen der Königlichen Akademie der Wissenschaften zu Berlin aus dem Jahre 1858:63–111
Schmidt WJ (1923) Bau und Bildung der Perlmuttermasse. Zoologische Jahrbücher Abteilung Anat Ontogenie Tiere 45:1–148
Schmidt WJ (1924) Die Bausteine des Tierkörpers in polarisiertem Lichte. F. Cohen, Bonn
Schmidt WJ (1932) Studien über Pinnaperlen. I. Über Prismenperlen von Pinna nobilis. Z Morph Ökol Tiere Abt A 25:235–277
Siletic T, Peharda M (2003) Population study of the fan shell *Pinna nobilis* L. in Malo and Veliko Jezero of the Mljet National Park (Adriatic Sea). Sci Mar 67:91–98
Simkiss K, Wilbur KM (1989) Biomineralization. Cell biology and mineral deposition. Academic, New York
Steiner G, Hammer S (2000) Molecular phylogeny of the Bivalvia inferred from 18 S rDNA sequences with particular reference to the Pteriomorphia. In: Harper EM, Taylor JD, Crame JA

(eds) Evolutionary biology of the Bivalvia. Geological Society, London, pp 11–29, Geological Society Special Publication, 177

Sudo S, Fujikawa T, Nagakura T, Ohkubo T, Sakagushi K, Tanaka M, Nakashima K (1997) Structures of mollusc shell framework proteins. Nature 387:563–564

Tabak LA (1995) In defense of the oral cavity: structure, biosynthesis, and function of salivary mucins. Annu Rev Physiol 57:547–564

Tabak LA, Levine MJ, Jain NK, Bryan AR, Cohen RE, Monte LD, Zawacki S, Nancollas GH, Slomiany A, Slomiany BL (1985) Adsorption of human salivary mucins to hydroxyapatite. Arch Oral Biol 30:423–427

Taylor JD, Kennedy WJ, Hall A (1969) The shell structure and mineralogy of the Bivalvia. Introduction. Nuculacea-Trigonacea. Bull Brit Mus Nat Hist Zool Lond supplem 3:1–125

Ubukata T (1994) Architectural constraints on the morphogenesis of prismatic structure in Bivalvia. Palaeontology 37:241–261

Vicente N (2003) La grande nacre de Méditerranée *Pinna nobilis*. Présentation générale. In: Vicente N (ed) Mémoires de l'Institut Océanographique Paul Ricard, 1er Séminaire International sur la grande Nacre de Méditerranée : Pinna nobilis, 10–12 Octobre 2002, Institut Océanographique Paul Ricard, pp 7–16

Vicente N, Riva A, Butler A (1992) Etude expérimentale préliminaire sur les échanges gazeux chez *Pinna nobilis*. In : Aspects Récents de la Biologie des Mollusques, Ifremer Brest, Actes de Colloques 13, pp 187

Wada K (1961) Crystal growth of molluscan shells. Bull Natl Pearl Res Lab 7:703–828

Wada K (1972) Nucleation and growth of aragonite crystals in the nacre of some bivalve molluscs. Biominer Res Rep 6:141–159

Wada K (1980) Initiation of mineralization in bivalve mollusc. In: Omori M, Watabe N (eds) The mechanism of biomineralization in animals and plants. Tokay University Press, Tokyo, pp 79–92

Weiner S (1983) Mollusk shell formation – Isolation of two organic matrix proteins associated with calcite deposition in the bivalve *Mytilus californianus*. Biochemistry 22:4139–4145

Weiner S, Hood L (1975) Soluble proteins of the organic matrix of mollusc shells: a potential template for shell formation. Science 190:987–989

Weiss IM, Tuross N, Addadi L, Weiner S (2002) Mollusc larval shell formation: amorphous calcium carbonate is a precursor phase for aragonite. J Exp Zool 293:478–491

Wetzel G (1900) Die organischen Substanzen der Schaalen von *Mytilus* und *Pinna*. Z. Phys Chem 29:386–410

Wheeler AP, Rusenko KW, Sikes CS (1988) Organic matrix from carbonate biomineral as a regulator of mineralization. In: Sikes CS, Wheeler AP (eds) Chemical aspects of regulation of mineralization. University of South Alabama Publication Service, Mobile, Alabama, pp 9–13

Wise SW (1970) Microarchitecture and mode of formation of nacre (mother-of-pearl) in pelecypods, gastropods, and cephalopods. Eclog Geol Helvet 63:775–797

Zavodnik D (1967) Contribution to the ecology of *Pinna nobilis* L. (Moll. Bivalvia) in the Northern Adriatic Sea. Thalass Yugol 3:93–102

Zolotoyabko E, Caspi EN, Fieramosca JS, Von Dreele RB, Marin F, Mor G, Addadi L, Weiner S, Politi Y (2010) Differences between bond lengths in biogenic and geological calcite. Cryst Growth Des 10:1207–1214

Index

W.E.G. Müller (ed.), *Molecular Biomineralization*, Progress in Molecular and Subcellular Biology 52, DOI 10.1007/978-3-642-21230-7,

M

N